Spacecraft Modeling, Attitude Determination, and Control
Quaternion-based Approach

Yaguang Yang
US Nuclear Regulatory Commission
Office of Research
Rockville, Maryland, USA

CRC Press
Taylor & Francis Group
Boca Raton London New York

CRC Press is an imprint of the
Taylor & Francis Group, an **informa** business
A SCIENCE PUBLISHERS BOOK

CRC Press
Taylor & Francis Group
6000 Broken Sound Parkway NW, Suite 300
Boca Raton, FL 33487-2742

First issued in paperback 2021

Version Date: 20181115

ISBN-13: 978-0-367-78035-7 (pbk)
ISBN-13: 978-1-138-33150-1 (hbk)

Library of Congress Cataloging-in-Publication Data

Names: Yang, Yaguang, author.
Title: Spacecraft modeling, attitude determination, and control : quaternion-based approach / Yaguang Yang (US Nuclear Regulatory Commission, Office of Research, Rockville, Maryland, USA).
Description: Boca Raton, FL : CRC Press, 2019. | "A science publishers book." | Includes bibliographical references and index.
Identifiers: LCCN 2018051045 | ISBN 9781138331501 (hardback)
Subjects: LCSH: Space vehicles--Attitude control systems. | Stability of space vehicles. | Rotational motion (Rigid dynamics) | Quaternions. | Vector analysis.
Classification: LCC TL3260 .Y36 2019 | DDC 629.47/42--dc23
LC record available at https://lccn.loc.gov/2018051045

Visit the Taylor & Francis Web site at
http://www.taylorandfrancis.com

and the CRC Press Web site at
http://www.crcpress.com

To my parents, my wife,
my son, and daughter

Preface

My interest in spacecraft modeling, attitude determination and control started at Orbital Science Corporation. At the end of the summer of 2005, I was looking for a job that would best use my background in controls and optimization. There was an open house for job applicants at Dulles campus of the company. That was the first time I visited Orbital Science Corporation. I was very fortunate to have a chance to talk to Dr. Brian Keller, the deputy director of GNC (guidance, navigation, and controls) at the time. I showed him my publications in controls and explained my work at previous companies, he listened and immediately promised to set up an interview for me. A few weeks later, my future manager at Orbital Science Corporation, Mr. James Bobbett, called me and an interview was scheduled. Both Brian and James knew that I did not have a background in spacecraft and launch vehicles, however they trusted my background in controls and believed that my prior experience to be beneficial in this work. They offered me the job! I joined Orbital Science Corporation in November 2005.

My time at Orbital Science Corporation was delightful. I was deeply involved in the control system designs for two spacecrafts and one launch vehicle. My first assignment was to review and learn the design of ROCSAT III in preparation for designing the next spacecraft. In a few weeks, I realized that the design could be improved and I proposed an alternative method. I was surprised that my manager, Mr. Bobbett, quickly replied to my email with his strong support for my proposal. The proposed changes were implemented and six satellites were launched in April, 2006, all achieving their design requirements.

During my time at Orbital Science Corporation, several textbooks on spacecraft controls, such as M.J. Sidi's book "Spacecraft Dynamics and Control: A Practical Engineering Approach", B. Wie's book "Space Vehicle Dynamics and Control", and J.R. Wertz's book "Spacecraft Attitude Determination and Control", were great source to me in understanding this topic. Although all these books are excellent, I believed that some materials could be improved, especially, the control system design methods. However, my work assignments at Orbital

Science Corporation were very challenging and I did not have time to think about the specific details of these improvements.

I left Orbital Science Corporation to join the US NRC in 2008. At NRC, I have had more free time, after eight hours in office, to think about these problems. I started to publish papers in 2010 on new methods for spacecraft control and algorithms to design spacecraft control systems, trying to address control related problems in different stages of different missions using different sensors and actuators to cover as many design problems as possible. After a few years, my publications covered a few important areas in spacecraft modeling, attitude determination and control.

On May 1, 2015, I received an email from Vijay Primlani from CRC Press, asking if I was interested in publishing a book with this established publisher. My immediate thought was: that is a cool idea. I said "yes, but it might take some time because I want to consider a few more design problems that I have not done yet, besides I had been working and would continue to work only in my spare time for this project." I did not know that the delay would be a few years but the promise has been the motivation for me to work continuously on this interesting project.

As this project approaches the finish line, I would like to thank a few people, who helped me along the way. First, I would like to thank Dr. Keller and Mr. Bobbett at Orbital Science Corporation for giving me the chance to work in this amazing area. Second, I would like to thank Mr. Primlani at CRC Press for his invitation to write a book with my choice of topic and for his patience with my slow progress. I am also indebted to my former colleague, Dr. Z. Zhou at NASA, who co-authored two papers which are included in this book. Last but not the least, I am grateful to my manager, Mr. Ronaldo Jenkins at the US NRC for his support and approval of writing this book in my spare time.

Contents

Chapter 1

Introduction

Spacecraft attitude determination and control is an important part of a spacecraft to achieve its designed mission. As of today, many spacecrafts have been successfully launched, and most of them have performed well as they were designed. Many research papers have been published to address the attitude determination and control design problems. Several textbooks are available for students to learn the technology and for engineers to use as references.

The most popular spacecraft models for attitude determination algorithms and control design methods are the *Euler angle models* and the *quaternion models*. The Euler angle models have been proved very efficient as the linearized models are controllable, therefore, all standard linear control system design methods are directly applicable. The drawbacks related to the Euler angle methods are: (a) the designs based on linearized models may not globally stabilize the original nonlinear spacecraft, i.e., the design may not work when the attitude of the spacecraft is far away from the point where the linearization is performed; (b) the models depend on the rotational sequences, this can be error prone if several teams work on the same project and they use different rotational sequences; (c) for any rotational sequence, there is a singular point where the model is not applicable; and (d) since most attitude determination methods use quaternion to represent the spacecraft attitude, there is a need to transform quaternion into Euler angles. On the other hand, for quaternion models, people have found controllers that can globally stabilize nonlinear spacecraft systems; the models do not depend on rotational sequences and they have no singular point; and the quaternion is provided by attitude determination system and is ready to use. The main problem with the quaternion model-based control system design is that the linearized quaternion model is not controllable. Therefore, most published design methods heavily rely on Lyapunov functions for the nonlinear spacecraft system. But there is no systematic way to obtain a desired Lyapunov functions. Moreover, the Lyapunov function-based designs focus on the closed-loop system stability but pay little attention to its performance.

In a series of papers, the author proposed some *reduced quaternion* models which lead to some controllable linearized spacecraft models. Therefore, all standard linear system theory can be directly applied to analyze and design the spacecraft control systems. We showed that, in some cases, the designed control system is not only optimal for the linearized system, but also globally stabilize the original *nonlinear system*. Clearly, the reduced quaternion models do not depend on rotational sequences. Due to the special structure of the linearized spacecraft model, some most important design methods, such as LQR design and robust pole assignment design are very simple, enjoy the analytical solutions for some problems, have direct connection to the performance measures, such as *settling time*, *rising time*, and *percentage of overshoot*. All these features are attractive for high quality control system designs.

The idea mentioned above is then extended to more spacecraft control problems using specific actuators, such as magnetic torque bars and control momentum gyroscopes. These types of actuators may not provide the exactly desired torque. Most existing methods use different conversions to get approximate solutions, meaning that these actuators may generate a torque close to but not equal to the desired one. Using the reduced quaternion models that incorporate the actuators into the system model, the control inputs are not torques but the operational parameters. The main benefit of this idea is that the control actions are not approximate but accurate. As all actuators have their operational limit, design with input constraints have also been considered in this book by using recently developed interior-point optimization techniques.

This book is a result of the author's research for over a decade on the spacecraft attitude determination and control design methods which are focused on using reduced quaternion models because of their merits as stated above. It provides all necessary background materials on orbital dynamics, rotations and quaternion, frequently used reference frames, transformations between reference frames, space environment and disturbance torques, ephemeris astronomical vector calculations and measurement instruments, spacecraft control actuators and their models, so that the readers get a global picture and can apply all this information into the spacecraft system modeling, attitude determination, and spacecraft control system designs, which is the main purpose of this book.

This book is different from existing books as it focuses on the quaternion-based spacecraft control system designs and considers only attitude control system design-related problems, from spacecraft modeling to attitude determination and estimation, to control system design method selection, to control algorithm development, and the simulation of the control system designs. Moreover, this book addresses different attitude control tasks in the spacecraft life cycle, including spacecraft maneuver, orbit raising, attitude control, and rendezvous. Finally, this book emphasizes the state-space design methods rather than the classical frequency design methods.

1.1 Organization of the Book

This book is organized as follows: Chapter 2 is a brief description of orbit dynamics and properties. The treatment is focused on two body systems, which provides necessary background to be used in other chapters, for example, Chapters 3, 11, and 15.

Chapter 3 discusses the frequently used coordinate system, the rotational sequences, and the quaternion mathematics. Similar to Chapter 2, this chapter provides readers the tools and background that will be repeatedly used in the rest of the chapters.

Chapter 4 introduces two spacecraft dynamical systems based on the spacecraft missions, and their representations using the reduced quaternion models. The merit of using reduced quaternion models is that their linearized spacecraft models are controllable while the spacecraft models using full quaternion are not. It is well known that all modern linear control system design methods require the systems to be controllable. This makes the reduced quaternion spacecraft model very attractive. The ultimate goal of this chapter is to establish a few linearized controllable spacecraft models for some mostly desired attitudes for spacecraft, i.e., the inertial pointing attitude and the nadir pointing attitude.

Chapter 5 explains the space environment and the major disturbance torques introduced in the space environment. Most of these torques are difficult to be included in the spacecraft models which are used in spacecraft attitude control system designs. This means that the designed controllers do not consider the effects of these disturbance torques. As a result, the designed controllers may not work in the real space environment because the control torques may not compensate these unmodeled torques. Because of this reason, there is a need to have some simulation test for the designed spacecraft feedback control system to make sure that the designed controller works in the space environment that includes these disturbance torques. This chapter provides the necessary information so that control engineers can build the simulated space environment to test the designed controller.

Chapter 6 discusses the quaternion-based attitude determination methods using vector measurements, including some recently proposed methods. In principle, spacecraft attitude can be determined by a set of observed (measured) astronomical vectors and corresponding ephemeris astronomical vectors at the given time. An important problem is to find some fast, accurate, and robust algorithms to calculate the spacecraft attitude. Though there are other attitude determination methods based on rotational matrix or Euler angle representation, it must be pointed out that quaternion-based attitude determination methods are the most efficient ones.

Chapter 7 explains how to measure the astronomical vectors and how to calculate the corresponding ephemeris astronomical vectors at any given time. The most widely used astronomical vectors are considered. Given the ephemeris information of the astronomical objects represented in reference frame and measured astronomical vectors represented in body frame, the spacecraft attitude can be obtained using the methods described in Chapter 6.

Since there always exist some random measurement noises, there is a need to have some filtering techniques to reduce the measurement noise effect. Kalman filter was developed in 1960s just for this purpose and this technique was widely used in spacecraft attitude determination. Chapter 8 discusses the attitude estimation problem using extended and traditional Kalman filters.

Chapter 9 talks about attitude control system designs with the desired torques as control variables. This chapter focuses on state-space Linear Quadratic Regulator (LQR) design method. For nadir pointing spacecraft, the solution described in Appendix B can be applied directly. But for inertial pointing spacecraft, which has an extremely simple linearized model, an analytic solution exists. For this case, the relation between the LQR design and the closed-loop pole positions is established. The analytical solution provides insight for engineers to trade off many conflict requirements. It is shown that the design globally stabilizes the nonlinear spacecraft system even though it is based on the linearized system. As a matter of fact, the LQR design discussed in this chapter is actually a robust pole assignment design. Therefore, the design is insensitive to the modeling error and is good for disturbance rejection.

All designs in Chapter 9 calculate the desired torques that are used to control the spacecraft attitude. These desired torques are supplied by using several different actuators or their combinations. Chapter 10 reviews some widely used spacecraft actuators, including reaction wheel and momentum wheel, control moment gyros, magnetic torque rods, and thrusters. This chapter reveals a fact that several types of actuators are not able to provide the desired torques in all directions. Therefore, the methods discussed in Chapter 9 (when these actuators are used) have a torque realization problem. A better design practice should include the actuators' models in the control system design. This consideration will be the topics in rest of the chapters.

Chapter 11 discusses system designs for spacecraft using magnetic torque rods. Although magnetic torque bars can provide torques only in a plane instead of three-dimensional space at any time, it is shown that the controllability of spacecraft using only magnetic torques is achievable under some mild conditions. Using the fact that the magnetic field is approximately a periodic function of the spacecraft orbit, periodic LQR design is considered in the controller design. Some efficient solutions for the algebraic periodic Riccati equation are proposed.

Chapter 12 discusses the spacecraft control system design using thrusters. A typical operation using thrusters, orbit-raising, is considered in this chapter. The control system models and controller designs depend on the thruster configurations. This chapter describes how to design the controller using the standard linear system theory. Although a particular thruster configuration is considered in this chapter, the idea can easily be used for any other thruster configurations.

Chapter 13 addresses Model Predictive Control (MPC) and its application to the spacecraft attitude control problems. Since MPC needs extensive on-board computation, it was not widely used in spacecraft control, as more powerful computers are installed on spacecraft. MPC is expected to find more applications in aerospace in the future. This chapter establishes the relation between constrained MPC and convex quadratic programming (QP) with box constraints. This formulation is directly applicable to the controller design problem when actuator saturation exists. An efficient interior-point algorithm specifically for this problem is proposed and its convergence is proved. The thruster control problem discussed in Chapter 12 is revisited and it is shown that the problem can be solved by the MPC control method proposed in this chapter.

Chapter 14 is dedicated to the spacecraft attitude control system design using control moment gyros. As it has already been explained in Chapter 10 that for given desired torques obtained in Chapter 9, there are singular points where one cannot find gimbal speeds of the CMGs to achieve the desired torques. This chapter presents a new operational concept for control moment gyros and proposes a MPC design method for this problem. Simulation test is used to demonstrate the feasibility of the proposed method.

Chapter 15 considers coupled orbit and attitude control that is the key technology for spacecraft rendezvous and soft docking. Coupled orbit and attitude control is an extensively studied problem with renewed interest because of installations of powerful on-board computers, availability of advanced theoretical results, and requirements for better performance in future missions. The method considered in this chapter addresses a fundamental requirements for soft docking, i.e., there is no oscillation crossing the horizontal line for relative position and relative attitude between chaser and target spacecraft to avoid collision during the docking process.

Three appendices are included for quick reference for the background used in the control system design methods discussed in this book. Appendix A is about the first order optimality conditions, which is used in several chapters and in Appendix B. Appendix B provides LQR problem formulation and numerical solutions. Appendix C summarizes the background and solutions for robust pole assignment design which has been used in several chapters. For readers who need to know more background information on optimization and control theory, they are referred to some standard textbooks [8, 41, 92, 108, 142, 168, 233] listed in the references.

1.2 Some Basic Notations and Identities

In this book, vectors are denoted by small case letters with bold font, for example, $\mathbf{a}$ is a vector. Vector magnitude is denoted by normal font, for example, a is the magnitude of $\mathbf{a}$. A n-dimensional linear space is denoted by $\mathbf{R}^n$. A collection of all real points is denoted by $\mathbf{R}$. Matrices are denoted by capital letters with bold font, for example, $\mathbf{A}$ is a matrix, its magnitude is denoted by 2-norm $\|\mathbf{A}\|$ unless it is explicitly indicated that other matrix norm is used. A $n \times m$ matrix space, or the collection of all $n \times m$ linear transformation, is denoted by $\mathbf{R}^{n\times m}$.

Throughout this book, we will use some common notations. For a column vector $\mathbf{x} = [x_1, x_2, \ldots, x_n]^{\mathrm{T}}$, we sometimes write it as $\mathbf{x} = (x_1, x_2, \ldots, x_n)$ to save space. For any two vectors $\mathbf{x}$ and $\mathbf{y}$, we will denote by $\mathbf{x} \cdot \mathbf{y} = \mathbf{x}^{\mathrm{T}}\mathbf{y}$ the dot product of $\mathbf{x}$ and $\mathbf{y}$, by $\mathbf{x} \times \mathbf{y}$ the cross product of $\mathbf{x}$ and $\mathbf{y}$, by $\mathbf{x} \circ \mathbf{y}$ the element-wise or Hadamard product of $\mathbf{x}$ and $\mathbf{y}$, by $\frac{\mathbf{x}}{\mathbf{y}}$ the element-wise division of $\mathbf{x}$ and $\mathbf{y}$ if all elements of $\mathbf{y}$ are not zero, by $\|\mathbf{x}\|$ the 2-norm of the vector of $\mathbf{x}$. For a vector $\mathbf{x}$, we use $\mathbf{X}$ to denote a matrix whose diagonal elements are the vector $\mathbf{x}$. Let $\mathbf{a}$, $\mathbf{b}$, and $\mathbf{c}$ be any three dimensional vectors, we will repeatedly use the following identities.

$$\mathbf{a} \times \mathbf{b} = -\mathbf{b} \times \mathbf{a} \tag{1.1}$$

$$(\mathbf{a} \times \mathbf{b}) \times \mathbf{c} = (\mathbf{a} \cdot \mathbf{c})\mathbf{b} - (\mathbf{b} \cdot \mathbf{c})\mathbf{a} \tag{1.2}$$

and

$$\mathbf{a} \times (\mathbf{b} \times \mathbf{c}) = (\mathbf{a} \cdot \mathbf{c})\mathbf{b} - (\mathbf{a} \cdot \mathbf{b})\mathbf{c} \tag{1.3}$$

and

$$(\mathbf{a} \times \mathbf{b}) \cdot \mathbf{a} = (\mathbf{a} \times \mathbf{b}) \cdot \mathbf{b} = 0 \tag{1.4}$$

We denote

$$\mathbf{i} = (1,0,0), \quad \mathbf{j} = (0,1,0), \quad \mathbf{k} = (0,0,1) \tag{1.5}$$

for the standard basis for $\mathbf{R}^3$, and $\mathbf{S}(\mathbf{x})$ a skew-symmetric matrix function of $\mathbf{x} = [x_1, x_2, x_3]^{\mathrm{T}}$ defined by

$$\mathbf{S}(\mathbf{x}) = \begin{bmatrix} 0 & -x_3 & x_2 \\ x_3 & 0 & -x_1 \\ -x_2 & x_1 & 0 \end{bmatrix} = \mathbf{x}^{\times} \tag{1.6}$$

The *cross product* of $\mathbf{x} \times \mathbf{y}$ can then be represented by a matrix multiplication $\mathbf{S}(\mathbf{x})\mathbf{y}$, i.e., $\mathbf{x} \times \mathbf{y} = \mathbf{S}(\mathbf{x})\mathbf{y} = \mathbf{x}^{\times}\mathbf{y}$. We will use $\bar{\mathbf{p}}$, $\bar{\mathbf{q}}$, and $\bar{\mathbf{r}}$ to denote quaternions which will be defined later on.

Chapter 2

Orbit Dynamics and Properties

This chapter introduces the necessary background about orbit dynamics and properties, which will be used in the remaining chapters. The presentation of this chapter follows closely the style of [37, 181, 205].

2.1 Orbit dynamics

Let $\mathbf{f}$ denote the *force* applied to a particle in space, m be the *mass* of the particle, $\mathbf{v}$ be the *velocity* of the particle in space, $\mathbf{p} = m\mathbf{v}$ be the *linear momentum*, and $\mathbf{a} = \frac{d\mathbf{v}}{dt}$ be *linear acceleration*. The most important Newton's law is

$$\mathbf{f} = \frac{d\mathbf{p}}{dt} = \frac{dm\mathbf{v}}{dt} = m\mathbf{a} \tag{2.1}$$

For any two particles in space with masses m_1 and m_2, respectively, their *distance in space* is expressed by a vector $\mathbf{r}$, and they attract to each other with a force given by the expression

$$\mathbf{f} = \frac{Gm_1m_2\mathbf{r}}{r^3} \tag{2.2}$$

where $G = 6.669 * 10^{-11} m^3/kg - s^2$ is the *universal constant of gravitation*. The magnitude of the force is $f = \frac{Gm_1m_2}{r^2}$. Note that for any force $\mathbf{f}_{12}$ exerted by particle 1 on particle 2, there must exist a force $\mathbf{f}_{21}$ exerted by particle 2 on particle 1 with the same magnitude but in opposite direction, i.e.,

$$\mathbf{f}_{21} = -\mathbf{f}_{12} \tag{2.3}$$

For a selected coordinate, let O be the coordinate origin. For a particle with mass m, its position can be defined by a vector $\mathbf{r}$ from origin O pointing to its location. Then, the *moment of the force* $\mathbf{f}$ about the origin (also known as the *torque*) is given by

$$\mathbf{t} = \mathbf{r} \times \mathbf{f} \tag{2.4}$$

The angular momentum about O is defined as

$$\mathbf{h} = m(\mathbf{r} \times \mathbf{v}) \tag{2.5}$$

Taking derivative on both side of the equation gives:

$$\frac{d\mathbf{h}}{dt} = \frac{d}{dt}(\mathbf{r} \times m\mathbf{v}) = \mathbf{v} \times (m\mathbf{v}) + \mathbf{r} \times \frac{d}{dt}(m\mathbf{v}) \ = \mathbf{0} + \mathbf{r} \times \mathbf{f} = \mathbf{t} \tag{2.6}$$

Equation (2.6) is very important, which will be used throughout the book. For two body system, if the mass of one particle is much larger than the other particle, since the attracting force $\mathbf{f}$ is collinear with $\mathbf{r}$, therefore, $\mathbf{r} \times \mathbf{f} = \mathbf{0} = \mathbf{t} = \frac{d\mathbf{h}}{dt}$, i.e., $\mathbf{h}$ is a constant, the *orbit* of the smaller particle is a plane.

Now, let's consider the motion of a small particle with mass of unit around a much large particle with mass M in the coordinate system as described in Figure 2.1.

In view of (2.5), $\mathbf{h} = \mathbf{r} \times \mathbf{v}$, one has

$$h = rv\sin(\alpha) = rv\cos(\beta) = r\left(r\frac{d\theta}{dt}\right) = r^2\frac{d\theta}{dt} \tag{2.7}$$

In Figure 2.1, $\mathbf{i}$ and $\mathbf{j}$ are unit length vectors. Therefore, $\mathbf{r} = r\mathbf{i}$, and we have

$$\frac{d\mathbf{i}}{dt} = \frac{d\mathbf{i}}{d\theta}\frac{d\theta}{dt} = \mathbf{j}\frac{d\theta}{dt}, \quad \frac{d\mathbf{j}}{dt} = \frac{d\mathbf{j}}{d\theta}\frac{d\theta}{dt} = -\mathbf{i}\frac{d\theta}{dt} \tag{2.8}$$

Hence,

$$\frac{d\mathbf{r}}{dt} = r\frac{d\mathbf{i}}{dt} + \mathbf{i}\frac{dr}{dt} = \mathbf{j}r\frac{d\theta}{dt} + \mathbf{i}\frac{dr}{dt} \tag{2.9}$$

Since the particle has the mass of unit, from (2.1), it follows

$$\begin{aligned}
\mathbf{f} &= \mathbf{a} = \frac{d^2\mathbf{r}}{dt^2} = \frac{d}{dt}\left(\mathbf{j}r\frac{d\theta}{dt} + \mathbf{i}\frac{dr}{dt}\right) \\
&= \frac{d\mathbf{j}}{dt}r\frac{d\theta}{dt} + \mathbf{j}\frac{dr}{dt}\frac{d\theta}{dt} + \mathbf{j}r\frac{d^2\theta}{dt^2} + \frac{d\mathbf{i}}{dt}\frac{dr}{dt} + \mathbf{i}\frac{d^2r}{dt^2} \\
&= -\mathbf{i}\frac{d\theta}{dt}r\frac{d\theta}{dt} + \mathbf{j}\frac{dr}{dt}\frac{d\theta}{dt} + \mathbf{j}r\frac{d^2\theta}{dt^2} + \mathbf{j}\frac{d\theta}{dt}\frac{dr}{dt} + \mathbf{i}\frac{d^2r}{dt^2} \\
&= \mathbf{i}\left(\frac{d^2r}{dt^2} - r\left(\frac{d\theta}{dt}\right)^2\right) + \mathbf{j}\left(r\frac{d^2\theta}{dt^2} + 2\frac{d\theta}{dt}\frac{dr}{dt}\right)
\end{aligned} \tag{2.10}$$

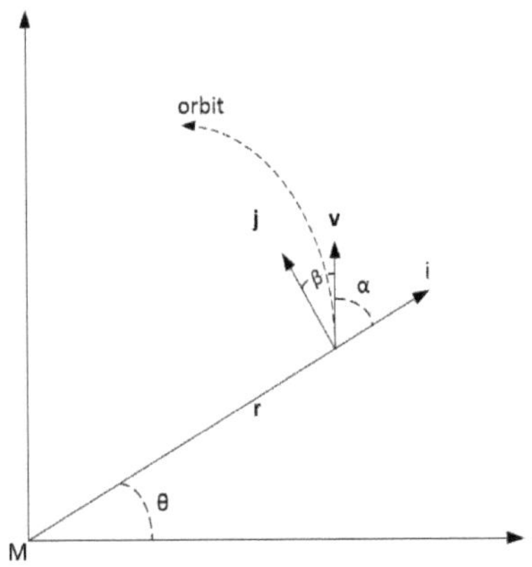

Figure 2.1: Radial and transverse components of motion in a plane.

Using (2.2) with $m_1 = 1$ unit and $m_2 = M$, we find

$$\mathbf{f} = \mathbf{a} = -\frac{GM}{r^3}\mathbf{i}r \tag{2.11}$$

Combining these two equations gives:

$$\frac{d^2r}{dt^2} - r\left(\frac{d\theta}{dt}\right)^2 = -\frac{GM}{r^2}, \quad r\frac{d^2\theta}{dt^2} + 2\frac{d\theta}{dt}\frac{dr}{dt} = 0 \tag{2.12}$$

The second equation implies

$$\frac{1}{r}\frac{d}{dt}\left(r^2\frac{d\theta}{dt}\right) = 0 \tag{2.13}$$

in view of (2.7), this implies

$$h = r^2\frac{d\theta}{dt} = constant \tag{2.14}$$

The first equation of (2.12) is a nonlinear differential equation and cannot be solved directly. Let $r = \frac{1}{u}$. Taking derivative on both sides yields

$$\frac{dr}{dt} = -\frac{1}{u^2}\frac{du}{dt} = -\frac{1}{u^2}\frac{du}{d\theta}\frac{d\theta}{dt} \tag{2.15}$$

Substituting $r = \frac{1}{u}$ into (2.14) yields

$$\frac{d\theta}{dt} = hu^2 \tag{2.16}$$

Substituting this equation into (2.15) gives

$$\frac{dr}{dt} = -\frac{1}{u^2}\frac{du}{d\theta}hu^2 = -h\frac{du}{d\theta} \tag{2.17}$$

Note that $\frac{dh}{dt} = 0$, taking the second derivative on both sides yields

$$\frac{d^2r}{dt^2} = -h\frac{d}{dt}\frac{du}{d\theta} = -h\frac{d}{d\theta}\frac{du}{d\theta}\frac{d\theta}{dt} = -h\frac{d^2u}{d\theta^2}\frac{d\theta}{dt} = -h^2u^2\frac{d^2u}{d\theta^2} \tag{2.18}$$

Denote the *standard gravitational parameter* $GM = \mu$ (μ is also known as the geocentric gravitational constant). Combining the first equations of (2.12), (2.14), and (2.18) yields

$$\begin{aligned}
& \frac{d^2r}{dt^2} = r\left(\frac{d\theta}{dt}\right)^2 - \frac{\mu}{r^2} \\
\Longleftrightarrow\quad & \frac{d^2r}{dt^2} = \frac{1}{u}\left(\frac{d\theta}{dt}\right)^2 - \mu u^2 \\
\Longleftrightarrow\quad & -h^2u^2\frac{d^2u}{d\theta^2} = \frac{1}{u}h^2u^4 - \mu u^2 \\
\Longleftrightarrow\quad & \frac{d^2u}{d\theta^2} = -u + \mu/h^2
\end{aligned} \tag{2.19}$$

The last equation is a second order linear differential equation of u which has the solution of the following form:

$$u = \frac{\mu}{h^2} + c\cos(\theta - \theta_0) \tag{2.20}$$

where c is a constant to be determined. Taking derivative of (2.20) yields

$$\frac{du}{d\theta} = -c\sin(\theta - \theta_0) \tag{2.21}$$

Let

$$E = v^2/2 - \mu/r \tag{2.22}$$

be the *total energy per unit mass*. The term of $v^2/2$ is the *kinetic energy* and μ/r is *potential energy* of the unit mass. Invoking (2.17), (2.9), and (2.16), one can write

$$v^2 = \left(\frac{dr}{dt}\right)^2 + \left(r\frac{d\theta}{dt}\right)^2 = \left(-h\frac{du}{d\theta}\right)^2 + \left(\frac{1}{u}hu^2\right)^2 = h^2\left[\left(\frac{du}{d\theta}\right)^2 + u^2\right] \tag{2.23}$$

Substituting (2.21) and (2.20) into this equation gives

$$v^2 = h^2\left[c^2 + \frac{2c\mu}{h^2}\cos(\theta - \theta_0) + \left(\frac{\mu}{h^2}\right)^2\right] = c^2h^2 + 2c\mu\cos(\theta - \theta_0) + (\mu/h)^2 \tag{2.24}$$

The principle of conservation of energy implies that $E = v^2/2 - \mu/r$ is a constant for any θ. Taking $\theta - \theta_0 = \frac{\pi}{2}$ and using (2.20) yields

$$E \;=\; v^2/2 - \mu/r$$

$$\begin{aligned} &= (ch)^2/2 + (\mu/h)^2/2 - u\mu \\ &= (ch)^2/2 + (\mu/h)^2/2 - \frac{\mu}{h^2}\mu \\ &= \frac{(ch)^2}{2} - \frac{\mu^2}{2h^2} \end{aligned} \tag{2.25}$$

This gives

$$c = \frac{\mu}{h^2}\sqrt{1 + 2E\frac{h^2}{\mu^2}} \tag{2.26}$$

Denote

$$e = \sqrt{1 + 2E\frac{h^2}{\mu^2}} \tag{2.27}$$

it can be seen later that e is the *eccentricity of the orbit*. Therefore, an important relationship between the eccentricity and the total energy of the orbit is given by

$$E = (e^2 - 1)\frac{\mu^2}{2h^2} \tag{2.28}$$

Substituting $c = \frac{\mu}{h^2}e$ and $r = 1/u$ into (2.20) yields one of the most important result so far.

$$r = \frac{h^2/\mu}{1 + e\cos(\theta - \theta_0)} = \frac{p}{1 + e\cos(\theta - \theta_0)} \tag{2.29}$$

where

$$p = h^2/\mu \tag{2.30}$$

is called the *semi-latus rectum*.

2.2 Conic Section and Different Orbits

Spacecraft orbits are closely related to *conic sections*. A conic section is the intersection of a plane and a right circular cone. Different intersections result in different orbital shapes: *circle*, *ellipse*, *parabola*, and *hyperbola* (see Figure 2.2). Since parabolic orbit is of no importance in the context of spacecrafts, therefore, only the circle, ellipse, and hyperbola orbits have been discussed here.

2.2.1 Circular Orbits

For *circular orbits*, the eccentricity meets the condition of $e = 0$ and r, the magnitude of the radius vector $\mathbf{r}$ of the orbit from the only *focus*, is a constant that meets the condition:

$$r = p = h^2/\mu = [rvcos(\beta)]^2/\mu \tag{2.31}$$

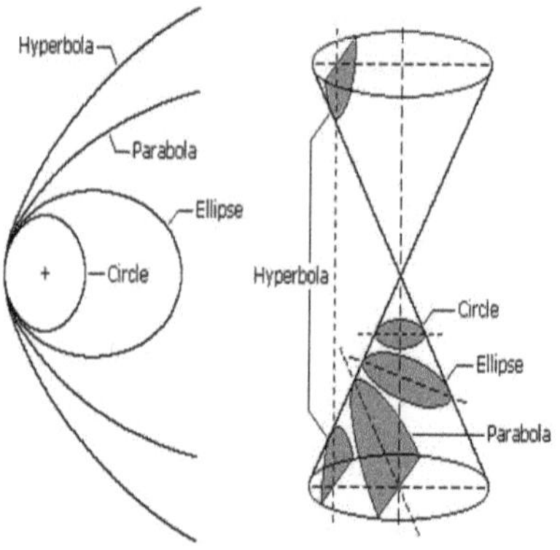

Figure 2.2: The orbits defined by the conic section.

In view of Figure 2.1, for circular orbit, it has $\beta = 0$ (the velocity of the body is perpendicular to the radius vector $\mathbf{r}$), therefore, it follows that

$$v^2 = \mu / r \tag{2.32}$$

This shows that the velocity v is a constant. Moreover, the energy is given by

$$E = -\mu^2/(2h^2) \tag{2.33}$$

2.2.2 Elliptic Orbits

For *elliptic orbit*, the eccentricity meets the condition of $0 < e < 1$, and from (2.28), its energy is given by $E < 0$. Representing the ellipse in a two-dimensional space, it is shown in Figure 2.3.

The point on the ellipse at $\theta = 0^o$ is called *perigee*, which corresponds to point A. The point on the ellipse at $\theta = 180^o$ is called *apogee*, which corresponds to point B. The foci are the points $F = (c, 0)$ and $F' = (-c, 0)$. The *prime focus* of the ellipse is F. For $\mathbf{r}$ at point A (the perigee, $\theta - \theta_0 = 0^o$), it follows from (2.29) that

$$r_p = \frac{p}{1+e} \tag{2.34}$$

For $\mathbf{r}$ at point B (the apogee, $\theta - \theta_0 = 180^o$), it follows from (2.29) that

$$r_a = \frac{p}{1-e} \tag{2.35}$$

Combining (2.34) and (2.35) gives

$$\frac{r_a}{r_p} = \frac{1+e}{1-e} \tag{2.36}$$

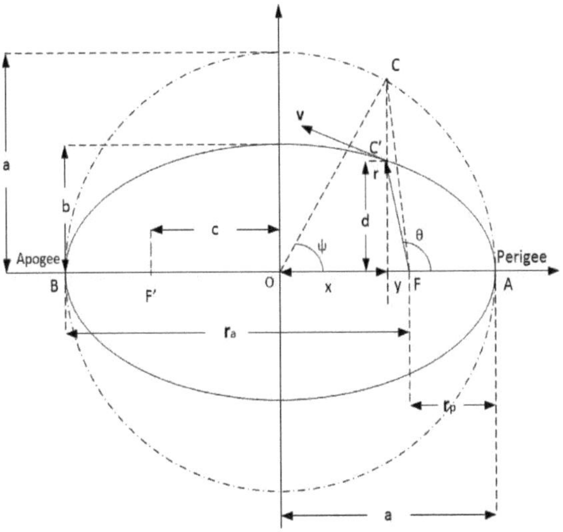

Figure 2.3: The ellipse orbit defined on a plane.

from which it follows that

$$e = \frac{r_a - r_p}{r_a + r_p} \tag{2.37}$$

In view of the Figure 2.3, the *major axis* of the ellipse is

$$2a = r_a + r_p = \frac{2p}{1 - e^2} \tag{2.38}$$

this yields

$$p = a(1 - e^2) = h^2/\mu \tag{2.39}$$

where a is called the *semi-major axis*. From (2.26) and (2.28), it follows that the total energy of a body with unit mass in the orbit is

$$E = \frac{v^2}{2} - \frac{\mu}{r} = \frac{(e^2-1)\mu^2}{2h^2} = \frac{(e^2-1)\mu}{2p} = \frac{(e^2-1)\mu}{2a(1-e^2)} = -\frac{\mu}{2a} \tag{2.40}$$

This yields

$$\frac{v^2}{2} = \frac{\mu}{r} - \frac{\mu}{2a} \tag{2.41}$$

Therefore, for an orbit to be elliptic, it must have

$$\frac{v^2}{2} < \frac{\mu}{r} \tag{2.42}$$

For an ellipse, it is known that $c = ae$. In view of (2.39), it follows that

$$b = \sqrt{a^2 - c^2} = a\sqrt{1-e^2} = \frac{p\sqrt{1-e^2}}{1-e^2} \frac{p}{\sqrt{1-e^2}} \tag{2.43}$$

where b is called *semi-minor axis* of the elliptic orbit. Combining (2.39) and $c = ae$ yields

$$c = \frac{pe}{1-e^2} \tag{2.44}$$

2.2.3 Hyperbolic Orbits

In this orbit, $e > 0$, in view of (2.28), it follows that $E > 0$. This means that the kinetic energy of the spacecraft is larger than its potential energy. Therefore, the spacecraft is about to leave the gravitational attraction field of the central body.

2.3 Property of Keplerian Orbits

This section discusses elliptic orbit. The location of the spacecraft in an orbit can be presented either by its angular deviation from the major axis or by the time elapsed from its passage at the perigee. In Figure 2.3, the *true anomaly* θ is defined as an angle between the major axis pointing to the perigee and the radius vector from the prime focus F to the spacecraft. To define the *eccentric anomaly*, an *auxiliary circle* with radius a is centered at the middle of the major axis. The eccentric anomaly ψ is the angle between OA and OC defined in Figure 2.3.

The relation between true anomaly and eccentric anomaly is derived as follows. Note

$$x + y = ae = c \tag{2.45a}$$

$$x = a\cos(\psi) \tag{2.45b}$$

$$y = r\cos(180 - \theta) = -r\cos(\theta) \tag{2.45c}$$

it follows

$$x + y = a\cos(\psi) - r\cos(\theta) = ae \tag{2.46}$$

From equations (2.29) and (2.39), it follows

$$\begin{aligned} x &= a\cos(\psi) = ae + r\cos(\theta) = ae + \frac{p\cos(\theta)}{1+e\cos(\theta)} \\ &= ae + \frac{a(1-e^2)\cos(\theta)}{1+e\cos(\theta)} = \frac{ae + a\cos(\theta)}{1+e\cos(\theta)} \end{aligned} \tag{2.47}$$

Therefore,

$$\cos(\psi) = \frac{e+\cos(\theta)}{1+e\cos(\theta)}, \quad \sin(\psi) = \sqrt{1-\cos^2(\psi)} = \frac{\sin(\theta)\sqrt{1-e^2}}{1+e\cos(\theta)} \tag{2.48}$$

This gives

$$\cos(\theta) = \frac{\cos(\psi) - e}{1 - e\cos(\psi)}, \quad \sin(\theta) = \frac{\sin(\psi)\sqrt{1-e^2}}{1-e\cos(\psi)} \tag{2.49}$$

Applying standard trigonometry yields

$$\tan\left(\frac{\theta}{2}\right) = \frac{\sin(\theta)}{1+\cos(\theta)} = \sqrt{\frac{1+e}{1-e}}\tan\left(\frac{\psi}{2}\right) \tag{2.50}$$

Substituting (2.39) and (2.49) into (2.29) yields

$$r = \frac{p}{1+e\cos(\theta)} = \frac{a(1-e^2)}{1+e\cos(\theta)} = \frac{a(1-e^2)}{1+e\frac{\cos(\psi)-e}{1-e\cos(\psi)}} = a(1-e\cos(\psi)) \tag{2.51}$$

Now, it is ready to derive *Kepler's second* and *third law*. In Figure 2.4, the spacecraft position vector **r** is swept in a differential period of time, the differential area $\Delta A = (\Delta\theta r^2)/2$. Therefore, it follows from (2.7) and (2.14) that

$$\frac{dA}{dt} = \frac{1}{2}\left(r^2\frac{d\theta}{dt}\right) = \frac{1}{2}h = constant \tag{2.52}$$

This proves Kepler's second law: the time rate of change in area is a constant. Integration of the above equation, the area swept in time t is given by

$$A = \frac{1}{2}ht \tag{2.53}$$

Because the area of a ellipse is $A = \pi ab$, if the time period of the orbit is $t = T$, from (2.53), (2.39), and (2.43) it follows that the *orbit period* of the spacecraft is given by

$$\begin{aligned} T &= \frac{2A}{h} = \frac{2\pi ab}{\sqrt{p\mu}} = \frac{2\pi ab}{\sqrt{a(1-e^2)\mu}} = \frac{2\pi a^2\sqrt{1-e^2}}{\sqrt{a(1-e^2)\mu}} \\ &= 2\pi\sqrt{\frac{a^3}{\mu}} = \frac{2\pi}{\omega_0} \end{aligned} \tag{2.54}$$

where

$$\omega_0 = \sqrt{\frac{\mu}{a^3}} = \frac{2\pi}{T} \tag{2.55}$$

is named the *mean motion*, and

$$M = \omega_0(t-t_p) = \frac{2\pi}{T}(t-t_p) \tag{2.56}$$

is named the *mean anomaly*, where t_p is the passing time from perigee. Equation (2.54) is the so-called Kepler's third law.

The last formula to be derived in this section is *Kepler's time equation*. Let the area (AFC') be denoted by S(AFC') and the area (AFC) be denoted by S(AFC) in Figure 2.3. Let $t_m = t - t_p$. Then, it follows from the law of the area that

$$\frac{t_m}{S(AFC')} = \frac{T}{\pi ab} \tag{2.57}$$

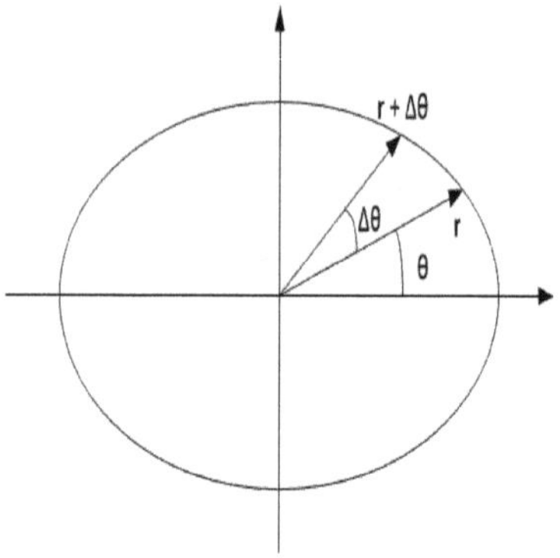

Figure 2.4: Geometry for deriving the law of area.

Since

$$S(AFC') = \frac{b}{a} S(AFC) \tag{2.58}$$

and

$$\begin{aligned} S(AFC) &= \frac{\psi}{2\pi}(\pi a^2) - S(OCF) \\ &= \frac{\psi a^2}{2} - \frac{1}{2} ac \sin(\psi) \\ &= \frac{\psi a^2}{2} - \frac{1}{2} a^2 e \sin(\psi) \end{aligned} \tag{2.59}$$

it follows from (2.57) and (2.58) that

$$t_m = \frac{b}{a} \frac{T}{\pi ab} \left(\frac{\psi a^2}{2} - \frac{1}{2} a^2 e \sin(\psi) \right) = \frac{T}{2\pi} [\psi - e \sin(\psi)] \tag{2.60}$$

In view of (2.56), this is equivalent to

$$t_m \frac{2\pi}{T} = (t - t_p)\omega_0 = M = \psi - e \sin(\psi) \tag{2.61}$$

2.4 Keplerian Orbits in Three-dimensional Space

In the previous section, Keplerian orbits in the orbital plane were discussed, which are easy to deal with. In the real world, a convenient spacecraft coordinate system is most likely in a three-dimensional space and the orbital plane is more likely a plane embedded in a three-dimensional space.

2.4.1 Celestial Inertial Coordinate System

For an Earth-orbiting spacecraft, it is convenient to define the center of mass of the Earth as its origin (a geocentric system). To make it easy to use the formulas developed in the previous sections of this chapter, the coordinate system should be an *inertial coordinate system* without acceleration or deceleration. Since the Earth moves in an almost circular orbit around the Sun with a long period, therefore, it is practically acceptable as an inertial system. Let **Z** be Earth's rotational axis, and this axis is selected as the **Z** axis of the inertial coordinate system. The **Z** direction is perpendicular to the Earth's equator which is in the **X**-**Y** plane of this coordinate system.

Next, we define the **X** axis of the *geocentric inertial coordinate system*. It is known that the Earth's equator plane is not on the same plane as of the *ecliptic plane*, which is the plane of the Earth orbiting around the Sun. The Earth's equator plane is inclined to the ecliptic plane at an angle of about 23.5°. The two planes intersect along a line that is called the *vernal equinox* vector (see Figure 2.5). While the Earth rotates around the Sun, it crosses this line twice a year. The point when Earth cross this line in March is called *vernal equinox*. The direction from the center of mass of the Sun to the vernal points is defined as the **X** direction of the geocentric inertial system. The third axis **Y** completes an orthogonal right-hand system. Both equator and ecliptic planes move slowly because of the force of attraction of astronomical bodies. The coordinate axes may need some corrections over the time.

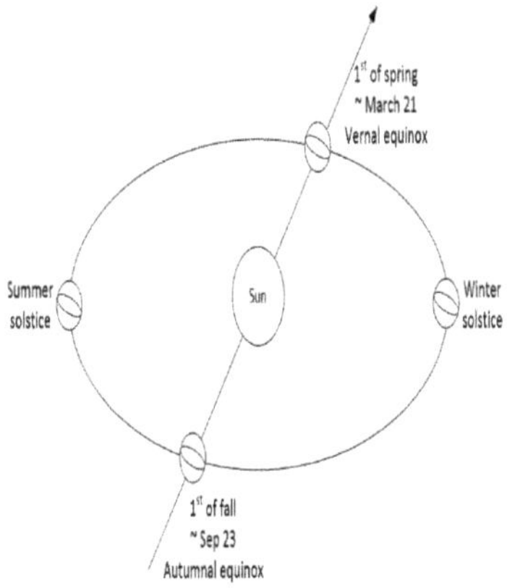

Figure 2.5: Vernal equinox description.

2.4.2 Orbital Parameters

Given the geocentric inertial coordinate system, the spacecraft orbit in this system can be described in Figure 2.6. As explained, the **X**-**Y** plane is the equator plane. **Z**-axis is the rotational axis of the Earth. The vector $\mathbf{r}_p$ is the vector from the center of the mass of the Earth pointing to the *perigee*. The vector **r** is a moving vector which moves from the center of the Earth to the position of the spacecraft, which moves along the direction **v**. The angle between $\mathbf{r}_p$ and **r**, θ, is called *true anomaly* which was defined in Figure 2.3 in a two-dimensional orbit plane. A coordinate system in the orbit plane is given by three vectors **P**, **Q**, and **W**, where **P** is the unit length vector from the *primary focus* (the center of the mass of the Earth) pointing to the perigee of the orbit. The unit length vector **Q** is on the orbit plane and 90° from **P** in the direction of the moving spacecraft. **W** is defined by $\mathbf{P} \times \mathbf{Q}$, which is the unit length vector along the *momentum axis* of the orbit. The angle between the orbit plane and the equator plane, i, is named as the *inclination* of the orbit. The orbit crosses the **X**-**Y** plane at two points, one is the *ascending node*, the other one is the *descending node*. The line that passes through the ascending node and descending node is called the *node line*. The angle between **X** axis and the node line pointing towards the ascending node is called the *right ascension*, Ω. The angle between the node line pointing towards the ascending node and **P** is ω which is called the *argument of perigee*. The three angles, i, Ω, and ω, plus three parameters discussed before, a, e, and $M = n(t - t_0)$, are known as classical orbit parameters. It is convenient to define a vector $\alpha = [a, e, i, \Omega, \omega, M]$ for the orbit parameters, which are summarized below:

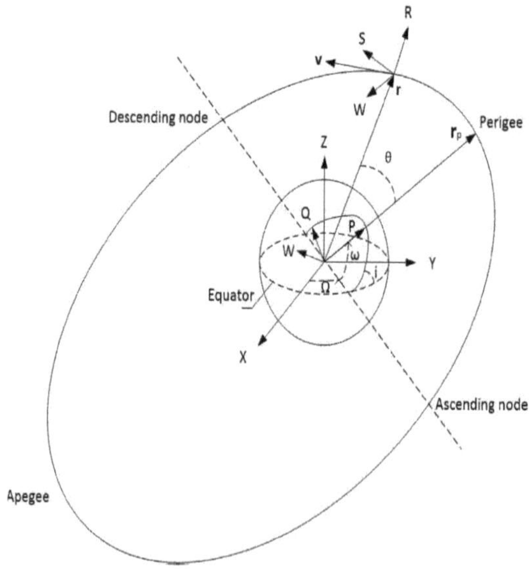

Figure 2.6: Parameters in orbit.

a, the semi-major axis;

e, the eccentricity;

i, the inclination;

Ω, the right ascension of the ascending node;

ω, the argument of the perigee; and

$M = \omega_0(t - t_0)$, the mean anomaly.

Clearly, there is another way to present the spacecraft moving around the orbit by given $(\mathbf{v}, \mathbf{r})$ at any time. Chapter 3 provides in detail the transformations between these two different presentations.

Chapter 3

Rotational Sequences and Quaternion

Based on the missions of a spacecraft, the attitude of the spacecraft represented by the body frame should be aligned with some desired frame. Spacecraft attitude determination is to provide the information of the difference between the spacecraft body frame and the desired frame. The desired spacecraft frame also depends on the spacecraft position and the current time, GPS signals may be used to determine the spacecraft current position and the time. Mostly the *universal time* (UT) is used in aerospace engineering [205]. The time and position can be used to calculate the ephemeris astronomical direction information, such as stars' directions, the Sun's direction, the Earth's direction, the Earth's magnetic field's direction, observed from the spacecraft position at the current time and represented in the desired frame. The body frame information can be obtained by the measurements of these directions by the spacecraft's on-board instruments. When the body frame is perfectly aligned with the desired frame, the calculated ephemeris stars' direction, the Sun's direction, the Earth's direction, and the Earth's magnetic field's direction at the given time should be identical or very close to the measurements by the spacecraft instruments. When the body frame is significantly different from the desired frame, the measured astronomical directions are significantly different from the ephemeris astronomical directions at the given time. This difference can be represented by a single rotation if quaternion is used or a series of rotations if Euler angles are used. In the latter case, the sequence of the rotations is very important. These rotations rotate some angle around certain rotational axis, thereby estimating the distance between the spacecraft body frame and the desired frame. Therefore, the mathematical definition of rotation and rotational sequences are the most important concepts in spacecraft attitude determination and control. There are many ways to characterize the rotation and rotational sequences. The quaternion representation is one of the best characterizations, and this chapter will focus on this representation. The presentation in this chapter follows the style of [99, 205, 219].

3.1 Some Frequently used Frames

Many coordinate frames are used in spacecraft related application. This section discusses some of the most important frames. For in-depth discussion, readers can refer to [205].

3.1.1 Body-fixed Frame

The body coordinate system is vehicle-carried and directly defined on the body of the spacecraft. Its origin is located at the center of the mass of the spacecraft. There may be different ways to define its axes. In this book, the axes are defined by using the so-called principal axes of rotation of the rigid body. Let $\mathbf{J}$ be the moment of inertia matrix of the spacecraft, which is a three-dimensional and real symmetric matrix. Because $\mathbf{J}$ is the real symmetric, it has three mutually orthogonal eigenvectors which are associated with three real eigenvalues, i.e., there are $\lambda_i, i = 1,2,3$, and $\mathbf{x}_i, i = 1,2,3$ such that

$$\mathbf{J}\mathbf{x}_i = \lambda_i \mathbf{x}_i \tag{3.1}$$

where, assuming that the spacecraft is in normal operation, $\mathbf{x}_1$ defines the axis $\mathbf{X}_b$ which points forward in the direction of the spacecraft velocity (but may not be identical unless the orbit is circular), $\mathbf{x}_2$ defines the axis $\mathbf{Z}_b$ which points downward and is on the orbit plane, and $\mathbf{x}_3$ defines the axis $\mathbf{Y}_b$ which complies with the right-hand rule.

3.1.2 The Earth Centered Inertial (ECI) Frame

The *Earth Centered Inertial* (ECI) frame is important because of two reasons. First, the Newton's laws of motion and gravity applied to the spacecraft are defined in the inertial frame. Second, many types of satellites are inertial pointing spacecraft. This frame is defined relative to the rotation axis of the Earth and the plane of the Earth's orbit (the ecliptic plane) about the Sun. The Earth's equator is perpendicular to the rotation axis of the Earth. As the Earth moves along the ecliptic orbit, the equator plane and the ecliptic have two cross points. These two cross points are special as the tilt of the Earth's rotational axis is inclined neither away nor towards the Sun (the center of the Sun being in the same plane as the

Earth's equator). The ECI frame is defined at one of these equinoxes, the *vernal equinox* (or March equinox). Because of many less significant (but may not be negligible) factors, such as the *precession of the equinoxes*, vernal equinox used by aerospace engineers is defined by 2000 coordinates and the true of date (TOD)[1]. The $\mathbf{X}_I$ of inertial frame is the direction from the Earth center to the vernal equinox. The $\mathbf{Z}_I$ axis is the Earth rotational axis. The $\mathbf{Y}_I$ follows the right-hand rule.

3.1.3 Local Vertical Local Horizontal Frame

The *local vertical local horizontal frame* (LVLH) is one of the most desired frames for many satellites because its $\mathbf{Z}_{lvlh}$ direction is always pointing to the center of the Earth (nadir pointing), which is a desired feature of many satellites. The origin of the local vertical local horizontal frame is the center of mass of an orbital spacecraft. The $\mathbf{X}_{lvlh}$ direction is along the spacecraft velocity direction and perpendicular to $\mathbf{Z}_{lvlh}$, and $\mathbf{Y}_{lvlh}$ is perpendicular to the orbit plan and follows the right-hand rule.

3.1.4 South-east Zenith (SEZ) Frame

The *South-east Zenith frame* is useful for ground stations to track a spacecraft. The location of the tracking instrument is the origin. $\mathbf{X}_{SEZ}$ is the direction pointing towards the south, $\mathbf{Y}_{SEZ}$ is the direction pointing towards the east, and $\mathbf{Z}_{SEZ}$ is the direction pointing towards the zenith. In this system, the azimuth is the angle measured from north, clockwise to the location beneath the object of interest. The elevation is measured from local horizon, positive up to the object of interest.

3.1.5 North-east Nadir (NED) Frame

The *North-east Nadir frame* is opposite to the SEZ frame which is defined by the local horizontal plane. The center of the horizontal plane is the origin. $\mathbf{X}_{NED}$ is the direction pointing towards the north, $\mathbf{Y}_{NED}$ is the direction pointing towards the east, and $\mathbf{Z}_{NED}$ is the nadir direction.

3.1.6 The Earth-centered Earth-fixed (ECEF) Frame

Like the Earth Centered Inertial (ECI) frame, the *Earth-centered Earth-fixed* (ECEF) frame is the Earth-based frame. The ECI frame is independent from the motion and the rotation of the Earth. However, it may not be convenient in some case as observatories on the ground rotate with the Earth. The center of ECEF

[1]For the rigorous and precise definition, please read [205].

frame is the center of the Earth. Using the convention adopted at the International Meridian Conference in Washington D.C. 1884, the primary meridian for the Earth is the meridian that the Royal Observatory at Greenwich lies on. The $\mathbf{X}_{ECEF}$ is the direction from the center of the Earth pointing towards the cross point of the primary meridian and equator. The $\mathbf{Z}_{ECEF}$ is the direction from the center of the Earth pointing towards the north pole. The $\mathbf{Y}_{ECEF}$ is the direction that follows the right-hand rule. The ECEF frame is sometimes called International Terrestrial Reference Frame (ITRF). Due to the plate tectonic motion, the frame may need some adjustment every year for certain applications.

3.1.7 The Orbit (Perifocal PQW) Frame

In *Perifocal PQW frame*, the fundamental plane is the spacecraft orbit, and the origin is at the center of the Earth (see Figure 2.6). The $\mathbf{P}_x$ axis points towards perigee, and the $\mathbf{Q}_y$ is 90^0 from $\mathbf{P}_x$ axis in the direction of the spacecraft motion. The $\mathbf{W}_z$ is normal to the orbit represented by $\mathbf{W}_z = \mathbf{P}_x \times \mathbf{Q}_y$.

3.1.8 The Spacecraft Coordinate (RSW) Frame

The *Spacecraft Coordinate (RSW) frame* is closely related to LVLH frame (see Figure 2.6). The $\mathbf{R}_x$ axis always points from the Earth's center towards the spacecraft as it moves through the orbit. The $\mathbf{S}_y$ axis points in the direction of (but not necessarily parallel to) the velocity vector and is perpendicular to the $\mathbf{R}_x$ axis, an important additional requirement. The $\mathbf{W}_z$ axis is normal to the orbital plane represented by $\mathbf{W}_z = \mathbf{R}_x \times \mathbf{S}_y$.

3.2 Rotation Sequences and Mathematical Representations

3.2.1 Representing a Fixed Point in a Rotational Frame

As discussed in the beginning of this chapter, the spacecraft attitude is determined by locating the astronomical objects in the sky from the spacecraft instruments which gives the directions in the body frame; from the ephemeris information represented in the desired frame. Therefore, there is information on some fixed (astronomical object) point in a rotational frame when the spacecraft body frame is different from the desired frame. This is equivalent to represent a fixed point in a rotational frame.

Let $(\mathbf{X}, \mathbf{Y}, \mathbf{Z})$ be the axes of a frame (see Figure 3.1, where $\mathbf{Z}$-axis points out of the paper), and $(\mathbf{x}, \mathbf{y}, \mathbf{z})$ be the axes of another frame which rotates an angle of θ about $\mathbf{Z}$ axis. Let P be a fixed point in $(\mathbf{X}, \mathbf{Y})$ plane. Assume that the distance

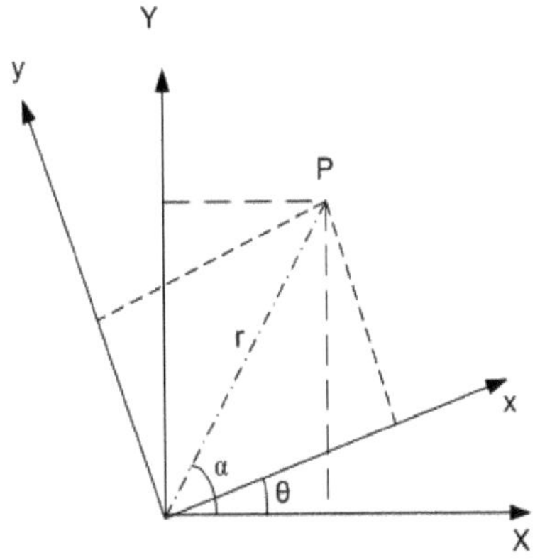

Figure 3.1: A fixed point in a rotational frame.

of P from the origin is r, then P can be expressed in the first frame coordinate as (x_1, y_1, z_1)

$$x_1 = r\cos(\alpha), \quad y_1 = r\sin(\alpha), \quad z_1 = 0 \tag{3.2}$$

and in the second frame coordinate as (x_2, y_2, z_2)

$$x_2 = r\cos(\alpha - \theta), \quad y_2 = r\sin(\alpha - \theta), \quad z_2 = 0$$

Thus, in view of (3.2), we have

$$\begin{aligned} x_2 &= r\cos(\alpha)\cos(\theta) + r\sin(\alpha)\sin(\theta) \\ &= x_1\cos(\theta) + y_1\sin(\theta) \\ y_2 &= r\sin(\alpha)\cos(\theta) - r\cos(\alpha)\sin(\theta) \\ &= y_1\cos(\theta) - x_1\sin(\theta) \\ z_2 &= 0 \end{aligned} \tag{3.3}$$

We can write this transformation in a matrix form

$$\begin{bmatrix} x_2 \\ y_2 \\ z_2 \end{bmatrix} = \begin{bmatrix} \cos(\theta) & \sin(\theta) & 0 \\ -\sin(\theta) & \cos(\theta) & 0 \\ 0 & 0 & 1 \end{bmatrix} \begin{bmatrix} x_1 \\ y_1 \\ z_1 \end{bmatrix} := Rot_3(\theta) \tag{3.4}$$

Similarly, for a fixed point, if the frame rotates about the **Y**-axis for an angle θ, then the transformation can be expressed as

$$\begin{bmatrix} x_2 \\ y_2 \\ z_2 \end{bmatrix} = \begin{bmatrix} \cos(\theta) & 0 & -\sin(\theta) \\ 0 & 1 & 0 \\ \sin(\theta) & 0 & \cos(\theta) \end{bmatrix} \begin{bmatrix} x_1 \\ y_1 \\ z_1 \end{bmatrix} := Rot_2(\theta) \tag{3.5}$$

For a fixed point, if the frame rotates about the **X**-axis for an angle θ, then the transformation can be expressed as

$$\begin{bmatrix} x_2 \\ y_2 \\ z_2 \end{bmatrix} = \begin{bmatrix} 1 & 0 & 0 \\ 0 & \cos(\theta) & \sin(\theta) \\ 0 & -\sin(\theta) & \cos(\theta) \end{bmatrix} \begin{bmatrix} x_1 \\ y_1 \\ z_1 \end{bmatrix} := Rot_1(\theta) \tag{3.6}$$

Rotational matrices of (3.4), (3.5), and (3.6) are all unitary matrices. By definition, the length of each column of a *unitary matrix* is one, each column is orthogonal to other columns. Unitary matrices have many useful properties. Let $\mathbf{C}_1$ and $\mathbf{C}_2$ be two unitary matrices and $\mathbf{v}$ be a vector. Some most important properties of the unitary matrix are (see [58]):

- $\|\mathbf{C}_1\mathbf{v}\| = \|\mathbf{C}_2\mathbf{v}\| = \|\mathbf{v}\|$, i.e., transformation by a unitary matrix does not change the vector length.
- $\mathbf{C}_2\mathbf{C}_1$ is a unitary matrix. For rotational matrices, it means that the consecutive rotations can be expressed by the product of the rotational matrices, where $\mathbf{C}_1$ is the first rotation and $\mathbf{C}_2$ is the second rotation.
- $\mathbf{C}_1^{-1} = \mathbf{C}_1^{\mathrm{T}}$, i.e., the inverse of a rotational matrix is simply a transpose of the rotational matrix.

3.2.2 Representing a Rotational Point in a Fixed Frame

When analyzing the relationship between frames, there is sometimes a need to represent a rotational point in a fixed frame. Let P_1 be a point obtained by rotating P at an angle of θ around the **Z**-axis (see Figure 3.2 where **Z**-axis points out of the paper). Then P_1 can be expressed as

$$x_2 = r\cos(\alpha+\theta), \quad y_2 = r\sin(\alpha+\theta), \quad z_2 = 0$$

Thus, in view of (3.2), we have

$$x_2 = x_1\cos(\theta) - y_1\sin(\theta), \quad y_2 = y_1\cos(\theta) + x_1\sin(\theta), \quad z_2 = 0$$

This transformation in a matrix form can be written as

$$\begin{bmatrix} x_2 \\ y_2 \\ z_2 \end{bmatrix} = \begin{bmatrix} \cos(\theta) & -\sin(\theta) & 0 \\ \sin(\theta) & \cos(\theta) & 0 \\ 0 & 0 & 1 \end{bmatrix} \begin{bmatrix} x_1 \\ y_1 \\ z_1 \end{bmatrix} := Rot_3(-\theta) \tag{3.7}$$

Similarly, for a rotational point, if it rotates about the **Y**-axis for an angle θ, then the transformation can be expressed as

$$\begin{bmatrix} x_2 \\ y_2 \\ z_2 \end{bmatrix} = \begin{bmatrix} \cos(\theta) & 0 & \sin(\theta) \\ 0 & 1 & 0 \\ -\sin(\theta) & 0 & \cos(\theta) \end{bmatrix} \begin{bmatrix} x_1 \\ y_1 \\ z_1 \end{bmatrix} := Rot_2(-\theta) \tag{3.8}$$

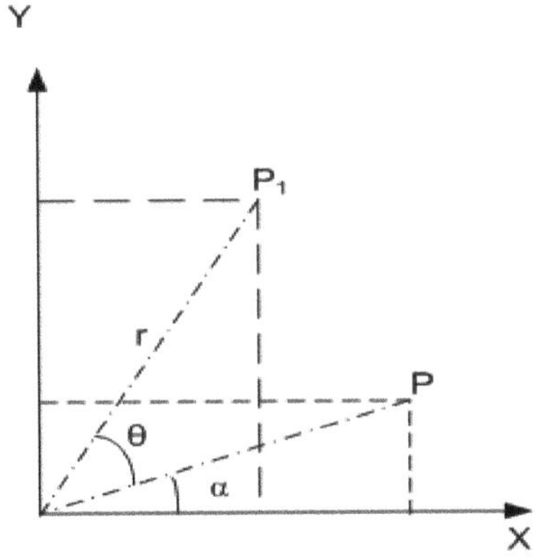

Figure 3.2: A rotational point in a fixed frame.

For a rotational point, if it rotates about the **X**-axis for an angle θ, then the transformation can be expressed as

$$\begin{bmatrix} x_2 \\ y_2 \\ z_2 \end{bmatrix} = \begin{bmatrix} 1 & 0 & 0 \\ 0 & \cos(\theta) & -\sin(\theta) \\ 0 & \sin(\theta) & \cos(\theta) \end{bmatrix} \begin{bmatrix} x_1 \\ y_1 \\ z_1 \end{bmatrix} := Rot_1(-\theta) \tag{3.9}$$

3.2.3 Rotations in Three-dimensional Space

The rotations discussed above are simple rotations in two-dimensional space. They are special cases where the rotational axis is one of the coordinates which is perpendicular to the plane spanned by vectors before and after the rotation. Spacecraft attitude determination and control involve general rotations in three-dimensional space. Considering the rotation described in Figure 3.3 where the axis **X** is rotated to the axis **x**. A popular method to represent this rotation is to use a series of rotations about the coordinate described in the previous subsections, i.e., first the frame is rotated at an α angle around $-$**Y**-axis, then the intermediate $\mathbf{x}'$ is rotated at a β angle around the new **Z**-axis ($\mathbf{z}'$ axis). The α and β angles are the so-called *Euler angles*. Therefore, the rotational matrix is given by

$$\begin{aligned} \mathbf{C} &= \begin{bmatrix} \cos(\beta) & \sin(\beta) & 0 \\ -\sin(\beta) & \cos(\beta) & 0 \\ 0 & 0 & 1 \end{bmatrix} \begin{bmatrix} \cos(\alpha) & 0 & \sin(\alpha) \\ 0 & 1 & 0 \\ -\sin(\alpha) & 0 & \cos(\alpha) \end{bmatrix} \\ &= \begin{bmatrix} \cos(\beta)\cos(\alpha) & \sin(\beta) & \cos(\beta)\sin(\alpha) \\ -\sin(\beta)\cos(\alpha) & \cos(\beta) & -\sin(\beta)\sin(\alpha) \\ -\sin(\alpha) & 0 & \cos(\alpha) \end{bmatrix} \\ &= \begin{bmatrix} C_{11} & C_{12} & C_{13} \\ C_{21} & C_{22} & C_{23} \\ C_{31} & C_{32} & C_{33} \end{bmatrix} \end{aligned} \tag{3.10}$$

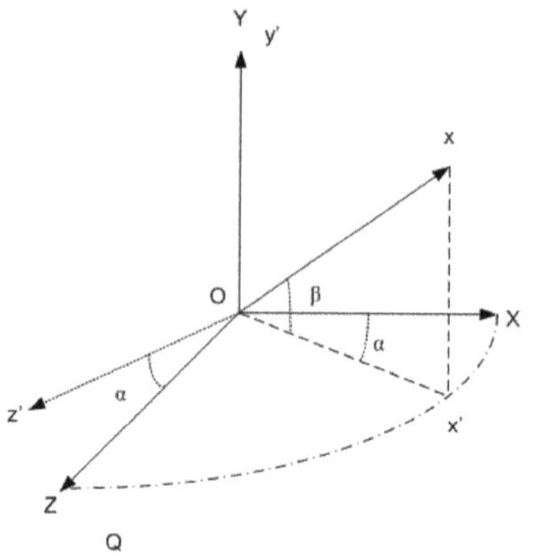

Figure 3.3: An axis rotation in three dimensional space.

which provides a different explanation of the rotation from **X**-axis to **x**-axis, i.e., the series of rotations can also be represented by a general rotational matrix (3.10). Let

$$\cos(\theta) = \frac{1}{2}(C_{11} + C_{22} + C_{33} - 1) \tag{3.11}$$

$$\hat{\mathbf{e}} = \frac{1}{2\sin(\theta)} \begin{bmatrix} C_{23} - C_{32} \\ C_{31} - C_{13} \\ C_{12} - C_{21} \end{bmatrix} = \begin{bmatrix} e_1 \\ e_2 \\ e_3 \end{bmatrix} \tag{3.12}$$

$$\mathbf{E} = \frac{1}{2\sin(\theta)}(\mathbf{C}^{\mathrm{T}} - \mathbf{C}) = \begin{bmatrix} 0 & -e_3 & e_2 \\ e_3 & 0 & -e_1 \\ -e_2 & e_1 & 0 \end{bmatrix}, \quad \theta \neq \pm k\pi, \quad k = 0, 1, 2, \ldots \tag{3.13}$$

the general rotational matrix (3.10) can be expressed as

$$\mathbf{C} = \cos(\theta)\mathbf{I} + (1 - \cos(\theta))\hat{\mathbf{e}}\hat{\mathbf{e}}^{\mathrm{T}} - \sin(\theta)\mathbf{E} \tag{3.14}$$

It can be verified that **C** is a rotational matrix, $\hat{\mathbf{e}}$ is the *rotational axis*, and θ is the rotational angle [74]. **C** is called the *direction cosine matrix*.

Actually, there may be infinitely many combinations of rotational axes and rotational angles that can rotate **X** to **x**. Moreover, Figure 3.4 and the following analysis show that in general case, the rotational axis of the direction cosine matrix may not be one of the coordinates. Let P be the middle point between **X** and **x** and ψ be the angle between O**x** and OP. Let OQ be the unit vector that is perpendicular to the plane spanned by **X** and **x** vectors. Obviously, the rotation can be achieved by rotating 2ψ around OQ. Alternatively, another rotation with rotational axis OP and rotational angle π can also rotate **X** to **x**. In fact, we can use any vector on the plane spanned by OP and OQ as the rotational axis and find an appropriate rotational angle which will rotate **X** to **x**. The first rotation

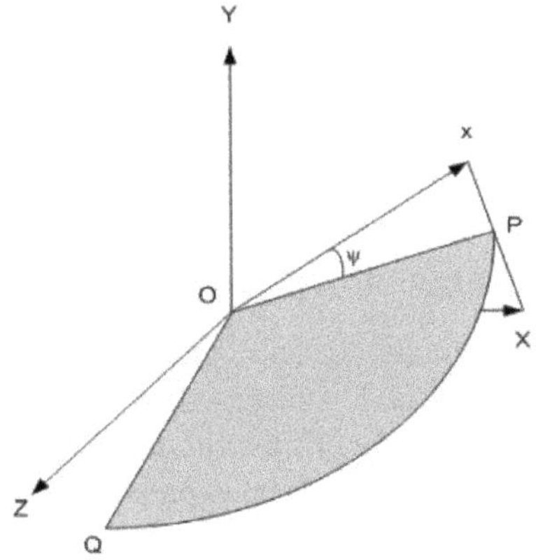

Figure 3.4: All possible rotations for one axis.

described above is sometimes called the *minimum-angle rotation*, and the second rotation described above is called the *maximum-angle rotation.*

3.2.4 Rotation from One Frame to Another Frame

In spacecraft attitude determination, oftentimes it is required to find a rotation that brings one frame to another one. This means that we need to find a rotational axis and an appropriate rotational angle that rotates one given frame $(\mathbf{X},\mathbf{Y},\mathbf{Z})$ to another given frame $(\mathbf{x},\mathbf{y},\mathbf{z})$. Let *S* be the middle point of $\mathbf{Y}$ and $\mathbf{y}$, *OR* be the unit length vector that is perpendicular to the plane spanned by $\mathbf{Y}$ and $\mathbf{y}$. The rotation that brings the frame $(\mathbf{X},\mathbf{Y},\mathbf{Z})$ to $(\mathbf{x},\mathbf{y},\mathbf{z})$ is described in Figure 3.5, where the plane OPQ spanned by *OP* and *OQ* defines all the rotational axes that can rotate

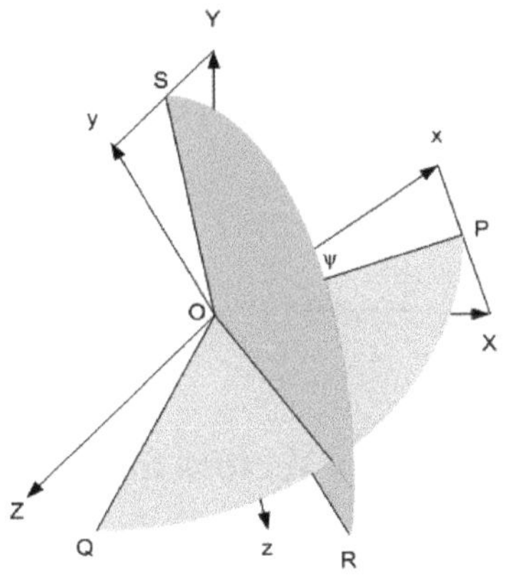

Figure 3.5: Rotation from one frame to another frame.

X to **x**; the plane OSR spanned by *OR* and *OS* defines all the rotational axes that can rotate **Y** to **y**. Therefore, the intersection of these two planes defines the unique rotational axis that can rotate **X** to **x** and **Y** to **y**, simultaneously. A rigorous derivation will be provided in Section 3.4.

3.2.5 Rate of Change of the Direction Cosine Matrix

In spacecraft dynamics modeling and controls, we need to not only know the attitude of the spacecraft, which is represented by the rotation from one frame to another frame, but also the *rate of this rotation*. The time dependence of the direction cosine matrix **A** at time t can be expressed by $\mathbf{A}(t)$. The time dependence of the direction cosine matrix **A** at time $t+\Delta t$ can be expressed by

$$\mathbf{A}(t+\Delta t) = \mathbf{C}\mathbf{A}(t)$$

where **C** is a rotation around $\hat{\mathbf{e}}$ with rotational angle $\theta = \Omega\Delta t$, and Ω is the *rate of the rotation* around the rotational axis. From (3.14),

$$\mathbf{C} = \cos(\Omega\Delta t)\mathbf{I} + (1-\cos(\Omega\Delta t))\hat{\mathbf{e}}\hat{\mathbf{e}}^{\mathrm{T}} - \sin(\Omega\Delta t)\mathbf{E}$$

As $\Delta t \to 0$, using the notation of (1.6),

$$\mathbf{C} \to \mathbf{I} - \mathbf{E}\Omega\Delta t = \mathbf{I} - \mathbf{S}(\omega)\Delta t = \mathbf{I} - \begin{bmatrix} 0 & -\omega_3 & \omega_2 \\ \omega_3 & 0 & -\omega_1 \\ -\omega_2 & \omega_1 & 0 \end{bmatrix} \Delta t$$

where $\omega = (\omega_1, \omega_2, \omega_3)$ is the rate vector along the rotational axis $\hat{\mathbf{e}}$, and

$$\mathbf{E}\Omega = \begin{bmatrix} 0 & -\omega_3 & \omega_2 \\ \omega_3 & 0 & -\omega_1 \\ -\omega_2 & \omega_1 & 0 \end{bmatrix} = \mathbf{S}(\omega)$$

This gives

$$\mathbf{A}(t+\Delta t) = (\mathbf{I} - \mathbf{S}(\omega)\Delta t)\mathbf{A}(t)$$

or

$$\mathbf{A}(t+\Delta t) - \mathbf{A}(t) = -\mathbf{S}(\omega)\mathbf{A}(t)\Delta t$$

therefore, we get

$$\frac{d\mathbf{A}}{dt} = -\mathbf{S}(\omega)\mathbf{A}(t) \tag{3.15}$$

3.2.6 Rate of Change of Vectors in Rotational Frame

In spacecraft dynamics modeling and controls, vectors and their rate of change are oftentimes represented in different frames. For the modeling and control purpose, we need to convert the vectors and their rate of change represented in different frames into a single frame. Therefore, the relationship between the time derivatives of an arbitrary vector along a coordinate axes of one system and the derivatives in a different system needs to be resolved. Let $\mathbf{a}'$ be the vector represented in a reference system and $\mathbf{a}$ be the same vector represented in the body frame. Then there is a rotational matrix $\mathbf{C}$ expressed in (3.14) such that

$$\mathbf{a} = \mathbf{C}\mathbf{a}'$$

The product rule for differentiation gives

$$\left(\frac{d\mathbf{a}}{dt}\right)\bigg|_b = \frac{d\mathbf{C}}{dt}\mathbf{a}' + \mathbf{C}\left(\frac{d\mathbf{a}'}{dt}\right)\bigg|_r$$

where the derivative $\left(\frac{d\mathbf{a}}{dt}\right)\Big|_b$ is represented in the body frame, and the derivative $\left(\frac{d\mathbf{a}'}{dt}\right)\Big|_r$ is represented in the reference frame. Since $\mathbf{C}$ is the rotation from reference frame to body frame, $C\left(\frac{d\mathbf{a}'}{dt}\right)\Big|_r = \left(\frac{d\mathbf{a}'}{dt}\right)\Big|_b$. From (3.15),

$$\begin{aligned}
\left(\frac{d\mathbf{a}}{dt}\right)\bigg|_b &= -\mathbf{S}(\omega)\mathbf{C}\mathbf{a}' + \mathbf{C}\frac{d\mathbf{a}'}{dt}\bigg|_r \\
&= -\mathbf{S}(\omega)\mathbf{a} + \left(\frac{d\mathbf{a}'}{dt}\right)\bigg|_b \\
&= -\omega \times \mathbf{a} + \left(\frac{d\mathbf{a}'}{dt}\right)\bigg|_b
\end{aligned} \tag{3.16}$$

where ω is the *rate of the rotation* between the reference frame and the body frame.

3.3 Transformation between Coordinate Systems

This section discusses some rotational matrix applications. It will focus on the transformation between different coordinate systems.

3.3.1 Transformation from ECI (XYZ) to PQW Coordinate

In view of Figure 2.6, one can see that the transformation of XYZ coordinate to PQW coordinate can be done by (a) rotate around the **Z**-axis by an angle Ω; (b) then rotate around the **X**-axis by an angle i, and (c) then rotate around the **Z**-axis by an angle ω. Let c be a short notation for cos and s be a short notation for sin. In mathematics formula, this transformation can be expressed as

$$\begin{bmatrix} P \\ Q \\ W \end{bmatrix}$$

$$= [Rot_3(\omega)][Rot_1(i)][Rot_3(\Omega)] \begin{bmatrix} X \\ Y \\ Z \end{bmatrix}$$

$$= \begin{bmatrix} c\omega & s\omega & 0 \\ -s\omega & c\omega & 0 \\ 0 & 0 & 1 \end{bmatrix} \begin{bmatrix} 1 & 0 & 0 \\ 0 & ci & si \\ 0 & -si & ci \end{bmatrix} \begin{bmatrix} c\Omega & s\Omega & 0 \\ -s\Omega & c\Omega & 0 \\ 0 & 0 & 1 \end{bmatrix} \begin{bmatrix} X \\ Y \\ Z \end{bmatrix} \tag{3.17}$$

3.3.2 Transformation from ECI (XYZ) to RSW Coordinate

In view of Figure 2.6, one can see that the transformation of XYZ coordinate to RSW coordinate can be done by (a) rotating around **Z**-axis by an angle Ω; (b) then rotating around the **X**-axis by an angle i, and (c) then rotating around **Z** axis by an angle $(\omega + \theta)$. Let c be a short notation for cos and s be a short notation for sin. In mathematics formula, this transformation can be expressed as

$$\begin{bmatrix} R \\ S \\ W \end{bmatrix}$$

$$= [Rot_3(\omega+\theta)][Rot_1(i)][Rot_3(\Omega)] \begin{bmatrix} X \\ Y \\ Z \end{bmatrix}$$

$$= \begin{bmatrix} c(\omega+\theta) & s(\omega+\theta) & 0 \\ -s(\omega+\theta) & c(\omega+\theta) & 0 \\ 0 & 0 & 1 \end{bmatrix} \begin{bmatrix} 1 & 0 & 0 \\ 0 & ci & si \\ 0 & -si & ci \end{bmatrix} \begin{bmatrix} c\Omega & s\Omega & 0 \\ -s\Omega & c\Omega & 0 \\ 0 & 0 & 1 \end{bmatrix} \begin{bmatrix} X \\ Y \\ Z \end{bmatrix} \tag{3.18}$$

where Ω is the *right ascension* of the *ascending node* of the orbit, i is the *inclination* of the orbit, ω is the *argument of perigee*, and θ is the *true anomaly*. The sum of ω and θ represents the location of the spacecraft relative to the ascending node.

3.3.3 Transformation from Six Classical Parameters to $(\mathbf{v}, \mathbf{r})$

This section will focus on finding the spacecraft position and speed in the ECI coordinate system given *six classical orbit parameters* $[a, e, i, \Omega, \omega, M]$. Since all Keplerian orbits are in a plane, a coordinate system **x**, **y** can be defined in a plane with $z = 0$. It follows from Figure 3.6 and (2.49) that

$$x = a\cos(\psi) - c = a(\cos(\psi) - e) \tag{3.19}$$

and

$$y = x\tan(\theta) = a(\cos(\psi) - e)\frac{\sin(\theta)}{\cos(\theta)}$$

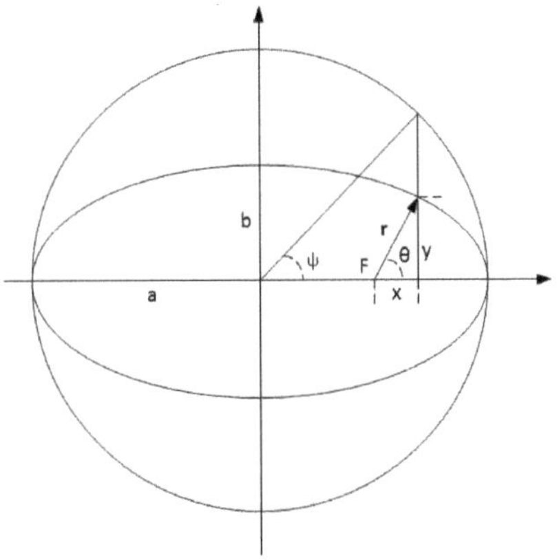

Figure 3.6: Transformation between orbit parameters and ECI frame.

$$= \quad a(\cos(\psi)-e)\frac{\sin(\psi)\sqrt{1-e^2}}{\cos(\psi)-e} = [a\sin(\psi)]\sqrt{1-e^2} \tag{3.20}$$

Given M and e, to find ψ, one can use Newton's method for the equation (2.61) which is provided again below

$$M = \psi - e\sin(\psi) \tag{3.21}$$

Given ψ, x and y are obtained from (3.19) and (3.20). In view of Figure 2.6, (x,y,z) determines the spacecraft location in the PQW coordinate frame with $z=0$. Therefore, to find $\mathbf{r}$ and $\mathbf{v}$, it follows that

$$\mathbf{r} = x\mathbf{P} + y\mathbf{Q} = a(\cos(\psi)-e)\mathbf{P} + a\sqrt{1-e^2}\sin(\psi)\mathbf{Q} \tag{3.22}$$

From this equation, the location of the spacecraft in ECI frame is given by the inverse transformation of Equation (3.17) which is given by

$$\begin{bmatrix} X \\ Y \\ Z \end{bmatrix} = [Rot_3(\Omega)]^{-1}[Rot_1(i)]^{-1}[Rot_3(\omega)]^{-1}\begin{bmatrix} x \\ y \\ 0 \end{bmatrix} \tag{3.23}$$

where (X,Y,Z) is the ECI coordinate of the spacecraft.

To calculate the velocity vector, one simply needs to use $\mathbf{v} = \frac{d\mathbf{r}}{dt}$ which gives

$$\mathbf{v} = \frac{d\mathbf{r}}{dt} = \frac{d\mathbf{r}}{d\psi}\frac{d\psi}{dt} \tag{3.24}$$

It follows from (3.21) and (2.56) that

$$\frac{dM}{dt} = \omega_0 = \frac{d\psi}{dt} - e\cos(\psi)\frac{d\psi}{dt} \tag{3.25}$$

which gives

$$\frac{d\psi}{dt} = \frac{\omega_0}{1 - e\cos(\psi)} = \frac{a\omega_0}{r} \tag{3.26}$$

The last equation follows from (2.51). Differentiating (3.22) and using (3.26) yield

$$\mathbf{v} = \frac{d\mathbf{r}}{dt} = \frac{a^2\omega_0}{r}\left[-\sin(\psi)\mathbf{P} + \sqrt{1-e^2}\cos(\psi)\mathbf{Q}\right] = v_p\mathbf{P} + v_q\mathbf{Q} \tag{3.27}$$

where $(v_p, v_q, 0)$ is the spacecraft velocity in PQW coordinate frame. Using the inverse transformation of Equation (3.17) gives

$$\begin{bmatrix} v_x \\ v_y \\ v_z \end{bmatrix} = [Rot_3(\Omega)]^{-1}[Rot_1(i)]^{-1}[Rot_3(\omega)]^{-1}\begin{bmatrix} v_p \\ v_q \\ 0 \end{bmatrix} \tag{3.28}$$

where (v_x, v_y, v_z) is the spacecraft velocity in the ECI coordinate frame.

3.3.4 Transformation from $(\mathbf{v}, \mathbf{r})$ to Six Classical Parameters

Now, we consider the inverse transformation, i.e., given $(\mathbf{v}, \mathbf{r})$ in Cartesian coordinates, **X**, **Y**, **Z**, v_x, v_y, and v_z, the task is to find the classical orbit parameters $\alpha = [a, e, i, \Omega, \omega, M]$. From (2.41), it follows immediately that

$$a = \frac{\mu}{2\left[\frac{\mu}{r} - \frac{v^2}{2}\right]} \tag{3.29}$$

Let $\mathbf{h} = [h_x, h_y, h_z]^{\mathrm{T}}$ be the *orbit momentum* represented in ECI frame and $h = |\mathbf{h}|$. Since $\mathbf{h} = \mathbf{r} \times \mathbf{v} = |\mathbf{h}|\mathbf{W}$ is a given, in view of Figure 2.6, it follows that

$$\cos(i) = h_z/h \tag{3.30}$$

From Figure 2.6 again, it follows that

$$\sin(\Omega) = \frac{h_x}{\sqrt{h_x^2 + h_y^2}}, \quad \cos(\Omega) = -\frac{h_y}{\sqrt{h_x^2 + h_y^2}} \tag{3.31}$$

In view of (2.39), it follows that

$$e = \sqrt{1 - \frac{h^2}{a\mu}} \tag{3.32}$$

From (3.21), to obtain $M = \psi - e\sin(\psi)$, one needs to know ψ. Given a, e, and $\mathbf{r}$, from (2.51), it follows that

$$\psi = \cos^{-1}\left(\frac{1 - |\mathbf{r}|}{ae}\right) \tag{3.33}$$

From (3.22), (3.27), and (2.51), it follows that

$$\begin{aligned}\mathbf{r}\cdot\mathbf{v} &= \frac{a^3\omega_0}{r}\sin(\psi)\left[-(\cos(\psi)-e)+(1-e^2)\cos(\psi)\right]\\ &= \frac{a^3\omega_0}{r}\sin(\psi)e(1-e)\cos(\psi)\\ &= \frac{a^2\omega_0}{r}\sin(\psi)er = a^2\omega_0 e\sin(\psi) \qquad (3.34)\end{aligned}$$

This yields, in view of (2.55), that

$$\sin(\psi) = \frac{\mathbf{r}\cdot\mathbf{v}}{a^2\omega_0 e} = \frac{\mathbf{r}\cdot\mathbf{v}}{e\sqrt{a\mu}} \qquad (3.35)$$

Equations (3.33) and (3.35) give ψ with correct sign. Therefore, M is obtained by using (2.61) which is given again below

$$M = \psi - e\sin(\psi) \qquad (3.36)$$

The last parameter is the *argument of perigee* ω. In view of Figure 3.6, in an orbit plane, we have

$$x = r\cos(\theta), \quad y = r\sin(\theta) \qquad (3.37)$$

Since $\mathbf{r} = [X,Y,Z]^{\mathrm{T}}$ is known in ECI frame, substituting x, y, X, Y, and Z into (3.18) gives

$$\sin(\omega+\theta) = \frac{Z}{r\sin(i)}, \quad \cos(\omega+\theta) = \frac{X\cos(\Omega)+Y\sin(\Omega)}{r} \qquad (3.38)$$

Since θ is given in (2.49), ω can be obtained from (3.38).

3.4 Quaternion and Its Properties

Unlike the Euler angles which represent a rotation by a series of rotations rotating around **X**, or **Y** or **Z** axes, *quaternion* represents a rotation by a rotational angle around a rotational axis, which is not necessarily around **X**, or **Y**, or **Z** axes. Quaternion was first introduced by the Irish mathematician William Rowan Hamilton in 1843 and applied to mechanics in three-dimensional space. A striking feature of quaternion is that the product of two quaternions is *non-commutative*, meaning that the product of two quaternions depends on which factor is to the left of the multiplication sign and which factor is to the right. Let the standard basis **i**, **j**, and **k** for the $\mathbf{R}^3$ satisfy the following condition

$$\mathbf{i}^2 = \mathbf{j}^2 = \mathbf{k}^2 = \mathbf{ijk} = -1 \qquad (3.39)$$

Let a 4-tuple of real numbers

$$\bar{\mathbf{q}} = (q_0, q_1, q_2, q_3) \tag{3.40}$$

we define a quaternion as the sum of a scalar and a vector

$$\bar{\mathbf{q}} = q_0 + \mathbf{i}q_1 + \mathbf{j}q_2 + \mathbf{k}q_3 = q_0 + \mathbf{q} \tag{3.41}$$

where q_0 is called the *scalar part of the quaternion* and

$$\mathbf{q} = \mathbf{i}q_1 + \mathbf{j}q_2 + \mathbf{k}q_3$$

is called the *vector part of the quaternion.* People use (3.40) and (3.41) interchangeably if no confusion is introduced. Though in aerospace engineering, one always uses a special *normalized quaternion* $q_0 = \cos(\frac{\alpha}{2})$, and $\mathbf{q} = \hat{\mathbf{e}}\sin(\frac{\alpha}{2})$, where $\hat{\mathbf{e}}$ is rotational axis, and α is the rotational angle. We will derive some useful properties for the general form of quaternion.

3.4.1 Equality and Addition

Let

$$\bar{\mathbf{p}} = p_0 + \mathbf{i}p_1 + \mathbf{j}p_2 + \mathbf{k}p_3$$

and

$$\bar{\mathbf{q}} = q_0 + \mathbf{i}q_1 + \mathbf{j}q_2 + \mathbf{k}q_3$$

be two quaternions, then the two quaternions are equal if and only if

$$p_0 = q_0, \quad p_1 = q_1, \quad p_2 = q_2, \quad p_3 = q_3$$

For the special normalized quaternion used in aerospace engineering, if two quaternions are equal, they will have the same rotational angle and the same rotational axis. The *sum of the two quaternions* is defined as

$$\bar{\mathbf{p}} + \bar{\mathbf{q}} = (p_0 + q_0) + \mathbf{i}(p_1 + q_1) + \mathbf{j}(p_2 + q_2) + \mathbf{k}(p_3 + q_3)$$

The *zero quaternion* has scalar part 0 and vector part $(0,0,0)$. The *negative or an additive inverse* of $\bar{\mathbf{q}}$ is $-\bar{\mathbf{q}}$.

3.4.2 Multiplication and the Identity

From (3.39), we have

$$\mathbf{ij} = \mathbf{k} = -\mathbf{ji}, \quad \mathbf{jk} = \mathbf{i} = -\mathbf{kj}, \quad \mathbf{ki} = \mathbf{j} = -\mathbf{ik} \tag{3.42}$$

Let $\bar{\mathbf{p}}$ and $\bar{\mathbf{q}}$ be defined as before, use (3.39) and (3.42), we define the *multiplication of two quaternions* $\bar{\mathbf{p}}$ and $\bar{\mathbf{q}}$ by

$$\bar{\mathbf{p}} \otimes \bar{\mathbf{q}} = p_0 q_0 - \mathbf{p} \cdot \mathbf{q} + p_0 \mathbf{q} + q_0 \mathbf{p} + \mathbf{p} \times \mathbf{q} \tag{3.43}$$

with the scalar part $p_0q_0 - \mathbf{p}\cdot\mathbf{q}$ and vector part $p_0\mathbf{q}+q_0\mathbf{p}+\mathbf{p}\times\mathbf{q}$. The quaternion *multiplicative identity* has scalar part 1 and vector part $(0,0,0)$.

The quaternion multiplication can be used to represent two consecutive rotations. Let $\bar{\mathbf{p}}$ and $\bar{\mathbf{q}}$ be the two consecutive rotations ($\bar{\mathbf{p}}$ represent the first rotation and $\bar{\mathbf{q}}$ represent the second rotation). The *composed rotation* is given by $\bar{\mathbf{r}} = \bar{\mathbf{p}}\otimes\bar{\mathbf{q}}$. The derivation is given in Section 3.4.4 (see also [220, pp. 319–320]).

3.4.3 Complex Conjugate, Norm, and Inverse

The *complex conjugate of* quaternion $\bar{\mathbf{q}}$ is denoted by

$$\bar{\mathbf{q}}^* = q_0 - \mathbf{q} = q_0 - \mathbf{i}q_1 - \mathbf{j}q_2 - \mathbf{k}q_3 \tag{3.44}$$

It is easy to see

$$\bar{\mathbf{q}} + \bar{\mathbf{q}}^* = (q_0+\mathbf{q}) + (q_0-\mathbf{q}) = 2q_0 \tag{3.45}$$

Given two quaternions $\bar{\mathbf{p}}$ and $\bar{\mathbf{q}}$, we have

$$(\bar{\mathbf{p}}\otimes\bar{\mathbf{q}})^* = \bar{\mathbf{q}}^*\otimes\bar{\mathbf{p}}^* \tag{3.46}$$

The *norm of* a quaternion is defined as $\|\bar{\mathbf{q}}\| = \sqrt{\bar{\mathbf{q}}^*\otimes\bar{\mathbf{q}}}$. It is also easy to verify that the norm satisfies

$$\|\bar{\mathbf{q}}\| = \sqrt{q_0^2+q_1^2+q_2^2+q_3^2} \tag{3.47}$$

We define the *inverse of* a quaternion by

$$\bar{\mathbf{q}}^{-1}\otimes\bar{\mathbf{q}} = \bar{\mathbf{q}}\otimes\bar{\mathbf{q}}^{-1} = 1$$

Pre- and post-multiplying by $\bar{\mathbf{q}}^*$ gives

$$\bar{\mathbf{q}}^{-1}\otimes\bar{\mathbf{q}}\otimes\bar{\mathbf{q}}^* = \bar{\mathbf{q}}^*\otimes\bar{\mathbf{q}}\otimes\bar{\mathbf{q}}^{-1} = \bar{\mathbf{q}}^*$$

Since $\bar{\mathbf{q}}^*\otimes\bar{\mathbf{q}} = \bar{\mathbf{q}}\otimes\bar{\mathbf{q}}^* = \|\bar{\mathbf{q}}\|^2$, we have

$$\bar{\mathbf{q}}^{-1} = \frac{\bar{\mathbf{q}}^*}{\|\bar{\mathbf{q}}\|^2} \tag{3.48}$$

For normalized quaternion which satisfies $\|\bar{\mathbf{q}}\| = \sqrt{q_0^2+q_1^2+q_2^2+q_3^2} = 1$

$$\bar{\mathbf{q}}^{-1} = \bar{\mathbf{q}}^* \tag{3.49}$$

Finally, the *norm of the product* of two quaternions $\bar{\mathbf{p}}$ and $\bar{\mathbf{q}}$ is the product of the individual norms because

$$\begin{aligned}\|\bar{\mathbf{p}}\otimes\bar{\mathbf{q}}\|^2 &= (\bar{\mathbf{p}}\otimes\bar{\mathbf{q}})\otimes(\bar{\mathbf{p}}\otimes\bar{\mathbf{q}})^* \\ &= \bar{\mathbf{p}}\otimes\bar{\mathbf{q}}\otimes\bar{\mathbf{q}}^*\otimes\bar{\mathbf{p}}^* \\ &= \bar{\mathbf{p}}\otimes\|\mathbf{q}\|^2\otimes\bar{\mathbf{p}}^* \\ &= \bar{\mathbf{p}}\otimes\bar{\mathbf{p}}^*\|\mathbf{q}\|^2 = \|\mathbf{p}\|^2\|\mathbf{q}\|^2\end{aligned} \tag{3.50}$$

3.4.4 Rotation by Quaternion Operator

This section will explain how to rotate a vector using quaternion operator. For this purpose, only the normalized quaternion $\bar{\mathbf{q}} = q_0 + \mathbf{q} = \cos(\frac{\alpha}{2}) + \hat{\mathbf{e}}\sin(\frac{\alpha}{2})$ is considered, where $\hat{\mathbf{e}}$ is the unit length rotational axis and α is the rotational angle. Clearly, quaternion does have the information about the rotational angle and the rotational axis. Similar to rotational matrices, the product of quaternions need to be able to represent consecutive rotations. Let $\bar{\mathbf{p}} = \cos(\frac{\alpha}{2}) + \hat{\mathbf{e}}\sin(\frac{\alpha}{2})$ and $\bar{\mathbf{q}} = \cos(\frac{\beta}{2}) + \hat{\mathbf{e}}\sin(\frac{\beta}{2})$, from (3.43), we have

$$\begin{aligned}
\bar{\mathbf{r}} &= \bar{\mathbf{p}} \otimes \bar{\mathbf{q}} = \left(\cos\left(\frac{\alpha}{2}\right) + \hat{\mathbf{e}}\sin\left(\frac{\alpha}{2}\right)\right) \otimes \left(\cos\left(\frac{\beta}{2}\right) + \hat{\mathbf{e}}\sin\left(\frac{\beta}{2}\right)\right) \\
&= \cos\left(\frac{\alpha}{2}\right)\cos\left(\frac{\beta}{2}\right) - \hat{\mathbf{e}}\sin\left(\frac{\alpha}{2}\right) \cdot \hat{\mathbf{e}}\sin\left(\frac{\beta}{2}\right) \\
&\quad + \cos\left(\frac{\alpha}{2}\right)\hat{\mathbf{e}}\sin\left(\frac{\beta}{2}\right) + \hat{\mathbf{e}}\sin\left(\frac{\alpha}{2}\right)\cos\left(\frac{\beta}{2}\right) \\
&\quad + \hat{\mathbf{e}}\sin\left(\frac{\alpha}{2}\right) \times \hat{\mathbf{e}}\sin\left(\frac{\beta}{2}\right) \\
&= \cos\left(\frac{\alpha}{2}\right)\cos\left(\frac{\beta}{2}\right) - \sin\left(\frac{\alpha}{2}\right)\sin\left(\frac{\beta}{2}\right) \\
&\quad + \hat{\mathbf{e}}\left(\sin\left(\frac{\alpha}{2}\right)\cos\left(\frac{\beta}{2}\right) + \cos\left(\frac{\alpha}{2}\right)\sin\left(\frac{\beta}{2}\right)\right) \\
&= \cos\left(\frac{\alpha+\beta}{2}\right) + \hat{\mathbf{e}}\sin\left(\frac{\alpha+\beta}{2}\right) \\
&= \cos(\gamma) + \hat{\mathbf{e}}\sin(\gamma)
\end{aligned} \tag{3.51}$$

This means that the product of two quaternions indeed represents two consecutive rotations. Parallel to the vector rotation using rotational matrix, it is expected that a quaternion rotation operator involves *multiplication of a quaternion and a vector*. Therefore, the multiplication of a quaternion and a vector should be defined. To this end, we consider a vector $\mathbf{v}$ as a pure quaternion in which the scalar part is zero and the vector part is $\mathbf{v}$, i.e., $\bar{\mathbf{v}} = 0 + \mathbf{v}$. For the sake of notational simplicity, $\bar{\mathbf{v}}$ and $\mathbf{v}$ are used interchangeably for both vector and pure quaternion. From (3.43), the multiplication of a vector and a quaternion is defined as

$$\bar{\mathbf{q}} \otimes \mathbf{v} = (q_0 + \mathbf{q}) \otimes (0 + \mathbf{v}) = -\mathbf{q} \cdot \mathbf{v} + q_0\mathbf{v} + \mathbf{q} \times \mathbf{v} \tag{3.52}$$

We also expect that the quaternion operator will rotate a vector into another vector, or a pure quaternion. Simple evaluation shows that neither $\mathbf{w} = \bar{\mathbf{q}} \otimes \mathbf{v}$ nor $\mathbf{w} = \mathbf{v} \otimes \bar{\mathbf{q}}$ is necessarily a pure vector. However, using (3.52) and (1.2), we have

$$\begin{aligned}
\mathbf{w} &= \bar{\mathbf{q}} \otimes \mathbf{v} \otimes \bar{\mathbf{q}}^* = (q_0 + \mathbf{q}) \otimes (0 + \mathbf{v}) \otimes (q_0 - \mathbf{q}) \\
&= (-\mathbf{q} \cdot \mathbf{v} + q_0\mathbf{v} + \mathbf{q} \times \mathbf{v}) \otimes (q_0 - \mathbf{q})
\end{aligned}$$

$$
\begin{aligned}
&= -q_0(\mathbf{q}\cdot\mathbf{v}) + q_0(\mathbf{v}\cdot\mathbf{q}) + (\mathbf{q}\times\mathbf{v})\cdot\mathbf{q} \\
&\quad +(\mathbf{q}\cdot\mathbf{v})\mathbf{q} + q_0^2\mathbf{v} + q_0(\mathbf{q}\times\mathbf{v}) - q_0(\mathbf{v}\times\mathbf{q}) - (\mathbf{q}\times\mathbf{v})\times\mathbf{q} \\
&= (\mathbf{q}\cdot\mathbf{v})\mathbf{q} + q_0^2\mathbf{v} + 2q_0(\mathbf{q}\times\mathbf{v}) - (\mathbf{q}\cdot\mathbf{q})\mathbf{v} + (\mathbf{v}\cdot\mathbf{q})\mathbf{q} \\
&= (2q_0^2-1)\mathbf{v} + 2(\mathbf{q}\cdot\mathbf{v})\mathbf{q} + 2q_0(\mathbf{q}\times\mathbf{v}) \\
&= \left(\cos^2\left(\frac{\alpha}{2}\right) - \sin^2\left(\frac{\alpha}{2}\right)\right)\mathbf{v} + 2(\mathbf{q}\cdot\mathbf{v})\mathbf{q} + 2q_0(\mathbf{q}\times\mathbf{v})
\end{aligned} \tag{3.53}
$$

which is a vector. In fact, the quaternion operator can be expressed by direction cosine matrix which may be more convenient in some cases. From (3.53), since

$$
2(q_0^2-1)\mathbf{v} = \begin{bmatrix} (2q_0^2-1) & 0 & 0 \\ 0 & (2q_0^2-1) & 0 \\ 0 & 0 & (2q_0^2-1) \end{bmatrix} \begin{bmatrix} v_1 \\ v_2 \\ v_3 \end{bmatrix}
$$

$$
2(\mathbf{v}\cdot\mathbf{q})\mathbf{q} = \begin{bmatrix} 2q_1^2 & 2q_1q_2 & 2q_1q_3 \\ 2q_1q_2 & 2q_2^2 & 2q_2q_3 \\ 2q_1q_3 & 2q_2q_3 & 2q_3^2 \end{bmatrix} \begin{bmatrix} v_1 \\ v_2 \\ v_3 \end{bmatrix}
$$

$$
2q_0(\mathbf{q}\times\mathbf{v}) = \begin{bmatrix} 0 & -2q_0q_3 & 2q_0q_2 \\ 2q_0q_3 & 0 & -2q_0q_1 \\ -2q_0q_2 & 2q_0q_1 & 0 \end{bmatrix} \begin{bmatrix} v_1 \\ v_2 \\ v_3 \end{bmatrix}
$$

we have

$$
\begin{bmatrix} w_1 \\ w_2 \\ w_3 \end{bmatrix} = \begin{bmatrix} 2q_0^2-1+2q_1^2 & 2q_1q_2-2q_0q_3 & 2q_1q_3+2q_0q_2 \\ 2q_1q_2+2q_0q_3 & 2q_2^2+2q_0^2-1 & 2q_2q_3-2q_0q_1 \\ 2q_1q_3-2q_0q_2 & 2q_2q_3+2q_0q_1 & 2q_3^2+2q_0^2-1 \end{bmatrix} \begin{bmatrix} v_1 \\ v_2 \\ v_3 \end{bmatrix} \tag{3.54}
$$

This means that either (3.53) or (3.54) can be used for quaternion rotation. They will be used for different applications in the rest of the book. It is worthwhile to note, in view of (3.53), that (3.54) defines a general rotational matrix as

$$
\mathbf{C} = (q_0^2 - \mathbf{q}^{\mathrm{T}}\mathbf{q})\mathbf{I} + 2\mathbf{q}\mathbf{q}^{\mathrm{T}} + 2q_0\mathbf{S}(\mathbf{q}) \tag{3.55}
$$

It can be seen that $\bar{\mathbf{q}}\otimes\mathbf{v}\otimes\bar{\mathbf{q}}^*$ is indeed the *quaternion operator* that rotates $\mathbf{v}$ an α angle around $\hat{\mathbf{e}}$. First, it is easy to verify that $\bar{\mathbf{q}}\otimes\mathbf{v}\otimes\bar{\mathbf{q}}^*$ is linear operator, i.e., for two vectors $\mathbf{a}$, $\mathbf{b}$, and a scalar k, the following relation holds.

$$
\bar{\mathbf{q}}\otimes(k\mathbf{a}+\mathbf{b})\otimes\bar{\mathbf{q}}^* = k\bar{\mathbf{q}}\otimes\mathbf{a}\otimes\bar{\mathbf{q}}^* + \bar{\mathbf{q}}\otimes\mathbf{b}\otimes\bar{\mathbf{q}}^* \tag{3.56}
$$

Then, vector $\mathbf{v}$ is decomposed into two components, $\mathbf{v} = \mathbf{v}_q + \mathbf{v}_n$, where $\mathbf{v}_q$ is parallel to $\mathbf{q}$ and $\mathbf{v}_n$ is perpendicular to $\mathbf{q}$. It is shown that (a) under quaternion operator $\bar{\mathbf{q}}\otimes\mathbf{v}\otimes\bar{\mathbf{q}}^*$, the first component $\mathbf{v}_q$ is invariant and (b) the second component $\mathbf{v}_n$ rotates at an angle of α. Since $\mathbf{v}_q = k\mathbf{q}$, where $k \le 1$ is a constant, from (3.56), (3.52), and (3.43), using the fact that $\bar{\mathbf{q}}$ is a normalized quaternion, we have

$$
\bar{\mathbf{q}}\otimes\mathbf{v}_q\otimes\bar{\mathbf{q}}^* = \bar{\mathbf{q}}\otimes(k\mathbf{q})\otimes\bar{\mathbf{q}}^* = k\bar{\mathbf{q}}\otimes(\mathbf{q})\otimes\bar{\mathbf{q}}^* = k(-\mathbf{q}\cdot\mathbf{q}+q_0\mathbf{q})\otimes(q_0-\mathbf{q}) = k\mathbf{q}
$$

This proves (a). Using the facts that

$$\mathbf{q} \cdot \mathbf{v}_n = 0$$

$$\cos(\alpha) = \cos^2\left(\frac{\alpha}{2}\right) - \sin^2\left(\frac{\alpha}{2}\right)$$

$$\sin(\alpha) = 2\cos\left(\frac{\alpha}{2}\right)\sin\left(\frac{\alpha}{2}\right)$$

$$q_0 = \cos\left(\frac{\alpha}{2}\right)$$

$$\|\mathbf{q}\| = \sin\left(\frac{\alpha}{2}\right)$$

$$\mathbf{q} \times \mathbf{v}_n = \|\mathbf{q}\|\|\mathbf{v}_n\|\sin\left(\frac{\pi}{2}\right)\mathbf{v}_\perp = \|\mathbf{q}\|\|\mathbf{v}_n\|\mathbf{v}_\perp$$

where $\mathbf{v}_\perp$ is a unit length vector perpendicular to both $\mathbf{q}$ and $\mathbf{v}_n$, and from (3.53), we have

$$\begin{aligned}
\bar{\mathbf{q}} \otimes (\mathbf{v}_n) \otimes \bar{\mathbf{q}}^* &= \left(\cos^2\left(\frac{\alpha}{2}\right) - \sin^2\left(\frac{\alpha}{2}\right)\right)\mathbf{v}_n + 2(\mathbf{q} \cdot \mathbf{v}_n)\mathbf{q} + 2q_0(\mathbf{q} \times \mathbf{v}_n) \\
&= \cos(\alpha)\mathbf{v}_n + 2q_0(\mathbf{q} \times \mathbf{v}_n) \\
&= \cos(\alpha)\mathbf{v}_n + 2\cos\left(\frac{\alpha}{2}\right)(\mathbf{q} \times \mathbf{v}_n) \\
&= \cos(\alpha)\mathbf{v}_n + 2\cos\left(\frac{\alpha}{2}\right)\sin\left(\frac{\alpha}{2}\right)\|\mathbf{v}_n\|\mathbf{v}_\perp \\
&= \cos(\alpha)\mathbf{v}_n + \sin(\alpha)\|\mathbf{v}_n\|\mathbf{v}_\perp
\end{aligned} \tag{3.57}$$

Since $\mathbf{v}_n$ and $\|\mathbf{v}_n\|\mathbf{v}_\perp$ have the same length, and they both are perpendicular to $\mathbf{v}_q$, equation (3.57) indicates that $\bar{\mathbf{q}} \otimes (\mathbf{v}_n) \otimes \bar{\mathbf{q}}^*$ rotates $\mathbf{v}_n$ an angle of α around axis $\mathbf{q}$. This proves (b).

A fact parallel to the rotational matrix is that $\bar{\mathbf{q}} \otimes (\mathbf{v}) \otimes \bar{\mathbf{q}}^*$ does not change the length of $\mathbf{v}$, which is a direct result of (3.50) and the fact that $\bar{\mathbf{q}}$ is a normalized quaternion.

$$\|\bar{\mathbf{q}} \otimes \mathbf{v} \otimes \bar{\mathbf{q}}^*\| = \|\bar{\mathbf{q}}\|\|\mathbf{v}\|\|\bar{\mathbf{q}}\|^* = \|\mathbf{v}\| \tag{3.58}$$

Similar to the rotational matrix, the *inverse of the quaternion operator* $\mathbf{w} = \bar{\mathbf{q}} \otimes (\mathbf{v}) \otimes \bar{\mathbf{q}}^*$ on $\mathbf{v}$ is simple and it is given by

$$\bar{\mathbf{q}}^* \otimes (\mathbf{w}) \otimes \bar{\mathbf{q}} = \bar{\mathbf{q}}^* \otimes (\bar{\mathbf{q}} \otimes (\mathbf{v}) \otimes \bar{\mathbf{q}}^*) \otimes \bar{\mathbf{q}} = (\bar{\mathbf{q}}^* \otimes \bar{\mathbf{q}}) \otimes \mathbf{v} \otimes (\bar{\mathbf{q}}^* \otimes \bar{\mathbf{q}}) = \mathbf{v}$$

which rotates $\mathbf{w}$ at an angle of α around $-\mathbf{q}$ and brings $\mathbf{w}$ back to $\mathbf{v}$. It is easy to verify that

$$\mathbf{v} = \bar{\mathbf{q}}^* \otimes \mathbf{w} \otimes \bar{\mathbf{q}} = (2q_0^2 - 1)\mathbf{w} + 2(\mathbf{q} \cdot \mathbf{w})\mathbf{q} - 2q_0(\mathbf{q} \times \mathbf{w}) \tag{3.59}$$

This gives

$$\begin{bmatrix} v_1 \\ v_2 \\ v_3 \end{bmatrix} = \begin{bmatrix} 2q_0^2-1+2q_1^2 & 2q_1q_2+2q_0q_3 & 2q_1q_3-2q_0q_2 \\ 2q_1q_2-2q_0q_3 & 2q_0^2-1+2q_2^2 & 2q_2q_3+2q_0q_1 \\ 2q_1q_3+2q_0q_2 & 2q_2q_3-2q_0q_1 & 2q_0^2-1+2q_3^2 \end{bmatrix} \begin{bmatrix} w_1 \\ w_2 \\ w_3 \end{bmatrix} \tag{3.60}$$

It is worthwhile to note, in view of (3.59), that (3.60) defines a general rotational matrix as

$$\mathbf{A} = (q_0^2 - \mathbf{q}^{\mathrm{T}}\mathbf{q})\mathbf{I} + 2\mathbf{q}\mathbf{q}^{\mathrm{T}} - 2q_0\mathbf{S}(\mathbf{q}) \tag{3.61}$$

Formula (3.61) is another form of the rotational matrix (3.14)

3.4.5 Matrix Form of Quaternion Production

It can be seen that in some applications, a *matrix form of quaternion production* is more convenient than the form of (3.43). Let $\bar{\mathbf{r}} = (r_0, r_1, r_2, r_3)$ be the composed quaternion of two consecutive quaternions of $\bar{\mathbf{p}}$ and $\bar{\mathbf{q}}$, i.e., $\bar{\mathbf{r}} = \bar{\mathbf{p}} \otimes \bar{\mathbf{q}}$. Expanding (3.43) gives

$$r_0 = p_0q_0 - p_1q_1 - p_2q_2 - p_3q_3 \tag{3.62a}$$

$$r_1 = p_0q_1 + p_1q_0 + p_2q_3 - p_3q_2 \tag{3.62b}$$

$$r_2 = p_0q_2 - p_1q_3 + p_2q_0 + p_3q_1 \tag{3.62c}$$

$$r_3 = p_0q_3 + p_1q_2 - p_2q_1 + p_3q_0 \tag{3.62d}$$

(3.62) can be written in the matrix form

$$\begin{bmatrix} r_0 \\ r_1 \\ r_2 \\ r_3 \end{bmatrix} = \begin{bmatrix} p_0 & -p_1 & -p_2 & -p_3 \\ p_1 & p_0 & -p_3 & p_2 \\ p_2 & p_3 & p_0 & -p_1 \\ p_3 & -p_2 & p_1 & p_0 \end{bmatrix} \begin{bmatrix} q_0 \\ q_1 \\ q_2 \\ q_3 \end{bmatrix} \tag{3.63a}$$

$$= \begin{bmatrix} q_0 & -q_1 & -q_2 & -q_3 \\ q_1 & q_0 & q_3 & -q_2 \\ q_2 & -q_3 & q_0 & q_1 \\ q_3 & q_2 & -q_1 & q_0 \end{bmatrix} \begin{bmatrix} p_0 \\ p_1 \\ p_2 \\ p_3 \end{bmatrix} \tag{3.63b}$$

3.4.6 Derivative of the Quaternion

The *derivative of quaternion* is obtained as follows. Let $\bar{\mathbf{q}}(t)$ be the quaternion to a reference frame at time t, $\bar{\mathbf{q}}(t+\Delta t)$ be the quaternion to the reference frame at $t+\Delta t$, and $\bar{\mathbf{p}}(t) = \cos(\frac{\Delta\alpha}{2}) + \hat{\mathbf{e}}(t)\sin(\frac{\Delta\alpha}{2})$ be the quaternion that brings $\bar{\mathbf{q}}(t)$ to $\bar{\mathbf{q}}(t+\Delta t)$, i.e., $\bar{\mathbf{p}}(t)$ is an incremental quaternion with rotational axis $\hat{\mathbf{e}}(t)$ and rotational angle $\Delta\alpha$. For $\Delta t \to 0$, $\cos(\frac{\Delta\alpha}{2}) \to 1$ and $\sin(\frac{\Delta\alpha}{2}) \to \frac{\Delta\alpha}{2}$, therefore,

$\bar{\mathbf{p}}(t) \approx 1+\hat{\mathbf{e}}(t)\frac{\Delta\alpha}{2}$. This gives

$$\bar{\mathbf{q}}(t+\Delta t) = \bar{\mathbf{q}}(t) \otimes \left(1+\hat{\mathbf{e}}(t)\frac{\Delta\alpha}{2}\right)$$

or

$$\bar{\mathbf{q}}(t+\Delta t) - \bar{\mathbf{q}}(t) = \bar{\mathbf{q}}(t) \otimes \left(0+\hat{\mathbf{e}}(t)\frac{\Delta\alpha}{2}\right)$$

Divide Δt at both sides and let $\Delta t \to 0$, we obtain

$$\frac{d\bar{\mathbf{q}}}{dt} = \bar{\mathbf{q}}(t) \otimes \left(0+\frac{1}{2}\hat{\mathbf{e}}(t)\Omega(t)\right) = \bar{\mathbf{q}}(t) \otimes \left(0+\frac{1}{2}\omega(t)\right)$$

where $\Omega(t) = \lim_{\Delta t \to 0} \frac{\Delta\alpha}{\Delta t}$ is a scalar, and $\omega(t) = \hat{\mathbf{e}}(t)\Omega(t)$ is a vector, and $(0+\frac{1}{2}\omega(t)) = \frac{1}{2}(0,\omega_1,\omega_2,\omega_3)$ is a quaternion. Using matrix expression (3.63) for the quaternion product, we obtain

$$\begin{aligned}\begin{bmatrix}\dot{q}_0\\ \dot{q}_1\\ \dot{q}_2\\ \dot{q}_3\end{bmatrix} &= \frac{1}{2}\begin{bmatrix}0 & -\omega_1 & -\omega_2 & -\omega_3\\ \omega_1 & 0 & \omega_3 & -\omega_2\\ \omega_2 & -\omega_3 & 0 & \omega_1\\ \omega_3 & \omega_2 & -\omega_1 & 0\end{bmatrix}\begin{bmatrix}q_0\\ q_1\\ q_2\\ q_3\end{bmatrix}\\ &= \frac{1}{2}\begin{bmatrix}q_0 & -q_1 & -q_2 & -q_3\\ q_1 & q_0 & -q_3 & q_2\\ q_2 & q_3 & q_0 & -q_1\\ q_3 & -q_2 & q_1 & q_0\end{bmatrix}\begin{bmatrix}0\\ \omega_1\\ \omega_2\\ \omega_3\end{bmatrix}\end{aligned} \tag{3.64}$$

Chapter 4

Spacecraft Dynamics and Modeling

The quaternion-based model has several advantages over Euler angle-based model. For example, the quaternion-based model is uniquely defined because it does not depend on a rotational sequence, while a Euler angle-based model can be different for different rotational sequences. Therefore, Euler angle-based models may be error-prone if different groups of people work on the same project but use different rotational sequences. In engineering design practice, an agreement is supposed to reach among different design groups working on the same project. Another attractive feature of quaternion-based model is that a full quaternion model does not have any singular point in any rotational sequence. Therefore, quaternion model-based control design methods using Lyapunov function have been discussed in many research papers, for example, [23, 214, 218]. Though Lyapunov function is a powerful tool in global stability analysis, obtaining a control law and the associated Lyapunov function for the nonlinear systems is postulated by intuition, as noted in [147]. Moreover, most of these designs focus on the global stability and do not pay much attention on the performance of the control system. In [147, 220], quaternion-based linear error dynamics are adapted to get desired performance for the attitude control system using classical frequency domain methods. However, state space time domain design methods, such as optimal control and pole assignment, are more attractive than the classical frequency domain design methods. In [273], a linearized state space quaternion model is derived. Unfortunately, the analysis shows that the linearized state space representation of the full quaternion model using all four components of the quaternion is uncontrollable. Therefore, pole assignment can only be achieved in some controllable subspace in the linearized state space quaternion model using all four components of the quaternion. In addition, the stability of the linearized closed loop system is unknown because an uncontrollable eigenvalue is at the origin of the complex plane.

In this chapter, firstly a controllable quaternion model for inertial pointing spacecraft has been described, the simplest one in many applications. To obtain a controllable quaternion model, only vector component of the quaternion is used in the model. The cost of using only three components of the quaternion in the model is that, similar to the Euler angle representation, the reduced model has a singular point at $\alpha = \pm\pi$, where α is the rotation angle around the rotation axis. However, this singular point is the farthest point to the point where the linearization is carried out. Therefore, the model and designed controller will work well in practice.

Secondly, a controllable quaternion model for nadir pointing spacecraft with momentum wheel(s) has been presented. This is a different model from the inertial pointing spacecraft without a momentum wheel discussed in many literatures. This model includes five important features of many low orbit nadir pointing spacecraft: (a) an additional term for the momentum wheels is incorporated to the nonlinear dynamic equations, (b) the local vertical local horizontal frame is used as the reference frame and the rotation between local vertical local horizontal frame and inertial frame is considered in the model similar to the treatment in [181] for the Euler angle-based models, (c) gravity gradient torque, a dominant and predictable disturbance for low orbit spacecraft, is included to improve the model accuracy, (d) unlike the Euler angle models, the reduced quaternion model does not depend on the rotational sequence, and (e) the singularity of the reduced quaternion model is at the farthest angle of π comparing to the singularity of Euler angle model at angle of $\pi/2$.

This chapter will show, by using only vector component of the quaternion, that these linearized spacecraft models are fully controllable. Therefore, it is easier to use these reduced models than the full quaternion models in controller design because all modern state space control system design methods can be applied directly. The stability of the designed closed-loop spacecraft system is guaranteed because the linearized control system is fully controllable. The justification of using reduced quaternion models and their benefits were fully discussed [243]. The similar strategy was used in [160, 265, 163] but the merits were not discussed.

4.1 The General Spacecraft System Equations

4.1.1 The Dynamics Equation

Let $\mathbf{J}$ be the *inertia matrix* of a spacecraft defined by

$$\mathbf{J} = \begin{bmatrix} J_{11} & J_{12} & J_{13} \\ J_{21} & J_{22} & J_{23} \\ J_{31} & J_{32} & J_{33} \end{bmatrix} \tag{4.1}$$

$\omega_I = [\omega_{I1}, \omega_{I2}, \omega_{I3}]^{\mathrm{T}}$ be the *angular velocity* vector of the spacecraft body with respect to the inertial frame, represented in the spacecraft body frame, $\mathbf{h}_I$ be the *angular momentum* vector of the spacecraft about its center of mass represented in the *inertial frame*, $\mathbf{h} = \mathbf{J}\omega_I$ be the same vector of $\mathbf{h}_I$ but represented in the *body frame*, $\mathbf{m}$ be the external torque acting on the body about its center of mass. Then, from [176], we have

$$\mathbf{m} = \left(\frac{d\mathbf{h}_I}{dt}\right)\bigg|_b$$

In view of (3.16), we have

$$\mathbf{m} = \left(\frac{d\mathbf{h}_I}{dt}\right)\bigg|_b = \left(\frac{d\mathbf{h}}{dt}\right) + \omega_I \times \mathbf{h} \tag{4.2}$$

This gives

$$\left(\frac{d\mathbf{h}}{dt}\right) = \mathbf{J}\dot{\omega}_I = -\omega_I \times \mathbf{J}\omega_I + \mathbf{m}$$

The external torques $\mathbf{m}$ are normally composed of (a) disturbance torques $\mathbf{t}_d$ due to gravitational, aerodynamic, solar radiation, and other environmental torques in body frame, and is expressed by

$$\mathbf{t}_d = [t_{d1}, t_{d2}, t_{d3}]^{\mathrm{T}} \tag{4.3}$$

and (b) the control torque $\mathbf{u}$ expressed by

$$\mathbf{u} = [u_1, u_2, u_3]^{\mathrm{T}} \tag{4.4}$$

Therefore,

$$\mathbf{J}\dot{\omega}_I = -\omega_I \times (\mathbf{J}\omega_I) + \mathbf{t}_d + \mathbf{u} = -\mathbf{S}(\omega_I)(\mathbf{J}\omega_I) + \mathbf{t}_d + \mathbf{u} \tag{4.5}$$

4.1.2 The Kinematics Equation

Denote the rotational axis of a body frame relative to a reference frame by a unit length vector $\hat{\mathbf{e}}$, the rotational angle around the rotational axis by α, the scalar component of the quaternion by $q_0 = \cos(\frac{\alpha}{2})$, the vector component of the quaternion by $\mathbf{q} = [q_1, q_2, q_3]^{\mathrm{T}} = \hat{\mathbf{e}} \sin(\frac{\alpha}{2})$, then, the quaternion that represents the rotation of the body frame relative to the reference frame is given by

$$\bar{\mathbf{q}} = [q_0, \mathbf{q}^{\mathrm{T}}]^{\mathrm{T}} = \left[\cos\left(\frac{\alpha}{2}\right), \hat{\mathbf{e}}^{\mathrm{T}} \sin\left(\frac{\alpha}{2}\right)\right]^{\mathrm{T}} \tag{4.6}$$

Let ω be the spacecraft *body rate* with respect to reference frame represented in the body frame. From (3.64), which is repeated below

$$\begin{aligned}
\begin{bmatrix} \dot{q}_0 \\ \dot{q}_1 \\ \dot{q}_2 \\ \dot{q}_3 \end{bmatrix} &= \frac{1}{2} \begin{bmatrix} 0 & -\omega_1 & -\omega_2 & -\omega_3 \\ \omega_1 & 0 & \omega_3 & -\omega_2 \\ \omega_2 & -\omega_3 & 0 & \omega_1 \\ \omega_3 & \omega_2 & -\omega_1 & 0 \end{bmatrix} \begin{bmatrix} q_0 \\ q_1 \\ q_2 \\ q_3 \end{bmatrix} \\
&= \frac{1}{2} \begin{bmatrix} q_0 & -q_1 & -q_2 & -q_3 \\ q_1 & q_0 & -q_3 & q_2 \\ q_2 & q_3 & q_0 & -q_1 \\ q_3 & -q_2 & q_1 & q_0 \end{bmatrix} \begin{bmatrix} 0 \\ \omega_1 \\ \omega_2 \\ \omega_3 \end{bmatrix}
\end{aligned} \tag{4.7}$$

the nonlinear spacecraft kinematics equations of motion can be represented by the quaternion as follows:

$$\begin{cases} \dot{\mathbf{q}} = -\frac{1}{2}\omega \times \mathbf{q} + \frac{1}{2} q_0 \omega \\ \dot{q}_0 = -\frac{1}{2}\omega^{\mathrm{T}} \mathbf{q} \end{cases} \tag{4.8}$$

In view of (4.7), using the fact that $q_0 = \sqrt{1 - q_1^2 - q_2^2 - q_3^2}$, we have

$$\begin{aligned}
\begin{bmatrix} \dot{q}_1 \\ \dot{q}_2 \\ \dot{q}_3 \end{bmatrix} &= \frac{1}{2} \begin{bmatrix} \sqrt{1-q_1^2-q_2^2-q_3^2} & -q_3 & q_2 \\ q_3 & \sqrt{1-q_1^2-q_2^2-q_3^2} & -q_1 \\ -q_2 & q_1 & \sqrt{1-q_1^2-q_2^2-q_3^2} \end{bmatrix} \begin{bmatrix} \omega_1 \\ \omega_2 \\ \omega_3 \end{bmatrix} \\
&= \frac{1}{2} \mathbf{Q}(q_1, q_2, q_3)\omega = \mathbf{g}(q_1, q_2, q_3, \omega)
\end{aligned} \tag{4.9}$$

It is easy to verify

$$\begin{aligned}
&\det \begin{pmatrix} \sqrt{1-q_1^2-q_2^2-q_3^2} & -q_3 & q_2 \\ q_3 & \sqrt{1-q_1^2-q_2^2-q_3^2} & -q_1 \\ -q_2 & q_1 & \sqrt{1-q_1^2-q_2^2-q_3^2} \end{pmatrix} \\
&= \det(\mathbf{Q}(q_1, q_2, q_3)) = \frac{1}{\sqrt{1-q_1^2-q_2^2-q_3^2}}
\end{aligned} \tag{4.10}$$

hence $\mathbf{Q}(q_1, q_2, q_3)$ is always a full rank matrix except for $\alpha = \pm\pi$. This means that unless $\alpha = \pm\pi$, the kinematics equation of motion using reduced quaternion representation can be simplified from (4.7) to (4.9).

The main advantages of using (4.9) instead of (4.7) is as follows: (a) the system dimension is reduced from 7 to 6, yielding a simpler model, (b) the linearized system is controllable, (c) the stability analysis can be directly conducted based on the linearized system (there is no uncontrollable unstable pole, see [273]), and (d) all closed loop eigenvalues can be assigned to any position by appropriate feedback control law because the linearized system is controllable. The results presented in this chapter are based on [243, 245].

4.2 The Inertial Pointing Spacecraft Model

4.2.1 The Nonlinear Inertial Pointing Spacecraft Model

The *inertial pointing spacecraft* is desired in many applications. The inertial pointing spacecraft model is one of the simplest spacecraft models. In this section, it is assumed that the spacecraft does not have a momentum wheel ($\mathbf{h}_w = 0$); therefore, the control torques are either thrusters or magnet torque rods or their combinations (more details about spacecraft control actuators will be discussed in Chapter 10). To simplify the model further, it is assumed that the disturbance torque is negligible. In this case, (4.5) is reduced to

$$\mathbf{J}\dot{\omega}_I = -\omega_I \times (\mathbf{J}\omega_I) + \mathbf{u} = -\mathbf{S}(\omega_I)(\mathbf{J}\omega_I) + \mathbf{u} \tag{4.11}$$

Let $\bar{\mathbf{q}}$ be the quaternion that represents the rotation of the spacecraft body frame relative to the inertial frame, the reduced kinematics equation is then the same as equation (4.9).

4.2.2 The Linearized Inertial Pointing Spacecraft Models

The linearized spacecraft system can be derived from (4.11) and (4.9) by using the first order Taylor expansion around the stationary point $q_1 = q_2 = q_3 = 0$ and $\omega_I = 0$ as follows:

$$\dot{\omega}_I \approx \mathbf{J}^{-1}\mathbf{u}$$

$$\left.\frac{\partial \mathbf{g}}{\partial \omega_I}\right|_{\substack{\omega_I \approx 0 \\ q_1=q_2=q_3\approx 0}} \approx \frac{1}{2}\mathbf{I}_3$$

$$\left.\frac{\partial \mathbf{g}}{\partial \mathbf{q}}\right|_{\substack{\omega_I \approx 0 \\ q_1=q_2=q_3\approx 0}} \approx \frac{1}{2}\mathbf{0}_3$$

Therefore,

$$\begin{bmatrix} \dot{\omega}_I \\ \dot{\mathbf{q}} \end{bmatrix} = \begin{bmatrix} \mathbf{0}_3 & \mathbf{0}_3 \\ \frac{1}{2}\mathbf{I}_3 & \mathbf{0}_3 \end{bmatrix} \begin{bmatrix} \omega_I \\ \mathbf{q} \end{bmatrix} + \begin{bmatrix} \mathbf{J}^{-1} \\ \mathbf{0}_3 \end{bmatrix} \mathbf{u} = \mathbf{A}\mathbf{x} + \mathbf{B}\mathbf{u} \tag{4.12}$$

where

$$\mathbf{A} = \begin{bmatrix} \mathbf{0}_3 & \mathbf{0}_3 \\ \frac{1}{2}\mathbf{I}_3 & \mathbf{0}_3 \end{bmatrix}, \quad \mathbf{x} = \begin{bmatrix} \omega_I \\ \mathbf{q} \end{bmatrix}, \quad \text{and} \quad \mathbf{B} = \begin{bmatrix} \mathbf{J}^{-1} \\ \mathbf{0}_3 \end{bmatrix} \tag{4.13}$$

It is easy to verify that this linearized spacecraft system equation is controllable.

4.3 Nadir Pointing Momentum Biased Spacecraft Model

4.3.1 The Nonlinear Nadir Pointing Spacecraft Model

Momentum biased spacecraft is widely used in practice, and is discussed extensively in [181, Chapter 8]. For momentum biased spacecraft, a momentum wheel is installed in the $\mathbf{Y}_b$ axis which is perpendicular to the orbit plane. Normally, the momentum wheel spins in a constant speed, but it may also be used to generate a control torque by changing the speed. Let

$$\mathbf{h} = [h_1, h_2, h_3]^{\mathrm{T}} = [0, h_2, 0]^{\mathrm{T}} \tag{4.14}$$

be the angular momentum of the momentum wheel in the body frame. The spacecraft model (4.5), therefore, becomes

$$\mathbf{J}\dot{\omega}_I = -\omega_I \times (\mathbf{J}\omega_I + \mathbf{h}) + \mathbf{t}_d + \mathbf{u} = -\mathbf{S}(\omega_I)(\mathbf{J}\omega_I + \mathbf{h}) + \mathbf{t}_d + \mathbf{u} \tag{4.15}$$

For a nadir pointing spacecraft, the attitude of the spacecraft is represented by the rotation of the spacecraft body frame relative to the local vertical and local horizontal (LVLH) frame. Therefore, the quaternion and spacecraft body rate will be represented in terms of the rotations of the spacecraft body frame relative to the LVLH frame. Let $\omega = [\omega_1, \omega_2, \omega_3]^{\mathrm{T}}$ be the body rate with respect to the LVLH frame represented in the body frame, $\omega_{lvlh} = [0, -\omega_0, 0]^{\mathrm{T}}$ be the *orbit rate* (or LVLH frame rate) with respect to the inertial frame, represented in the LVLH frame. Let v be the speed of the spacecraft, r be the distance from the spacecraft to the center of the Earth, p be the orbit period, then for circular orbit spacecraft, we have (see also the definition of mean motion of (2.55))

$$\omega_0 = \frac{v}{r} = \frac{2\pi}{p} \tag{4.16}$$

Let $\mathbf{A}_l^b$ represent the *transformation matrix* from the LVLH frame to the spacecraft body frame. Then, ω_I can be expressed by

$$\omega_I = \omega + \mathbf{A}_l^b \omega_{lvlh} = \omega + \omega_{lvlh}^b \tag{4.17}$$

where $\omega_{lvlh}^b = \mathbf{A}_l^b \omega_{lvlh}$ is the rate of the LVLH frame with respect to the inertial frame, represented in the body frame. From (3.15), $\dot{A}_l^b = -\omega \times A_l^b$, therefore, $\dot{\omega}_I$ is given by

$$\dot{\omega}_I = \dot{\omega} + \dot{\mathbf{A}}_l^b \omega_{lvlh} + \mathbf{A}_l^b \dot{\omega}_{lvlh} = \dot{\omega} - \omega \times \mathbf{A}_l^b \omega_{lvlh} = \dot{\omega} - \omega \times \omega_{lvlh}^b \tag{4.18}$$

where it is assumed that $\dot{\omega}_{lvlh}$ is small and can be neglected.[1] Using equations (4.17) and (4.18), equation (4.15) can be rewritten as

$$\mathbf{J}\dot{\omega} \quad = \quad \mathbf{J}(\omega \times \omega_{lvlh}^b) - \omega \times (\mathbf{J}\omega) - \omega \times (\mathbf{J}\omega_{lvlh}^b) - \omega_{lvlh}^b \times (\mathbf{J}\omega)$$

[1] This assumption is true for most satellites as long as the orbit eccentricity is small, i.e., the orbit is close to a circle.

$$
\begin{aligned}
& -\omega^b_{lvlh} \times (\mathbf{J}\omega^b_{lvlh}) - \omega \times \mathbf{h} - \omega^b_{lvlh} \times \mathbf{h} + \mathbf{t}_d + \mathbf{u} \\
= \; & \mathbf{f}(\omega, \omega^b_{lvlh}, \mathbf{h}) + \mathbf{t}_d + \mathbf{u}
\end{aligned} \quad (4.19)
$$

where

$$
\begin{aligned}
& \mathbf{f}(\omega, \omega^b_{lvlh}, \mathbf{h}) \\
= \; & \mathbf{J}(\omega \times \omega^b_{lvlh}) - \omega \times (\mathbf{J}\omega) - \omega \times (\mathbf{J}\omega^b_{lvlh}) - \omega^b_{lvlh} \times (\mathbf{J}\omega) \\
& -\omega^b_{lvlh} \times (\mathbf{J}\omega^b_{lvlh}) - \omega \times \mathbf{h} - \omega^b_{lvlh} \times \mathbf{h}
\end{aligned} \quad (4.20)
$$

Let $\bar{\mathbf{q}} = [q_0, q_1, q_2, q_3]^{\mathrm{T}} = [q_0, \mathbf{q}^{\mathrm{T}}]^{\mathrm{T}} = [\cos(\frac{\alpha}{2}), \hat{\mathbf{e}}^{\mathrm{T}} \sin(\frac{\alpha}{2})]^{\mathrm{T}}$ be the quaternion representing the rotation of the body frame relative to the LVLH frame, where $\hat{\mathbf{e}}$ is the unit length rotational axis and α is the rotation angle about $\hat{\mathbf{e}}$. Therefore, the reduced kinematics equation is given by (4.9). From (3.60), $\mathbf{A}^b_l$ can be written as

$$
\mathbf{A}^b_l = \begin{bmatrix} 2q_0^2 - 1 + 2q_1^2 & 2q_1q_2 + 2q_0q_3 & 2q_1q_3 - 2q_0q_2 \\ 2q_1q_2 - 2q_0q_3 & 2q_0^2 - 1 + 2q_2^2 & 2q_2q_3 + 2q_0q_1 \\ 2q_1q_3 + 2q_0q_2 & 2q_2q_3 - 2q_0q_1 & 2q_0^2 - 1 + 2q_3^2 \end{bmatrix}
$$

4.3.2 The Linearized Nadir Pointing Spacecraft Model

It is difficult to design a controller with specified performance (such as settling time, rising time, and percentage of overshoot) using the nonlinear spacecraft system model as described by equations (4.19) and (4.9). The common practice is to design the controller using a linearized system and then check if the designed controller works for the original nonlinear system using simulation. For a nadir pointing spacecraft system, the closed-loop spacecraft system needs to have the following features: (a) the spacecraft body rate with respect to the LVLH frame should be as small as possible, ideally, $\omega = 0$; and (b) the spacecraft body frame should be aligned with the LVLH frame, i.e., the error is as small as possible, ideally, $q_1 = q_2 = q_3 = 0$. Since the rotation axis length is always 1, this implies that the rotation angle $\alpha = 0$. Therefore, the linearized model is the first order model of Taylor expansion of the nonlinear system (4.19) and (4.9) about $\omega = 0$ and $q_1 = q_2 = q_3 = 0$. By using quaternion representation of $\mathbf{A}^b_l$, assuming $\mathbf{J}$ is almost diagonal (which is almost always true in real spacecraft designs), and neglecting high order terms of q_1, q_2, and q_3, we have the following relations.

$$
\omega^b_{lvlh} = \mathbf{A}^b_l \omega_{lvlh} = \begin{bmatrix} 2q_1q_2 + 2q_0q_3 \\ 2q_0^2 - 1 + 2q_2^2 \\ 2q_2q_3 - 2q_0q_1 \end{bmatrix} (-\omega_0) \Bigg|_{\substack{\omega \approx 0 \\ q_1 = q_2 = q_3 \approx 0}} \approx \begin{bmatrix} -2q_3 \\ -1 \\ 2q_1 \end{bmatrix} \omega_0 \quad (4.21)
$$

Using (1.6) and

$$
\mathbf{J}\omega^b_{lvlh} \approx \begin{bmatrix} -2J_{11}q_3\omega_0 \\ -J_{22}\omega_0 \\ 2J_{33}q_1\omega_0 \end{bmatrix}
$$

we have

$$
\begin{aligned}
&\omega^b_{lvlh} \times (\mathbf{J}\omega^b_{lvlh})\Big|_{\substack{\omega\approx 0\\ q_1=q_2=q_3\approx 0}} \\
&= \begin{bmatrix} 0 & 2q_1\omega_0 & \omega_0 \\ -2q_1\omega_0 & 0 & -2q_3\omega_0 \\ -\omega_0 & 2q_3\omega_0 & 0 \end{bmatrix} \begin{bmatrix} 2J_{11}q_3\omega_0 \\ J_{22}\omega_0 \\ -2J_{33}q_1\omega_0 \end{bmatrix}\Bigg|_{\substack{\omega\approx 0\\ q_1=q_2=q_3\approx 0}} \\
&\approx \omega_0^2 \begin{bmatrix} 2(J_{22}-J_{33})q_1 \\ 0 \\ 2(J_{22}-J_{11})q_3 \end{bmatrix}
\end{aligned}
\tag{4.22}
$$

and

$$
\begin{aligned}
\omega^b_{lvlh} \times \mathbf{h}\Big|_{\substack{\omega\approx 0\\ q_1=q_2=q_3\approx 0}} &= -\begin{bmatrix} 0 & 2q_1\omega_0 & \omega_0 \\ -2q_1\omega_0 & 0 & -2q_3\omega_0 \\ -\omega_0 & 2q_3\omega_0 & 0 \end{bmatrix} \begin{bmatrix} 0 \\ h_2 \\ 0 \end{bmatrix}\Bigg|_{\substack{\omega\approx 0\\ q_1=q_2=q_3\approx 0}} \\
&\approx -\omega_0 \begin{bmatrix} 2h_2q_1 \\ 0 \\ 2h_2q_3 \end{bmatrix}
\end{aligned}
\tag{4.23}
$$

Using (4.21), (4.22), and (4.23), we have

$$
\frac{\partial \mathbf{f}}{\partial \omega}\Big|_{\substack{\omega\approx 0\\ q_1=q_2=q_3\approx 0}} \approx -\mathbf{J}\mathbf{S}(\omega^b_{lvlh}) + \mathbf{S}(\mathbf{J}\omega^b_{lvlh}) - \mathbf{S}(\omega^b_{lvlh})\mathbf{J} + \mathbf{S}(\mathbf{h}) \tag{4.24}
$$

$$
\begin{aligned}
\frac{\partial \mathbf{f}}{\partial \mathbf{q}}\Big|_{\substack{\omega\approx 0\\ q_1=q_2=q_3\approx 0}} &= \frac{\partial(-\omega^b_{lvlh} \times (\mathbf{J}\omega^b_{lvlh}) - \omega^b_{lvlh} \times \mathbf{h})}{\partial \mathbf{q}}\Bigg|_{\substack{\omega\approx 0\\ q_1=q_2=q_3\approx 0}} \\
&\approx \begin{bmatrix} 2\omega_0^2(J_{33}-J_{22}) + 2h_0\omega_0 & 0 & 0 \\ 0 & 0 & 0 \\ 0 & 0 & 2\omega_0^2(J_{11}-J_{22}) + 2h_0\omega_0 \end{bmatrix}
\end{aligned}
\tag{4.25}
$$

$$
\frac{\partial \mathbf{g}}{\partial \omega}\Big|_{\substack{\omega\approx 0\\ q_1=q_2=q_3\approx 0}} \approx \frac{1}{2}\mathbf{I}_3 \tag{4.26}
$$

$$
\frac{\partial \mathbf{g}}{\partial \mathbf{q}}\Big|_{\substack{\omega\approx 0\\ q_1=q_2=q_3\approx 0}} \approx \frac{1}{2}\mathbf{0}_3 \tag{4.27}
$$

where $\mathbf{I}_3$ is a 3×3 dimensional identity matrix, $\mathbf{0}_3$ is a 3×3 dimensional zero matrix. Equation (4.24) can be simplified further as follows.

$$
\mathbf{J}\mathbf{S}(\omega^b_{lvlh}) = -\begin{bmatrix} -J_{13}\omega_0 & 0 & J_{11}\omega_0 \\ -J_{23}\omega_0 & 0 & J_{21}\omega_0 \\ -J_{33}\omega_0 & 0 & J_{31}\omega_0 \end{bmatrix} = \begin{bmatrix} 0 & 0 & -J_{11}\omega_0 \\ 0 & 0 & 0 \\ J_{33}\omega_0 & 0 & J0 \end{bmatrix} \tag{4.28}
$$

$$\mathbf{S}(\mathbf{J}\omega_{lvlh}^{b}) = -\begin{bmatrix} 0 & -J_{32}\omega_0 & J_{22}\omega_0 \\ J_{32}\omega_0 & 0 & -J_{12}\omega_0 \\ -J_{22}\omega_0 & J_{12}\omega_0 & 0 \end{bmatrix}$$

$$= \begin{bmatrix} 0 & 0 & -J_{22}\omega_0 \\ 0 & 0 & 0 \\ J_{22}\omega_0 & 0 & 0 \end{bmatrix} \tag{4.29}$$

$$\mathbf{S}(\omega_{lvlh}^{b})\mathbf{J} = -\begin{bmatrix} J_{31}\omega_0 & J_{32}\omega_0 & J_{33}\omega_0 \\ 0 & 0 & 0 \\ -J_{11}\omega_0 & -J_{12}\omega_0 & -J_{13}\omega_0 \end{bmatrix}$$

$$= \begin{bmatrix} 0 & 0 & -J_{33}\omega_0 \\ 0 & 0 & 0 \\ J_{11}\omega_0 & 0 & 0 \end{bmatrix} \tag{4.30}$$

$$\mathbf{S}(\mathbf{h}) = \begin{bmatrix} 0 & 0 & h_2 \\ 0 & 0 & 0 \\ -h_2 & 0 & 0 \end{bmatrix} \tag{4.31}$$

Therefore,

$$\left.\frac{\partial \mathbf{f}}{\partial \omega}\right|_{\substack{\omega\approx 0 \\ q_1=q_2=q_3\approx 0}} = \begin{bmatrix} 0 & 0 & (J_{11}-J_{22}+J_{33})\omega_0+h_2 \\ 0 & 0 & 0 \\ -(J_{11}-J_{22}+J_{33})\omega_0-h_2 & 0 & 0 \end{bmatrix} \tag{4.32}$$

For many nadir pointing satellites, we need to model disturbance torque in the linearized model. For low Earth orbit spacecraft, aerodynamic torque and gravity gradient torque are the dominant disturbance torques. It is difficult to model the aerodynamic torque because it is related to solar activity, geomagnetic index, spacecraft geometry, spacecraft attitude, spacecraft altitude, and many other factors, but it is known that the gravity gradient torque can be modeled by (see derivation in Chapter 5 or [181, 66])

$$\mathbf{t}_{gg} = \begin{bmatrix} 3\omega_0^2(J_{33}-J_{22})\phi \\ 3\omega_0^2(J_{33}-J_{11})\theta \\ 0 \end{bmatrix} \tag{4.33}$$

where ϕ and θ are the Euler angles for the roll and the pitch. For small Euler angles (see [219]), $\phi = 2q_1$ and $\theta = 2q_2$, this gives

$$\mathbf{t}_{gg} = \begin{bmatrix} 6\omega_0^2(J_{33}-J_{22})q_1 \\ 6\omega_0^2(J_{33}-J_{11})q_2 \\ 0 \end{bmatrix}$$

$$= \begin{bmatrix} 6\omega_0^2(J_{33}-J_{22}) & 0 & 0 \\ 0 & 6\omega_0^2(J_{33}-J_{11}) & 0 \\ 0 & 0 & 0 \end{bmatrix} \begin{bmatrix} q_1 \\ q_2 \\ q_3 \end{bmatrix} \tag{4.34}$$

From (4.19),

$$\mathbf{J}\dot{\omega} \approx \frac{\partial \mathbf{f}}{\partial \omega}\omega + \frac{\partial \mathbf{f}}{\partial \mathbf{q}}\mathbf{q} + \mathbf{t}_d + \mathbf{u} \tag{4.35}$$

Assuming $\mathbf{t}_d = \mathbf{t}_{gg}$, and combining equations (4.35), (4.25), (4.26), (4.27), (4.32), and (4.34), we have the quaternion based linearized spacecraft system described by

$$\begin{bmatrix} 1 & 0 & 0 & 0 & 0 & 0 \\ 0 & 1 & 0 & 0 & 0 & 0 \\ 0 & 0 & 1 & 0 & 0 & 0 \\ 0 & 0 & 0 & J_{11} & J_{12} & J_{13} \\ 0 & 0 & 0 & J_{21} & J_{22} & J_{23} \\ 0 & 0 & 0 & J_{31} & J_{32} & J_{33} \end{bmatrix} \begin{bmatrix} \dot{q}_1 \\ \dot{q}_2 \\ \dot{q}_3 \\ \dot{\omega}_1 \\ \dot{\omega}_2 \\ \dot{\omega}_3 \end{bmatrix}$$
$$= \begin{bmatrix} 0 & 0 & 0 & .5 & 0 & 0 \\ 0 & 0 & 0 & 0 & .5 & 0 \\ 0 & 0 & 0 & 0 & 0 & .5 \\ f_{41} & 0 & 0 & 0 & 0 & f_{46} \\ 0 & f_{52} & 0 & 0 & 0 & 0 \\ 0 & 0 & f_{63} & f_{64} & 0 & 0 \end{bmatrix} \begin{bmatrix} q_1 \\ q_2 \\ q_3 \\ \omega_1 \\ \omega_2 \\ \omega_3 \end{bmatrix} + \begin{bmatrix} 0 \\ 0 \\ 0 \\ u_x \\ u_y \\ u_z \end{bmatrix} \tag{4.36}$$

where $f_{41} = 8(J_{33}-J_{22})\omega_0^2 + 2h_2\omega_0$, $f_{46} = (J_{11}-J_{22}+J_{33})\omega_0 + h_2$, $f_{64} = -f_{46}$, $f_{52} = 6(J_{33}-J_{11})\omega_0^2$, and $f_{63} = 2(J_{11}-J_{22})\omega_0^2 + 2h_2\omega_0$. It is easy to check that the linearized spacecraft model is fully controllable. Therefore, all modern control design methods in linear system theory can be applied directly, and the designed linear system is guaranteed to be stable. Clearly, it is easy to modify the model to include three reaction wheels.

Chapter 5

Space Environment and Disturbance Torques

The previous chapter briefly mentioned that *disturbance torques* affect spacecraft's attitude. The gravitational torque was considered in the modeling process because this torque is predictable and is easy to calculate. There are several other disturbance torques induced by the space environment. These torques can significantly affect the attitude of a spacecraft if the attitude control system is not well designed because these torques are difficult to predict and they are likely not incorporated into the spacecraft dynamics models used for the control system design. These unmodeled torques introduce uncertainties. Although these disturbance torques are normally not considered in the analytical models that are used to design the controllers, in engineering design practice, the designed controller should be able to compensate these unmodeled disturbance torques to make sure a spacecraft's attitude is aligned with its desired frame.

On the other hand, given the information, such as the geometry, the electrical and the mechanical properties of the spacecraft, the attitude, the altitude, the coordinate, the speed of the spacecraft, the current time, etc., it is still possible to model the space environment and to approximately calculate these disturbance torques. Therefore, in engineering practice, the designed controllers' performances should be verified or tested in a simulation system that includes both the space environment models and the disturbance torques omitted in the design stage. This chapter will discuss the models of the most significant space environment phenomena and the associated unmodeled disturbance torques.

5.1 Gravitational Torques

The study of a rigid body in a gravitational field is based on Newton's laws. The problem has been studied for hundreds of years. A good historical review in this field can be found in [66]. The importance of *gravitational torques* on spacecraft were quickly realized in the early stage of the spacecraft development. For example, a detailed analysis of various disturbance torques acted on Sputnik 3 has shown that the gravitational torques were the major disturbance torques and were larger, by a factor of six, than the next largest disturbance torque, the magnetic torque acted on the spacecraft [11]. This large disturbance torque caused some operational problems for some spacecraft when the designs did not consider this disturbance torque. For example, the first Canadian spacecraft, Alouette 1, was spin stabilized and employed four long antennas. The long booms caused a large inertia difference which introduced a comparatively rapid precession [146]. The adversary effect of the gravitational torques was carefully studied and the formula of the gravitational torque was derived. An experiment was conducted in the spacecraft Explorer 11 where angular momentum vector was determined by radio signals and spacecraft's motion was checked against calculated gravitational torque acting on the spacecraft. A good match between calculated torque and measured torque was obtained [140]. The knowledge about the gravitational torques is sometimes used in the spacecraft design to stabilize some spacecraft [181]. Now, it has become a widely accepted engineering practice to include the gravitational torque in spacecraft models whenever it is appropriate. But still, in some applications, gravitational torques are treated as unmodeled disturbance.

Our description about gravitational torques in this section follows the style of [66, 181]. Let $\mathbf{r}$ be a vector of length r along the line connecting the centers of mass of two objects whose masses are m_1 and m_2. Let $G = 6.669 * 10^{-11} m^3/kg - s^2$ be the universal constant of gravitation. The force attracting the two objects to each other is given by (2.2) (see also [176])

$$\mathbf{f} = \frac{G m_1 m_2 \mathbf{r}}{|\mathbf{r}|^3}$$

If the first object is the Earth, and the second object is the spacecraft, since the mass of the Earth m_1 is a constant, the formula can be simplified as

$$\mathbf{f} = \frac{\mu m \mathbf{r}}{|\mathbf{r}|^3}$$

where $\mu = G m_1$ is the geocentric gravitational constant of the Earth and $m = m_2$ is the mass of the spacecraft. Let dm be a small element of the spacecraft, the vector from the center of the mass of the spacecraft to dm be $\mathbf{p}$, the vector from the center of Earth to the center of the mass of the spacecraft be $\mathbf{R}$. Since $\mathbf{r} = \mathbf{R} + \mathbf{p}$ and $d\mathbf{f} = -\frac{\mu dm}{|\mathbf{r}|^3}\mathbf{r}$, the gravitational torque or the moment induced by dm about the center of the mass of the spacecraft is given by

$$d\mathbf{t}_g = \mathbf{p} \times d\mathbf{f} = -\mathbf{p} \times \frac{\mu dm}{|\mathbf{r}|^3}\mathbf{r} = -\frac{\mu dm}{|\mathbf{r}|^3}\mathbf{p} \times \mathbf{r} \approx -\frac{\mu dm}{|\mathbf{r}|^3}\mathbf{p} \times \mathbf{R} \tag{5.1}$$

Since $|\mathbf{p}| << |\mathbf{R}|$ and for small x, $(1+x)^{-k} \approx 1 - kx$

$$\begin{aligned} |\mathbf{r}|^{-3} &= \left((\mathbf{R}+\mathbf{p})^{\mathrm{T}}(\mathbf{R}+\mathbf{p})\right)^{-\frac{3}{2}} = (|\mathbf{R}|^2 + 2\mathbf{R}\cdot\mathbf{p} + |\mathbf{p}|^2)^{-\frac{3}{2}} \\ &\approx |\mathbf{R}|^{-3}\left(1 + \frac{2\mathbf{R}\cdot\mathbf{p}}{|\mathbf{R}|^2}\right)^{-\frac{3}{2}} \approx |\mathbf{R}|^{-3}\left(1 - \frac{3\mathbf{R}\cdot\mathbf{p}}{|\mathbf{R}|^2}\right) \end{aligned} \tag{5.2}$$

Integration of (5.1) over the entire spacecraft body mass and using (5.2) yield

$$\mathbf{t}_g = \int -\frac{\mu dm}{|\mathbf{R}|^3}\left(1 - \frac{3\mathbf{R}\cdot\mathbf{p}}{|\mathbf{R}|^2}\right)\mathbf{p} \times \mathbf{R} \tag{5.3}$$

Because $\int \mathbf{p}dm = 0$ by the definition of the center of mass, the gravitational torque or gravity gradient torque is given by

$$\mathbf{t}_g = \frac{3\mu}{|\mathbf{R}|^5}\int (\mathbf{R}\cdot\mathbf{p})(\mathbf{p}\times\mathbf{R})dm = -\frac{3\mu}{|\mathbf{R}|^5}\mathbf{R} \times \int \mathbf{p}(\mathbf{p}dm\cdot\mathbf{R}) \tag{5.4}$$

Using the definition of *inertia dyadic* (see for example [220, page 335])

$$\mathbf{J} = \int (\rho^2\mathbf{I} - \mathbf{pp})dm$$

or

$$\int \mathbf{pp}dm = \int \rho^2\mathbf{I}dm - \mathbf{J}$$

(5.4) can be reduced as

$$\mathbf{t}_g = -\frac{3\mu}{|\mathbf{R}|^5}\mathbf{R} \times \int (\rho^2\mathbf{I}dm - \mathbf{J})\cdot\mathbf{R} = \frac{3\mu}{|\mathbf{R}|^5}\mathbf{R} \times \mathbf{J}\mathbf{R} \tag{5.5}$$

where the last relation uses the fact that $\mathbf{R} \times \rho^2\mathbf{I}\mathbf{R} = \rho^2\mathbf{R} \times \mathbf{R} = 0$. The *gravity gradient torque* needs to be represented in the body frame. Notice that in local vertical local horizontal frame

$$\mathbf{R} = \begin{bmatrix} 0 \\ 0 \\ -|\mathbf{R}| \end{bmatrix}$$

Let $\bar{\mathbf{q}}$ be the quaternion transformation between body frame and local vertical local horizontal frame. Then, using (3.60), $\mathbf{R}$ can be represented in body frame as

$$\mathbf{R} = \begin{bmatrix} 2q_0^2 - 1 + 2q_1^2 & 2q_1q_2 + 2q_0q_3 & 2q_1q_3 - 2q_0q_2 \\ 2q_1q_2 - 2q_0q_3 & 2q_0^2 - 1 + 2q_2^2 & 2q_2q_3 + 2q_0q_1 \\ 2q_1q_3 + 2q_0q_2 & 2q_2q_3 - 2q_0q_1 & 2q_0^2 - 1 + 2q_3^2 \end{bmatrix} \begin{bmatrix} 0 \\ 0 \\ -|\mathbf{R}| \end{bmatrix}$$

When body frame is close to local vertical local horizontal frame, $q_0 \approx 1$, $q_1 \approx 0$, $q_2 \approx 0$, and $q_3 \approx 0$, this means

$$\mathbf{R} = |\mathbf{R}| \begin{bmatrix} 2q_2 \\ -2q_1 \\ -1 \end{bmatrix}$$

Assuming that $\mathbf{J} = \text{diag}(J_{11}, J_{22}, J_{33})$, we have

$$\mathbf{R} \times \mathbf{J}\mathbf{R} \approx |\mathbf{R}|^2 \begin{bmatrix} 2q_1(J_{33} - J_{22}) \\ 2q_2(J_{33} - J_{11}) \\ 0 \end{bmatrix} \tag{5.6}$$

Since the lateral velocity of a body in a circular orbit of radius $|\mathbf{R}|$ is given in (2.32) (see also [220, page 221])

$$v = \sqrt{\frac{\mu}{|\mathbf{R}|}} \tag{5.7}$$

and angular orbital velocity of the body is given by (2.55)

$$\omega_0 = \frac{v}{|\mathbf{R}|} = \sqrt{\frac{\mu}{|\mathbf{R}|^3}} \tag{5.8}$$

substituting (5.6) and (5.8) into (5.5) yields

$$\mathbf{t}_g = \begin{bmatrix} 6\omega_0^2(J_{33} - J_{22})q_1 \\ 6\omega_0^2(J_{33} - J_{11})q_2 \\ 0 \end{bmatrix} \tag{5.9}$$

which is identical to (4.34) used in the linearized model for the controller design. To verify the controller design in a simulation system, the more accurate formula (5.5) should be used.

5.2 Atmosphere-induced Torques

Atmospheric condition is the source that causes one of the major disturbance torques for spacecraft. The atmospheric condition is determined by many factors. The most significant one is the air density that directly affects the torques which result from the aerodynamic interaction between the spacecraft and the atmosphere. A simple conservative estimate of the aerodynamic force that involves only the density is given in [117].

$$\mathbf{f} = -\rho V^2[(2 - \sigma_n - \sigma_t)(\mathbf{e}_v \cdot \mathbf{e}_n)^2\mathbf{e}_n + \sigma_t(\mathbf{e}_v \cdot \mathbf{e}_n)\mathbf{e}_v]dA \tag{5.10}$$

where $\mathbf{f}$ is the aerodynamic force on an element area dA, dA is the projected area of spacecraft element normal to the incident flow which is related to the spacecraft geometry and attitude, V is the spacecraft velocity which is related to the altitude of the spacecraft, ρ is the atmospheric density, σ_n is the normal momentum exchange coefficient, σ_t is the tangential momentum exchange coefficient, $\mathbf{e}_v$ is the unit spacecraft velocity vector, and $\mathbf{e}_n$ is the outward unit vector normal to dA. The momentum exchange coefficients are generally considered to be functions of the surface material of the spacecraft. An empirical value of 0.8 has been used for σ_t and σ_n in applications. For some simple geometric figures, formulas of aerodynamic force are given in [219, page 575, table 17-3].

Having the aerodynamic force, the *aerodynamic torque* can be evaluated by

$$\mathbf{t}_a = \mathbf{r} \times \mathbf{f} \tag{5.11}$$

where $\mathbf{r}$ is the moment arm.

The density is varied due to a lot of the factors, but a very simple graph that represents density as a function of altitude can be used for the purpose of a coarse estimation [219, page 107].

More accurate modeling atmospheres have been developed based on both physical relationships and observed phenomena [210, 211]. A detailed description of the theory and observations are beyond the scope of this book. In [112], seven different effects other than altitude that result in variations of density, temperature, and composition of the upper atmosphere are listed as follows:

variations with solar activity

diurnal variation

variations with geomagnetic activity

semiannual variation

seasonal-latitudinal variations of the lower thermo-sphere

seasonal-latitudinal variations of helium

rapid density fluctuations probably associated with tidal and gravity waves

These effects are discussed in details and many references are provided in [112]. To compute more accurate atmospheric density that take these effects into account, a set of formulas that use 10.7-cm solar flux and geomagnetic activity as inputs are also provided in Appendix A of [112].

It is easy to see that the density model is not simple but involves many factors. Therefore, the aerodynamic disturbance torque are most likely not incorporated into spacecraft dynamic models that are used for the controller design purpose. This requires that the spacecraft attitude controller designs have good disturbance rejection performance. Furthermore, the designed controller should be verified in a simulation model that includes atmospheric density and aerodynamic torque estimations.

5.3 Magnetic Field-induced Torques

Similar to the gravitational torques, the *magnetic field induced torques* can adversely affect on-board equipment and can change the spacecraft's drag, attitude, and direction of motion. A description on the degradation of the performance of the attitude control system due to magnetic field induced torques was reported in [173]. On the other hand, people quickly realized that the magnetic field induced torques can be used with the *magnet torque rods* to control the spacecraft attitude [3]. Many control algorithms are specifically designed for control systems using only magnet torque rods, for example, [159, 182, 165].

Magnetic disturbance torques are a result of the interaction between the spacecraft's residual magnetic field and the geomagnetic field. The dominant source of the magnetic disturbance torque is the spacecraft's magnetic moment because the material selection in a spacecraft design makes other magnetic disturbance sources negligible [10, 43]. The magnetic moment induced torque is given by

$$\mathbf{t}_m = \mathbf{m} \times \mathbf{r}_m \tag{5.12}$$

where $\mathbf{m}$ (in $A \cdot m^2$) is the sum of the individual magnetic moments caused by permanent and induced magnetism and the spacecraft-generated current loops, and $\mathbf{r}_m$ is the *geocentric magnetic flux density* (in Wb/m^2). The description of geocentric magnetic field is discussed in [65, 46, 141]. Given the spacecraft geocentric spherical polar coordinates (r, θ, ϕ), where r is the spacecraft geocentric distance pointing down in nadir direction, θ is the *co-elevation* pointing to the north direction, and ϕ is the *east longitude* from Greenwich pointing to the east (this information can be provided by GPS installed on spacecraft), the geomagnetic flux density vector $\mathbf{r}_m = -grad(V) := \bigtriangledown \times V$ is obtained by taking gradient of $V(r, \theta, \phi)$. The *scalar potential function* $V(r, \theta, \phi)$ is given by the following formula [65, 174, 39, 141]:

$$V(r,\theta,\phi) = a \sum_{n=1}^{\infty} \sum_{m=0}^{n} \left(\frac{a}{r}\right)^{n+1} P_n^m \cos(\theta) \left(g_n^m \cos(m\phi) + h_n^m \sin(m\phi)\right) \tag{5.13}$$

where $a = 6378km$, is the *equatorial radius* of the Earth, $P_n^m(\theta)$ are *Schmidt semi-normalized Legendre polynomials* of degree n and order m (the input to these polynomials are actually in $\cos(\theta)$, rather than θ, but this has been dropped for brevity), g_n^m and h_n^m are *Gauss coefficients* in unit nanotesla (nT). The set of Gaussian coefficients used in the analytical models are called the *International Geomagnetic Reference Field* (IGRF). These coefficients are updated every five years by a group of scientists from the International Association of Geomagnetism and Aeronomy (IAGA). The recent one, which takes advantage of a comprehensive set of observation data, including satellite measurements from the CHAMP, Orsted and SAC-C missions, was published in 2015 [75, 199]. This version of IGRF remains valid until 2020.

By using the conservative of the magnetic field ($\bigtriangledown \times \mathbf{B} = 0$), we have the geomagnetic vector $\mathbf{r}_m = -grad(V)$ by taking minus gradient of V for (r, θ, ϕ) [219].

$$B_r = \frac{-\partial V}{\partial r} = \sum_{n=1}^{\infty} \left(\frac{a}{r}\right)^{n+2} (n+1) \sum_{m=0}^{n} (g_n^m \cos(m\phi) + h_n^m \sin(m\phi)) P_n^m(\theta) \quad (5.14a)$$

$$B_\theta = \frac{-1}{r} \frac{\partial V}{\partial \theta} = \sum_{n=1}^{\infty} \left(\frac{a}{r}\right)^{n+2} \sum_{m=0}^{n} (g_n^m \cos(m\phi) + h_n^m \sin(m\phi)) \frac{\partial P_n^m(\theta)}{\partial \theta} \quad (5.14b)$$

$$B_\phi = \frac{-1}{r\sin(\theta)} \frac{\partial V}{\partial \phi} = \frac{-1}{\sin(\theta)} \sum_{n=1}^{\infty} \left(\frac{a}{r}\right)^{n+2} \sum_{m=0}^{n} m(-g_n^m \sin(m\phi) + h_n^m \cos(m\phi)) P_n^m(\theta) \quad (5.14c)$$

In order to calculate the magnetic field, one must first calculate the associated Legendre polynomials. Legendre polynomials are a set of orthogonal polynomials that also satisfy the zero mean condition. The following equations for the Legendre polynomials and associated Legendre polynomials are provided in [174]. The regular Legendre polynomials $P_n(v)$ are calculated to satisfy the following equation:

$$(1 - 2vx + x^2) - 1/2 = \sum_{n=0}^{\infty} P_n(v) x^n \quad (5.15)$$

Solving this equation gives

$$P_n(v) = \frac{1}{2^n n!} \left(\frac{d}{dv}\right)^n (v^2 - 1)^n \quad (5.16)$$

The above Legendre polynomials are related to the associated Legendre polynomials through the following equation:

$$P_{n,m}(v) = \left(1 - v^2\right)^{1/2m} \frac{d^m}{dv^m} (P_n(v)) \quad (5.17)$$

Note that for all $m > n$, the associated Legendre polynomial is equal to zero. The formulas in Equation (5.17) represent traditional associated Legendre polynomials that have not been normalized. There are two commonly used normalizations. The first is the Gaussian normalized associated Legendre polynomials, $P^{n,m}$, which is related to the non-normalized set by the following equation

$$P^{n,m}(v) = \frac{2^n!(n-m)!}{(2n)!} P_{n,m}(v) \quad (5.18)$$

The second is the Schmidt semi-normalized form, P_n^m, which is related to the non-normalized set by the following equation

$$P_n^m = \left(\frac{2(n-m)!}{(n+m)!}\right)^{1/2} P_{n,m} \tag{5.19}$$

The two Gaussian normalized associated Legendre polynomials are related as [219]:

$$P_n^m = S_{n,m} P^{n,m} \tag{5.20}$$

where $S_{n,m}$ is defined by

$$S_{n,m} = \left(\frac{(2-\delta_m^0)(n-m)!}{(n+m)!}\right)^{1/2} \frac{(2n-1)!!}{(n-m)!} \tag{5.21}$$

where the Kronecker delta is defined as $\delta_i^j = 1$ if $i = j$ and $\delta_i^j = 0$ if $i \neq j$, and $(2n-1)!! := 1 \cdot 3 \cdot \cdot (2n-1)$. Due to the fact that these normalization values can be calculated irrespective of the value of θ at which the associated Legendre polynomials are calculated, it is much simpler to instead normalize the model coefficients, g_n^m and h_n^m, such that

$$g^{n,m} = S_{n,m} g_n^m \tag{5.22}$$

and

$$h^{n,m} = S_{n,m} h_n^m \tag{5.23}$$

In order to produce efficient computer code, the preceding formulas should be decomposed into recursive formulas as seen in [39, 219]. The following recursive relationships is used in MATLAB® code of [39]. First, the recursive formulas for the Gaussian normalized associated Legendre polynomials are as follows:

$$P^{0,0} = 1 \tag{5.24a}$$

$$P^{n,n} = \sin(\theta) P^{n-1,n-1} \tag{5.24b}$$

$$P^{n,m} = \cos(\theta) P^{n-1,m} - K^{n,m} P^{n-2,m} \tag{5.24c}$$

$$K^{n,m} = 0, \quad n = 1 \tag{5.24d}$$

$$K^{n,m} = \frac{(n-1)^2 - m^2}{(2n-1)(2n-3)}, \quad n > 1 \tag{5.24e}$$

The recursive formulas for the Gaussian normalized derivatives of the associated Legendre polynomials are

$$\frac{\partial P^{0,0}}{\partial \theta} = 0 \tag{5.25a}$$

$$\frac{\partial P^{n,n}}{\partial \theta} = \sin(\theta) \frac{\partial P^{n-1,n-1}}{\partial \theta} + \cos(\theta) P^{n-1,n-1} \tag{5.25b}$$

$$\frac{\partial P^{n,m}}{\partial \theta} = \cos(\theta)\frac{\partial P^{n-1,m}}{\partial \theta} - \sin(\theta)P^{n-1,m} - K^{n,m}\frac{\partial P^{n-2,m}}{\partial \theta} \tag{5.25c}$$

Using mathematical induction, one can get the recursive formulas for $S_{n,m}$ as follows:

$$S_{0,0} = 1 \tag{5.26a}$$

$$P_{n,0} = P_{n-1,0}\left(\frac{2n-1}{n}\right), \quad n \geq 1 \tag{5.26b}$$

$$S_{n,m} = S_{n,m-1}\sqrt{\frac{(n-m+1)(\delta_m^1+1)}{n+m}} \quad m \geq 1 \tag{5.26c}$$

The procedure to calculate (B_r, B_θ, B_ϕ) is summarized as follows:

Algorithm 5.1

1. *Get the Gauss coefficients g_n^m and h_n^m from IGRF table.*
2. *Calculate $S_{n,m}$ from (5.26).*
3. *Calculate $P^{n,m}$ from (5.24).*
4. *Calculate P_n^m from (5.20).*
5. *Calculate $\frac{\partial P^{n,m}}{\partial \theta}$ from (5.25).*
6. *Calculate $\frac{\partial P_n^m}{\partial \theta} = S_{n,m}\frac{\partial P^{n,m}}{\partial \theta}$.*
7. *(B_r, B_θ, B_ϕ) is given by (5.14).*

Similar to ECEF frame, the geocentric spherical polar coordinates (r, θ, ϕ) rotates with the Earth (relatively with ECI frame as described in [219, Appendix H]). In order for the results of Equation (5.14) to be effective in spacecraft application, they must be converted to geocentric inertial frame (ECI frame). This is done by the following transformation [219, (H-14), page782].

$$B_x^I = (B_r\cos(\delta) + B_\theta\sin(\delta))\cos(\alpha) - B_\phi\sin(\alpha) \tag{5.27a}$$

$$B_y^I = (B_r\cos(\delta) + B_\theta\sin(\delta))\sin(\alpha) + B_\phi\cos(\alpha) \tag{5.27b}$$

$$B_z^I = (B_r\sin(\delta) - B_\theta\cos(\delta)) \tag{5.27c}$$

where δ is the latitude measured positive North from the equator (declination), and α is the local sidereal time of the location in question (celestial time in Greenwich). The details on the computation of (5.27) is provided in [39] and a Matlab code is attached there.

The next step is to transform the magnetic field to the orbit (PQW) frame using the following equation [205, Figure 2-16 and (3.28)].

$$\mathbf{B}^o = Rot_3(\omega)Rot_1(i)Rot_3(\Omega)\mathbf{B}^I \tag{5.28}$$

where ω is the argument of perigee, Ω is the right ascension of the ascending node, and i the inclination. Let $s\cdot$ and $c\cdot$ denote for $\sin(\cdot)$ and $\cos(\cdot)$. Expanding (5.28) gives:

$$\begin{bmatrix} B_x^o \\ B_y^o \\ B_z^o \end{bmatrix} = \begin{bmatrix} c\omega & s\omega & 0 \\ -s\omega & c\omega & 0 \\ 0 & 0 & 1 \end{bmatrix} \begin{bmatrix} 1 & 0 & 0 \\ 0 & ci & si \\ 0 & -si & ci \end{bmatrix} \begin{bmatrix} c\Omega & s\Omega & 0 \\ -s\Omega & c\Omega & 0 \\ 0 & 0 & 1 \end{bmatrix} \begin{bmatrix} B_x^I \\ B_y^I \\ B_z^I \end{bmatrix} \tag{5.29}$$

Then, a transformation from orbit frame to spacecraft coordinate (RSW) frame is needed. This transformation is given by (3.17) (see also [205, Figure 2-16 and (3.29)]):

$$\mathbf{B}^s = Rot_3(\theta)\mathbf{B}^o \tag{5.30}$$

where θ is the true anomaly. Combining (5.29) and (5.30) gives (3.18) (see also [181, (2.6.4), pages 25–26]):

$$\begin{aligned} &\mathbf{B}^s \\ = \; & Rot_3(\omega+\theta)Rot_1(i)Rot_3(\Omega)\mathbf{B}^I \\ = \; & \begin{bmatrix} c(\omega+\theta)c\Omega - cis(\omega+\theta)s\Omega & c(\omega+\theta)s\Omega + s(\omega+\theta)cic\Omega & s(\omega+\theta)si \\ -s(\omega+\theta)c\Omega - cis\Omega c(\omega+\theta) & -s(\omega+\theta)s\Omega + c(\omega+\theta)cic\Omega & c(\omega+\theta)si \\ sis\Omega & -sic\Omega & ci \end{bmatrix} \begin{bmatrix} B_x^I \\ B_y^I \\ B_z^I \end{bmatrix} \end{aligned} \tag{5.31}$$

From spacecraft coordinate frame (see Figure 2.6), one can determine the magnetic field vector in LVLH coordinate

$$\mathbf{B}^L = \begin{bmatrix} 0 & 1 & 0 \\ 0 & 0 & 1 \\ 1 & 0 & 0 \end{bmatrix} Rot_2(\pi)\mathbf{B}^s \tag{5.32}$$

Finally, to calculate the magnetic field vector described in (5.14) in body frame, (B_x^L, B_y^L, B_z^L) needs to be transformed to the spacecraft body frame as $\mathbf{r}_m$, one may use 1-2-3 rotational sequence [219, Table E-1, page 764], the formula is given by

$$\begin{aligned} r_{m_x} &= \cos(\psi)\cos(\theta)B_x^L \\ &+(\cos(\psi)\sin(\theta)\sin(\phi)+\sin(\psi)\cos(\phi))B_y^L \\ &+(-\cos(\psi)\cos(\phi)\sin(\theta)+\sin(\psi)\sin(\phi))B_z^L \end{aligned} \tag{5.33a}$$

$$\begin{aligned} r_{m_y} &= -\sin(\psi)\cos(\theta)B_x^L \\ &+(-\sin(\psi)\sin(\theta)\sin(\phi)+\cos(\psi)\cos(\phi))B_y^L \end{aligned}$$

$$+(\sin(\psi)\sin(\theta)\cos(\phi)+\cos(\psi)\sin(\phi))B_z^L \quad (5.33b)$$

$$r_{m_z}=\sin(\theta)B_x^L-\cos(\theta)\sin(\phi)B_y^L+\cos(\theta)\cos(\phi)B_z^L \quad (5.33c)$$

where ϕ, θ, and ψ are roll, pitch, and yaw angles, respectively. When these angles are small, equation (5.33) can be simplified as

$$r_{m_x}=B_x^L+\psi B_y^L+\theta B_z^L \quad (5.34a)$$

$$r_{m_y}=-\psi B_x^L+B_y^L+\phi B_z^L \quad (5.34b)$$

$$r_{m_z}=\theta B_x^L-\phi B_y^L+B_z^L \quad (5.34c)$$

5.4 Solar Radiation Torques

Solar radiation acting on the spacecraft surface generates radiation force or pressure on the surface of the spacecraft. The magnitude of this force or pressure depends on several factors, such as the intensity and spectral distribution of the incident radiation, the geometry of the surface and its optical properties, and the orientation of the Sun vector relative to the spacecraft [219, Section 17.2.2]. The mean momentum flux pressure acting on the surface normal to the Sun's radiation is $P=4.563\times10^{-6}N/m^2$ $1AU$ from the Sun. Let A be the surface area, $\mathbf{n}$ be a unit vector normal to the surface and opposite to the vector of incoming photons $\mathbf{q}$, $\mathbf{t}$ be the transverse unit vector perpendicular to the $\mathbf{n}$ and in the plane spanned by $\mathbf{q}$ and $\mathbf{n}$, α be the photon incident angle between $\mathbf{q}$ and $-\mathbf{n}$, ρ_s be the fraction of specularly reflected photons, ρ_d be the fraction of diffusely reflected photons, and ρ_a be the fraction of absorbed photons ($\rho_s+\rho_d+\rho_a=1$), then the solar radiation pressure induced force is given by [222]

$$\mathbf{f}=F_n\mathbf{n}+F_t\mathbf{t} \quad (5.35)$$

where

$$F_n=PA\left[(1+\rho_s)\cos^2(\alpha)+\frac{2}{3}\rho_d\cos(\alpha)\right]$$

and

$$F_t=PA(1-\rho_s)\cos(\alpha)\sin(\alpha)$$

For other simple geometric figures other than the flat plate, the solar radiation pressure induced force is given in [219, Table 17.2]. Given $\mathbf{f}$ in (5.35), the *solar pressure induced torque* is given by [217]

$$\mathbf{t}_s=\mathbf{r}\times\mathbf{f} \quad (5.36)$$

where $\mathbf{r}$ is the vector from the body center of mass to the optical center of pressure.

5.5 Internal Torques

Internal torques can be generated by moving parts of the spacecraft, the astronauts inside a manned space station, or the leak of gas or liquid in thrusters. When these leaks, motions, or rotations happen, they generate torques. It is relatively easier to model these torques than the torques mentioned in the previous sections. Some of these motion-induced torques are relatively large, such as the deployments of the solar panels or booms. These torques must be incorporated at least in the simulation systems to check if the designed controller can compensate these torques or not. If not, these torques may have to be incorporated into spacecraft dynamical models for the controller design purpose. If it is impossible to design a controller based on a high fidelity physics model that includes these large disturbance torques. Spacecraft design may have to be modified. For example, it may require to reduce the forces or the torques generated by the instrument deployments or increase the capacity of the actuators. This issue has not been addressed in this chapter because it is based on specific spacecraft designs.

Chapter 6

Spacecraft Attitude Determination

Spacecraft attitude determination is very important for two reasons. First, control engineers need to know if the spacecraft attitude is in the desired orientation. Second, if the spacecraft attitude is not in the perfect position, the attitude information will be compared automatically to the desired attitude, and the error information is then used to calculate how much action is needed for each actuator to bring the spacecraft to the desired attitude.

From Section 3.2.4, it can be seen that to determine the frame rotation, one needs to know the coordinates of at least two vector pairs in the body frame and the desired reference frame. Given this coordinate information, one can determine the rotational axis and the rotational angle, which represent the attitude deviation of the body frame from the desired reference frame. This intuition has been used by many researchers to develop their attitude determination methods, such as [9, 22, 124, 152, 164, 179, 180, 213, 234]. In this chapter, we will first introduce *Wahba's problem* [213], then *Davenport's formula* [38], followed by a well-known method *QUEST* [180], an *analytic solution for* a special case of Wahba's problem developed in [124], and an analytic solution to the general Wahba's problem. Noticing that QUEST and the analytic solution divide the computation of the spacecraft attitude into two steps: (a) compute the largest eigenvalue of Davenport's **K**-*matrix* and (b) compute the corresponding eigenvector, and the second step is sensitive to the accuracy of the first step, therefore, some numerical method that combines the two steps into one, i.e., directly solve the largest eigenvalue and its corresponding eigenvector of the **K**-matrix, is proposed. Some simple analysis is performed and some simulation results are presented to show the potential advantages of the direct method.

6.1 Wahba's Problem

Suppose there are measurements of two directions represented by two unit vectors $\mathbf{b}_1$ and $\mathbf{b}_2$ in the spacecraft body frame. These measurements can be unit vectors of some observed objects, such as stars, or the Sun, or the Earth, or some ambient vector field, such as the Earth's magnetic field or gravity vector. The engineers consider only unit vectors because the length of the vectors has no information relevant to the attitude determination and unit length makes expression simpler. As pointed out earlier, engineers also need to know the representations of these two unit vectors in some reference frame $\mathbf{r}_1$ and $\mathbf{r}_2$. Depending on the mission of the spacecraft, the reference frame is usually the inertial frame or the local vertical local horizontal frame. The attitude to be determined is the rotational matrix or the quaternion that rotates the reference frame to the spacecraft body frame. Therefore, one can find an attitude matrix $\mathbf{A}$ such that

$$\mathbf{A}\mathbf{r}_1 = \mathbf{b}_1 \tag{6.1a}$$

$$\mathbf{A}\mathbf{r}_2 = \mathbf{b}_2 \tag{6.1b}$$

Since a rotational matrix is also orthogonal, equation (6.1) implies

$$\mathbf{b}_1 \cdot \mathbf{b}_2 = (\mathbf{A}\mathbf{r}_1) \cdot (\mathbf{A}\mathbf{r}_2) = \mathbf{r}_1^{\mathrm{T}} \mathbf{A}^{\mathrm{T}} \mathbf{A}\mathbf{r}_2 = \mathbf{r}_1 \cdot \mathbf{r}_2 \tag{6.2}$$

In general, given two sets of m known reference vectors $\{\mathbf{r}_1, \ldots, \mathbf{r}_m\}$ and m observation vectors $\{\mathbf{b}_1, \ldots, \mathbf{b}_m\}$, $m \geq 2$, find the proper rotational matrix $\mathbf{A}$ which brings the first set into the best least squares coincidence with the second, i.e.,

$$\min_{\mathbf{A}} \frac{1}{2} \sum_{i=1}^{m} \|\mathbf{b}_i - \mathbf{A}\mathbf{r}_i\|^2 \tag{6.3}$$

This problem was first defined by Wahba and is called Wahba's problem [213] which is the base of the most attitude determination methods.

A slightly more general assumption is that there is a set of weights a_i, each is associated with a corresponding observation $\mathbf{b}_i$, and $\sum_i a_i = 1$. Then Wahba's problem takes the following form:

$$\min_{\mathbf{A}} \frac{1}{2} \sum_{i=1}^{m} a_i \|\mathbf{b}_i - \mathbf{A}\mathbf{r}_i\|^2 \tag{6.4}$$

6.2 Davenport's Formula

All popular methods, such as QUEST [180], ESOQ [136], and FOMA [123], use Davenport's q-method [38] (**K**-matrix derivation is accessible in [91]). Rewriting (6.3) by using equations (6.1) and (6.2), then using the facts: (a) $\mathbf{b}_i$ and $\mathbf{r}_i$ are unit vectors, and (b) $\mathbf{A}$ is orthogonal matrix, we have

$$\begin{aligned} \frac{1}{2}\sum_{i=1}^{m} \|\mathbf{b}_i - \mathbf{A}\mathbf{r}_i\|^2 &= \frac{1}{2}\sum_{i=1}^{m} (\mathbf{b}_i^{\mathrm{T}}\mathbf{b}_i - 2\mathbf{b}_i^{\mathrm{T}}\mathbf{A}\mathbf{r}_i + \mathbf{r}_i^{\mathrm{T}}\mathbf{A}^{\mathrm{T}}\mathbf{A}\mathbf{r}_i) \\ &= \left(m - \sum_{i=1}^{m} \mathbf{b}_i^{\mathrm{T}}\mathbf{A}\mathbf{r}_i\right) = \left(m - Tr(\mathbf{W}^{\mathrm{T}}\mathbf{A}\mathbf{V})\right) \end{aligned} \tag{6.5}$$

where $\mathbf{W} = [\mathbf{b}_1, \ldots, \mathbf{b}_m]$, $\mathbf{V} = [\mathbf{r}_1, \ldots, \mathbf{r}_m]$, and $Tr(\cdot)$ represents the trace of the matrix in the argument. Using (3.61) and the fact that $Tr(\mathbf{AB}) = Tr(\mathbf{BA})$ for any matrices $\mathbf{A}$ and $\mathbf{B}$ with appropriate dimensions, we have

$$\begin{aligned} & Tr(\mathbf{W}^{\mathrm{T}}\mathbf{A}\mathbf{V}) \\ = \; & Tr(\mathbf{W}^{\mathrm{T}}\left((q_0^2 - \mathbf{q}^{\mathrm{T}}\mathbf{q})\mathbf{I} + 2\mathbf{q}\mathbf{q}^{\mathrm{T}} - 2q_0\mathbf{q}^{\times}\right)\mathbf{V}) \\ = \; & (q_0^2 - \mathbf{q}^{\mathrm{T}}\mathbf{q})Tr(\mathbf{W}^{\mathrm{T}}\mathbf{V}) + 2Tr(\mathbf{q}\mathbf{q}^{\mathrm{T}}\mathbf{V}\mathbf{W}^{\mathrm{T}}) - 2q_0 Tr(\mathbf{W}^{\mathrm{T}}\mathbf{q}^{\times}\mathbf{V}) \end{aligned} \tag{6.6}$$

Let $\mathbf{B} = \mathbf{W}\mathbf{V}^{\mathrm{T}}$, $\sigma = Tr(\mathbf{B})$, $\mathbf{H} = \mathbf{B} + \mathbf{B}^{\mathrm{T}}$, and $\mathbf{z}^{\mathrm{T}} = [B_{23} - B_{32}, B_{31} - B_{13}, B_{12} - B_{21}]$. The second term of (6.6) can be rewritten as

$$2Tr(\mathbf{q}\mathbf{q}^{\mathrm{T}}\mathbf{V}\mathbf{W}^{\mathrm{T}}) = 2\mathbf{q}^{\mathrm{T}}\mathbf{V}\mathbf{W}^{\mathrm{T}}\mathbf{q} = \mathbf{q}^{\mathrm{T}}(\mathbf{V}\mathbf{W}^{\mathrm{T}} + \mathbf{W}\mathbf{V}^{\mathrm{T}})\mathbf{q} = \mathbf{q}^{\mathrm{T}}\mathbf{H}\mathbf{q} \tag{6.7}$$

Since $\mathbf{z}^{\times} = \mathbf{B}^{\mathrm{T}} - \mathbf{B}$, $\mathbf{q}^{\times^{\mathrm{T}}} = -\mathbf{q}^{\times}$, and $Tr(\mathbf{q}^{\times}\mathbf{z}^{\times}) = -2\mathbf{q}^{\mathrm{T}}\mathbf{z}$, the third term of (6.6) can be rewritten as

$$\begin{aligned} & 2q_0 Tr(\mathbf{q}^{\times}\mathbf{V}\mathbf{W}^{\mathrm{T}}) \\ = \; & q_0 Tr(\mathbf{q}^{\times}\mathbf{B}^{\mathrm{T}} + \mathbf{B}\mathbf{q}^{\times^{\mathrm{T}}}) \\ = \; & q_0 Tr(\mathbf{q}^{\times}\mathbf{B}^{\mathrm{T}} + \mathbf{q}^{\times^{\mathrm{T}}}\mathbf{B}) \\ = \; & q_0 Tr(\mathbf{q}^{\times}(\mathbf{B}^{\mathrm{T}} - \mathbf{B})) \\ = \; & q_0 Tr(\mathbf{q}^{\times}\mathbf{z}^{\times}) = -2q_0\mathbf{q}^{\mathrm{T}}\mathbf{z} \end{aligned} \tag{6.8}$$

Substituting (6.7) and (6.8) into (6.6) produces

$$\begin{aligned} & Tr(\mathbf{W}^{\mathrm{T}}\mathbf{A}\mathbf{V}) \\ = \; & (q_0^2 - \mathbf{q}^{\mathrm{T}}\mathbf{q})\sigma + \mathbf{q}^{\mathrm{T}}\mathbf{H}\mathbf{q} + 2q_0\mathbf{q}^{\mathrm{T}}\mathbf{z} \\ = \; & \begin{bmatrix} q_0 & \mathbf{q}^{\mathrm{T}} \end{bmatrix} \begin{bmatrix} \sigma & \mathbf{z}^{\mathrm{T}} \\ \mathbf{z} & \mathbf{H} - \sigma\mathbf{I} \end{bmatrix} \begin{bmatrix} q_0 \\ \mathbf{q} \end{bmatrix} \\ := \; & \bar{\mathbf{q}}^{\mathrm{T}}\mathbf{K}\bar{\mathbf{q}} \end{aligned} \tag{6.9}$$

where

$$\mathbf{K} = \begin{bmatrix} \sigma & \mathbf{z}^{\mathrm{T}} \\ \mathbf{z} & \mathbf{H} - \sigma\mathbf{I} \end{bmatrix} \tag{6.10}$$

Therefore,

$$\min_{\mathbf{A}} \frac{1}{2} \sum_{i=1}^{m} \|\mathbf{b}_i - \mathbf{A}\mathbf{r}_i\|^2 = \left(m - \max_{\mathbf{A}} Tr(\mathbf{W}^{\mathrm{T}}\mathbf{A}\mathbf{V}) \right) = \left(m - \max_{\bar{\mathbf{q}}=1} \bar{\mathbf{q}}^{\mathrm{T}}\mathbf{K}\bar{\mathbf{q}} \right) \tag{6.11}$$

Introducing the Lagrange multiplier λ for the unit length constraint of $\|\bar{\mathbf{q}}\| = 1$ reduces Wahba's problem to Davenport's problem

$$\max_{\lambda,\bar{\mathbf{q}}} \bar{\mathbf{q}}^{\mathrm{T}}\mathbf{K}\bar{\mathbf{q}} - \lambda(\bar{\mathbf{q}}^{\mathrm{T}}\bar{\mathbf{q}} - 1) \tag{6.12}$$

Taking the derivative of (6.12) gives the optimal solution which satisfies

$$\mathbf{K}\bar{\mathbf{q}} = \lambda\bar{\mathbf{q}} \tag{6.13}$$

The optimization problem is reduced to find the largest eigenvalue of **K** and its corresponding eigenvector, which is Davenport's formula.

6.3 Attitude Determination Using QUEST and FOMA

In the early of 1980s, the computation of the largest eigenvalue and its corresponding eigenvector of the **K**-matrix in an on-board computer was a burden. Shuster [180] developed QUEST algorithm to approximately solve (6.13). By using the *Cayley-Hamilton theorem* (cf. [169, pages 4–5]), Shuster [180] derived the first analytic formula of the characteristic polynomial of the **K**-matrix which is a polynomial of degree 4, given as

$$f(\lambda) = \lambda^4 - (a+b)\lambda^2 - c\lambda + (ab + c\sigma - d) = 0 \tag{6.14}$$

where $\sigma = 0.5Tr(\mathbf{H}) = Tr(\mathbf{B})$, $\kappa = Tr(adj(\mathbf{H}))$, $\Delta = \det(\mathbf{H})$, $a = \sigma^2 - \kappa$, $b = \sigma^2 + \mathbf{z}^{\mathrm{T}}\mathbf{z}$, $c = \Delta + \mathbf{z}^{\mathrm{T}}\mathbf{H}\mathbf{z}$, and $d = \mathbf{z}^{\mathrm{T}}\mathbf{H}^2\mathbf{z}$.

For many applications, the largest eigenvalue may be approximated by $\lambda \approx 1$. Shuster [180] suggested using *Newton-Raphson iteration* to find the λ using the initial guess $\lambda^0 = 1$. To calculate the eigenvector using λ, Shuster used the *Rodriguez parameters* defined as follows:

$$\mathbf{p} = \frac{\mathbf{q}}{q_0} = \mathbf{q}\tan\left(\frac{\alpha}{2}\right)$$

Since $\mathbf{K}\bar{\mathbf{q}} = \lambda\bar{\mathbf{q}}$, from the **K**-matrix, it is easy to see that

$$[(\lambda + \sigma)\mathbf{I} - \mathbf{H}]\mathbf{p} = \mathbf{z}$$

$\mathbf{p}$ can be obtained by solving linear system equations. Once $\mathbf{p}$ is available, the quaternion is given by

$$\bar{\mathbf{q}} = \frac{1}{\sqrt{1+\mathbf{p}^{\mathrm{T}}\mathbf{p}}} \begin{bmatrix} \mathbf{p} \\ 1 \end{bmatrix} \tag{6.15}$$

To avoid the possible singularity in Rodriguez parameter, Shuster and Oh developed a method of sequential rotations which avoids the singularity. This method is widely recognized and is refereed to as the QUEST method. The operation count for QUEST method was analyzed in [251] and is listed as follows.

1. Constructing the characteristic polynomial (6.14): 67 flops in total
2. In each iteration of Newton method: 18 flops
3. Constructing the quaternion (6.15): 33 flops

This flop count shows that QUEST needs very small number of flops in every iteration. The construction of the characteristic polynomial and the quaternion may be the main effort in QUEST.

Markley [123] derived an equivalent characteristic polynomial for the $\mathbf{K}$-matrix and also used Newton's method for his expression of the polynomial to find the largest eigenvalue λ iteratively. Using this largest eigenvalue, Markley's method finds the rotational matrix explicitly. This method is now referred to as the FOMA algorithm. This method is more expensive than QUEST, and similar to QUEST, is sensitive to the accuracy of the solution of the largest eigenvalue.

6.4 Analytic Solution of Two Vector Measurements

Though QUEST is very efficient, if the attitude determination is based on only two vector measurements, there is a simpler method which is an analytic solution [124].

6.4.1 The Minimum-angle Rotation Quaternion

First, it is worthwhile to notice that for the quaternion which maps the reference vector $\mathbf{r}_1$ to the body frame vector $\mathbf{b}_1$, the minimal rotational angle α is determined by $\cos(\alpha) = \mathbf{b}_1 \cdot \mathbf{r}_1$. Using the minimum-angle rotation quaternion (see Figure 3.4), the rotational axis must be perpendicular to $\mathbf{r}_1$ and $\mathbf{b}_1$ and satisfy the right-hand rule, which means that the unit length rotational axis is given by $\hat{\mathbf{e}} = \frac{\mathbf{b}_1 \times \mathbf{r}_1}{\sin(\alpha)}$. Using the following identities of the trigonometry [157]

$$\frac{1-\cos(\alpha)}{2} = \sin^2\left(\frac{\alpha}{2}\right)$$

$$\cot\left(\frac{\alpha}{2}\right) = \frac{1+\cos(\alpha)}{\sin(\alpha)}$$

we can verify that the minimum-angle rotation quaternion is given by

$$\begin{aligned}
&(1+\mathbf{b}_1\cdot\mathbf{r}_1,\mathbf{b}_1\times\mathbf{r}_1)\frac{1}{\sqrt{2(1+\mathbf{b}_1\cdot\mathbf{r}_1)}} \\
=\ &(1+\cos(\alpha),\mathbf{b}_1\times\mathbf{r}_1)\sqrt{\frac{1}{2(1+\cos(\alpha))}} \\
=\ &(1+\cos(\alpha),\mathbf{b}_1\times\mathbf{r}_1)\sqrt{\frac{1-\cos(\alpha)}{2(1-\cos^2(\alpha))}} \\
=\ &(1+\cos(\alpha),\mathbf{b}_1\times\mathbf{r}_1)\frac{\sin\left(\frac{\alpha}{2}\right)}{\sin(\alpha)} \\
=\ &\left(\frac{1+\cos(\alpha)}{\sin(\alpha)},\frac{\mathbf{b}_1\times\mathbf{r}_1}{\sin(\alpha)}\right)\sin\left(\frac{\alpha}{2}\right) \\
=\ &\left(\cot\left(\frac{\alpha}{2}\right),\hat{\mathbf{e}}\right)\sin\left(\frac{\alpha}{2}\right) \\
=\ &\left(\cos\left(\frac{\alpha}{2}\right),\hat{\mathbf{e}}\sin\left(\frac{\alpha}{2}\right)\right) = \bar{\mathbf{q}}_{min}
\end{aligned} \tag{6.16}$$

6.4.2 The General Rotation Quaternion

Denote $\bar{\mathbf{q}}(\hat{\mathbf{e}},\alpha)$ as the quaternion that has rotational axis $\hat{\mathbf{e}}$ and rotational angle α. Then, the most general rotation that maps $\mathbf{r}_1$ to $\mathbf{b}_1$ is given by

$$\bar{\mathbf{q}}_1 = \bar{\mathbf{q}}(\mathbf{r}_1,\phi_r)\otimes\bar{\mathbf{q}}_{min}\otimes\bar{\mathbf{q}}(\mathbf{b}_1,\phi_b) \tag{6.17}$$

where ϕ_b and ϕ_r are arbitrary angles of rotation about $\mathbf{b}_1$ and $\mathbf{r}_1$, respectively. Using (1.2), (1.3), (1.4), (3.43), and the facts that

$$\sin(\alpha+\beta) = \sin(\alpha)\cos(\beta)+\cos(\alpha)\sin(\beta) \tag{6.18}$$

and

$$\cos(\alpha+\beta) = \cos(\alpha)\cos(\beta)-\sin(\alpha)\sin(\beta) \tag{6.19}$$

(6.17) can be reduced by using (3.43) and (1.1), as follows:

$$\begin{aligned}
&(1+\mathbf{b}_1\cdot\mathbf{r}_1,\mathbf{b}_1\times\mathbf{r}_1)\otimes\left(\cos\left(\frac{\phi_b}{2}\right),\mathbf{b}_1\sin\left(\frac{\phi_b}{2}\right)\right) \\
=\ &\left((1+\mathbf{b}_1\cdot\mathbf{r}_1)\cos\left(\frac{\phi_b}{2}\right)-(\mathbf{b}_1\times\mathbf{r}_1)\cdot\mathbf{b}_1\sin\left(\frac{\phi_b}{2}\right)\right. \\
&\left.+(1+\mathbf{b}_1\cdot\mathbf{r}_1)\mathbf{b}_1\sin\left(\frac{\phi_b}{2}\right)+(\mathbf{b}_1\times\mathbf{r}_1)\cos\left(\frac{\phi_b}{2}\right)+(\mathbf{b}_1\times\mathbf{r}_1)\times\mathbf{b}_1\sin\left(\frac{\phi_b}{2}\right)\right)
\end{aligned}$$

$$
\begin{aligned}
&= \left((1+\mathbf{b}_1\cdot\mathbf{r}_1)\cos\left(\frac{\phi_b}{2}\right)\right. \\
&\quad \left. +(1+\mathbf{b}_1\cdot\mathbf{r}_1)\mathbf{b}_1\sin\left(\frac{\phi_b}{2}\right)+(\mathbf{b}_1\times\mathbf{r}_1)\cos\left(\frac{\phi_b}{2}\right)+(\mathbf{r}_1-(\mathbf{b}_1\cdot\mathbf{r}_1)\mathbf{b}_1)\sin\left(\frac{\phi_b}{2}\right)\right) \\
&= \left((1+\mathbf{b}_1\cdot\mathbf{r}_1)\cos\left(\frac{\phi_b}{2}\right),(\mathbf{b}_1+\mathbf{r}_1)\sin\left(\frac{\phi_b}{2}\right)+(\mathbf{b}_1\times\mathbf{r}_1)\cos\left(\frac{\phi_b}{2}\right)\right)
\end{aligned} \tag{6.20}
$$

Thus, we have

$$
\begin{aligned}
&\left(\cos\left(\frac{\phi_r}{2}\right),\mathbf{r}_1\sin\left(\frac{\phi_r}{2}\right)\right) \\
&\otimes(1+\mathbf{b}_1\cdot\mathbf{r}_1,\mathbf{b}_1\times\mathbf{r}_1)\otimes\left(\cos\left(\frac{\phi_b}{2}\right),\mathbf{b}_1\sin\left(\frac{\phi_b}{2}\right)\right) \\
&= \left(\cos\left(\frac{\phi_r}{2}\right),\mathbf{r}_1\sin\left(\frac{\phi_r}{2}\right)\right) \\
&\otimes\left((1+\mathbf{b}_1\cdot\mathbf{r}_1)\cos\left(\frac{\phi_b}{2}\right),(\mathbf{b}_1+\mathbf{r}_1)\sin\left(\frac{\phi_b}{2}\right)+(\mathbf{b}_1\times\mathbf{r}_1)\cos\left(\frac{\phi_b}{2}\right)\right)
\end{aligned} \tag{6.21}
$$

Let q_0 and $\mathbf{q}$ be the scalar part and vector part of the quaternion defined by (6.21). Using (3.43), (1.4), and (6.19), and the fact that $\|\mathbf{r}_1\|=1=\|\mathbf{b}_1\|$, we have

$$
\begin{aligned}
q_0 &= (1+\mathbf{b}_1\cdot\mathbf{r}_1)\cos\left(\frac{\phi_r}{2}\right)\cos\left(\frac{\phi_b}{2}\right) \\
&\quad -\mathbf{r}_1\cdot(\mathbf{b}_1+\mathbf{r}_1)\sin\left(\frac{\phi_r}{2}\right)\sin\left(\frac{\phi_b}{2}\right)-\mathbf{r}_1\cdot(\mathbf{b}_1\times\mathbf{r}_1)\sin\left(\frac{\phi_r}{2}\right)\cos\left(\frac{\phi_b}{2}\right) \\
&= (1+\mathbf{b}_1\cdot\mathbf{r}_1)\cos\left(\frac{\phi_r}{2}\right)\cos\left(\frac{\phi_b}{2}\right)-(1+\mathbf{b}_1\cdot\mathbf{r}_1)\sin\left(\frac{\phi_r}{2}\right)\sin\left(\frac{\phi_b}{2}\right) \\
&= (1+\mathbf{b}_1\cdot\mathbf{r}_1)\left(\cos\left(\frac{\phi_r}{2}\right)\cos\left(\frac{\phi_b}{2}\right)-\sin\left(\frac{\phi_r}{2}\right)\sin\left(\frac{\phi_b}{2}\right)\right) \\
&= (1+\mathbf{b}_1\cdot\mathbf{r}_1)\cos\left(\frac{\phi_r+\phi_b}{2}\right)
\end{aligned} \tag{6.22}
$$

From (6.21), using (3.43), (1.3), (6.18), and (6.19), we have

$$
\begin{aligned}
\mathbf{q} &= (1+\mathbf{b}_1\cdot\mathbf{r}_1)\mathbf{r}_1\cos\left(\frac{\phi_b}{2}\right)\sin\left(\frac{\phi_r}{2}\right) \\
&\quad +(\mathbf{b}_1+\mathbf{r}_1)\sin\left(\frac{\phi_b}{2}\right)\cos\left(\frac{\phi_r}{2}\right)+(\mathbf{b}_1\times\mathbf{r}_1)\cos\left(\frac{\phi_b}{2}\right)\cos\left(\frac{\phi_r}{2}\right) \\
&\quad +\mathbf{r}_1\times(\mathbf{b}_1+\mathbf{r}_1)\sin\left(\frac{\phi_r}{2}\right)\sin\left(\frac{\phi_b}{2}\right)+\mathbf{r}_1\times(\mathbf{b}_1\times\mathbf{r}_1)\sin\left(\frac{\phi_r}{2}\right)\cos\left(\frac{\phi_b}{2}\right)
\end{aligned}
$$

$$\begin{aligned}
&= \mathbf{r}_1 \cos\left(\frac{\phi_b}{2}\right)\sin\left(\frac{\phi_r}{2}\right) + (\mathbf{b}_1\cdot\mathbf{r}_1)\mathbf{r}_1\cos\left(\frac{\phi_b}{2}\right)\sin\left(\frac{\phi_r}{2}\right)\\
&\quad +(\mathbf{b}_1+\mathbf{r}_1)\sin\left(\frac{\phi_b}{2}\right)\cos\left(\frac{\phi_r}{2}\right) + (\mathbf{b}_1\times\mathbf{r}_1)\cos\left(\frac{\phi_r}{2}\right)\cos\left(\frac{\phi_b}{2}\right)\\
&\quad -(\mathbf{b}_1\times\mathbf{r}_1)\sin\left(\frac{\phi_r}{2}\right)\sin\left(\frac{\phi_b}{2}\right)\\
&\quad +\mathbf{b}_1\sin\left(\frac{\phi_r}{2}\right)\cos\left(\frac{\phi_b}{2}\right) - (\mathbf{b}_1\cdot\mathbf{r}_1)\mathbf{r}_1\cos\left(\frac{\phi_b}{2}\right)\sin\left(\frac{\phi_r}{2}\right)\\
&= (\mathbf{b}_1+\mathbf{r}_1)\sin\left(\frac{\phi_b}{2}\right)\cos\left(\frac{\phi_r}{2}\right) + (\mathbf{b}_1+\mathbf{r}_1)\sin\left(\frac{\phi_r}{2}\right)\cos\left(\frac{\phi_b}{2}\right)\\
&\quad +(\mathbf{b}_1\times\mathbf{r}_1)\left(\cos\left(\frac{\phi_r}{2}\right)\cos\left(\frac{\phi_b}{2}\right) - \sin\left(\frac{\phi_r}{2}\right)\sin\left(\frac{\phi_b}{2}\right)\right)\\
&= (\mathbf{b}_1+\mathbf{r}_1)\sin\left(\frac{\phi_r+\phi_b}{2}\right) + (\mathbf{b}_1\times\mathbf{r}_1)\cos\left(\frac{\phi_r+\phi_b}{2}\right)\\
&= (\mathbf{b}_1+\mathbf{r}_1)\sin\left(\frac{\phi}{2}\right) + (\mathbf{b}_1\times\mathbf{r}_1)\cos\left(\frac{\phi}{2}\right) \qquad (6.23)
\end{aligned}$$

where $\phi = \phi_r + \phi_b$. Combining (6.17), (6.16), (6.21), (6.22), and (6.23) yields

$$\bar{\mathbf{q}}_1 = \frac{1}{\sqrt{2(1+\mathbf{b}_1\cdot\mathbf{r}_1)}}\left((1+\mathbf{b}_1\cdot\mathbf{r}_1)\cos\left(\frac{\phi}{2}\right), (\mathbf{b}_1\times\mathbf{r}_1)\cos\left(\frac{\phi}{2}\right) + (\mathbf{b}_1+\mathbf{r}_1)\sin\left(\frac{\phi}{2}\right)\right) \quad (6.24)$$

Similarly, the most general rotation that maps $\mathbf{r}_2$ to $\mathbf{b}_2$ is given by

$$\bar{\mathbf{q}}_2 = \frac{1}{\sqrt{2(1+\mathbf{b}_2\cdot\mathbf{r}_2)}}\left((1+\mathbf{b}_2\cdot\mathbf{r}_2)\cos\left(\frac{\psi}{2}\right), (\mathbf{b}_2\times\mathbf{r}_2)\cos\left(\frac{\psi}{2}\right) + (\mathbf{b}_2+\mathbf{r}_2)\sin\left(\frac{\psi}{2}\right)\right) \quad (6.25)$$

for some angle ψ.

6.4.3 Attitude Determination Using Two Vector Measurements

As every quaternion in the family of $\bar{\mathbf{q}}_1(\phi)$ maps $\mathbf{r}_1$ to $\mathbf{b}_1$ and every quaternion in the family of $\bar{\mathbf{q}}_2(\psi)$ maps $\mathbf{r}_2$ to $\mathbf{b}_2$, we need to find a quaternion $\bar{\mathbf{q}}$ which is in both families so that it can map $\mathbf{r}_1$ to $\mathbf{b}_1$ and $\mathbf{r}_2$ to $\mathbf{b}_2$ simultaneously. This means that both the scalar part and the vector part of $\bar{\mathbf{q}}_1$ and $\bar{\mathbf{q}}_2$ are equal for some ϕ and ψ. For the scalar part, we need

$$\begin{aligned}
\frac{(1+\mathbf{r}_1\cdot\mathbf{b}_1)}{\sqrt{2(1+\mathbf{r}_1\cdot\mathbf{b}_1)}}\cos\left(\frac{\phi}{2}\right) &= \frac{(1+\mathbf{r}_2\cdot\mathbf{b}_2)}{\sqrt{2(1+\mathbf{r}_2\cdot\mathbf{b}_2)}}\cos\left(\frac{\psi}{2}\right)\\
\Longrightarrow \qquad \cos\left(\frac{\psi}{2}\right) &= \sqrt{\frac{1+\mathbf{r}_1\cdot\mathbf{b}_1}{1+\mathbf{r}_2\cdot\mathbf{b}_2}}\cos\left(\frac{\phi}{2}\right) \qquad (6.26a)
\end{aligned}$$

$$\Longrightarrow \qquad \sin\left(\frac{\psi}{2}\right) = \sqrt{\frac{1+\mathbf{r}_2\cdot\mathbf{b}_2-(1+\mathbf{r}_1\cdot\mathbf{b}_1)\cos^2\left(\frac{\phi}{2}\right)}{1+\mathbf{r}_2\cdot\mathbf{b}_2}} \tag{6.26b}$$

For the vector part, we need

$$\begin{aligned} & \frac{(\mathbf{b}_1\times\mathbf{r}_1)}{\sqrt{(1+\mathbf{b}_1\cdot\mathbf{r}_1)}}\cos\left(\frac{\phi}{2}\right)+\frac{(\mathbf{b}_1+\mathbf{r}_1)}{\sqrt{(1+\mathbf{b}_1\cdot\mathbf{r}_1)}}\sin\left(\frac{\phi}{2}\right) \\ = & \frac{(\mathbf{b}_2\times\mathbf{r}_2)}{\sqrt{(1+\mathbf{b}_2\cdot\mathbf{r}_2)}}\cos\left(\frac{\psi}{2}\right)+\frac{(\mathbf{b}_2+\mathbf{r}_2)}{\sqrt{(1+\mathbf{b}_2\cdot\mathbf{r}_2)}}\sin\left(\frac{\psi}{2}\right) \end{aligned} \tag{6.27}$$

Substituting (6.26a) and (6.26b) into (6.27) yields

$$\begin{aligned} & (\mathbf{b}_1\times\mathbf{r}_1)\cos\left(\frac{\phi}{2}\right)+(\mathbf{b}_1+\mathbf{r}_1)\sin\left(\frac{\phi}{2}\right) \\ = & \frac{1+\mathbf{b}_1\cdot\mathbf{r}_1}{1+\mathbf{b}_2\cdot\mathbf{r}_2}\cos\left(\frac{\phi}{2}\right)(\mathbf{b}_2\times\mathbf{r}_2)+(\mathbf{b}_2+\mathbf{r}_2)\frac{\sqrt{(1+\mathbf{b}_1\cdot\mathbf{r}_1)}}{(1+\mathbf{b}_2\cdot\mathbf{r}_2)}\sqrt{1+\mathbf{b}_2\cdot\mathbf{r}_2-(1+\mathbf{b}_1\cdot\mathbf{r}_1)\cos^2\left(\frac{\phi}{2}\right)} \end{aligned}$$

Applying dot product of $\mathbf{b}_2-\mathbf{r}_2$ on both the sides, the right-hand side vanishes because $(\mathbf{b}_2+\mathbf{r}_2)\cdot(\mathbf{b}_2-\mathbf{r}_2)=0$, and from (1.4), $(\mathbf{b}_2\times\mathbf{r}_2)\cdot(\mathbf{b}_2-\mathbf{r}_2)=0$. Therefore, we have

$$(\mathbf{b}_1\times\mathbf{r}_1)\cdot(\mathbf{b}_2-\mathbf{r}_2)\cos\left(\frac{\phi}{2}\right)+(\mathbf{b}_1+\mathbf{r}_1)\cdot(\mathbf{b}_2-\mathbf{r}_2)\sin\left(\frac{\phi}{2}\right)=0 \tag{6.28}$$

or

$$\frac{\sin\left(\frac{\phi}{2}\right)}{\cos\left(\frac{\phi}{2}\right)}=-\frac{(\mathbf{b}_1\times\mathbf{r}_1)\cdot(\mathbf{b}_2-\mathbf{r}_2)}{(\mathbf{b}_1+\mathbf{r}_1)\cdot(\mathbf{b}_2-\mathbf{r}_2)} \tag{6.29}$$

For any two vectors **a** and **b**, if **a** is proportional to **b**, we denote this relation as $\mathbf{a}\propto\mathbf{b}$. Clearly, if $\mathbf{a}\propto\mathbf{b}$, and $\mathbf{b}\propto\mathbf{c}$, then $\mathbf{a}\propto\mathbf{c}$. From (6.24) and (6.29), we have

$$\begin{aligned} \bar{\mathbf{q}} & \propto \left(1+\mathbf{b}_1\cdot\mathbf{r}_1,\mathbf{b}_1\times\mathbf{r}_1+\frac{\sin\left(\frac{\phi}{2}\right)}{\cos\left(\frac{\phi}{2}\right)}(\mathbf{b}_1+\mathbf{r}_1)\right) \\ & \propto \left((1+\mathbf{b}_1\cdot\mathbf{r}_1),\mathbf{b}_1\times\mathbf{r}_1-\frac{(\mathbf{b}_1\times\mathbf{r}_1)\cdot(\mathbf{b}_2-\mathbf{r}_2)}{(\mathbf{b}_1+\mathbf{r}_1)\cdot(\mathbf{b}_2-\mathbf{r}_2)}(\mathbf{b}_1+\mathbf{r}_1)\right) \\ & \propto \left((\mathbf{b}_1+\mathbf{r}_1)\cdot(\mathbf{b}_2-\mathbf{r}_2),\frac{(\mathbf{b}_1\times\mathbf{r}_1)((\mathbf{b}_1+\mathbf{r}_1)\cdot(\mathbf{b}_2-\mathbf{r}_2))-((\mathbf{b}_1\times\mathbf{r}_1)\cdot(\mathbf{b}_2-\mathbf{r}_2))(\mathbf{b}_1+\mathbf{r}_1)}{(1+\mathbf{b}_1\cdot\mathbf{r}_1)}\right) \end{aligned} \tag{6.30}$$

In view of (6.2), the scalar part of (6.30) implies

$$(\mathbf{b}_1+\mathbf{r}_1)\cdot(\mathbf{b}_2-\mathbf{r}_2)=\mathbf{b}_2\cdot\mathbf{r}_1-\mathbf{b}_1\cdot\mathbf{r}_2 \tag{6.31}$$

For the numerator of the vector part of (6.30), using (1.3), (1.2), and the fact that $\mathbf{b}_1$ and $\mathbf{r}_1$ are unit vectors, we have

$$\begin{aligned}
& (\mathbf{b}_1 \times \mathbf{r}_1)((\mathbf{b}_1 + \mathbf{r}_1) \cdot (\mathbf{b}_2 - \mathbf{r}_2)) - ((\mathbf{b}_1 \times \mathbf{r}_1) \cdot (\mathbf{b}_2 - \mathbf{r}_2))(\mathbf{b}_1 + \mathbf{r}_1) \\
= \; & (\mathbf{b}_2 - \mathbf{r}_2) \times ((\mathbf{b}_1 \times \mathbf{r}_1) \times (\mathbf{b}_1 + \mathbf{r}_1)) \\
= \; & (\mathbf{b}_2 - \mathbf{r}_2) \times (\mathbf{r}_1 - (\mathbf{r}_1 \cdot \mathbf{b}_1)\mathbf{b}_1 + (\mathbf{b}_1 \cdot \mathbf{r}_1)\mathbf{r}_1 - \mathbf{b}_1) \\
= \; & (\mathbf{b}_2 - \mathbf{r}_2) \times (\mathbf{r}_1 - \mathbf{b}_1 + (\mathbf{r}_1 - \mathbf{b}_1)(\mathbf{r}_1 \cdot \mathbf{b}_1)) \\
= \; & ((\mathbf{b}_2 - \mathbf{r}_2) \times (\mathbf{r}_1 - \mathbf{b}_1))(1 + \mathbf{r}_1 \cdot \mathbf{b}_1) \\
= \; & ((\mathbf{b}_1 - \mathbf{r}_1) \times (\mathbf{b}_2 - \mathbf{r}_2))(1 + \mathbf{r}_1 \cdot \mathbf{b}_1)
\end{aligned} \tag{6.32}$$

Combining (6.30), (6.31), and (6.32) yields

$$\bar{\mathbf{q}} \propto (\mathbf{b}_2 \cdot \mathbf{r}_1 - \mathbf{b}_1 \cdot \mathbf{r}_2, (\mathbf{b}_1 - \mathbf{r}_1) \times (\mathbf{b}_2 - \mathbf{r}_2))$$

Normalizing the right-hand side gives

$$\bar{\mathbf{q}} = \frac{(\mathbf{b}_2 \cdot \mathbf{r}_1 - \mathbf{b}_1 \cdot \mathbf{r}_2, (\mathbf{b}_1 - \mathbf{r}_1) \times (\mathbf{b}_2 - \mathbf{r}_2))}{\sqrt{(\mathbf{b}_2 \cdot \mathbf{r}_1 - \mathbf{b}_1 \cdot \mathbf{r}_2)^2 + \|(\mathbf{b}_1 - \mathbf{r}_1) \times (\mathbf{b}_2 - \mathbf{r}_2)\|^2}} \tag{6.33}$$

Therefore, given known ephemeris $\mathbf{r}_1$ and $\mathbf{r}_2$, observations $\mathbf{b}_1$ and $\mathbf{b}_2$, the attitude quaternion is uniquely defined. The attitude quaternion is extremely simple though the derivation is tedious. It is worthwhile to note that this solution does not need to compute the largest eigenvalue and its corresponding eigenvector. The operation count is very low. In fact, the calculation of $\mathbf{b}_2 \cdot \mathbf{r}_1 - \mathbf{b}_1 \cdot \mathbf{r}_2$ needs 11 flops and the calculation of $(\mathbf{b}_1 - \mathbf{r}_1) \times (\mathbf{b}_2 - \mathbf{r}_2)$ needs 15 flops. Given these two quantities, the calculation of the square root needs seven flops. Therefore, the total flops is $11 + 15 + 7 + 4 = 37$ flops.

6.5 Analytic Formula for General Case

Although all flight experiences were successful for QUEST method, using a specific example, Markley and Mortari [127] showed that QUEST does not always converge. In fact, it is well known that Newton's method (used in QUEST to find zeros of a polynomial) is inadequate for general use since it may fail to converge to a solution. Cheng and Shuster [27] find a fix for the specific problem raised by Markley and Mortari [127]. But even if Newton's method converge, its behavior may be erratic in regions where the function is not convex [142]. On the other hand, equation (6.14) is a polynomial of degree 4 which admits analytic solutions.

6.5.1 Analytic Formula

Since the characteristic polynomial of (6.14) has order of four, it admits analytic solution. Mortari noticed this and proposed a closed-form solution which is now referred to as the ESOQ algorithm [136]. This solution, however, was known to be not numerically stable by experts for a long time but this issue was not discussed openly in literatures.

In this section, we provide a different but more robust analytic solution based on the characteristic polynomial of the **K**-matrix presented in [136] which is given as follows.

$$p(x) = x^4 + ax^3 + bx^2 + cx + d = 0 \tag{6.34}$$

where $a = 0$, $b = -2(tr[\mathbf{B}])^2 + tr[adj(\mathbf{H})] - \mathbf{z}^{\mathrm{T}}\mathbf{z}$, $\mathbf{H} = \mathbf{B} + \mathbf{B}^{\mathrm{T}}$, $adj(\mathbf{H})$ the adjugate matrix of $\mathbf{H}$, $c = -tr[adj(\mathbf{K})]$, and $d = \det(\mathbf{K})$ are all known parameters. Several different methods were proposed in the last several hundred years [71] to solve (6.34). A latest effort was by Shmakov [178] who found a universal method to find the roots of the general quartic polynomial. A special case of this method is simpler than all previous methods and it can be directly adopted to solve (6.34). The steps can be summarized as follows (see [259]).

First, equation (6.34) can be factorized as the product of two quadratic polynomials as

$$\begin{aligned} & (x^2 + g_1 x + h_1)(x^2 + g_2 + h_2) \\ = \; & x^4 + (g_1 + g_2)x^3 + (g_1 g_2 + h_1 + h_2)x^2 \\ & +(g_1 h_2 + g_2 h_1)x + h_1 h_2 = 0 \end{aligned} \tag{6.35}$$

Moreover, g_1, g_2, h_1, and h_2 are solutions of two quadratic equations defined by

$$g^2 - ag + \frac{2}{3}b - y = 0 \tag{6.36a}$$

$$h^2 - \left(y + \frac{b}{3}\right)h + d = 0 \tag{6.36b}$$

where y is the real root(s) of the following cubic polynomial

$$y^3 + py + q = 0 \tag{6.37a}$$

$$p = ac - \frac{b^2}{3} - 4d \tag{6.37b}$$

$$q = \frac{abc}{3} - a^2 d - \frac{2}{27}b^3 - c^2 + \frac{8}{3}bd \tag{6.37c}$$

The roots of the cubic equation can be obtained by the famous *Cardano's formula* [157]

$$y_1 = \sqrt[3]{-\frac{q}{2} + \sqrt{\left(\frac{q}{2}\right)^2 + \left(\frac{p}{3}\right)^3}} + \sqrt[3]{-\frac{q}{2} - \sqrt{\left(\frac{q}{2}\right)^2 + \left(\frac{p}{3}\right)^3}} \tag{6.38a}$$

$$y_2 = \omega_1 \sqrt[3]{-\frac{q}{2} + \sqrt{\left(\frac{q}{2}\right)^2 + \left(\frac{p}{3}\right)^3}} + \omega_2 \sqrt[3]{-\frac{q}{2} - \sqrt{\left(\frac{q}{2}\right)^2 + \left(\frac{p}{3}\right)^3}} \quad (6.38b)$$

$$y_3 = \omega_2 \sqrt[3]{-\frac{q}{2} + \sqrt{\left(\frac{q}{2}\right)^2 + \left(\frac{p}{3}\right)^3}} + \omega_1 \sqrt[3]{-\frac{q}{2} - \sqrt{\left(\frac{q}{2}\right)^2 + \left(\frac{p}{3}\right)^3}} \quad (6.38c)$$

where $\omega_1 = \frac{-1+i\sqrt{3}}{2}$ and $\omega_2 = \frac{-1-i\sqrt{3}}{2}$. It is well known that (6.37) has either one real solution or three real solutions. If the discriminate

$$\Delta = \left(\frac{q}{2}\right)^2 + \left(\frac{p}{3}\right)^3 > 0$$

then (6.37) has a real solution given by (6.38a), and a pair of complex conjugate solutions given by (6.38b) and (6.38c). If $\Delta = 0$, (6.37) has three zero solutions. If $\Delta < 0$, then (6.37) has three distinct real solutions. In this case, to avoid complex operations, the solutions can be given in a different form. Let $r = \sqrt{-\left(\frac{p}{3}\right)^3}$, $\theta = \frac{1}{3}\arccos\left(-\frac{q}{2r}\right)$, then the three real solutions are given by

$$y_1 = 2r^{\frac{1}{3}} \cos(\theta) \quad (6.39a)$$

$$y_2 = 2r^{\frac{1}{3}} \cos\left(\theta + \frac{2\pi}{3}\right) \quad (6.39b)$$

$$y_3 = 2r^{\frac{1}{3}} \cos\left(\theta + \frac{4\pi}{3}\right) \quad (6.39c)$$

Given a real y, from (6.36), we have

$$g_{1,2} = \pm\sqrt{y - \frac{2}{3}b} \quad (6.40a)$$

$$h_{1,2} = \frac{y + \frac{b}{3} \pm \sqrt{(y+b/3)^2 - 4d}}{2} \quad (6.40b)$$

In view of (6.35), it is worthwhile to notice that the following relations must be held

$$(g_1 + g_2) = a \quad (6.41a)$$

$$g_1 g_2 + h_1 + h_2 = b \quad (6.41b)$$

$$g_1 h_2 + g_2 h_1 = c \quad (6.41c)$$

$$h_1 h_2 = d \quad (6.41d)$$

where (6.41a), (6.41b), and (6.41d) do not depend on the selections of g_1, g_2, h_1, and h_2 (these relations always hold), but (6.41c) does depend on the choices of g_1, g_2, h_1, and h_2. In practice, it can always be assumed that g_1 takes a positive sign in (6.40a) and g_2 takes minus sign in (6.40a); it can then be tried that h_1

takes a positive sign in (6.40b) and h_2 takes minus sign in (6.40b); if (6.41c) holds, the correct selection is obtained; otherwise, h_1 takes minus sign in (6.40b) and h_2 takes positive sign in (6.40b) so that (6.41c) holds. Finally, the roots of the quartic (6.34) are given by

$$x_{1,2} = \frac{-g_1 \pm \sqrt{g_1^2 - 4h_1}}{2} \tag{6.42a}$$

$$x_{3,4} = \frac{-g_2 \pm \sqrt{g_2^2 - 4h_2}}{2} \tag{6.42b}$$

A MATLAB® code of this method can be downloaded from MATLAB file exchange website https://www.mathworks.com/matlabcentral/fileexchange/54255-quartic-roots-m.

6.5.2 Numerical Test

The proposed analytic method and QUEST method have been implemented in MATLAB and tested against each other.

A simple problem: The first simple test is the following problem.

$$p(x) = x^4 + ax^3 + bx^2 + cx + d = 0 \tag{6.43}$$

where $a = 0$, $b = -2$, $c = 0$, and $d = 1$. The problem has two positive solutions of $x = 1$ and two negative solutions of $x = -1$. The analytic method finds all solutions without numerical error. Starting from $x = 1.1$, the QUEST method finds the largest positive solution $x = 1.00000001251746$ after 23 iterations.

Randomly generated problems: The simple problem shows that the analytic method may be promising. Extensive numerical test were conducted for tens of thousands randomly generated problems. These test problems are generated as follows. First, Euler angles $\alpha \in [0,\pi]$, $\beta \in [0,\pi]$, and $\gamma \in [0,\pi]$ are randomly generated. This gives the true rotational matrix $\mathbf{A}$ which is converted as the true rotational quaternion $\bar{\mathbf{q}}_{t_i}$ for each randomly generated problem. Then three unit vectors representing the astronomic objectives $\mathbf{r}_i$, $i = 1,2,3$, are randomly generated. It is then assumed that the measurement vectors $\mathbf{b}_i$ is the rotation of $\mathbf{r}_i$ with measurement noise given by

$$\mathbf{A}\mathbf{r}_i = \mathbf{b}_i + \mathbf{n}_i$$

where $\mathbf{n}_i \in [0,N]$ are random noise whose maximum magnitude N varies in our test. The relative weight associated with each measurement is taken as $a_i = \frac{1}{n}$, where n is the total number of measurements. For each prescribed N, 1000 randomly generated Wahba's problems are solved by both analytic and QUEST methods, the results are denoted as $\bar{\mathbf{q}}_{a_i}$ and $\bar{\mathbf{q}}_{n_i}$, respectively. The cumulative errors between the true quaternions and estimated quaternions are calculated as

$$E_a = \sum_{i=1}^{1000} \|\bar{\mathbf{q}}_{t_i} - \bar{\mathbf{q}}_{a_i}\|_2, \qquad E_n = \sum_{i=1}^{1000} \|\bar{\mathbf{q}}_{t_i} - \bar{\mathbf{q}}_{n_i}\|_2$$

The results are given in Table 6.1.

This test result shows that if the upper bound of the noise is greater than 10^{-8}, the estimation accuracies for both analytic method and QUEST method are very similar. For very small noise (the maximum magnitude is less than 10^{-8}), QUEST method is slightly better. The MATLAB code for calculating the roots of the quartic equation can be downloaded from [76].

Table 6.1: Comparison of analytic method and QUEST method.

Noise size	Analytic method E_a	QUEST method E_n
N=0.01	4.50344692811497	4.50336882243908
N=0.001	0.46355508921313	0.46356102302689
N=0.0001	0.04633308474148	0.04636056974745
N=0.00001	0.00464952990550	0.00462173419676
N=0.000001	0.46855497718417E-3	0.45068048617712E-3
N=0.0000001	0.48374024654480E-4	0.46367084520959E-4
N=0.00000001	0.32071390174853E-4	0.04635740127652E-4
N=0.000000001	0.67150605970535E-5	0.04666503538671E-5
N=0.0000000001	0.93419725779054E-5	0.00465660360757E-5

6.6 Riemann-Newton Method

For problems with more than two measurements, both QUEST method and the analytic method described in the previous section solve Davenport's problem in two steps. First, find the largest eigenvalue of the **K** matrix; then find the quaternion using the analytic formula. It has been noticed that the second step is sensitive to the accuracy of the the largest eigenvalue of the **K**-matrix but directly solves Davenport's method is much more robust, which was also observed in [123]. Since $\bar{\mathbf{q}}$ is a unit length vector, maximizing (6.11) is equivalent to solving *Rayleigh quotient problem* [73]:

$$\lambda_{\max} = \max_{\|\bar{\mathbf{q}}\|=1} \frac{1}{2} \bar{\mathbf{q}}^{\mathrm{T}} \mathbf{K} \bar{\mathbf{q}} \tag{6.44}$$

where $\bar{\mathbf{q}}$ is also the eigenvector associated with the largest eigenvalue $\lambda_{\max}$ of the **K**-matrix. Problem (6.44) is an optimization problem with a sphere constraint $\|\bar{\mathbf{q}}\| = 1$ which is much simpler than Wahba's problem.

As the size of the problem (6.44) is small, Newton's method should be considered. Noticing that both Euclidean space and smooth algebraic equation systems are Riemannian manifolds, Smith [186] extended unconstrained Newton's method in Euclidean space to include all Riemannian manifolds (smoothly constrained optimization problem). The method derived from the idea is not only mathematically elegant, but also turns out, for some cases including the unit sphere constraint in (6.44), to be extremely efficient [186, 241]. In the following discussion, a slightly different but more efficient method is proposed to solve the problem defined in (6.44).

Instead of searching along straight line, optimization on sphere (or in general on manifolds) searches along geodesics on the sphere (or in general on manifolds). The first important result is, therefore, to find the geodesic defined by the current point on sphere and a descent direction. Let $\mathbf{B}\mathcal{S}^{n-1} := \{\bar{\mathbf{q}} \in \mathbf{R}^n : \|\bar{\mathbf{q}}\| = 1\}$ be a sphere in n-dimensional space, let $\mathbf{y}$ be a descent direction and the tangent space of $\mathbf{B}\mathcal{S}^{n-1}$ at $\bar{\mathbf{q}}$ be denoted as $\mathcal{T}_{\bar{\mathbf{q}}}(\mathbf{B}\mathcal{S}^{n-1})$, then we have (see [242]) the following

Theorem 6.1
Let $\bar{\mathbf{q}} \in \mathbf{B}\mathcal{S}^3$, $\mathbf{y} \in \mathcal{T}_{\bar{\mathbf{q}}}(\mathbf{B}\mathcal{S}^3)$ *be any tangent vector at* $\bar{\mathbf{q}}$, *and* $\|\mathbf{y}\| = 1$. *Then, the unique geodesic* $\mathbf{g}(t)$ *on* $\mathbf{B}\mathcal{S}^3$ *emanating from* $\bar{\mathbf{q}}$ *along the direction of* $\mathbf{y}$ *is given by*

$$\mathbf{g}(t) = \bar{\mathbf{q}}\cos(t) + \mathbf{y}\sin(t) \tag{6.45}$$

where $t \in [0, \frac{\pi}{2}]$

The main steps of the original Riemann-Newton method in [186] are: (a) from current iterate $\bar{\mathbf{q}}$, calculate the Newton direction (a vector) in $\mathbf{R}^n$, (b) project the vector onto the tangent space $\mathcal{T}_{\bar{\mathbf{q}}}(\mathbf{B}\mathcal{S}^{n-1})$, (c) normalize the vector in the tangent space to get $\mathbf{y}$, and (d) search the optimizer along the geodesic (6.45) to a new iterate $\bar{\mathbf{q}}$. Repeat Steps (a) to (d) until an optimal solution is obtained. Using the simple structure of spheres and fixed step size, steps (a) and (b) can be simplified as follows. Let $\mathbf{P}_{\bar{\mathbf{q}}_k} = (\mathbf{I} - \bar{\mathbf{q}}_k\bar{\mathbf{q}}_k^{\mathrm{T}})$ be the orthogonal projection from $\mathbf{R}^4$ to $\mathcal{T}_{\bar{\mathbf{q}}}(\mathbf{B}\mathcal{S}^3)$. Since the gradient of $\frac{1}{2}\bar{\mathbf{q}}^{\mathrm{T}}\mathbf{K}\bar{\mathbf{q}}$ is $\mathbf{P}_{\bar{\mathbf{q}}_k}\mathbf{K}\bar{\mathbf{q}}$, and the Hessian of $\frac{1}{2}\bar{\mathbf{q}}^{\mathrm{T}}\mathbf{K}\bar{\mathbf{q}}$ on the sphere manifold can be expressed as $\mathbf{P}_{\bar{\mathbf{q}}_k}\mathbf{K}\mathbf{P}_{\bar{\mathbf{q}}_k} - \bar{\mathbf{q}}_k^{\mathrm{T}}\mathbf{K}\bar{\mathbf{q}}_k\mathbf{I}$. The Newton equation for (6.44) is given by

$$(\mathbf{P}_{\bar{\mathbf{q}}_k}\mathbf{K}\mathbf{P}_{\bar{\mathbf{q}}_k} - \bar{\mathbf{q}}_k^{\mathrm{T}}\mathbf{K}\bar{\mathbf{q}}_k\mathbf{I})\mathbf{y}_k = -\mathbf{P}_{\bar{\mathbf{q}}_k}\mathbf{K}\bar{\mathbf{q}}_k \tag{6.46}$$

Steps (c) and (d) can be approximated in a much more efficient way described as follows. As $\mathbf{y}_k$ must be on the tangent plane $\mathcal{T}_{\bar{\mathbf{q}}}(\mathbf{B}\mathcal{S}^{n-1})$, the Newton full size update on the tangent plane is $\bar{\mathbf{q}}_k + \mathbf{y}_k$. Because of the special structure of sphere, searching along geodesic can be replaced by

$$\bar{\mathbf{q}}_{k+1} = \frac{\bar{\mathbf{q}}_k + \mathbf{y}_k}{\|\bar{\mathbf{q}}_k + \mathbf{y}_k\|} \tag{6.47}$$

The algorithm is, therefore, given as follows.

Algorithm 6.1

Select $\bar{\mathbf{q}}_0 \in \mathbf{R}^4$ *such that* $\|\bar{\mathbf{q}}_0\| = 1$
for $k = 0, 1, 2, \ldots$

Solve linear systems $\mathbf{P}_{\bar{\mathbf{q}}_k}\mathbf{K}\mathbf{P}_{\bar{\mathbf{q}}_k}\mathbf{y}_k - \mathbf{y}_k\bar{\mathbf{q}}_k^{\mathrm{T}}\mathbf{K}\bar{\mathbf{q}}_k = -\mathbf{P}_{\bar{\mathbf{q}}_k}\mathbf{K}\bar{\mathbf{q}}_k$ *and* $\bar{\mathbf{q}}_k^{\mathrm{T}}\mathbf{y}_k = 0$

Set $\bar{\mathbf{q}}_{k+1} = \frac{\bar{\mathbf{q}}_k + \mathbf{y}_k}{\|\bar{\mathbf{q}}_k + \mathbf{y}_k\|}$ *and* $k = k+1$

end (for)

For general problem, Riemann-Newton method in [186] does not have a useful rule to choose a good initial point. For attitude determination problem, however, Shuster observed [180] that the largest eigenvalue of $\mathbf{K}$-matrix is very close to one. Therefore, the initial point $\bar{\mathbf{q}}_0$ can be determined as follows. Let $\bar{\mathbf{K}} = \mathbf{K} - \mathbf{I}$. Since $\mathbf{K}\bar{\mathbf{q}} \approx \bar{\mathbf{q}}$, or equivalently $\bar{\mathbf{K}}\bar{\mathbf{q}} \approx \mathbf{0}$, using MATLAB notation, this gives

$$\bar{\mathbf{K}}(:,2:4)\mathbf{q} = -\bar{\mathbf{K}}(:,1) \tag{6.48}$$

and set $\bar{\mathbf{q}}_0 = \frac{[1,\mathbf{q}^{\mathrm{T}}]^{\mathrm{T}}}{\|[1,\mathbf{q}^{\mathrm{T}}]\|}$. Numerical experience shows that this selection of $\bar{\mathbf{q}}_0$ is very close to the solution of (6.44). In many cases, there is no need for any iteration. Another possible way to select the initial point is to use (6.33) for two vector observations, which is slightly cheaper than the method of solving linear system equations (6.48). Numerical test in [251] demonstrated the efficiency and robustness of this method. The MATLAB code of the method can be downloaded in [77].

6.7 Rotation Rate Determination Using Vector Measurements

The information of the rotation rate of the spacecraft may be needed in the feedback controller design. Many spacecrafts are equipped with on-board three axis rate-gyros to measure the angular rate [53]. But some spacecrafts do not install the rate-gyros because of the economical consideration. In this case, angular rate can be estimated using vector measurements, for example, the method published in [184]. This section presents a very simple method. Let

$$\mathbf{E} = \begin{bmatrix} -q_1 & q_0 & q_3 & -q_2 \\ -q_2 & -q_3 & q_0 & q_1 \\ -q_3 & q_2 & -q_1 & q_0 \end{bmatrix} \tag{6.49}$$

Pre-multiplying $2\mathbf{E}$ on both sides of (3.64) gives

$$\begin{bmatrix} \omega_1 \\ \omega_2 \\ \omega_3 \end{bmatrix} = 2 \begin{bmatrix} -q_1 & q_0 & q_3 & -q_2 \\ -q_2 & -q_3 & q_0 & q_1 \\ -q_3 & q_2 & -q_1 & q_0 \end{bmatrix} \begin{bmatrix} \frac{dq_0}{dt} \\ \frac{dq_1}{dt} \\ \frac{dq_2}{dt} \\ \frac{dq_3}{dt} \end{bmatrix} = 2\mathbf{E}\frac{d\bar{\mathbf{q}}}{dt} \tag{6.50}$$

In theory, after getting the quaternion, then taking the differences $\Delta\bar{\mathbf{q}} = \bar{\mathbf{q}}(t_i) - \bar{\mathbf{q}}(t_{i-1})$, $\Delta t = t_i - t_{i-1}$, and the division of $\frac{\Delta\bar{\mathbf{q}}}{\Delta t}$, we can approximate $\frac{d\bar{\mathbf{q}}}{dt}$ and get the angular rate. However, in practical application, due to the measurement noise, this angular rate determination based on the differentiation may not be reliable because of the high frequency noise. A low pass Butterworth digital filter [144], whose input is the ω obtained from (6.50) and the output is the refined angular rate, will significantly suppress the noise and thereby improve the angular rate determination. Furthermore, this angular rate can be further refined by a Kalman filter that will be discussed in Chapter 8.

The next problem for spacecraft attitude determination is about how to get ephemeris and observation vectors. These vector pairs can be any astronomical vectors, such as the Sun vector pairs, the Earth vector pairs, the Earth's magnetic vector pairs, or any star vector pair. There are a lot of literatures that discuss these topics. For example, for the sun direction measurement, one can read [111]. For the ephemeris sun direction, the formula is given in [205]. For geomagnetic vector measurement, a magnetometer can be used [78]. For the ephemeris geomagnetic vector, the formula is given in [219]. For star tracker and algorithms, one can read [81]. These topics will be discussed in the next Chapter.

Chapter 7

Astronomical Vector Measurements

As seen in Chapter 6, the attitude determination depends primarily on the calculations of known reference vectors and the measurements of the astronomical vectors. The most frequently used astronomical vector measurements are the Sun's vector, the Earth's vector, the Earth's magnetic vector, and stars' vectors. In this chapter, vectors $\mathbf{r}_i$, $i = o,m,s$, are used for reference vectors; subscript o for the astronomical objects, m for geomagnetic field, and s for the Sun. Similarly, $\mathbf{b}_i$, $i = o,m,s$, are used for measured vectors of the astronomical objects, the geomagnetic field, and the Sun. It will be further discussed how these vectors are obtained in principle.

7.1 Stars' Vectors

Using stars in navigation and attitude determination has a long history. On the *celestial sphere* (an imaginary sphere of arbitrarily large radius, concentric with the Earth, with *celestial equator* on the same plane as Earth's equator and *celestial poles* in the same directions as the Earth's poles), all objects in the sky can be projected upon the celestial sphere and they all have essentially fixed positions on the celestial sphere. Therefore, if a spacecraft attitude is perfectly aligned with LVLH frame, the $-\mathbf{Z}$ direction will point to a certain astronomical object, which is a known direction vector $\mathbf{r}_o$ in the reference frame. If a *Charge Coupled Device* (CCD) camera mounted on the spacecraft with the *field of view* (FOV) in the $-\mathbf{Z}$ direction of the body detects some astronomical object, then a measured vector $\mathbf{b_o}$ is obtained. To make this idea work, several things are needed. First, a map is needed that gives the information about the star and their location and position in the celestial sphere (the spacecraft position is determined by a GPS mounted on the spacecraft). Several requirements are needed for this map: (a) stars in this map should be bright enough for CCD camera to see them, (b) stars in this map should be uniformly distributed everywhere so that CCD camera is always pointing to certain stars. This kind of map is called a *star catalog*. People have created many star catalogs for the purpose of attitude determination, for example, [171]. Second, after CCD detects some stars, we need to know where these stars are located in the star catalogs. There are numerous methods to use, see the survey paper by [188]. Based on the ideas described above, *star trackers* can be built (see [172]). Therefore, the observation vector and measurement vector are obtained as follows. Giving the spacecraft position, the $\mathbf{r_o}$ is immediately available from the star catalog; using CCD camera, stars are found, using star identification algorithm, stars observed on CCD are identified in the star catalog, thereby measured vector $\mathbf{b_o}$ is obtained.

7.2 Earth's Magnetic Field Vectors

To use Earth's *magnetic field* vectors in attitude determination, given the spacecraft position, we need to know ephemeris Earth's magnetic field vector in the reference frame, for example, in ECI frame or in LVLH frame; and the measured Earth's magnetic field vector in the body frame.

7.2.1 Ephemeris Earth's Magnetic Field Vector

The geomagnetic vector is based on the International Geomagnetic Reference Field (IGRF) model which is propagated by the flight software. Given the spacecraft *geocentric spherical polar coordinates* (r, θ, ϕ) (spacecraft *geocentric distance*, co-elevation, and east longitude from Greenwich) by GPS, the ephemeris Earth's magnetic field vector $\mathbf{r}_m$ is related to the scalar magnetic potential function V

$$V(r,\theta,\phi) = a\sum_{n=1}^{\infty}\sum_{m=0}^{n}\left(\frac{a}{r}\right)^{n+1} P_n^m \cos(\theta)\left(g_n^m \cos(m\phi) + h_n^m \sin(m\phi)\right) \tag{7.1}$$

and $\mathbf{r}_m = -\text{grad}(V)$ is given in (5.14), this geomagnetic flux density should then be expressed in reference frame, i.e., ECI frame or LVLH frame. The transformations are discussed in Chapter 5 (see also [219, 141]).

7.2.2 Measured Earth's Magnetic Field Vector

There are many different magnetic sensors for various applications [107]. Among these sensors, flux-gate type *magnetometer* is the one used most for spacecraft to measure the Earth's magnetic field vector. The sensor is installed on the spacecraft with the known orientation. The geomagnetic vector in the body frame $\mathbf{b_m}$ can be then obtained from the magnetometer (TAM) measurement without any signal processing. A digital filter may also be used to reduce the measurement noise, but that may introduce some signal delay. Since the measurement noise of TAM is relatively small, a digital filter is likely not used. For some recent development in magnetometer design, readers are referred to [33] and the references therein.

7.3 Sun Vector

To use *sun vector* in attitude determination, given the spacecraft position, we need to know the sun vector in the reference frame, for example, in ECI frame or in LVLH frame; and the Sun vector in the body frame.

7.3.1 Ephemeris Sun Vector

The Sun vector in ECI frame is the vector from the center of ECI frame to the Sun, which is described in Figure 7.1. In this frame, it can be seen that the Sun rotates around the Earth in the ecliptic plane which is tilted at an angle of ε to the plane of celestial equator. In this figure, (x,y,z) are the coordinators of the ECI frame. (x',y',z') are the coordinators of a different frame in which x' coincides with x, and z' is perpendicular to the ecliptic plane, y' is in the ecliptic plane and completes the right-hand rule. The λ is the angle between the Sun vector and the x-axis. Clearly, the angle is time-dependent. The ε is nearly a constant (≈ 23.44) but changes over time. The Sun vector is clearly determined by λ and ε and it can be expressed in ECI frame as follows

$$\mathbf{r_s} = \begin{bmatrix} \cos(\lambda) \\ \cos(\varepsilon)\sin(\lambda) \\ \sin(\varepsilon)\sin(\lambda) \end{bmatrix} \tag{7.2}$$

The λ and ε can be calculated based on the mathematic model described in [219, page 141], [204], or [205]. The formulas of [205] is as follows. First, given year, month, day (January first is the first day), hour, minute, and second, the *Julian date* JD is given as [205, page 186]

$$\begin{aligned} JD \ &= 367(year) - floor\left(\frac{7\left(year+floor(\frac{month+9}{12})\right)}{4}\right) + floor\left(\frac{275month}{9}\right) \\ &+day + 1721013.5 + \frac{hour}{24} + \frac{minute}{1440} + \frac{second}{86400} \end{aligned} \tag{7.3}$$

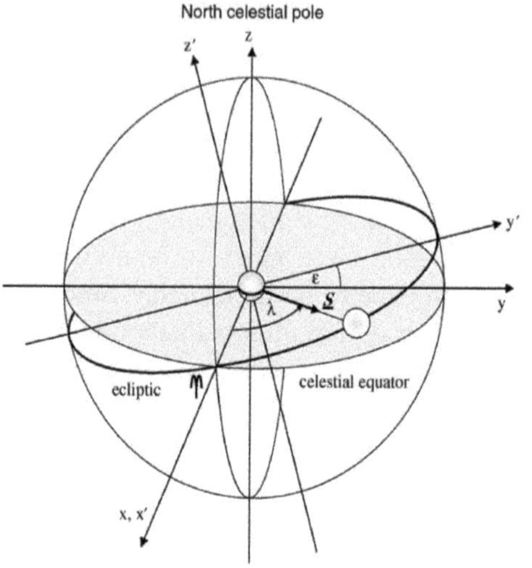

Figure 7.1: Sun vector represented in ECI frame.

where floor is the greatest integer smaller than its argument. From JD, we need to convert the date to J2000 which is given by [205, page 188]

$$T_{UT1} = \frac{JD - 2541545.0}{36525}$$

Then, the *mean longitude L of the Sun* is given by [205, pages 365–368]

$$L = 280.4606184 + 36000.77005361 \times T_{UT1} \tag{7.4}$$

Assume that *Barycentric dynamical time* $T_{TBD} = T_{UT1}$. The *mean anomaly* g of the Sun is given by

$$g = 357.5277233 + 35999.05034 \times T_{TBD} \tag{7.5}$$

The *ecliptic longitude* (λ) of the sun is given by

$$\lambda = L + 1.914666471 \times \sin(g) + 0.019994643 \times \sin(2 \times g) \tag{7.6}$$

The *tilted angle* is given by

$$\varepsilon = 23.439291^o - 0.0130042 \times T_{TBD} \tag{7.7}$$

Substituting (7.6) and (7.7) into (7.2) gives the Sun vector in ECI frame. The Sun vector in LVLH frame can be calculated by applying rotational matrix that transforms ECI frame to LVLH frame, which has been discussed in Chapter 5.

7.3.2 Sun Vector Measurement

Unlike geomagnetic vector, the Sun vector cannot be directly measured from the *Coarse Sun Sensors* (CSS) and some signal processing is necessary. Based on the specification of the *view angle* of CSS, a total of n CSS are needed to guarantee that at least two Sun sensors are available at any orientation when the spacecraft is not in *eclipse*. Each CSS measures the current proportional to the projection of the Sun vector onto the sensor *bore-sight*. Let the measured current of the ith sun sensor $\mathbf{n_i}$ be

$$b_i = I_i/I_0 = (\mathbf{n_i} \cdot \mathbf{b_s}) \tag{7.8}$$

where, $i = 1, 2, \ldots, m < n$, m is the number of Sun sensors that receive the Sun light at current spacecraft attitude, I_i is the measured current of the ith CSS, $\mathbf{n_i}$ is the known boresight unit vector of the ith CSS in body frame, $\mathbf{b_s}$ is the Sun direction vector to be determined, and I_0 is the known maximum CSS current.

There are two different cases that need two different methods to solve equation (7.8). In the first case, a valid current is measured from at least 3 of the n Sun sensors. The CSS processing algorithm computes a measured Sun vector by solving the system of equations (7.8) for $\mathbf{b}_s$ using a pseudo-inverse. The unit Sun vector $\hat{\mathbf{b}}_s$ is then obtained by normalizing $\mathbf{b_s}$. All vectors $\mathbf{n_i}$, $\mathbf{b_s}$, and $\hat{\mathbf{b}}_s$ are expressed in body frame.

In the second case, a valid current is measured from only 2 of the n Sun sensors. The resulting two linear system equations and quadratic constraint over the unit sphere has two possible solutions, and some extra information is needed to decide which solution is the true sun vector (a solution that is closer to the previous valid solution is a reasonable guess but it can be wrong).

Clearly, the solution obtained in the first case gives better estimation in general than the solution obtained in the second case. To avoid the second case, one needs more CSS.

Sun Vector Measurement

Chapter 8

Spacecraft Attitude Estimation

The previous two chapters discussed spacecraft attitude determination methods based on the knowledge of astronomical object vectors $\mathbf{r}_i$ at current time and the location of the spacecraft, and the vector measurements $\mathbf{b}_i$ at current time. However, due to various reasons, these measurements are normally noise signals, which oftentimes result in inaccurate attitude determination. In 1960, Kalman published his famous *Kalman filter* [87] and this technique quickly found its use in some high-profile missions in the aerospace industry, such as the Apollo project [132]. The success of the Apollo project made Kalman filter a widely known method that has been used in many applications where measurement signals are noisy. Spacecraft attitude estimation has been a major research area since the Kalman filter was invented [106]. Because both quaternion kinematics and spacecraft dynamics are nonlinear, for spacecraft attitude estimation, *extended Kalman filter* was developed by Smith et al. [187] and is now widely used in spacecraft attitude estimation.

In 2000, Julier et al. [82] proposed a different filtering and estimation method, the *unscented Kalman filter*, for nonlinear system estimation problem. This estimation method has attracted a lot of attention. Many research papers, for example [31, 34, 35] and references therein, were published. Many reports claim that the unscented Kalman filter produces better estimation result than extended Kalman filter. But some simulation comparison between the two methods leads to different opinions about the potential advantages of unscented Kalman filter [104]. We will not discuss the unscented Kalman filter method in this chapter. The readers interested in this method and its application to the spacecraft attitude estimation are directed to [35] which includes a lot of references.

8.1 Extended Kalman Filter Using Reduced Quaternion Model

Although many different methods have been proposed, most models suggest using only quaternion kinematics equations of motion for the attitude estimation without considering spacecraft dynamics, for example, some widely cited survey papers [35, 106] and references therein. This model reduces the problem size but discards useful spacecraft attitude information available in the spacecraft dynamics equation. The drawbacks of this simplified model are: (a) when gyros measurements have significant noise, the spacecraft dynamics information is not used to prevent the degradation of the attitude estimation, and (b) when gyro measurements are not available (as a matter of fact, gyros are not used in most small spacecrafts for economical consideration), the simplified model cannot be used to estimate the spacecraft attitude. There are some papers that consider models including the spacecraft dynamics in Kalman filter designs, for example, [93, 115], but comparison about which model is a better fit for the application of spacecraft attitude estimation was not carried out for a long time. In a recent research [260], the performance comparison for Kalman filters using two different models was performed. The result shows that the Kalman filter should include spacecraft dynamics. This section is based on [260].

The spacecraft model with Gaussian noise considered in this section can be expressed as follows [243, 249]:

$$\dot{\omega} = -\mathbf{J}^{-1}\omega \times (\mathbf{J}\omega) + \mathbf{J}^{-1}\mathbf{u} + \phi_1 \tag{8.1a}$$

$$\dot{\mathbf{q}} = \frac{1}{2}\Omega(\omega + \phi_2) \tag{8.1b}$$

where $\mathbf{q}$ is the vector part of the quaternion ($\mathbf{q}$ is referred as the reduced quaternion in this book), ω is the spacecraft rotational rate with respect to the inertial frame, $\phi = [\phi_1, \phi_2]^{\mathrm{T}}$ is the process Gaussian noise, $\mathbf{J}$ is the inertia matrix of the spacecraft, and Ω is a matrix given by

$$\Omega = \begin{bmatrix} g(\mathbf{q}) & -q_3 & q_2 \\ q_3 & g(\mathbf{q}) & -q_1 \\ -q_2 & q_1 & g(\mathbf{q}) \end{bmatrix} \tag{8.2}$$

with $g(\mathbf{q}) = \sqrt{1 - q_1^2 - q_2^2 - q_3^2}$.

It is worthwhile to note that unlike ϕ_1, the noise ϕ_2 is added to ω so that the kinematic equations are consistent with the form of (4.8). Depending on the design, we may have angular rate measurements ω_y and quaternion measurement $\mathbf{q}_y$; or we may have only quaternion measurement $\mathbf{q}_y$. Assuming that three gyros and quaternion measurement sensors are installed on board, then the measurement equation can be written as [34]

$$\dot{\beta} = \phi_3 \tag{8.3a}$$

$$\omega_y = \omega + \beta + \psi_1 \tag{8.3b}$$

$$\mathbf{q}_y = \mathbf{q} + \psi_2 \tag{8.3c}$$

where β is a drift in the angular rate measurement, ϕ_3 is the process noise, ω_y is the angular rate measurement, $\mathbf{q}_y$ is the quaternion measurement, and ψ_1 and ψ_2 are measurement noise. The overall system equations are given as follows:

$$\dot{\omega} = -\mathbf{J}^{-1}\omega \times (\mathbf{J}\omega) + \mathbf{J}^{-1}\mathbf{u} + \phi_1 \tag{8.4a}$$

$$\dot{\mathbf{q}} = \frac{1}{2}\Omega(\omega + \phi_2) \tag{8.4b}$$

$$\dot{\beta} = \phi_3 \tag{8.4c}$$

$$\omega_y = \omega + \beta + \psi_1 \tag{8.4d}$$

$$\mathbf{q}_y = \mathbf{q} + \psi_2 \tag{8.4e}$$

which can be rewritten as a standard state space model as follows:

$$\dot{\mathbf{x}} = \mathbf{f}(\mathbf{x}, \mathbf{u}, \phi) \tag{8.5a}$$

$$\mathbf{y} = \mathbf{H}\mathbf{x} + \psi \tag{8.5b}$$

where $\mathbf{x} = [\omega^{\mathrm{T}}, \mathbf{q}^{\mathrm{T}}, \beta^{\mathrm{T}}]^{\mathrm{T}}$, $\mathbf{y} = [\omega_y^{\mathrm{T}}, \mathbf{q}_y^{\mathrm{T}}]^{\mathrm{T}}$, $\phi = [\phi_1^{\mathrm{T}}, \phi_2^{\mathrm{T}}, \phi_3^{\mathrm{T}}]^{\mathrm{T}}$, $\psi = [\psi_1^{\mathrm{T}}, \psi_2^{\mathrm{T}}]^{\mathrm{T}}$, and

$$\mathbf{H} = \begin{bmatrix} \mathbf{I}_3 & \mathbf{0}_3 & \mathbf{I}_3 \\ \mathbf{0}_3 & \mathbf{I}_3 & \mathbf{0}_3 \end{bmatrix}$$

Some noticeable difference between this model and other popular models are (a) it is a reduced quaternion model and (b) it uses the additive noise expression rather than the multiplicative noise expression.

The reduced quaternion geometry of $\mathbf{q}_y$ can be seen from the following argument. First, the noise ψ_2 can be viewed as a reduced rotational quaternion whose rotational axis is $\frac{\psi_2}{\|\psi_2\|}$ and rotational angle δ meets the condition $\sin\left(\frac{\delta}{2}\right) = \|\psi_2\|$. For small noise ψ_2 and the quaternion $\mathbf{q} = \hat{\mathbf{e}}\sin(\frac{\alpha}{2})$ which is bounded away from a singular point ($\|\mathbf{q}\| < 1$), we can see that $\mathbf{q}_y = \mathbf{q} + \psi_2 = \frac{\mathbf{q}_y}{\|\mathbf{q}_y\|}\sin(\frac{\alpha+\Delta}{2})$ is a reduced quaternion whose rotational axis is a perturbation of $\mathbf{q}$ satisfying $\|\mathbf{q}_y\| \leq \|\mathbf{q}\| + \|\psi_2\|$ and $\|\mathbf{q}_y\| \leq 1$ (where $\|\psi_2\|$ is small), and the rotational angle around $\mathbf{q}_y$ is $\alpha + \Delta$ and Δ is small. Therefore, the mathematical treatment for this model is much easier than the multiplicative perturbation model.

Let dt be the sampling time period. The discrete version of (8.4) is given by

$$\begin{bmatrix} \omega_{k+1} \\ \mathbf{q}_{k+1} \\ \beta_{k+1} \end{bmatrix} = \left(\begin{bmatrix} \omega_k \\ \mathbf{q}_k \\ \beta_k \end{bmatrix} + \begin{bmatrix} -\mathbf{J}^{-1}\omega_k \times (\mathbf{J}\omega_k) + \mathbf{J}^{-1}\mathbf{u}_k \\ \frac{1}{2}\,\Omega_k\omega_k \\ 0 \end{bmatrix} dt \right) + \begin{bmatrix} \phi_{1_k} \\ \frac{1}{2}\,\Omega_k\phi_{2_k} \\ \phi_{3_k} \end{bmatrix} dt$$

$$= \mathbf{F}(\mathbf{x}_k, \mathbf{u}_k) + \mathbf{G}(\mathbf{x}_k, \mathbf{u}_k)\phi_k \tag{8.6a}$$

$$\begin{bmatrix} \omega_{\mathbf{y}_k} \\ \mathbf{q}_{\mathbf{y}_k} \end{bmatrix} = \begin{bmatrix} \mathbf{I}_3 & \mathbf{0}_3 & \mathbf{I}_3 \\ \mathbf{0}_3 & \mathbf{I}_3 & \mathbf{0}_3 \end{bmatrix} \begin{bmatrix} \omega_k \\ \mathbf{q}_k \\ \beta_k \end{bmatrix} + \begin{bmatrix} \psi_{1_k} \\ \psi_{2_k} \end{bmatrix} = \mathbf{H}\mathbf{x}_k + \psi_k \tag{8.6b}$$

where

$$\Omega_k = \begin{bmatrix} \sqrt{1-q_{1_k}^2-q_{2_k}^2-q_{3_k}^2} & -q_{3_k} & q_{2_k} \\ q_{3_k} & \sqrt{1-q_{1_k}^2-q_{2_k}^2-q_{3_k}^2} & -q_{1_k} \\ -q_{2_k} & q_{1_k} & \sqrt{1-q_{1_k}^2-q_{2_k}^2-q_{3_k}^2} \end{bmatrix} \tag{8.7}$$

Note that for two vectors $\mathbf{w} = [w_1, w_2, w_3]^{\mathrm{T}}$ and $\mathbf{v} = [v_1, v_2, v_3]^{\mathrm{T}}$, the cross product of $\mathbf{w} \times \mathbf{v}$ can be written as the product of matrix $\mathbf{w}^{\times}$ and vector $\mathbf{v}$ where

$$\mathbf{w}^{\times} = \begin{bmatrix} 0 & -w_3 & w_2 \\ w_3 & 0 & -w_1 \\ -w_2 & w_1 & 0 \end{bmatrix}$$

We also assume ϕ_k and ψ_k are white noise signals satisfying the following equations

$$E(\phi_k) = 0, \quad E(\psi_k) = 0, \quad \forall k \tag{8.8a}$$

$$E(\phi_k \phi_k^{\mathrm{T}}) = \mathbf{Q}_k, \quad E(\psi_k \psi_k^{\mathrm{T}}) = \mathbf{R}_k, \quad E(\psi_j \phi_i^{\mathrm{T}}) = 0, \quad \forall i, j, k \tag{8.8b}$$

$$E(\phi_j \phi_i^{\mathrm{T}}) = 0, \quad E(\psi_j \psi_i^{\mathrm{T}}) = 0, \quad \forall i \neq j \tag{8.8c}$$

For

$$\mathbf{F}_1(\mathbf{x}, \mathbf{u}) = \left(-\mathbf{J}^{-1}\omega_k \times (\mathbf{J}\omega_k) + \mathbf{J}^{-1}\mathbf{u}_k\right) dt + \omega_k$$

we have

$$\frac{\partial \mathbf{F}_1}{\partial \mathbf{x}} = \begin{bmatrix} \mathbf{I} - \mathbf{J}^{-1}(\omega_k^{\times}\mathbf{J} - (\mathbf{J}\omega_k)^{\times})dt & \mathbf{0}_3 & \mathbf{0}_3 \end{bmatrix}$$

For $\mathbf{F}_2(\mathbf{x}, \mathbf{u}) = \frac{1}{2}\,\Omega_k \omega_k dt + \mathbf{q}_k$, we have

$$\frac{\partial \mathbf{F}_2}{\partial \mathbf{x}} = \begin{bmatrix} \frac{\partial \mathbf{F}_2}{\partial \omega} & \frac{\partial \mathbf{F}_2}{\partial \mathbf{q}} & \mathbf{0}_3 \end{bmatrix}$$

with

$$\frac{\partial \mathbf{F}_2}{\partial \omega} = \begin{bmatrix} \frac{g}{2} & -\frac{q_3}{2} & \frac{q_2}{2} \\ \frac{q_3}{2} & \frac{g}{2} & -\frac{q_1}{2} \\ -\frac{q_2}{2} & \frac{q_1}{2} & \frac{g}{2} \end{bmatrix} dt = \frac{1}{2}\Omega dt \tag{8.9}$$

and

$$\frac{\partial \mathbf{F}_2}{\partial \mathbf{q}} = \begin{bmatrix} \frac{1}{dt} - \frac{q_1\omega_1}{2g(q)} & \frac{\omega_3}{2} - \frac{q_2\omega_1}{2g(q)} & -\frac{\omega_2}{2} - \frac{q_3\omega_1}{2g(q)} \\ -\frac{\omega_3}{2} - \frac{q_1\omega_2}{2g(q)} & \frac{1}{dt} - \frac{q_2\omega_2}{2g(q)} & \frac{\omega_1}{2} - \frac{q_3\omega_2}{2g(q)} \\ \frac{\omega_2}{2} - \frac{q_1\omega_3}{2g(q)} & -\frac{\omega_1}{2} - \frac{q_2\omega_3}{2g(q)} & \frac{1}{dt} - \frac{q_3\omega_3}{2g(q)} \end{bmatrix} dt \tag{8.10}$$

For $\mathbf{F}_3(\mathbf{x},\mathbf{u}) = \beta_k$, we have

$$\frac{\partial \mathbf{F}_3}{\partial \mathbf{x}} = \begin{bmatrix} \mathbf{0}_3 & \mathbf{0}_3 & \mathbf{I}_3 \end{bmatrix}$$

Therefore,

$$\begin{aligned} \mathbf{F}_{k-1} &:= \left.\frac{\partial \mathbf{F}}{\partial \mathbf{x}}\right|_{\hat{\mathbf{x}}_{k-1|k-1},\mathbf{u}_{k-1}} \\ &= \left.\begin{bmatrix} \mathbf{I}-\mathbf{J}^{-1}(\omega^{\times}J-(\mathbf{J}\omega)^{\times})dt & \mathbf{0}_3 & \mathbf{0}_3 \\ \frac{\partial \mathbf{F}_2}{\partial \omega} & \frac{\partial \mathbf{F}_2}{\partial \mathbf{q}} & \mathbf{0}_3 \\ \mathbf{0}_3 & \mathbf{0}_3 & \mathbf{I}_3 \end{bmatrix}\right|_{\hat{\mathbf{x}}_{k-1|k-1}} \end{aligned} \tag{8.11}$$

Similarly,

$$\mathbf{L}_{k-1} = \left.\frac{\partial \mathbf{G}}{\partial \phi_k}\right|_{\hat{\mathbf{x}}_{k-1|k-1},\mathbf{u}_{k-1}} = \begin{bmatrix} \mathbf{I}_3 & \mathbf{0}_3 & \mathbf{0}_3 \\ \mathbf{0}_3 & \frac{1}{2}\Omega_{k-1} & \mathbf{0}_3 \\ \mathbf{0}_3 & \mathbf{0}_3 & \mathbf{I}_3 \end{bmatrix} dt \tag{8.12}$$

The extended Kalman filter iteration is as follows:

$$\hat{\mathbf{x}}_{k|k-1} = \mathbf{F}(\hat{\mathbf{x}}_{k-1|k-1},\mathbf{u}_{k-1}) \tag{8.13a}$$

$$\mathbf{P}_{k|k-1} = \mathbf{F}_{k-1}\mathbf{P}_{k-1|k-1}\mathbf{F}_{k-1}^{\mathrm{T}} + \mathbf{L}_{k-1}\mathbf{Q}_{k-1}\mathbf{L}_{k-1}^{\mathrm{T}} \tag{8.13b}$$

$$\tilde{\mathbf{y}}_k = \mathbf{y}_k - \mathbf{H}\hat{\mathbf{x}}_{k|k-1} \tag{8.13c}$$

$$\mathbf{S}_k = \mathbf{H}\mathbf{P}_{k|k-1}\mathbf{H}^{\mathrm{T}} + \mathbf{R}_k \tag{8.13d}$$

$$\mathbf{K}_k = \mathbf{P}_{k|k-1}\mathbf{H}^{\mathrm{T}}\mathbf{S}_k^{-1} \tag{8.13e}$$

$$\hat{\mathbf{x}}_{k|k} = \hat{\mathbf{x}}_{k|k-1} + \mathbf{K}_k\tilde{\mathbf{y}}_k \tag{8.13f}$$

$$\mathbf{P}_{k|k} = (\mathbf{I} - \mathbf{K}_k\mathbf{H})\mathbf{P}_{k|k-1} \tag{8.13g}$$

The beauty of the Kalman filter using spacecraft dynamics can be seen from (8.13f). The best estimation is composed of two parts. The first part is a prediction $\hat{\mathbf{x}}_{k|k-1}$ which includes the spacecraft dynamics and the inertia matrix information for the specific spacecraft. The second part is a correction $\tilde{\mathbf{y}}_k$ which is based on observations. The filter gain $\mathbf{K}_k$ is constantly adjusted such that: (a) if the noise is higher, the gain is reduced so that the estimation depends more on the information of the system dynamics, and (b) if the noise is lower, the gain is increased so that the estimation depends more on the measurement. That is the reason why spacecraft dynamics should be included in the attitude estimation problem even if angular rate measurements are available.

The simulation test in [260] shows that the extended Kalman filter is robust to the modeling errors, in particular, when the spacecraft inertia matrix is not accurate, the estimation is still accurate enough for the practical application.

As mentioned before, the Kalman filter with spacecraft dynamics works without the (gyro) measurement of spacecraft angular velocity vector with respect to the inertial frame. In this case, gyro measurement drift β does not exist. Therefore, the continuous system (8.4) is reduced to

$$\dot{\omega} = -\mathbf{J}^{-1}\omega \times (\mathbf{J}\omega) + \mathbf{J}^{-1}\mathbf{u} + \phi_1 \tag{8.14a}$$

$$\dot{\mathbf{q}} = \frac{1}{2}\Omega(\omega + \phi_2) \tag{8.14b}$$

$$\mathbf{q}_y = \mathbf{q} + \psi \tag{8.14c}$$

We still use (8.5) for this system but $\mathbf{x} = [\omega^{\mathrm{T}}, \mathbf{q}^{\mathrm{T}}]^{\mathrm{T}}$, $\mathbf{y} = \mathbf{q}_y$, $\phi = [\phi_1^{\mathrm{T}}, \phi_2^{\mathrm{T}}]^{\mathrm{T}}$, and $\mathbf{C} = \begin{bmatrix} \mathbf{0}_3 & \mathbf{I}_3 \end{bmatrix}$. The discrete version of (8.14) is given by

$$\begin{bmatrix} \omega_{k+1} \\ \mathbf{q}_{k+1} \end{bmatrix} = \left(\begin{bmatrix} \omega_k \\ \mathbf{q}_k \end{bmatrix} + \begin{bmatrix} -\mathbf{J}^{-1}\omega_k \times (\mathbf{J}\omega_k) + \mathbf{J}^{-1}\mathbf{u}_k \\ \frac{1}{2}\Omega_k\omega_k \end{bmatrix} dt \right) + \begin{bmatrix} \phi_{1_k} \\ \frac{1}{2}\Omega_k\phi_{2_k} \end{bmatrix} dt$$

$$= \mathbf{F}(\mathbf{x}_k, \mathbf{u}_k) + \mathbf{G}(\mathbf{x}_k, \mathbf{u}_k)\phi_k \tag{8.15a}$$

$$\mathbf{q}_{y_k} = \begin{bmatrix} \mathbf{0}_3 & \mathbf{I}_3 \end{bmatrix} \begin{bmatrix} \omega_k \\ \mathbf{q}_k \end{bmatrix} + \psi_k = \mathbf{H}\mathbf{x}_k + \psi_k \tag{8.15b}$$

where Ω_k is the same as in (8.7). We also assume that ϕ_k and ψ_k are white noise signals satisfying equations (8.8).

For $\mathbf{F}_1(\mathbf{x}, \mathbf{u}) = \left(-\mathbf{J}^{-1}\omega_k \times (\mathbf{J}\omega_k) + \mathbf{J}^{-1}\mathbf{u}_k\right) dt + \omega_k$, we have

$$\frac{\partial \mathbf{F}_1}{\partial \mathbf{x}} = \begin{bmatrix} \mathbf{I} - \mathbf{J}^{-1}(\omega^{\times} J - (\mathbf{J}\omega)^{\times})dt & \mathbf{0}_3 \end{bmatrix}$$

For $\mathbf{F}_2(\mathbf{x}, \mathbf{u}) = \frac{1}{2}\Omega_k\omega_k dt + \mathbf{q}_k$, we have

$$\frac{\partial \mathbf{F}_2}{\partial \mathbf{x}} = \begin{bmatrix} \frac{\partial \mathbf{F}_2}{\partial \omega} & \frac{\partial \mathbf{F}_2}{\partial \mathbf{q}} \end{bmatrix}$$

with $\frac{\partial \mathbf{F}_2}{\partial \omega}$ and $\frac{\partial \mathbf{F}_2}{\partial \mathbf{q}}$ the same as (8.9) and (8.10). Therefore,

$$\mathbf{F}_{k-1} := \left.\frac{\partial \mathbf{F}}{\partial \mathbf{x}}\right|_{\hat{\mathbf{x}}_{k-1|k-1}, \mathbf{u}_{k-1}} = \left.\begin{bmatrix} \mathbf{I} - \mathbf{J}^{-1}(\omega^{\times} J - (\mathbf{J}\omega)^{\times})dt & \mathbf{0}_3 \\ \frac{\partial \mathbf{F}_2}{\partial \omega} & \frac{\partial \mathbf{F}_2}{\partial \mathbf{q}} \end{bmatrix}\right|_{\hat{\mathbf{x}}_{k-1|k-1}} \tag{8.16}$$

Let

$$\mathbf{L}_{k-1} = \left.\frac{\partial \mathbf{G}}{\partial \phi_k}\right|_{\hat{\mathbf{x}}_{k-1|k-1}, \mathbf{u}_{k-1}} = \begin{bmatrix} \mathbf{I}_3 & \mathbf{0}_3 \\ \mathbf{0}_3 & \frac{1}{2}\Omega_{k-1} \end{bmatrix} dt \tag{8.17}$$

The extended Kalman filter will be the same as (8.13).

8.2 Kalman Filter Using Reduced Quaternion Model

The idea of the extended Kalman filter is to use as much (nonlinear) information as possible and hopefully to improve the estimation performance. Therefore, part of the iteration uses the nonlinear equation (8.13a). But linearization has to be done in (8.11) and (8.12) and the linear approximation is used in (8.13b). The drawbacks of this method are: (a) in general, the extended Kalman filter is not an optimal estimator [4], (b) if the initial estimate of the state is wrong, the filter may diverge [68, 153], and (c) the estimated covariance matrix tends to underestimate the true covariance matrix and therefore risks becoming inconsistent in the statistical sense [153].

On the other hand, if the nonlinear spacecraft system equations are linearized and Kalman filter for the linear system is used, the accuracy in state prediction may be lost. In exchange, some benefits will be gained: (a) the estimate is optimal for the linearized system, (b) the initial guess is not as crucial as the extended Kalman filter, (c) the numerically stable algorithms are fully investigated, and (d) Kalman filter design and linear quadratic optimal control system design can be separated [230].

Therefore, this section briefly discusses a Kalman filter implementation for the spacecraft estimation problem using a reduced quaternion model proposed in [243]. Unlike most models [35] used in the spacecraft attitude estimation problem, we will include the spacecraft dynamics discussed in the previous section to make full use of the available information. We also adopt a simple additive noise model as suggested in the previous section rather than a more complex multiplicative noise model used in [34, 35, 106, 125, 126, 154]. Another benefit of using the reduced model is that the unit norm constraint for quaternion is not required as in [31, 49, 266], which greatly simplifies the problem and reduces the cost of computation. Other merits of using reduced quaternion model can be found in [245, 250].

As discussed in the previous sections, we can first linearize the nonlinear system equation and then use (linear) Kalman filter for the attitude estimation problem. Using exactly the same method as the previous section, to simplify the discussion, assuming that there is no measurement drafting, the linearized system is given as follows.

$$\begin{bmatrix} \dot{\omega} \\ \dot{\mathbf{q}} \end{bmatrix} = \begin{bmatrix} \mathbf{0}_3 & \mathbf{0}_3 \\ \frac{1}{2}\mathbf{I}_3 & \mathbf{0}_3 \end{bmatrix} \begin{bmatrix} \omega \\ \mathbf{q} \end{bmatrix} + \begin{bmatrix} \mathbf{J}^{-1} \\ \mathbf{0}_3 \end{bmatrix} \mathbf{u} \tag{8.18a}$$

$$\begin{bmatrix} \omega_y \\ \mathbf{q}_y \end{bmatrix} = \begin{bmatrix} \mathbf{I}_3 & \mathbf{0}_3 \\ \mathbf{0}_3 & \mathbf{I}_3 \end{bmatrix} \begin{bmatrix} \omega \\ \mathbf{q} \end{bmatrix} \tag{8.18b}$$

The corresponding discrete system with added noise is, therefore, as follows.

$$\begin{aligned} \mathbf{x}_{k+1} &= \begin{bmatrix} \omega_{k+1} \\ \mathbf{q}_{k+1} \end{bmatrix} = \begin{bmatrix} \mathbf{I}_3 & \mathbf{0}_3 \\ \frac{1}{2}\mathbf{I}_3 dt & \mathbf{I}_3 \end{bmatrix} \begin{bmatrix} \omega_k \\ \mathbf{q}_k \end{bmatrix} + \begin{bmatrix} \mathbf{J}^{-1} dt \\ \mathbf{0}_3 \end{bmatrix} \mathbf{u}_k + \begin{bmatrix} \phi_{1_k} \\ \phi_{2_k} \end{bmatrix} \\ &= \mathbf{A}\mathbf{x}_k + \mathbf{B}\mathbf{u}_k + \phi_k \end{aligned} \tag{8.19a}$$

$$\mathbf{y}_{k+1} = \begin{bmatrix} \omega_{y_k} \\ \mathbf{q}_{y_k} \end{bmatrix} = \begin{bmatrix} \mathbf{I}_3 & \mathbf{0}_3 \\ \mathbf{0}_3 & \mathbf{I}_3 \end{bmatrix} \begin{bmatrix} \omega_k \\ \mathbf{q}_k \end{bmatrix} + \begin{bmatrix} \psi_{1_k} \\ \psi_{2_k} \end{bmatrix}$$

$$= \mathbf{C}\mathbf{x}_k + \psi_k \tag{8.19b}$$

Assume $\hat{\mathbf{x}}_{0|0} = E(\mathbf{x}_0)$ and $\mathbf{P}_0 = E([\mathbf{x}_0 - E(\mathbf{x}_0)]^{\mathrm{T}}[\mathbf{x}_0 - E(\mathbf{x}_0)])$, the conventional Kalman filter for this linear system is updated as follows [4]:

$$\hat{\mathbf{x}}_{k|k-1} = \mathbf{A}\mathbf{x}_{k-1|k-1} + \mathbf{B}\mathbf{u}_{k-1} \tag{8.20a}$$

$$\mathbf{P}_{k|k-1} = \mathbf{A}\mathbf{P}_{k-1|k-1}\mathbf{A}^{\mathrm{T}} + \mathbf{Q}_{k-1} \tag{8.20b}$$

$$\tilde{\mathbf{y}}_k = \mathbf{y}_k - \mathbf{C}\hat{\mathbf{x}}_{k|k-1} \tag{8.20c}$$

$$\mathbf{S}_k = \mathbf{C}\mathbf{P}_{k|k-1}\mathbf{C}^{\mathrm{T}} + \mathbf{R}_k \tag{8.20d}$$

$$\mathbf{K}_k = \mathbf{P}_{k|k-1}\mathbf{C}^{\mathrm{T}}\mathbf{S}_k^{-1} \tag{8.20e}$$

$$\hat{\mathbf{x}}_{k|k} = \hat{\mathbf{x}}_{k|k-1} + \mathbf{K}_k\tilde{\mathbf{y}}_k \tag{8.20f}$$

$$\mathbf{P}_{k|k} = (\mathbf{I} - \mathbf{K}_k\mathbf{C})\mathbf{P}_{k|k-1} \tag{8.20g}$$

Although Kalman filter in the conventional form (8.20) has been very successful in applications, people noticed that formulas in (8.20) are sometimes not stable because of numerical errors which lead to the loss of positive definiteness of $\mathbf{P}_{k|k}$ by using (8.20g). Various alternative schemes were proposed. For example, *Joseph-form stabilized Kalman filter*[1] was discussed in [55], *root square filter* was proposed by Potter, Stern, and Carlson in [26, 158], and *Chandrasekhar square root filter* was introduced by Morf and Kailath in [122]. Some detailed numerical analysis and test were conducted by Verhaegen and Van Dooren [209] in which a root square fitler algorithm described in [4] was recommended because of its overall performance and robustness. When $\mathbf{R}_k$ matrix is diagonal, Bierman [18] suggested U-D factorization method which sequentially calculates the Kalman gain matrix $\mathbf{K}_k$ in (8.20e) and covariance matrix $\mathbf{P}_{k|k}$ in (8.20g) (one observation at a time).

8.3 A Short Comment

In this chapter, two Kalman filters have been presented to estimate the spacecraft attitude and body rate. In the aerospace industry, the extended Kalman filter is widely used. But there are pros and cons from the theoretical point of view for both methods. However, to the best of the author's knowledge, it is not clear which one is a perfect fit for application and niether has anyone done a test comparison.

[1]Joseph-form update replaces (8.20g) by $\mathbf{P}_{k|k} = (\mathbf{I} - \mathbf{K}_k\mathbf{C})\mathbf{P}_{k|k-1}(\mathbf{I} - \mathbf{K}_k\mathbf{C})^{\mathrm{T}} + \mathbf{K}_k\mathbf{R}_k\mathbf{K}_k^{\mathrm{T}}$, which is computationally more expensive but numerically more stable because $\mathbf{P}_{k|k}$ is guaranteed to be positive semidefinite.

Chapter 9

Spacecraft Attitude Control

Control design methods based on the quaternion spacecraft model have been investigated for decades. Most quaternion-based design methods use Lyapunov functions and focus on the global stability; these methods pay little attention to the control system performance which is important in practical system design. Not many researchers have considered the performance of the quaternion-based control systems. Using the classical frequency domain method, Paielli and Bach [147] adopted quaternion-based linear error dynamics to get the desired performance for the attitude control system; Wie, Weiss, and Arapostathis [224] showed that there exists some state feedback that globally stabilizes the nonlinear spacecraft system and the feedback matrix assigns the closed-loop poles for the dynamics described by the rotational angle about the rotational axis. These methods are in classical domain and they are not easy to extend to modern designs. Zhou and Colgren [273] obtained a linearized state space model with all components of the quaternion in the state variables. However, this linearized state space model is not fully controllable. This explains why many powerful design methods in linear control system theory, such as *pole assignment*, *linear quadratic regulator* (LQR) control, and $\mathbf{H}_\infty$ *control* were not directly applied to the spacecraft control system design if full quaternion based linearized model is used.

On the other hand, although the Euler angle representation has a singular point and the representation depends on the rotational sequential, the linearized Euler angle-based spacecraft model has been proved to be fully controllable. Therefore, all linear system design methods can be directly applied to spacecraft control system design for the Euler angle model and these methods are described in many standard textbooks, for example, [181, 219, 220]. More importantly, there are many successful applications of these powerful control design methods, for example [190, 229].

It is given in Chapter 4 that the reduced quaternion model that uses only vector components of the quaternion is fully controllable. Also the linearized reduced quaternion models have some simple and special structures. The design methods based on the reduced quaternion models have been considered in the rest of the book. For nadir pointing spacecraft, one can directly use standard linear control system design methods, such as LQR design [8], *robust pole assignment* design [90, 201, 241], $\mathbf{H}_\infty$ design [42], for the linearized system. The designed controller can then be checked by simulation with the original nonlinear spacecraft system in the space environment as discussed in Chapter 5. For inertial pointing spacecraft, since the linearized system has a very simple structure, using this linearized reduced quaternion model, one can derive an analytical formula for LQR optimal control that is explicitly related to the cost matrices $\mathbf{Q}$ and $\mathbf{R}$. Moreover, it can be shown that under some mild restriction, the LQR feedback controller globally stabilizes the original nonlinear spacecraft. In addition, the LQR controller has a diagonal structure in the state feedback matrices $\mathbf{D}$ and $\mathbf{K}$. Using this structure, it can be proved that the LQR design is actually a robust pole assignment design. The main results presented here are based on [8, 245, 249].

9.1 LQR Design for Nadir Pointing Spacecraft

We first consider the general linear system described as follows.

$$\begin{aligned} \dot{\mathbf{x}} &= \mathbf{A}\mathbf{x} + \mathbf{B}\mathbf{u} \\ \mathbf{y} &= \mathbf{C}\mathbf{x} \end{aligned} \tag{9.1}$$

The LQR design is to find a state feedback matrix

$$\mathbf{u} = -[\mathbf{D}, \mathbf{K}]\mathbf{x} = -\mathbf{G}\mathbf{x}$$

to minimize the following cost function

$$L = \frac{1}{2}\int_0^\infty \left(\mathbf{x}^{\mathrm{T}}\mathbf{Q}\mathbf{x} + \mathbf{u}^{\mathrm{T}}\mathbf{R}\mathbf{u}\right) dt \tag{9.2}$$

where $\mathbf{Q}$ and $\mathbf{R}$ are positive definite matrices, $\mathbf{x}^{\mathrm{T}}\mathbf{Q}\mathbf{x}$ represents the cost of the deviation from the desired equilibrium point, $\mathbf{u}^{\mathrm{T}}\mathbf{R}\mathbf{u}$ represents the cost of the energy consumption. The LQR control problem was first considered by Hall [64] and Wiener [225], but Kalman [88] provided a much better solution and popularized the design. If Kalman filter [87] is used as part of the feedback loop, then the control design method is the LQG control. Surprisingly, Kalman filter and

LQR control law can be designed separately because of the *separation theorem* obtained by Wonham [230].

For nadir pointing spacecraft system given by (4.36), the optimal control of LQR design is uniquely given by (see Appendix B or a comprehensive treatment of [8])

$$\mathbf{u}(t) = -\mathbf{R}^{-1}\mathbf{B}^{\mathrm{T}}\mathbf{F}\mathbf{x}(t) = -\mathbf{G}\mathbf{x} \tag{9.3}$$

where $\mathbf{F}$ is a constant positive definite matrix which is the solution of the *algebraic Riccati matrix equation*

$$-\mathbf{F}\mathbf{A} - \mathbf{A}^{\mathrm{T}}\mathbf{F} + \mathbf{F}\mathbf{B}\mathbf{R}^{-1}\mathbf{B}^{\mathrm{T}}\mathbf{F} - \mathbf{Q} = \mathbf{0} \tag{9.4}$$

This control law can be directly used for the nadir pointing spacecraft without any modification. For inertial pointing spacecraft, due to the simple structure of the linearized reduced quaternion model, analytic solution to LQR design can be obtained.

9.2 The LQR Design for Inertial Pointing Spacecraft

In this section, LQR design is considered for inertial pointing spacecraft for which $\mathbf{A}$ and $\mathbf{B}$ are defined in (4.13). It is further assumed that the constant inertia matrix of the spacecraft $\mathbf{J}$ defined in (4.1) is diagonal. This assumption is reasonable because in practical spacecraft design, $\mathbf{J}$ is always designed close to a diagonal matrix. In the rest of the discussion of this subsection, it is assumed that $\mathbf{Q}$, and $\mathbf{R}$ are diagonal matrices because $\mathbf{Q}$ and $\mathbf{R}$ are oftentimes selected to be diagonal in engineering design practice. With these assumptions, the problem can greatly be simplified.

9.2.1 The Analytic Solution

It is well known that the LQR feedback based on (9.3) and (9.4) guarantees the *stability* of the linearized closed-loop system and minimizes the cost function of (9.2) that is a combined cost of cumulative control system error and cumulative energy consumption.

First, we derive the analytical solution for the spacecraft model (4.12). Let

$$\mathbf{F} = \begin{bmatrix} \mathbf{F}_{11} & \mathbf{F}_{12} \\ \mathbf{F}_{21} & \mathbf{F}_{22} \end{bmatrix}, \quad \mathbf{Q} = \begin{bmatrix} \mathbf{Q}_{11} & \mathbf{0} \\ \mathbf{0} & \mathbf{Q}_{22} \end{bmatrix} \tag{9.5}$$

where the elements of $\mathbf{F}$ and $\mathbf{Q}$ in (9.5) are all 3 by 3 matrices. Substituting $\mathbf{A}$ and $\mathbf{B}$ defined in (4.13), $\mathbf{F}$ and $\mathbf{Q}$ defined in (9.5) into (9.4), after simple manipulations, we get

$$\begin{bmatrix} \mathbf{F}_{11}\mathbf{J}^{-1}\mathbf{R}^{-1}\mathbf{J}^{-1}\mathbf{F}_{11} & \mathbf{F}_{11}\mathbf{J}^{-1}\mathbf{R}^{-1}\mathbf{J}^{-1}\mathbf{F}_{12} \\ \mathbf{F}_{12}^{-1}\mathbf{J}^{-1}\mathbf{R}^{-1}\mathbf{J}^{-1}\mathbf{F}_{11} & \mathbf{F}_{12}^{\mathrm{T}}\mathbf{J}^{-1}\mathbf{R}^{-1}\mathbf{J}^{-1}\mathbf{F}_{12} \end{bmatrix} = \begin{bmatrix} \frac{1}{2}\left(\mathbf{F}_{12}^{\mathrm{T}} + \mathbf{F}_{12}\right) + \mathbf{Q}_{11} & \frac{1}{2}\mathbf{F}_{22} \\ \frac{1}{2}\mathbf{F}_{22} & \mathbf{Q}_{22} \end{bmatrix} \tag{9.6}$$

Since **J**, **Q** and **R** are positive definite, noticing that $\mathbf{F}_{21}^{\mathrm{T}} = \mathbf{F}_{12}$, comparing the (2,2) block on both sides of (9.6) yields

$$\mathbf{F}_{12} = \mathbf{J}\mathbf{R}^{\frac{1}{2}}\mathbf{Q}_{22}^{\frac{1}{2}} \tag{9.7}$$

Since **J**, $\mathbf{Q}_{11} = \mathrm{diag}(q_{1i})$, $\mathbf{Q}_{22} = \mathrm{diag}(q_{2i})$, and $\mathbf{R} = \mathrm{diag}(r_i)$ are diagonal, it can be concluded that $\mathbf{F}_{12}$ is diagonal. Substituting (9.7) into the (1,1) block of (9.6) gives

$$\mathbf{F}_{11} = \mathbf{J}\mathbf{R}^{\frac{1}{2}}\left(\mathbf{Q}_{11} + \frac{1}{2}\left(\mathbf{J}\mathbf{R}^{\frac{1}{2}}\mathbf{Q}_{22}^{\frac{1}{2}} + \mathbf{Q}_{22}^{\frac{1}{2}}\mathbf{R}^{\frac{1}{2}}\mathbf{J}\right)\right)^{\frac{1}{2}} \tag{9.8}$$

Therefore, $\mathbf{F}_{11}$ is diagonal. Substituting (9.7) and (9.8) into the (2,1) block of (9.6) gives

$$\mathbf{F}_{22} = 2\mathbf{Q}_{22}^{\frac{1}{2}}\left(\mathbf{Q}_{11} + \mathbf{J}\mathbf{R}^{\frac{1}{2}}\mathbf{Q}_{22}^{\frac{1}{2}}\right)^{\frac{1}{2}} \tag{9.9}$$

which is also diagonal. Equations (9.7), (9.8), and (9.9) give a complete solution of Riccati matrix equation (9.4). Therefore, (9.3) can be rewritten as

$$\mathbf{u}(t) = -\mathbf{R}^{-1}\mathbf{B}^{\mathrm{T}}\mathbf{F}\mathbf{x}(t) = -[\mathbf{R}^{-1}\mathbf{J}^{-1}\mathbf{F}_{11}, \mathbf{R}^{-1}\mathbf{J}^{-1}\mathbf{F}_{12}]\mathbf{x} = -[\mathbf{D}, \mathbf{K}]\mathbf{x} \tag{9.10}$$

Clearly, matrices **D** and **K** are diagonal.

9.2.2 The Global Stability of the Design

To show the global stability of the design, first the definition of global stability for nonlinear systems is reviewed [92, page 111].

Definition 9.1 Let $\mathbf{x}(t)$ be the solution of the nonlinear inertial pointing spacecraft system defined by (4.11) and (4.9). If for any initial state $\mathbf{x}(0)$, the trajectory $\mathbf{x}(t)$ approaches the origin as $t \to \infty$, no matter how large $\|\mathbf{x}(0)\|$ is, then the region of attraction (also called the region of asymptotic stability) is the entire space $\mathbf{R}^n$. If an asymptotically stable equilibrium point at the origin has this property, it is said to be globally asymptotically stable.

A theorem on *globally asymptotically stable* is given in [92, Corollary 3.2][1], which is restated below.

Theorem 9.1
Let $\mathbf{x} = \mathbf{0}$ *be an equilibrium point for the system defined by (4.11) and (4.9). Let a Lyapunov function* $V : \mathbf{R}^n \to \mathbf{R}$ *be a continuously differentiable, radially unbounded, positive definite function such that* $\dot{V}(\mathbf{x}) \leq 0$ *for all* $\mathbf{x} \in \mathbf{R}^n$. *Let* $S = \{\mathbf{x} \in \mathbf{R}^n | \dot{V}(\mathbf{x}) =$

[1] The original result is applicable to a much more general case.

0}, and suppose that no solution can stay forever in S, other than the trivial solution (equilibrium). Then, the origin is globally asymptotically stable.

Next, it is shown that under some additional conditions, the LQR optimal control given by (9.10) globally stabilizes the nonlinear system described by (4.11) and (4.8). Let $\mathbf{P} = \mathbf{Q}_{22}^{-\frac{1}{2}}\mathbf{R}^{\frac{1}{2}}\mathbf{J}$, and the Lyapunov function be

$$V = \frac{1}{2}\omega_I^{\mathrm{T}}\mathbf{P}\omega_I + q_1^2 + q_2^2 + q_3^2 + (1-q_0)^2 \tag{9.11}$$

It is easy to check, in view of (4.8), that

$$\begin{aligned}
&\frac{d}{dt}\left(q_1^2+q_2^2+q_3^2+(1-q_0)^2\right)\\
=\ & 2\mathbf{q}^{\mathrm{T}}\dot{\mathbf{q}} - 2(1-q_0)\dot{q}_0\\
=\ & 2\mathbf{q}\cdot\left(-\frac{1}{2}\omega_I\times\mathbf{q}+\frac{1}{2}q_0\omega_I\right)+2(1-q_0)\left(\frac{1}{2}\mathbf{q}^{\mathrm{T}}\omega_I\right)\\
=\ & q_0\mathbf{q}^{\mathrm{T}}\omega_I + (1-q_0)\mathbf{q}^{\mathrm{T}}\omega_I\\
=\ & \mathbf{q}^{\mathrm{T}}\omega_I
\end{aligned} \tag{9.12}$$

Using definition of $\mathbf{P}$ and (9.7), it is easy to see that

$$\begin{aligned}
&\omega_I^{\mathrm{T}}\mathbf{P}\mathbf{J}^{-1}\mathbf{R}^{-1}\mathbf{J}^{-1}\mathbf{F}_{12}\mathbf{q}\\
=\ & \omega_I^{\mathrm{T}}(\mathbf{Q}_{22}^{-\frac{1}{2}}\mathbf{R}^{\frac{1}{2}}\mathbf{J})\mathbf{J}^{-1}\mathbf{R}^{-1}\mathbf{J}^{-1}(\mathbf{J}\mathbf{R}^{\frac{1}{2}}\mathbf{Q}_{22}^{\frac{1}{2}})\mathbf{q}\\
=\ & \omega_I^{\mathrm{T}}\mathbf{q}
\end{aligned} \tag{9.13}$$

Therefore, using (9.12), (9.13), (9.3), and (4.13), the derivative of the Lyapunov function along the trajectory described by the nonlinear system equations (4.11) and (4.8) is given by

$$\begin{aligned}
\frac{dV}{dt} =\ & \frac{d}{dt}\left(\frac{1}{2}\omega_I^{\mathrm{T}}\mathbf{P}\omega_I+q_1^2+q_2^2+q_3^2+(1-q_0)^2\right)\\
=\ & \omega_I^{\mathrm{T}}\mathbf{P}\left(-\mathbf{J}^{-1}\omega_I\times\mathbf{J}\omega_I-\mathbf{J}^{-1}\mathbf{R}^{-1}\begin{bmatrix}\mathbf{J}^{-1} & \mathbf{0}\end{bmatrix}\begin{bmatrix}\mathbf{F}_{11} & \mathbf{F}_{12}\\ \mathbf{F}_{12}^{\mathrm{T}} & \mathbf{F}_{22}\end{bmatrix}\begin{bmatrix}\omega_I\\ \mathbf{q}\end{bmatrix}\right)\\
& +\omega_I^{\mathrm{T}}\mathbf{q}\\
=\ & -\omega_I^{\mathrm{T}}\mathbf{P}\mathbf{J}^{-1}\omega_I\times\mathbf{J}\omega_I-\omega_I^{\mathrm{T}}\mathbf{P}\mathbf{J}^{-1}\mathbf{R}^{-1}\mathbf{J}^{-1}\mathbf{F}_{11}\omega_I-\omega_I^{\mathrm{T}}\mathbf{P}\mathbf{J}^{-1}\mathbf{R}^{-1}\mathbf{J}^{-1}\mathbf{F}_{12}\mathbf{q}\\
& +\omega_I^{\mathrm{T}}\mathbf{q}\\
=\ & -\omega_I^{\mathrm{T}}\mathbf{P}\mathbf{J}^{-1}\omega_I\times\mathbf{J}\omega_I-\omega_I^{\mathrm{T}}\mathbf{Q}_{22}^{-\frac{1}{2}}\left(\mathbf{Q}_{11}+\frac{1}{2}\left(\mathbf{J}\mathbf{R}^{\frac{1}{2}}\mathbf{Q}_{22}^{\frac{1}{2}}+\mathbf{Q}_{22}^{\frac{1}{2}}\mathbf{R}^{\frac{1}{2}}\mathbf{J}\right)\right)^{\frac{1}{2}}\omega_I\\
=\ & -\omega_I^{\mathrm{T}}\mathbf{Q}_{22}^{-\frac{1}{2}}\mathbf{R}^{\frac{1}{2}}\omega_I\times\mathbf{J}\omega_I-\omega_I^{\mathrm{T}}\mathbf{Q}_{22}^{-\frac{1}{2}}\left(\mathbf{Q}_{11}+\frac{1}{2}\left(\mathbf{J}\mathbf{R}^{\frac{1}{2}}\mathbf{Q}_{22}^{\frac{1}{2}}+\mathbf{Q}_{22}^{\frac{1}{2}}\mathbf{R}^{\frac{1}{2}}\mathbf{J}\right)\right)^{\frac{1}{2}}\omega_I
\end{aligned} \tag{9.14}$$

Since **P**, **Q**, **R**, and **J** are all diagonal positive definite matrices, the second term of the last expression is negative definite.

If $\mathbf{Q}_{22}^{-1}\mathbf{R} = c\mathbf{I}$, i.e.,

$$\mathbf{R} = c\mathbf{Q}_{22} \tag{9.15}$$

or $\mathbf{Q}_{22}^{-1}\mathbf{R} = c\mathbf{J}$, i.e.,

$$\mathbf{R} = c\mathbf{Q}_{22}\mathbf{J} \tag{9.16}$$

where c is a constant, then the first term of (9.14) vanishes; therefore $\frac{dV}{dt}$ is negative semidefinite, and the nonlinear system described by (4.11) and (4.8) is globally stable with the optimal controller given by (9.10). To show that the closed-loop nonlinear system is asymptotically stable, we define $S = \{\mathbf{x}|\dot{V}(\mathbf{x}) = \mathbf{0}\}$. Since **J**, **Q**, and **R** are positive definite matrices, clearly, equation (9.14) indicates that $S = \{\mathbf{x}|\mathbf{x} = (\omega_I, \mathbf{q}) = (\mathbf{0}, \mathbf{q})\}$. From (4.11), since **D** and **K** are full rank matrices and $\mathbf{u} = -\mathbf{D}\omega_I - \mathbf{Kq} \neq \mathbf{0}$ if $\mathbf{q} \neq \mathbf{0}$, no solution can always stay in S except a subset $S_1 = \{\mathbf{x} = (\omega_I, \mathbf{q}) = (\mathbf{0}, \mathbf{0})\} \subset S$. Using Theorem 9.1, the origin is globally asymptotically stable. Therefore, the region of attraction (see [92]) of the nonlinear system is the whole space spanned by $\mathbf{R}^n$.

Remark 9.1 Spacecraft rotation is a special case of the attitude motion of a rigid body which can be expressed mathematically by SO(3), the group of rotational matrices. Bhat and Bernstein [17] showed that there is an intrinsic windup problem associated with the attitude motion of a rigid body when $q_0 < 0$. But many researchers realize that there are designs that eventually stabilize the system at $\bar{q} = [1,0,0,0]$. Tayebi in his paper [198] referred this type of design to "almost global asymptotic stability" design. ■

In system design practice, if the performance and the local stability are the only design considerations, **Q** and **R** can be chosen without any restriction; if the global stability is also required for the nonlinear spacecraft system, some restriction, though it is mild, should be placed on **Q** and **R**, i.e., either $\mathbf{R} = c\mathbf{Q}_{22}$ or $\mathbf{R} = c\mathbf{Q}_{22}\mathbf{J}$, where c is any positive constant.

9.2.3 The Closed-loop Poles

To establish the relationship between the closed-loop poles and the design matrices **Q** and **R**, we can simplify (9.10) further as follows.

$$\begin{aligned} \mathbf{D} &= \mathbf{R}^{-\frac{1}{2}}\left(\mathbf{Q}_{11} + \frac{1}{2}\left(\mathbf{J}\mathbf{R}^{\frac{1}{2}}\mathbf{Q}_{22}^{\frac{1}{2}} + \mathbf{Q}_{22}^{\frac{1}{2}}\mathbf{R}^{\frac{1}{2}}\mathbf{J}\right)\right)^{\frac{1}{2}} \\ &= \text{diag}(d_i) = \text{diag}\left(\sqrt{\frac{q_{1i}}{r_i} + J_{ii}\sqrt{\frac{q_{2i}}{r_i}}}\right) \end{aligned} \tag{9.17}$$

with

$$d_i = \sqrt{\frac{q_{1i}}{r_i} + J_{ii}\sqrt{\frac{q_{2i}}{r_i}}}$$

and

$$\mathbf{K} = \mathbf{R}^{-\frac{1}{2}}\mathbf{Q}_{22}^{\frac{1}{2}} = \text{diag}(k_i) = \text{diag}\left(\sqrt{\frac{q_{2i}}{r_i}}\right) \tag{9.18}$$

with

$$k_i = \sqrt{\frac{q_{2i}}{r_i}}$$

Therefore, (9.10) becomes

$$\begin{aligned} \mathbf{u}(x) &= -[\mathbf{D},\mathbf{K}]\mathbf{x} \\ &= -\begin{bmatrix} d_1 & 0 & 0 & k_1 & 0 & 0 \\ 0 & d_2 & 0 & 0 & k_2 & 0 \\ 0 & 0 & d_3 & 0 & 0 & k_3 \end{bmatrix} \begin{bmatrix} \omega_{I1} \\ \omega_{I2} \\ \omega_{I3} \\ q_1 \\ q_2 \\ q_3 \end{bmatrix} \end{aligned} \tag{9.19}$$

From (4.12), it is straightforward to write the closed-loop system as follows:

$$\begin{aligned} \begin{bmatrix} \frac{d\omega_I}{dt} \\ \frac{d\mathbf{q}}{dt} \end{bmatrix} &= \begin{bmatrix} -\mathbf{J}^{-1}\mathbf{R}^{-\frac{1}{2}}\left(\mathbf{Q}_{11} + \frac{1}{2}\left(\mathbf{J}\mathbf{R}^{\frac{1}{2}}\mathbf{Q}_{22}^{\frac{1}{2}} + \mathbf{Q}_{22}^{\frac{1}{2}}\mathbf{R}^{\frac{1}{2}}\mathbf{J}\right)\right)^{\frac{1}{2}} & -\mathbf{J}^{-1}\mathbf{R}^{-\frac{1}{2}}\mathbf{Q}_{22}^{\frac{1}{2}} \\ \frac{1}{2}\mathbf{I}_3 & \mathbf{0}_3 \end{bmatrix} \begin{bmatrix} \omega_I \\ \mathbf{q} \end{bmatrix} \\ &= \begin{bmatrix} -\frac{d_1}{J_{11}} & 0 & 0 & -\frac{k_1}{J_{11}} & 0 & 0 \\ 0 & -\frac{d_2}{J_{22}} & 0 & 0 & -\frac{k_2}{J_{22}} & 0 \\ 0 & 0 & -\frac{d_3}{J_{33}} & 0 & 0 & -\frac{k_3}{J_{33}} \\ 0.5 & 0 & 0 & 0 & 0 & 0 \\ 0 & 0.5 & 0 & 0 & 0 & 0 \\ 0 & 0 & 0.5 & 0 & 0 & 0 \end{bmatrix} \begin{bmatrix} \omega_{I1} \\ \omega_{I2} \\ \omega_{I3} \\ q_1 \\ q_2 \\ q_3 \end{bmatrix} \\ &= \bar{\mathbf{A}}\mathbf{x} \end{aligned} \tag{9.20}$$

For i=1, 2, and 3, let $s_i = \frac{d_i}{J_{ii}}$, $t_i = \frac{k_i}{J_{ii}}$, and

$$C_i = \frac{\frac{d_i}{J_{ii}} + \sqrt{\left(\frac{d_i}{J_{ii}}\right)^2 - 2\frac{k_i}{J_{ii}}}}{2\frac{k_i}{J_{ii}}} = \frac{s_i + \sqrt{s_i^2 - 2t_i}}{2t_i} \tag{9.21}$$

Then, we have

$$\bar{\mathbf{A}} = \begin{bmatrix} -s_1 & 0 & 0 & -t_1 & 0 & 0 \\ 0 & -s_2 & 0 & 0 & -t_2 & 0 \\ 0 & 0 & -s_3 & 0 & 0 & -t_3 \\ 0.5 & 0 & 0 & 0 & 0 & 0 \\ 0 & 0.5 & 0 & 0 & 0 & 0 \\ 0 & 0 & 0.5 & 0 & 0 & 0 \end{bmatrix} \tag{9.22}$$

Let the linear matrix transformation $\mathbf{T}_{ij}(C)$ be a matrix with the following properties: (a) the (i,j) element of $\mathbf{T}_{ij}(C)$ is C, (b) the diagonal elements are ones, (c) all the remaining elements are zeros. It is well known that the inverse of $\mathbf{T}_{ij}(C)$ is $\mathbf{T}_{ij}^{-1}(C) = \mathbf{T}_{ij}(-C)$. Pre-multiplying $\mathbf{T}_{41}(C_1)$ to $\bar{\mathbf{A}}$ is equivalent to multiply the first row of $\bar{\mathbf{A}}$ by C_1 and add this result to the 4th row of the matrix. This gives

$$\mathbf{T}_{41}(C_1)\bar{\mathbf{A}} = \begin{bmatrix} -s_1 & 0 & 0 & -t_1 & 0 & 0 \\ 0 & -s_2 & 0 & 0 & -t_2 & 0 \\ 0 & 0 & -s_3 & 0 & 0 & -t_3 \\ -\frac{s_1^2+s_1\sqrt{s_1^2-2t_1}}{2t_1}+0.5 & 0 & 0 & -\frac{s_1+\sqrt{s_1^2-2t_1}}{2} & 0 & 0 \\ 0 & 0.5 & 0 & 0 & 0 & 0 \\ 0 & 0 & 0.5 & 0 & 0 & 0 \end{bmatrix} \tag{9.23}$$

Post-multiplying $\mathbf{T}_{41}(-C_1)$ to this matrix is equivalent to multiply the 4th column by $-C_1$ and add this result to the first column of the matrix. Since

$$-s_1 + \frac{t_1\left(s_1+\sqrt{s_1^2-2t_1}\right)}{2t_1} = \frac{-s_1+\sqrt{s_1^2-2t_1}}{2}$$

and

$$\frac{s_1+\sqrt{s_1^2-2t_1}}{2}\,\frac{s_1+\sqrt{s_1^2-2t_1}}{2t_1} = \frac{s_1^2+s_1\sqrt{s_1^2-2t_1}}{2t_1} - 0.5$$

this gives

$$\mathbf{T}_{41}(C_1)\bar{\mathbf{A}}\mathbf{T}_{41}(-C_1) = \begin{bmatrix} \frac{-s_1+\sqrt{s_1^2-2t_1}}{2} & 0 & 0 & -t_1 & 0 & 0 \\ 0 & -s_2 & 0 & 0 & -t_2 & 0 \\ 0 & 0 & -s_3 & 0 & 0 & -t_3 \\ 0 & 0 & 0 & \frac{-s_1-\sqrt{s_1^2-2t_1}}{2} & 0 & 0 \\ 0 & 0.5 & 0 & 0 & 0 & 0 \\ 0 & 0 & 0.5 & 0 & 0 & 0 \end{bmatrix} \tag{9.24}$$

Repeating the similar manipulation, we have

$$\mathbf{T}_{63}(C_3)\mathbf{T}_{52}(C_2)\mathbf{T}_{41}(C_1)\bar{\mathbf{A}}\mathbf{T}_{41}(-C_1)\mathbf{T}_{52}(-C_2)\mathbf{T}_{63}(-C_3)$$

$$= \begin{bmatrix} \operatorname{diag}\left(\frac{-s_i+\sqrt{s_i^2-2t_i}}{2}\right) & \operatorname{diag}(-t_i) \\ \mathbf{0} & \operatorname{diag}\left(\frac{-s_i-\sqrt{s_i^2-2t_i}}{2}\right) \end{bmatrix} \tag{9.25}$$

Since

$$s_i = \frac{d_i}{J_{ii}} = \frac{1}{J_{ii}}\sqrt{\frac{q_{1i}}{r_i} + J_{ii}\sqrt{\frac{q_{2i}}{r_i}}} = \sqrt{\frac{q_{1i}}{J_{ii}^2 r_i} + \frac{1}{J_{ii}}\sqrt{\frac{q_{2i}}{r_i}}}$$

and

$$\begin{aligned} s_i^2 - 2t_i &= \frac{q_{1i}}{J_{ii}^2 r_i} + \frac{1}{J_{ii}}\sqrt{\frac{q_{2i}}{r_i}} - 2\frac{k_i}{J_{ii}} = \frac{q_{1i}}{J_{ii}^2 r_i} + \frac{1}{J_{ii}}\sqrt{\frac{q_{2i}}{r_i}} - \frac{2}{J_{ii}}\sqrt{\frac{q_{2i}}{r_i}} \\ &= \frac{q_{1i}}{J_{ii}^2 r_i} - \frac{1}{J_{ii}}\sqrt{\frac{q_{2i}}{r_i}} \end{aligned} \tag{9.26}$$

the closed-loop eigenvalues of the linear system (9.20) using LQR design are given by, for i=1, 2, and 3,

$$\lambda_i, \lambda_{i+3} = \frac{-s_i \pm \sqrt{s_i^2 - 2t_i}}{2} = \frac{-\sqrt{\frac{1}{J_{ii}}\sqrt{\frac{q_{2i}}{r_i}} + \frac{q_{1i}}{J_{ii}^2 r_i}} \pm \sqrt{\frac{q_{1i}}{J_{ii}^2 r_i} - \frac{1}{J_{ii}}\sqrt{\frac{q_{2i}}{r_i}}}}{2} \tag{9.27}$$

Equation (9.27) provides a lot of useful information for the LQR design. First, as $r_i \to 0$, the corresponding pair of eigenvalues go to minus infinity of the complex plane; as $r_i \to \infty$, the corresponding pair of eigenvalues go to the origin of the complex plane. Second, as long as $q_{1i} > \sqrt{q_{2i} r_i} J_{ii}$, the corresponding pair of eigenvalues are real and unequal; since $\frac{d_i}{J_{ii}} > \sqrt{\left(\frac{d_i}{J_{ii}}\right)^2 - 2\frac{k_i}{J_{ii}}}$, these two eigenvalues are always negative. Third, if $q_{1i} = \sqrt{q_{2i} r_i} J_{ii}$, there are two equal real negative eigenvalues. Fourth, if $q_{1i} < \sqrt{q_{2i} r_i} J_{ii}$, there is a pair of complex eigenvalues with negative real part. Therefore, increasing q_{1i} and decreasing q_{2i} will increase the dumping ratio; otherwise, it will decrease the dumping ratio. Finally, increasing q_{2i} and decreasing r_i will increase the natural frequency; otherwise, it will decrease the natural frequency. This information can be useful in the spacecraft system design.

Using the LQR design, we implicitly assign the closed-loop poles as defined by (9.20) and we can balance the requirements on accumulative control error and power consumption (both are important in the practical design).

9.2.4 The Simulation Result

An example from [273] is used to illustrate the design procedure. The spacecraft inertia matrix is given by

$$\mathbf{J} = \begin{bmatrix} 1200 & 100 & -200 \\ 100 & 2200 & 300 \\ -200 & 300 & 3100 \end{bmatrix} \tag{9.28}$$

It is clear that the diagonal elements of the matrix are significantly larger than off-diagonal elements. Assuming that the spacecraft inertia matrix can be approximate by a diagonal matrix whose diagonal elements are equal to those of $\mathbf{J}$, let $\mathbf{Q} = \text{diag}(5,5,5,5,5,5)$ and $\mathbf{R} = \text{diag}(8,8,8)$, the closed-loop poles are then given as in Table 9.1 and the feedback matrix $\mathbf{D}$ and $\mathbf{K}$ are as follows

$$\mathbf{D} = \begin{bmatrix} 31.06637549427606 & 0 & 0 \\ 0 & 41.71184140316478 & 0 \\ 0 & 0 & 49.51151569716377 \end{bmatrix} \tag{9.29}$$

$$\mathbf{K} = \begin{bmatrix} 0.7905694150429 & 0 & 0 \\ 0 & 0.7905694150429 & 0 \\ 0 & 0 & 0.7905694150429 \end{bmatrix} \tag{9.30}$$

We apply the designed feedback controller to the nonlinear spacecraft system described by (4.5) and (4.8) with the full Monte Carlo perturbation model described as follows: (a) in inertia matrix J, the off-diagonal elements are randomly selected between [0, 310], (b) the initial Euler angle errors of the nonlinear spacecraft system are randomly selected between $[0, \pi]$ and these initial Euler angles are converted into quaternion, and (c) the initial angular rates are randomly selected between [0, 0.1] deg/second, and 300 Monte Carlo simulation runs are conducted; the simulated runs are all asymptotically stable. This result is shown in Figure 9.1.

Table 9.1: Required closed-loop poles.

–0.01273212110421 +/– 0.01272387326295i
–0.00798572833825 +/– 0.00798369205833i
–0.00947996395486 +/– 0.00947655794419i

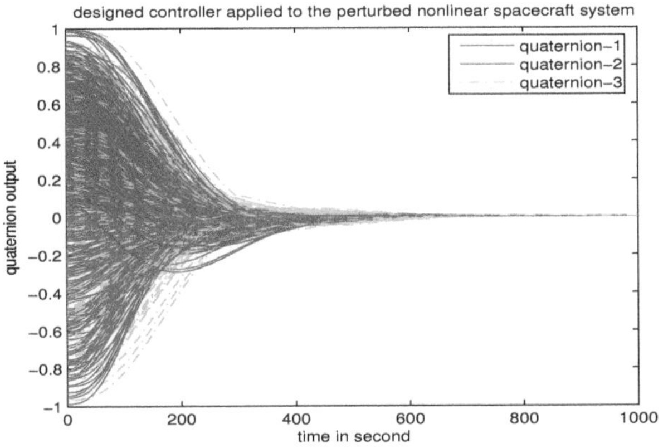

Figure 9.1: Monte Carlo simulation for the nonlinear spacecraft model with perturbation.

9.3 The LQR Design is a Robust Pole Assignment

9.3.1 Robustness of the Closed-loop Poles

In the previous section, a simple analytic LQR control design method was derived. The closed-loop eigenvalues are explicitly related to the spacecraft inertia matrix and the selected **Q** and **R** matrices. Therefore, the LQR design is equivalent to the pole assignment design. This section shows that the pole assignment design is a robust pole assignment design which is insensitive to the modeling error.

First, we have seen that the closed-loop system eigenvalues for the LQR design are

$$\lambda_i, \lambda_{i+3} = \frac{-\frac{d_i}{J_{ii}} \pm \sqrt{\left(\frac{d_i}{J_{ii}}\right)^2 - 2\frac{k_i}{J_{ii}}}}{2} \tag{9.31}$$

Let the desired spacecraft closed-loop eigenvalues be expressed as

$$\lambda_i, \lambda_{i+3} = -\zeta_i \omega_{in} \pm j\omega_{in}\sqrt{1-\zeta_i^2} = -\zeta_i \omega_{in} \pm j\omega_{id} \tag{9.32}$$

Comparing (9.31) and (9.32) yields the analytic feedback controller

$$k_i = 2\omega_{in}^2 J_{ii} \tag{9.33}$$

$$d_i = 2\zeta_i \omega_{in} J_{ii} \tag{9.34}$$

Therefore, for any LQR design which minimizes (9.2), there is an implicit set of desired spacecraft closed-loop eigenvalues defined by (9.27) or (9.31) or (9.32), the diagonal feedback matrices **D** and **K** with diagonal elements given by (9.33) and (9.34) assign the prescribed closed-loop eigenvalues. It is shown in the previous section that the closed-loop nonlinear system is globally asymptotically stable if some additional condition holds.

It is well known that for any controllable linear system and for any prescribed closed-loop pole locations, one can always find a state feedback controller such that the closed-loop system has the prescribed pole locations. For multi-input systems, the solution that achieves the closed-loop pole positions is not unique. As an example, let $(\mathbf{A},\mathbf{B})$ be a linear system with

$$\mathbf{A}=\begin{bmatrix} 0 & 0 \\ 0 & 1 \end{bmatrix}, \quad \mathbf{B}=\begin{bmatrix} 1 & 0 \\ 0 & 1 \end{bmatrix}$$

The open-loop system has two eigenvalues (0, 1) and the system is not stable. Assuming that the desired close-loop eigenvalues are (–1,–1), one may select two different feedback matrices

$$\mathbf{G}_1=\begin{bmatrix} -1 & 0 \\ 0 & -2 \end{bmatrix}, \quad \mathbf{G}_2=\begin{bmatrix} 1 & 4*10^{10} \\ -10^{-10} & -4 \end{bmatrix}$$

such that

$$\mathbf{A}+\mathbf{B}\mathbf{G}_1=\begin{bmatrix} -1 & 0 \\ 0 & -1 \end{bmatrix}, \quad \mathbf{A}+\mathbf{B}\mathbf{G}_2=\begin{bmatrix} 1 & 4*10^{10} \\ -10^{-10} & -3 \end{bmatrix}$$

It is easy to verify that $\det(\lambda\mathbf{I}-(\mathbf{A}+\mathbf{B}\mathbf{G}_1))=\det(\lambda\mathbf{I}-(\mathbf{A}+\mathbf{B}\mathbf{G}_2))=(\lambda+1)^2$. Both feedbacks achieve the desired closed-loop poles. The first system is robust because any small perturbation will not destabilize the system. However, the second system is not robust as a small perturbation of 10^{-10} in the left low corner of the matrix $\mathbf{A}+\mathbf{B}\mathbf{G}_2$ will change the closed-loop eigenvalues to $(1,-3)$. We show that the LQR defined pole assignment is a robust pole assignment.

9.3.2 The Robust Pole Assignment

For readers who are not familiar with the robust pole assignment, they should refer to Appendix C.

The robust pole assignment design makes full use of the extra degrees of freedom in a multi-input system to find the most robust controller from indefinitely many solutions of the pole assignment feedback matrices. Since the spacecraft attitude control system is a typical multi-input system that has three control torque inputs (roll, pitch, and yaw), getting a robust pole assignment design that is insensitive to the modeling error is very attractive and desirable. This section will show that the controller with diagonal **D** and **K** proposed in the previous subsection is a robust pole assignment design.

There are many different robust metrics that can be used in a robust pole assignment (see Appendix C or [227] [90]. The robust measurement proposed in

[239] will be adopted here as the design criterion because some algorithms based on this robust measurement lead to some efficient and effective design [183]. These design algorithms extend a well-known algorithm proposed by Kautsky, Nichols, and Van Dooren (KNV) [90], in which the angles of closed-loop eigenvectors are intuitively maximized one by one in a cyclic manner. Let $\mathbf{X}$ be the matrix whose columns are the unit length closed-loop eigenvectors. The robustness of the closed-loop eigenvalues (poles) can be measured by the absolute value of the determinant of $\mathbf{X}$. Geometrically, this determinant measures how close the matrix $\mathbf{X}$ is to an orthogonal matrix. Yang and Tits [258] showed that one of the KNV algorithm is equivalent to maximizing the absolute value of the determinant of $\mathbf{X}$. The greater the absolute value of the determinant, the more robust the closed-loop eigenvalues will be (see detailed discussions in Appendix C or [239, 241]). By maximizing the absolute value of the determinant under some constraints, we are guaranteed that the closed-loop poles obtained by the robust pole assignment design are insensitive to the modeling errors [249]. For a controllable linear system $(\mathbf{A}, \mathbf{B})$, where $\mathbf{B}$ is a full column rank, and any given set of desired closed-loop eigenvalues λ_i, the corresponding closed-loop eigenvectors $\mathbf{x}_i$ must be in the subspace (see Appendix C)

$$\mathcal{S}_i = \{\mathbf{x} : (\mathbf{A} - \lambda \mathbf{I})\mathbf{x} \in \mathbf{R}_c(\mathbf{B})\} \tag{9.35}$$

where

$$\mathbf{R}_c(\mathbf{B}) = \{\mathbf{B}\mathbf{y} : \mathbf{y} \in \mathbf{C}^m\}$$

m is the rank of $\mathbf{B}$, and $\mathbf{C}^m$ is a m-dimensional complex space. First, using *QR* decomposition on $\mathbf{B}$, we have

$$\mathbf{B} = \begin{bmatrix} \mathbf{U}_0 & \mathbf{U}_1 \end{bmatrix} \begin{bmatrix} \mathbf{V} \\ \mathbf{0} \end{bmatrix}$$

Let Λ be the diagonal matrix whose diagonal elements are the desired closed-loop eigenvalues, and $\mathbf{X}$ be the matrix whose columns are composed of the eigenvectors corresponding to the desired eigenvalues. Then,

$$\mathbf{B}\mathbf{G} = \mathbf{U}_0\mathbf{V}\mathbf{G} = \mathbf{X}\Lambda\mathbf{X}^{-1} - \mathbf{A} \tag{9.36}$$

Pre-multiplication of $\mathbf{U}_0^{\mathrm{T}}$ and $\mathbf{U}_1^{\mathrm{T}}$ gives

$$\mathbf{V}\mathbf{G} = \mathbf{U}_0^{\mathrm{T}}(\mathbf{X}\Lambda\mathbf{X}^{-1} - \mathbf{A}) \tag{9.37a}$$

$$\mathbf{0} = \mathbf{U}_1^{\mathrm{T}}(\mathbf{X}\Lambda\mathbf{X}^{-1} - \mathbf{A}) \tag{9.37b}$$

The first relation gives the closed-loop feedback matrix as

$$\mathbf{G} = \mathbf{V}^{-1}\mathbf{U}_0^{\mathrm{T}}(\mathbf{A} - \mathbf{X}\Lambda\mathbf{X}^{-1}) \tag{9.38}$$

The second relation shows that $\mathbf{x}_i$ must be in the subspace $\mathcal{S}_i$, or

$$\mathbf{U}_1^{\mathrm{T}}(\mathbf{A} - \lambda_i \mathbf{I})\mathbf{x}_i = \mathbf{0}$$

Therefore, $\mathbf{x}_i$ must be in the null space of $(\mathbf{A}^{\mathrm{T}} - \lambda_i \mathbf{I})\mathbf{U}_1$. Using QR decomposition again on $(\mathbf{A}^{\mathrm{T}} - \lambda_i \mathbf{I})\mathbf{U}_1$ gives

$$(\mathbf{A}^{\mathrm{T}} - \lambda_i \mathbf{I})\mathbf{U}_1 = \begin{bmatrix} \mathbf{W}_{1i} & \mathbf{W}_{2i} \end{bmatrix} \begin{bmatrix} \mathbf{R} \\ \mathbf{0} \end{bmatrix}$$

$\mathbf{W}_{2i}$ forms the basis of $\mathcal{S}_i$. We now apply the similar procedure to the linearized spacecraft system (4.12). Since $\mathbf{B}$ can be written as

$$\mathbf{B} = \begin{bmatrix} \mathbf{I} & \mathbf{0} \\ \mathbf{0} & \mathbf{I} \end{bmatrix} \begin{bmatrix} \mathbf{J}^{-1} \\ \mathbf{0} \end{bmatrix}$$

therefore and

$$\mathbf{U}_0 = \begin{bmatrix} \mathbf{I} \\ \mathbf{0} \end{bmatrix}, \quad \mathbf{U}_1 = \begin{bmatrix} \mathbf{0} \\ \mathbf{I} \end{bmatrix}, \quad \mathbf{V} = \mathbf{J}^{-1} \tag{9.39}$$

Since $\mathbf{A}$ is defined as in (4.13), we can write a similar decomposition of $(\mathbf{A}^{\mathrm{T}} - \lambda_i \mathbf{I})\mathbf{U}_1$ as

$$\begin{aligned} (\mathbf{A}^{\mathrm{T}} - \lambda_i \mathbf{I})\mathbf{U}_1 &= \begin{bmatrix} -\lambda_i \mathbf{I} & \frac{1}{2}\mathbf{I} \\ \mathbf{0} & -\lambda_i \mathbf{I} \end{bmatrix} \begin{bmatrix} \mathbf{0} \\ \mathbf{I} \end{bmatrix} \\ &= \begin{bmatrix} \frac{1}{2}\mathbf{I} \\ -\lambda_i \mathbf{I} \end{bmatrix} = \begin{bmatrix} 0.5\mathbf{I} & -\lambda_i \mathbf{I} \\ -\lambda_i \mathbf{I} & -0.5\mathbf{I} \end{bmatrix} \begin{bmatrix} \mathbf{I} \\ \mathbf{0} \end{bmatrix} \end{aligned} \tag{9.40}$$

therefore,

$$\mathbf{W}_{1i} = \begin{bmatrix} 0.5\mathbf{I} \\ -\lambda_i \mathbf{I} \end{bmatrix}$$

which is orthogonal to the subspace

$$\mathbf{W}_{2i} = \begin{bmatrix} -\lambda_i \mathbf{I} \\ -0.5\mathbf{I} \end{bmatrix}$$

Though $\begin{bmatrix} \mathbf{W}_{1i} & \mathbf{W}_{2i} \end{bmatrix}$ may not be a unitary matrix, it is clear that $\mathbf{W}_{2i}$ forms the basis of $\mathcal{S}_i$ (and we can always normalize $\mathbf{W}_{2i}$ to make it orthonormal). For the sake of simplicity, we prove, only for the case where all eigenvalues are real, that the design given by (9.19) is a robust pole assignment. For robust pole assignment design, since $\mathbf{x}_i \in \mathcal{S}_i$, we can write $\mathbf{x}_i = \mathbf{W}_{2i}\mathbf{p}_i$, where $\mathbf{p}_i = [p_{i1}, p_{i2}, p_{i3}]^{\mathrm{T}}$, therefore, the closed-loop eigenvector matrix must have the form

$$\mathbf{X} = \begin{bmatrix} \lambda_1 p_{11} & \lambda_2 p_{21} & \lambda_3 p_{31} & \lambda_4 p_{41} & \lambda_5 p_{51} & \lambda_6 p_{61} \\ \lambda_1 p_{12} & \lambda_2 p_{22} & \lambda_3 p_{32} & \lambda_4 p_{42} & \lambda_5 p_{52} & \lambda_6 p_{62} \\ \lambda_1 p_{13} & \lambda_2 p_{23} & \lambda_3 p_{33} & \lambda_4 p_{43} & \lambda_5 p_{53} & \lambda_6 p_{63} \\ 0.5 p_{11} & 0.5 p_{21} & 0.5 p_{31} & 0.5 p_{41} & 0.5 p_{51} & 0.5 p_{61} \\ 0.5 p_{12} & 0.5 p_{22} & 0.5 p_{32} & 0.5 p_{42} & 0.5 p_{52} & 0.5 p_{62} \\ 0.5 p_{13} & 0.5 p_{23} & 0.5 p_{33} & 0.5 p_{43} & 0.5 p_{53} & 0.5 p_{63} \end{bmatrix}$$

where p_{ij}, $i = 1,2,3,4,5,6$ and $j = 1,2,3$, are the real parameters that will be used to optimize the objective function. Therefore, the robust pole assignment design for linearized spacecraft system (11) becomes[2]

$$\begin{aligned} \max \quad & \det(\mathbf{X}) \\ s.t. \quad & \sum_{j=1}^{3} \left(|\lambda_i|^2 + 0.5^2\right) p_{ij}^2 = 1, \quad i = 1,2,3,4,5,6 \end{aligned} \tag{9.41}$$

It is well-known that an optimal solution for a general optimization problem has to satisfy the KKT conditions (see Appendix A). For (9.41), let the μ_i $i = 1,2,3,4,5,6$ be the Lagrangian multipliers, the Lagrangian function of (9.41) is given by

$$\begin{aligned} \mathcal{L} &= \det(\mathbf{X}) - \mu_1 \left(\sum_{j=1}^{3} \left(|\lambda_1|^2 + 0.5^2\right) p_{1j}^2 - 1 \right) - \mu_2 \left(\sum_{j=1}^{3} \left(|\lambda_2|^2 + 0.5^2\right) p_{2j}^2 - 1 \right) \\ &- \mu_3 \left(\sum_{j=1}^{3} \left(|\lambda_3|^2 + 0.5^2\right) p_{3j}^2 - 1 \right) - \mu_4 \left(\sum_{j=1}^{3} \left(|\lambda_4|^2 + 0.5^2\right) p_{4j}^2 - 1 \right) \\ &- \mu_5 \left(\sum_{j=1}^{3} \left(|\lambda_5|^2 + 0.5^2\right) p_{5j}^2 - 1 \right) - \mu_6 \left(\sum_{j=1}^{3} \left(|\lambda_6|^2 + 0.5^2\right) p_{6j}^2 - 1 \right) \end{aligned}$$

The corresponding KKT conditions are as follows (see Appendix A):

$$\frac{\partial \mathcal{L}}{\partial p_{ij}} = 0, \quad i = 1,2,3,4,5,6, \quad j = 1,2,3 \tag{9.42a}$$

$$-\frac{\partial \mathcal{L}}{\partial \mu_1} = \sum_{j=1}^{3} \left(|\lambda_1|^2 + 0.5^2\right) p_{1j}^2 - 1 = 0 \tag{9.42b}$$

$$-\frac{\partial \mathcal{L}}{\partial \mu_2} = \sum_{j=1}^{3} \left(|\lambda_2|^2 + 0.5^2\right) p_{2j}^2 - 1 = 0 \tag{9.42c}$$

$$-\frac{\partial \mathcal{L}}{\partial \mu_3} = \sum_{j=1}^{3} \left(|\lambda_3|^2 + 0.5^2\right) p_{3j}^2 - 1 = 0 \tag{9.42d}$$

$$-\frac{\partial \mathcal{L}}{\partial \mu_4} = \sum_{j=1}^{3} \left(|\lambda_4|^2 + 0.5^2\right) p_{4j}^2 - 1 = 0 \tag{9.42e}$$

$$-\frac{\partial \mathcal{L}}{\partial \mu_5} = \sum_{j=1}^{3} \left(|\lambda_5|^2 + 0.5^2\right) p_{5j}^2 - 1 = 0 \tag{9.42f}$$

[2] In [241], $|\det(\mathbf{X})|$ is used as the measurement of the robustness. If the maximum of $|\det(\mathbf{X})|$ is achieved at $-\det(\mathbf{X}^*)$, let $\mathbf{X}^0$ be the matrix obtained by changing the sign of some column of $\mathbf{X}^*$, $|\det(\mathbf{X})|$ is also achieved at $\mathbf{X}^0$. Therefore, $\det(\mathbf{X})$ can be simply used here as the objective function in the problem.

$$-\frac{\partial \mathcal{L}}{\partial \mu_6} = \sum_{j=1}^{3} \left(|\lambda_6|^2 + 0.5^2\right) p_{6j}^2 - 1 = 0 \tag{9.42g}$$

It is tedious but straightforward to verify that the following solution satisfies the KKT conditions:

$$\begin{cases} p_{i,i} = \sqrt{\frac{1}{|\lambda_i|^2+0.5^2}}, & i=j, \ \ i=1,2,3, \ \ j=1,2,3 \\ p_{i+3,j} = \sqrt{\frac{1}{|\lambda_{i+3}|^2+0.5^2}}, & i=j, \ \ i=1,2,3, \ \ j=1,2,3 \\ p_{i,j} = 0, & i \neq j, \ \ i \neq j+3, \ \ i=1,2,3,4,5,6, \ \ j=1,2,3 \end{cases} \tag{9.43}$$

Clearly, this set of $p_{i,j}$ meets (9.42b), (9.42c), (9.42d), (9.42e), (9.42f), and (9.42g). To show that the set of $p_{i,i}$ satisfies (9.42a), we use the observation that $\frac{\partial \det(\mathbf{X})}{\partial p_{ij}} = 0$ for all p_{ij} defined in (9.43) except $p_{11}, p_{22}, p_{33}, p_{41}, p_{52}, p_{63}$; therefore, $\frac{\partial \mathcal{L}}{\partial p_{ij}} = 0$ for all $p_{ij} \notin \{p_{11}, p_{22}, p_{33}, p_{41}, p_{52}, p_{63}\}$. As an example, let us consider $\frac{\partial \mathcal{L}}{\partial p_{12}}$, since

$$\begin{aligned} \frac{\partial \det(\mathbf{X})}{\partial p_{12}} &= \lambda_1 \begin{vmatrix} 0 & 0 & \lambda_4 p_{41} & 0 & 0 \\ 0 & \lambda_3 p_{33} & 0 & 0 & \lambda_6 p_{63} \\ 0 & 0 & 0.5 p_{41} & 0 & 0 \\ 0.5 p_{22} & 0 & 0 & 0.5 p_{52} & 0 \\ 0 & 0.5 p_{33} & 0 & 0 & 0.5 p_{63} \end{vmatrix} \\ &+ 0.5 \begin{vmatrix} 0 & 0 & \lambda_4 p_{41} & 0 & 0 \\ \lambda_2 p_{22} & 0 & 0 & \lambda_5 p_{52} & 0 \\ 0 & \lambda_3 p_{33} & 0 & 0 & \lambda_6 p_{63} \\ 0 & 0 & 0.5 p_{41} & 0 & 0 \\ 0 & 0.5 p_{33} & 0 & 0 & 0.5 p_{63} \end{vmatrix} = 0 \end{aligned}$$

(the last equation holds because the first row and the third row are proportional in the first determinant and the first row and the fourth row are proportional in the second determinant), we have

$$\frac{\partial \mathcal{L}}{\partial p_{12}} = \frac{\partial \det(\mathbf{X})}{\partial p_{12}} - 2\mu_1 p_{12} \left(|\lambda_1|^2 + 0.5^2\right)\Big|_{p_{12}=0} = 0 \tag{9.44}$$

Similarly, for all $p_{ij} \notin \{p_{11}, p_{22}, p_{33}, p_{41}, p_{52}, p_{63}\}$, the same way can be used to check that equation (9.42a) is valid. For each of these 6 $p_{ij} \in \{p_{11}, p_{22}, p_{33}, p_{41}, p_{52}, p_{63}\}$, $\frac{\partial \det(\mathbf{X})}{\partial p_{ij}} \neq 0$, one can select one of the multipliers μ_1, μ_2, μ_3, μ_4, μ_5, μ_6 to make $\frac{\partial \mathcal{L}}{\partial p_{ij}} = 0$. Therefore, the set of p_{ij} satisfying (9.43) is a candidate of the optimal solution of (9.41). This proves that the closed-loop eigenvector matrix has the form as

$$\mathbf{X} = \begin{bmatrix} \lambda_1 p_{1,1} & 0 & 0 & \lambda_4 p_{4,1} & 0 & 0 \\ 0 & \lambda_2 p_{2,2} & 0 & 0 & \lambda_5 p_{5,2} & 0 \\ 0 & 0 & \lambda_3 p_{3,3} & 0 & 0 & \lambda_6 p_{6,3} \\ 0.5 p_{1,1} & 0 & 0 & 0.5 p_{4,1} & 0 & 0 \\ 0 & 0.5 p_{2,2} & 0 & 0 & 0.5 p_{5,2} & 0 \\ 0 & 0 & 0.5 p_{3,3} & 0 & 0 & 0.5 p_{6,3} \end{bmatrix}$$
$$= \begin{bmatrix} \mathrm{diag}(\lambda_i p_{i,i}) & \mathrm{diag}(\lambda_{i+3} p_{i+3,i}) \\ \mathrm{diag}(0.5 p_{i,i}) & \mathrm{diag}(0.5 p_{i+3,i}) \end{bmatrix}, \quad i = 1,2,3 \tag{9.45}$$

It is easy to verify that

$$\mathbf{X}^{-1} = \begin{bmatrix} \mathrm{diag}\left(\frac{1}{(\lambda_i - \lambda_{i+3}) p_{i,i}}\right) & \mathrm{diag}\left(\frac{-\lambda_{i+3}}{0.5(\lambda_i - \lambda_{i+3}) p_{i,i}}\right) \\ \mathrm{diag}\left(\frac{-1}{(\lambda_i - \lambda_{i+3}) p_{i+3,i}}\right) & \mathrm{diag}\left(\frac{\lambda_i}{0.5(\lambda_i - \lambda_{i+3}) p_{i+3,i}}\right) \end{bmatrix} \tag{9.46}$$

Substituting (9.39), (9.45), and (9.46) into (9.38) gives the robust pole assignment state feedback

$$\begin{aligned} \mathbf{G} &= \mathbf{J} \begin{bmatrix} \mathbf{I} & \mathbf{0} \end{bmatrix} \\ & \left(\begin{bmatrix} \mathbf{0} & \mathbf{0} \\ 0.5\mathbf{I} & \mathbf{0} \end{bmatrix} - \begin{bmatrix} \mathrm{diag}(\lambda_i p_{i,i}) & \mathrm{diag}(\lambda_{i+3} p_{i+3,i}) \\ \mathrm{diag}(0.5 p_{i,i}) & \mathrm{diag}(0.5 p_{i+3,i}) \end{bmatrix} \begin{bmatrix} \mathrm{diag}(\lambda_i) & \mathbf{0} \\ \mathbf{0} & \mathrm{diag}(\lambda_{i+3}) \end{bmatrix} \mathbf{X}^{-1} \right) \\ &= -\mathbf{J} \begin{bmatrix} \mathrm{diag}(\lambda_i p_{i,i}) & \mathrm{diag}(\lambda_{i+3} p_{i+3,i}) \end{bmatrix} \begin{bmatrix} \mathrm{diag}(\lambda_i) & \mathbf{0} \\ \mathbf{0} & \mathrm{diag}(\lambda_{i+3}) \end{bmatrix} \mathbf{X}^{-1} \\ &= -\mathbf{J} \begin{bmatrix} \mathrm{diag}(\lambda_i^2 p_{i,i}) & \mathrm{diag}(\lambda_{i+3}^2 p_{i+3,i}) \end{bmatrix} \begin{bmatrix} \mathrm{diag}\left(\frac{1}{(\lambda_i - \lambda_{i+3}) p_{i,i}}\right) & \mathrm{diag}\left(\frac{-\lambda_{i+3}}{0.5(\lambda_i - \lambda_{i+3}) p_{i,i}}\right) \\ \mathrm{diag}\left(\frac{-1}{(\lambda_i - \lambda_{i+3}) p_{i+3,i}}\right) & \mathrm{diag}\left(\frac{\lambda_i}{0.5(\lambda_i - \lambda_{i+3}) p_{i+3,i}}\right) \end{bmatrix} \\ &= -\mathbf{J} \left[\mathrm{diag}\left(\frac{\lambda_i^2 - \lambda_{i+3}^2}{\lambda_i - \lambda_{i+3}}\right), \quad \mathrm{diag}\left(\frac{\lambda_{i+3}^2 \lambda_i - \lambda_i^2 \lambda_{i+3}}{0.5(\lambda_i - \lambda_{i+3})}\right) \right] \\ &= -\mathbf{J} \left[\mathrm{diag}(\lambda_i + \lambda_{i+3}), \quad \mathrm{diag}(-2\lambda_i \lambda_{i+3}) \right] \end{aligned} \tag{9.47}$$

or

$$\mathbf{G} = \left[\mathrm{diag}(-J_{ii}(\lambda_i + \lambda_{i+3})), \quad \mathrm{diag}(2J_{ii}(\lambda_i \lambda_{i+3})) \right] \tag{9.48}$$

Substituting (9.27) into (9.48) yields (9.19). Therefore, we conclude that the LQR design method is actually a robust pole assignment design for the linearized system (4.12), and the feedback matrix $\mathbf{G} = -[\mathbf{D}, \mathbf{K}]$ is composed of two diagonal matrices $\mathbf{D}$ and $\mathbf{K}$. With the same restriction as discussed before, the robust pole assignment controller globally stabilizes the nonlinear spacecraft system.

9.3.3 Disturbance Rejection of Robust Pole Assignment

In Appendix C, it is shown that maximizing $\det(\mathbf{X})$ amounts to minimizing an upper bound of the condition number κ_2, which improves the robustness of the closed-loop eigenvalues to modeling uncertainties (see [227] and [189]). It is further shown that minimizing the upper bound of the condition number also reduces the impact of disturbance torques on the system output. It is easy to see that the spacecraft system with disturbance torques can be modeled as

$$\dot{\mathbf{x}} = \mathbf{A}\mathbf{x} + \mathbf{B}\mathbf{u} + \mathbf{t}_d, \qquad \mathbf{y} = \mathbf{C}\mathbf{x} \tag{9.49}$$

where $\mathbf{t}_d$ is the vector of disturbance torques. Since $\mathbf{u} = \mathbf{G}\mathbf{x}$, taking Laplace transformation, we have

$$s\mathbf{x}(s) = \mathbf{A}\mathbf{x}(s) + \mathbf{B}\mathbf{u}(s) + \mathbf{t}_d(s), \qquad \mathbf{y}(s) = \mathbf{C}\mathbf{x}(s) \tag{9.50}$$

In view of (9.36), this gives

$$\mathbf{Y}(s) = \mathbf{C}(s\mathbf{I} - (\mathbf{A} + \mathbf{B}\mathbf{G}))^{-1}\mathbf{t}_d(s) = \mathbf{C}\mathbf{X}(s\mathbf{I} - \Lambda)^{-1}\mathbf{X}^{-1}\mathbf{t}_d(s)$$

Therefore,

$$\|\mathbf{Y}(s)\| = \|\mathbf{C}\|\|\mathbf{X}\|\|(s\mathbf{I} - \Lambda)^{-1}\|\|\mathbf{X}^{-1}\|\|\mathbf{t}_d(s)\|$$

Since Λ is a diagonal matrix whose elements are the prescribed closed-loop eigenvalues and $\mathbf{C}$ is fixed by a spacecraft design, minimizing condition number ($\kappa_2 = \|\mathbf{X}\|\|\|\mathbf{X}^{-1}\|$) will reduce the impact of the disturbance torques on the system output.

9.3.4 A Design Example

The same example used in the previous subsection is used to describe the pole assignment design procedure. The spacecraft inertia matrix is given in (9.28). The spacecraft inertia matrix is approximated by a diagonal matrix whose diagonal elements are equal to the diagonal elements of $\mathbf{J}$. Assuming that the desired closed-loop linear system has a fast settling time of $T_s \le 10$ seconds, and a small percentage of overshoot (smaller than 5%), the system is designed by first considering the dominant pole positions and then loosely assigning the remaining poles to certain desired regions such that their real parts are smaller than the real parts of the dominant poles. Since the settling time is (see for example, [41, pages 84–85])

$$T_s = \frac{4}{\zeta_3 \omega_{3n}}$$

$\zeta_3\omega_{3n} = 0.4$. We select $\zeta_3 = 0.8$ to meet the requirement of low percentage of overshoot (smaller than 5%). This gives $\omega_{3n} = 0.5$. Therefore, the dominant poles are at $-0.4 + j0.3$. To ensure that the design is globally asymptotically stable (see (9.33) and (9.15)), we use

$$\alpha = \frac{1}{\omega_{3n}^2 J_{33}} = \frac{1}{0.25 * 3100} = \frac{4}{3100} = \frac{1}{\omega_{2n}^2 J_{22}} = \frac{1}{\omega_{1n}^2 J_{11}}$$

Similarly, we select

$$\omega_{2n} = \frac{1}{\sqrt{\alpha J_{22}}} = \sqrt{\frac{3100}{4 * 2200}} = 0.5935$$

$$\omega_{1n} = \frac{1}{\sqrt{\alpha J_{11}}} = \sqrt{\frac{3100}{4 * 1200}} = 0.8036$$

Clearly, by selecting $\zeta_1 = \zeta_2 = 1$, we have two closed-loop poles at -0.5935 and two closed-loop poles at -0.8036. All of these poles have smaller real parts than the real part of the dominant poles. Therefore, from (9.34), the feedback matrices are given by

$$d_1 = \frac{2\zeta_1\sqrt{J_{11}}}{\sqrt{\alpha}} = 1928.73, \quad d_2 = \frac{2\zeta_2\sqrt{J_{22}}}{\sqrt{\alpha}} = 2611.513, \quad d_3 = \frac{2\zeta_3\sqrt{J_{33}}}{\sqrt{\alpha}} = 2480$$

and from (9.33),

$$k_1 = k_2 = k_3 = \frac{2}{\alpha} = 1550$$

Noticing that $\mathbf{K} = \mathrm{diag}(k_1, k_2, k_3) = 1550\mathbf{I}$, from (9.18), we have $\mathbf{K}^2 = 1500^2\mathbf{I} = \mathbf{R}^{-1}\mathbf{Q}_{22}$, i.e., $\mathbf{R} = c\mathbf{Q}_{22}$, which is the condition of (9.15). Therefore, the designed system is globally asymptotically stable.

Applying the designed feedback controller to the linearized system (4.12) with diagonal inertia matrix (9.28), assuming that the initial Euler angle errors of the linearized system are 10 degrees in roll, pitch, and yaw, and converting these initial Euler angles into quaternion, we have the simulated quaternion response as shown by Figure 9.2. It is clear that the designed control system meets the design criteria, i.e., the settling time is less than 10 seconds and the percentage overshoot is smaller than 5% even though the design is focused on the dominant poles while the remaining poles are loosely placed left to the dominant poles. The closed-loop system is globally asymptotically stable as expected.

Applying the same designed feedback controller to the original nonlinear system with non-diagonal matrix $\mathbf{J}$ given by (9.28), again assuming that the initial Euler angle errors of the linearized system are 10 degrees in roll, pitch, and yaw, and converting these initial Euler angles into quaternion, we have the simulated quaternion response as shown by Figure 9.3. This simulation result shows that the robust pole assignment design is insensitive to the perturbation in off-diagonal elements of $\mathbf{J}$.

As real spacecraft control torques are normally restricted by the solar panel size, energy consumption of the on-board instruments, fuel, etc., it is preferred to have a slow response with a low percentage overshoot to reduce energy consumption. Therefore, a different but a representative design is considered. We choose $\mathbf{Q} = \mathrm{diag}(5,5,5,5,5,5)$ and $\mathbf{R} = \mathrm{diag}(8,8,8)$. This is equivalent to selecting the closed-loop poles as

$$-0.0127 + / - 0.0127i; -0.0080 + / - 0.0080i; -0.0095 + / - 0.0095i$$

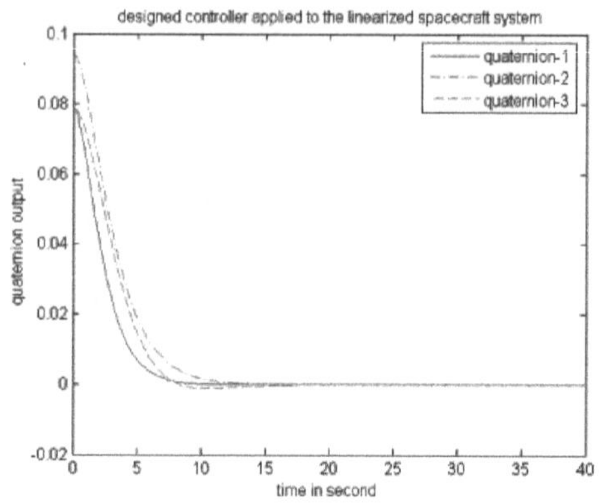

Figure 9.2: Designed controller applied to the linear spacecraft model.

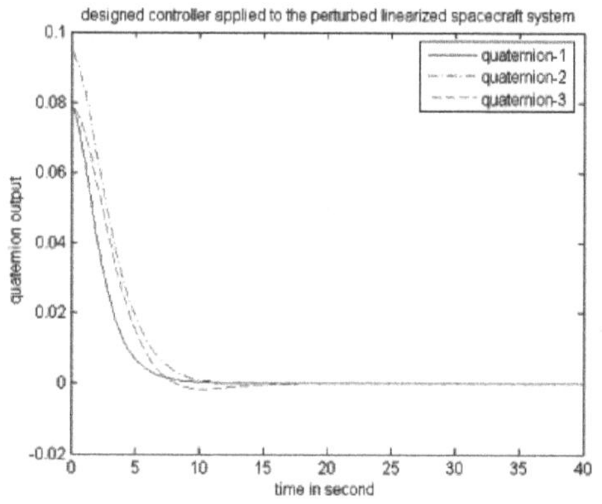

Figure 9.3: Designed controller applied to the nonlinear spacecraft model.

Notice that this is the same design of the LQR as described in the previous subsection. The feedback matrices **D** and **K** in this design are given in (9.29) and (9.30) which are significantly smaller than the ones in the previous design.

Applying the designed feedback controller to the linearized system (4.12) with diagonal inertia matrix (9.28), and assuming that the initial Euler angle errors of the linearized system are 10 degrees in roll, pitch, and yaw, converting these initial Euler angles into quaternion, the simulation result is shown in Figure 9.4, the rising time, settling time, and overshoot of the three quaternion components for the nominal linearized system are given in Table 9.2.

A very aggressive test has been conducted for this design (see the previous subsection, i.e., apply the same designed feedback controller to the nonlinear spacecraft system described by (4.11) and (4.8) with the full Monte Carlo perturbation model described as follows: (a) in inertia matrix **J**, the off-diagonal ele-

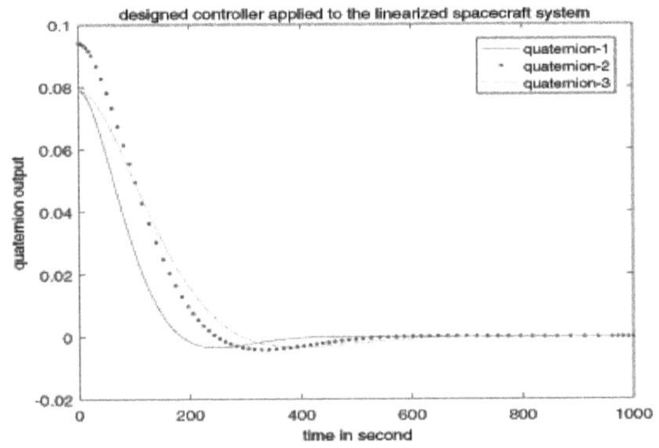

Figure 9.4: Designed controller applied to the linear spacecraft model.

Table 9.2: Performance of the nominal linearized system.

	Quaternion 1	Quaternion 2	Quaternion 3
Rising time (seconds)	140	196	222
Settling time (seconds)	310	430	500
Overshoot (percentage)	3.4%	4%	3.4%

ments are randomly selected between $[0,310]$, (b) the initial Euler angle errors of the nonlinear spacecraft system are randomly selected between $[0,\pi]$ and these initial Euler angles are converted into quaternion, and (c) the initial angular rates are randomly selected between $[0,0.1]$ deg/second. Three-hundred Monte Carlo simulation runs were conducted. The simulated quaternion response is given in Figure 9.1. This simulation result shows that although the designed robust pole assignment controller is obtained from the linearized system with diagonal inertia matrix, it actually stabilizes the original nonlinear spacecraft system with any initial Euler angles, any small initial angular rates (less than 0.1deg/second), and any perturbation in off-diagonal elements whose magnitudes are smaller than 10% of the magnitude of the largest element in the inertia matrix. Table 9.3 provides the means and standard deviations of the rising time, settling time, and overshoot of the perturbed nonlinear systems. Although these standard deviations appear somewhat large, the design meets the most important design target which is to stabilize the system in a few hours under all uncertainties related to the modeling error and initial conditions. A similar simulation is done for the Euler angle controller. The system is first designed for a linearized Euler angle model (see [181]) using LQR method and exactly the same set of closed-loop eigenvalues

$$-0.0127+/-0.0127i; -0.0080+/-0.0080i; -0.0095+/-0.0095i$$

Table 9.3: Performance of the perturbed nonlinear system.

	Quaternion 1	Quaternion 2	Quaternion 3
Mean rising time (seconds)	225	227	259
Std rising time	88	71	107
Mean settling time (seconds)	430	612	666
Std settling time	64	86	93
Mean overshoot (percentage)	15%	45%	30%
Std overshoot	18	37	29

to get the feedback control matrices **D** and **K**. Use the same Monte Carlo perturbation model described as above with perturbed nonlinear system (a) in inertia matrix J, the off-diagonal elements are randomly selected between $[0, 310]$, (b) the initial Euler angle errors of the nonlinear spacecraft system are randomly selected between $[0, \pi]$, and (c) the initial angular rates are randomly selected between $[0, 0.1]$ deg/second. In 300 Monte Carlo runs, the Euler angle controller stabilizes only 132 cases. The comparison is clearly in favor of the quaternion design described in this section. Sidi [181, page 156–158] has done some interesting comparisons of Euler angle design and quaternion design for maneuvers operation. The result shows that for small maneuvers, both designs have similar performance, but for large maneuvers, the quaternion design is clearly superior.

Chapter 10

Spacecraft Actuators

Spacecraft actuators are components that produce the control torques to achieve the desired attitude. The desired control torques can be calculated using the methods proposed in the previous chapter. The most frequently used actuators are *reaction wheel*, *momentum wheel*, *control moment gyros* (CMG), *magnetic torque rods*, and *thrusters*. This chapter will discuss these actuators. It will also show that given the designed torques, some actuators, such as reaction wheels and thruster, can easily provide the desired torques. But some other actuators, such as magnetic torque bars and CMGs, may not be able to provide the desired torques, at least in some situations, which means that we need to have alternative design methods specifically for those actuators. These topics will be discussed in Chapters 11 and 14.

10.1 Reaction Wheel and Momentum Wheel

Reaction wheel and momentum wheel are very similar. They both have *fly-wheel(s)* and are driven by electric motors. Both of these are used for attitude control. A reaction wheel is spun up and down to create the torque to either compensate disturbance torque to stabilize the spacecraft or to create a torque and force the spacecraft to rotate for attitude manipulation. A momentum wheel is always spinning at a very high speed, which creates a momentum bias, making it resistant to changing its attitude. But a momentum wheel can also be used as a reaction wheel, meaning that the acceleration and deceleration is near a momentum biased high speed instead of near the zero speed. The torques of both reaction wheel and the momentum wheel are generated from acceleration or deceleration of the rotational flywheel and torque can be calculated by the following relation [109]

$$\mathbf{u} = -\dot{\mathbf{h}}_w = -\mathbf{J}_w \dot{\omega} \tag{10.1}$$

where the $\mathbf{h}_w$ is the angular momentum vector of the flywheel, $\mathbf{J}_w$ is the moment of inertia about the flywheel rotation axis, ω is the angular velocity vector of the flywheel. The electricity that drives a flywheel of reaction wheel or momentum wheel, comes from the batteries which are charged by solar panels.

Both reaction wheel and momentum wheel are normally aligned with body axes. [181] has a chapter to discuss momentum biased spacecraft attitude stabilization. Since flywheels have maximum speed, once the maximum speed is reached, from (10.1), one cannot get the torque by increasing the flywheel speed. Therefore, the momentum management control, which makes sure that the flywheel speed does not approach to the maximum speed, is necessary. The momentum management control uses magnetic torque rods or thrusters to balance the total torques required by the attitude control, thereby maintaining the flywheel's speed within its limit. There are many papers which discuss this topic, for example, [29, 53]. This issue will be discussed later in Chapter 11.

10.2 Control Moment Gyros

Like a reaction wheel, a control moment gyro has spinning flywheels controlled by an electrical motor. Unlike a reaction wheel, which has a fixed rotational axis, the spinning axis of a control moment gyros changes as the flywheel is suspended in a *gimbal* and a second motor controls the gimbal axis. Another difference between a reaction wheel and a control moment gyro is that the torque of a reaction wheel is produced by changing the flywheel speed, while flywheel in a CMG rotate in a constant speed, the torque of a CMG is obtained by changing the gimbal's rotational speed. There are two different CMGs. One is *single gimbal CMG* and the other is *double gimbal CMG*. The advantage of the single CMG is the well-known torque amplification property, i.e., a rate about the gimbal axis can produce an output torque orthogonal to both the gimbal and spin axes which is much greater than the gimbal axis torque [50]. But CMG is more complicated to model and more expensive. Only the single gimbal control moment gyro is discussed in this book because it is the most effective CMG. Some good references about CMG are [100, 101]. A thorough performance comparison between CMGs and reaction wheels is discussed in [212].

Three mutually orthogonal unit vectors are shown in Figure 10.1 and defined as follows: Let $\hat{\mathbf{g}}$ be the unit-length *gimbal vector*, $\mathbf{h}$ be the *angular momentum vector* of the flywheel, $\mathbf{c} = \hat{\mathbf{g}} \times \mathbf{h}$ be the normalized *CMG torque vector*, then the torque of the CMG is given by

$$\mathbf{t}_c = \mathbf{c}\omega_g = (\hat{\mathbf{g}} \times \mathbf{h})\omega_g = \mathbf{g} \times \mathbf{h} \tag{10.2}$$

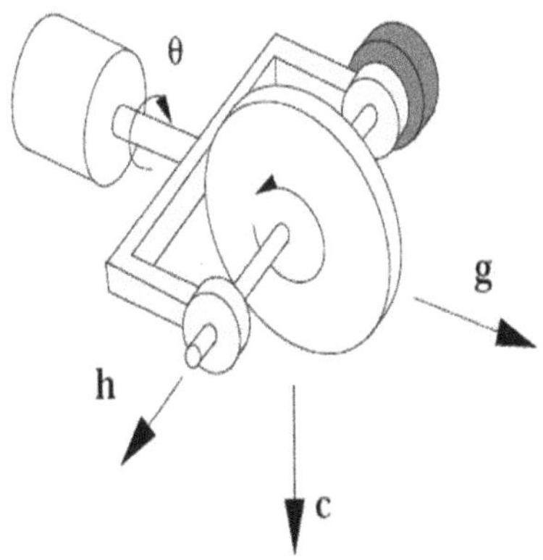

Figure 10.1: Orthonormal vectors of a CMG unit.

where the ω_g is the rotational speed of the gimbal and $\mathbf{g} = \hat{\mathbf{g}}\omega_g$. Therefore, the control variable is ω_g. If n identical single control moment gimbals are used, the total torque is given by

$$\mathbf{t}_c = [\mathbf{c}_1, \ldots, \mathbf{c}_n][\omega_{g_1}, \ldots, \omega_{g_n}]^{\mathrm{T}} = \mathbf{C}\omega_g \tag{10.3}$$

where $\mathbf{c}_i$ is the ith CMG's torque vector. Using the control system design method described in Chapter 9, we can find the desired control torque $\mathbf{t}_c$. Then the gimbal rotational speed ω_g is given by

$$\omega_g = \mathbf{C}^{\mathrm{T}}(\mathbf{C}\mathbf{C}^{\mathrm{T}})^{-1}\mathbf{t}_c \tag{10.4}$$

A solution does not exist when $\det(\mathbf{C}\mathbf{C}^{\mathrm{T}}) = 0$. This is the so-called *gimbal singularity*.

It is worthwhile to note that although the gimbal vector $\hat{\mathbf{g}}$ is a constant in body frame, the angular momentum vector $\mathbf{h}$ and therefore the normalized CMG torque vector $\mathbf{c}$ depend on the gimbal angle θ. Several methods are proposed to deal with the gimbal singularity problem, for example, [139, 143, 52, 223, 221]. An experimental comparison for these methods is given in [83]. Chapter 14 will discuss a novel method of CMG control.

10.3 Magnetic Torque Rods

Magnetic torque rods have been used in most low orbit earth satellites. Magnetic torque rods are generally planar coils of uniform wire rigidly placed along the spacecraft body axes. When electricity passes through the coils, a *magnetic dipole* is created. The strength of the dipole depends on several factors, such as amount of electricity and total area enclosed by the coils, etc. This dipole interacts with Earth's magnetic field, causing the *coils* to attempt to align their own magnetic field in the direction opposite to that of Earth's.

The advantages of magnetic torque rods are that they are lightweight, reliable, and energy-efficient. The electricity comes from battery which is charged by *solar panels*. Unlike reaction wheel and momentum wheel, magnetic torque rods do not have moving parts; therefore, they are much more reliable.

The disadvantages are that the magnetic torques generated by the magnetic tongue rods depends not only on the electricity applied, but also on the spacecraft location or the orbit, i.e., depends on Earth's magnetic field strength and direction. It is also impossible to control attitude in all three axes at any time even if the full three coils are used because the torque can be generated only perpendicular to the Earth's magnetic field vector.

Let $\mathbf{m}$ be the magnetic moment created by the magnetic torque rods, $\mathbf{r}_m$ be the *Earth's magnetic field intensity*, the mechanical torque $\mathbf{t}_m$ applied to the spacecraft, due to the interaction between $\mathbf{m}$ and $\mathbf{r}_m$ is given by

$$\mathbf{t}_m = \mathbf{m} \times \mathbf{r}_m \tag{10.5}$$

which should be equal to the required torque $\mathbf{u}$ obtained by attitude controller design described in Chapter 9. But in the implementation of attitude control, given the desired $\mathbf{u}$ and the geomagnetic field $\mathbf{r}_m$ which is given by (5.14) (we need to represent the field $\mathbf{r}_m$ in the body frame), one can only select $\mathbf{m}$ such that $\|\mathbf{u} - \mathbf{m} \times \mathbf{r}_m\|$ is minimized. Since $\mathbf{t}_m$ can be generated only perpendicular to the Earth's magnetic field vector, it is very likely that $\mathbf{u} \neq \mathbf{t}_m$. The best we can do is to find a minimum norm solution to the least squared problem. Denote $\hat{\mathbf{r}}_m$ the normalized vector of $\mathbf{r}_m$, since $\mathbf{t}_m = \mathbf{m} \times \mathbf{r}_m$, we have

$$\begin{aligned}
\mathbf{r}_m \times \mathbf{t}_m &= \mathbf{r}_m \times (\mathbf{m} \times \mathbf{r}_m) = \mathbf{r}_m^{\mathrm{T}} \mathbf{r}_m \mathbf{m} - \mathbf{r}_m \mathbf{r}_m^{\mathrm{T}} \mathbf{m} \\
&= \mathbf{r}_m^{\mathrm{T}} \mathbf{r}_m \mathbf{m} - \mathbf{r}_m^{\mathrm{T}} \mathbf{r}_m \frac{\mathbf{r}_m}{\|\mathbf{r}_m\|} \frac{\mathbf{r}_m^{\mathrm{T}}}{\|\mathbf{r}_m\|} \mathbf{m} = \mathbf{r}_m^{\mathrm{T}} \mathbf{r}_m (\mathbf{I} - \hat{\mathbf{r}}_m \hat{\mathbf{r}}_m^{\mathrm{T}}) \mathbf{m}
\end{aligned}$$

This gives

$$\frac{\mathbf{r}_m \times \mathbf{t}_m}{\mathbf{r}_m^{\mathrm{T}} \mathbf{r}_m} = (\mathbf{I} - \hat{\mathbf{r}}_m \hat{\mathbf{r}}_m^{\mathrm{T}}) \mathbf{m}$$

It is clear that

$$\mathbf{m} = \frac{\mathbf{r}_m \times \mathbf{t}_m}{\mathbf{r}_m^{\mathrm{T}} \mathbf{r}_m} \tag{10.6}$$

is the solution of the above equation because $\mathbf{r}_m \times \mathbf{t}_m$ is orthogonal to $\mathbf{r}_m$. Therefore, from the vector $\mathbf{m}$, the current applied to each magnetic torque rods can be obtained.

10.4 Thrusters

Thrusters are another type of actuators. They can be used for attitude control for any spacecraft. Fuels have to be loaded to thrusters and fuel budget is a major limitation on the use of thrusters. Thrusters use different propellants, such as cold gas propellant, solid chemical propellant, liquid chemical propellant, and electrical propellant. The same basic equation of propulsion holds for all kinds of propellants. The thrust force F is related to the exhaust velocity V_e relative to the satellite body, the fuel consumption rate $\frac{dm}{dt}$, the gas and ambient pressures P_e and P_a, and the area of the nozzle exit A_e. More specifically (see [181]),

$$F = V_e \frac{dm}{dt} + A_e(P_e - P_a) \tag{10.7}$$

Given the force and the thruster mounting information, the torques generated by thrusters can be obtained. This will be discussed later in Chapters 12 and 15.

Chapter 11

Spacecraft Control Using Magnetic Torques

In principle, the control system design methods presented in Chapter 9 can be implemented using any control actuators. But as seen in Chapter 10 that this may not be a good idea for magnetic torque control because given a desired control torque vector $\mathbf{u}$, one can only obtain an approximate solution $\mathbf{t}_m = \mathbf{m} \times \mathbf{r}_m$ given by $\mathbf{m}$ which minimizes the norm of $\|\mathbf{u} - \mathbf{t}_m\|$. For thrust control system, the torques generated by thruster(s) depend on the selection of thrusters and their configuration design. For control system using CMGs, given the desired torques, there are singular points where the desired torques are not achievable by any CMG gimbals' speeds. Therefore, to improve the control system design involving actuators other than reaction wheels only, we need to use models with more detailed information, such as *geomagnetic field* in magnetic torque control system design and thrusters' installation information in thrust control system design. This chapter focuses on the control system design involving magnetic torque bars/coils[1]. The materials of this chapter are mainly based on [252, 253, 254, 256].

Spacecraft attitude control using *magnetic torque* is a very attractive technique because the implementation is seamless, the system is reliable (without moving mechanical parts), the torque coils are inexpensive, and their weights are light. The main issue of using only magnetic torques to control the attitude is that the magnetic torques generated by magnetic coils are not available in all desired axes at any time [181]. However, because of the constant change of the Earth's

[1]Since the functions of magnetic torque bar and magnetic torque coils are the same, we use these names interchangeably.

magnetic field as a spacecraft circles around the Earth, the controllable subspace changes all the time. Many researchers believe that a spacecraft's attitude is actually controllable by using only magnetic torques. Numerous spacecraft attitude control designs were proposed in the last 25 years exploring the features of the time-varying systems [160, 265, 163, 159, 138, 156, 228, 113, 182, 114, 237, 30]. Some of these papers tried Euler angle model and Linear Quadratic Regulator (LQR) formulations [159, 138, 156, 228, 30] which are explicitly or implicitly assume that the *controllability* for the *linear time-varying* system holds so that the optimal solutions exist [87]. Therefore, the controllability conditions need to be established for the problem of spacecraft attitude control using only magnetic torque.

Other researchers [113, 182, 114] proposed direct design methods using *Lyapunov stability* theory. The existence of the solutions for these methods implicitly depends on the controllability for the nonlinear time-varying system. Therefore, Bhat [16] investigated controllability of the nonlinear time-varying systems. However, the condition for the controllability of the nonlinear time-varying systems obtained by Bhat is hard to be verified and is a sufficient condition.

A *reduced quaternion model* was discussed in previous chapters and its merits over *Euler angle model* were discussed (see also in [243, 245, 249]). The reduced quaternion model was also used for the design of spacecraft attitude control system using magnetic torque [160, 163, 265]. Because the controllability of the linear time-varying (LTV) systems was not established, the existence of the solutions was not guaranteed.

This chapter first considers the reduced linear quaternion model proposed in [243] for the case that magnetic torques are the only control torques. We establish the conditions of the controllability for this linear time-varying system. The same strategy can easily be used to prove the controllability of the Euler angle based linear time-varying system considered in [159]. However, we will not derive the similar result because of the merits of the reduced quaternion model as discussed in [243, 245, 249]. In Section 11.3, the LQR design is discussed for the *linear periodic* system. Instead of directly applying a well-known algorithm, the author has proposed a different algorithm that makes full use of the feature that only input matrix of the system is a periodic matrix. Then, a combined method is suggested in Section 11.4 to design the attitude control and the *momentum management* system at the same time, which were normally considered as two different problems in separate designs. In the last section of this Chapter, a different LQR design for the linear periodic system is discussed. This design uses a novel lifting method to convert the linear periodic system into an augmented linear time-invariant system and then proposed a new method to solve the Riccati equation. Numerical simulation is performed to show the efficiency of the new method.

11.1 The Linear Time-varying Model

The focus of the discussion in this section is on the nadir pointing spacecraft using a reduced quaternion model.[2] Therefore, the attitude of the spacecraft is represented by the rotation of the spacecraft *body fixed frame* relative to the *local vertical and local horizontal* (LVLH) frame. Let $\omega = [\omega_1, \omega_2, \omega_3]^{\mathrm{T}}$ be the body rate with respect to the LVLH frame represented in the body frame, ω_0 be the orbit (and LVLH frame) rate with respect to the inertial frame, represented in the LVLH frame. Let $\bar{\mathbf{q}} = [q_0, q_1, q_2, q_3]^{\mathrm{T}} = [q_0, \mathbf{q}^{\mathrm{T}}]^{\mathrm{T}} = [\cos(\frac{\alpha}{2}), \hat{\mathbf{e}}^{\mathrm{T}} \sin(\frac{\alpha}{2})]^{\mathrm{T}}$ be the quaternion representing the rotation of the body frame relative to the LVLH frame, where $\hat{\mathbf{e}}$ is the unit length rotational axis and α is the rotational angle about $\hat{\mathbf{e}}$. Therefore, the reduced quaternion-based kinematics equation can be expressed as (4.9).

Assume that the inertia matrix of the spacecraft is diagonal which is approximately correct for real systems, let the control torque vector be $\mathbf{u} = [u_x, u_y, u_z]^{\mathrm{T}}$, then the linearized nadir pointing spacecraft model with gravity gradient disturbance torque is a special case of (4.36) and is given as follows:

$$\begin{bmatrix} \dot{q}_1 \\ \dot{q}_2 \\ \dot{q}_3 \\ \dot{\omega}_1 \\ \dot{\omega}_2 \\ \dot{\omega}_3 \end{bmatrix} = \begin{bmatrix} 0 & 0 & 0 & .5 & 0 & 0 \\ 0 & 0 & 0 & 0 & .5 & 0 \\ 0 & 0 & 0 & 0 & 0 & .5 \\ f_{41} & 0 & 0 & 0 & 0 & f_{46} \\ 0 & f_{52} & 0 & 0 & 0 & 0 \\ 0 & 0 & f_{63} & f_{64} & 0 & 0 \end{bmatrix} \begin{bmatrix} q_1 \\ q_2 \\ q_3 \\ \omega_1 \\ \omega_2 \\ \omega_3 \end{bmatrix} + \begin{bmatrix} 0 \\ 0 \\ 0 \\ u_x/J_{11} \\ u_y/J_{22} \\ u_z/J_{33} \end{bmatrix} \tag{11.1}$$

where

$$f_{41} = [8(J_{33} - J_{22})\omega_0^2]/J_{11} \tag{11.2a}$$

$$f_{46} = (J_{11} - J_{22} + J_{33})\omega_0/J_{11} \tag{11.2b}$$

$$f_{64} = (-J_{11} + J_{22} - J_{33})\omega_0/J_{33} \tag{11.2c}$$

$$f_{52} = [6(J_{33} - J_{11})\omega_0^2]/J_{22} \tag{11.2d}$$

$$f_{63} = [2(J_{11} - J_{22})\omega_0^2]/J_{33} \tag{11.2e}$$

The control torques generated by magnetic coils interacting with the Earth's magnetic field is given by (see [181])

$$\mathbf{u} = \mathbf{m} \times \mathbf{b}$$

where the vector of the Earth's magnetic field represented in spacecraft coordinates, $\mathbf{b}(t) = [b_1(t), b_2(t), b_3(t)]^{\mathrm{T}}$, is computed using the spacecraft position, the spacecraft attitude, and a spherical harmonic model of the Earth's magnetic field as discussed in Section 5.3 (see also [219]); and $\mathbf{m} = [m_1, m_2, m_3]^{\mathrm{T}}$ is the

[2] The same idea can be used to derive the controllability condition for inertial pointing spacecraft and/or using Euler angle model.

spacecraft magnetic coils' induced magnetic moment in the spacecraft body coordinates.

The time-variation of the system is an approximate periodic function of $\mathbf{b}(t) = \mathbf{b}(t+T)$ where $T = \frac{2\pi}{\omega_0}$ is the orbital period (see (2.55)). This magnetic field $\mathbf{b}(t)$ can be approximately expressed as follows [159]:

$$\begin{bmatrix} b_1(t) \\ b_2(t) \\ b_3(t) \end{bmatrix} = \frac{\mu_f}{a^3} \begin{bmatrix} \cos(\omega_0 t)\sin(i_m) \\ -\cos(i_m) \\ 2\sin(\omega_0 t)\sin(i_m) \end{bmatrix} \tag{11.3}$$

where i_m is the inclination of the spacecraft orbit with respect to the magnetic equator, $\mu_f = 7.9 \times 10^{15}$ Wb-m is the field's dipole strength, and a is the orbit's semi-major axis. The time $t = 0$ is measured at the ascending node crossing of the magnetic equator. Therefore, the *reduced quaternion* linear time-varying system is given as follows:

$$\begin{aligned} \begin{bmatrix} \dot{q}_1 \\ \dot{q}_2 \\ \dot{q}_3 \\ \dot{\omega}_1 \\ \dot{\omega}_2 \\ \dot{\omega}_3 \end{bmatrix} &= \begin{bmatrix} 0 & 0 & 0 & .5 & 0 & 0 \\ 0 & 0 & 0 & 0 & .5 & 0 \\ 0 & 0 & 0 & 0 & 0 & .5 \\ f_{41} & 0 & 0 & 0 & 0 & f_{46} \\ 0 & f_{52} & 0 & 0 & 0 & 0 \\ 0 & 0 & f_{63} & f_{64} & 0 & 0 \end{bmatrix} \begin{bmatrix} q_1 \\ q_2 \\ q_3 \\ \omega_1 \\ \omega_2 \\ \omega_3 \end{bmatrix} \\ &+ \begin{bmatrix} 0 & 0 & 0 \\ 0 & 0 & 0 \\ 0 & 0 & 0 \\ 0 & \frac{b_3(t)}{J_{11}} & -\frac{b_2(t)}{J_{11}} \\ -\frac{b_3(t)}{J_{22}} & 0 & \frac{b_1(t)}{J_{22}} \\ \frac{b_2(t)}{J_{33}} & -\frac{b_1(t)}{J_{33}} & 0 \end{bmatrix} \begin{bmatrix} m_1 \\ m_2 \\ m_3 \end{bmatrix} \\ &:= \begin{bmatrix} \mathbf{0}_3 & \frac{1}{2}\mathbf{I}_3 \\ \Lambda_1 & \Sigma_1 \end{bmatrix} \begin{bmatrix} \mathbf{q} \\ \omega \end{bmatrix} + \begin{bmatrix} \mathbf{0}_3 \\ \mathbf{B}_2(t) \end{bmatrix} \mathbf{m} \\ &= \mathbf{A}\mathbf{x} + \mathbf{B}(t)\mathbf{m} \end{aligned} \tag{11.4}$$

Substituting (11.3) into (11.4) yields

$$\mathbf{B}_2(t) = \begin{bmatrix} 0 & b_{42}(t) & b_{43}(t) \\ b_{51}(t) & 0 & b_{53}(t) \\ b_{61}(t) & b_{62}(t) & 0 \end{bmatrix} \tag{11.5}$$

where

$$b_{42}(t) = \frac{2\mu_f}{a^3 J_{11}} \sin(i_m)\sin(\omega_0 t) \tag{11.6a}$$

$$b_{43}(t) = \frac{\mu_f}{a^3 J_{11}} \cos(i_m) \tag{11.6b}$$

$$b_{53}(t) = \frac{\mu_f}{a^3 J_{22}} \sin(i_m)\cos(\omega_0 t) \tag{11.6c}$$

$$b_{51}(t) = -\frac{2\mu_f}{a^3 J_{22}} \sin(i_m)\sin(\omega_0 t) = -b_{42}\frac{J_{11}}{J_{22}} \tag{11.6d}$$

$$b_{61}(t) = -\frac{\mu_f}{a^3 J_{33}} \cos(i_m) = -b_{43}\frac{J_{11}}{J_{33}} \tag{11.6e}$$

$$b_{62}(t) = -\frac{\mu_f}{a^3 J_{33}} \sin(i_m)\cos(\omega_0 t) = -b_{53}\frac{J_{22}}{J_{33}} \tag{11.6f}$$

Therefore, taking the first order and second order derivatives, we have

$$b'_{42}(t) = \frac{2\mu_f \omega_0}{a^3 J_{11}} \sin(i_m)\cos(\omega_0 t) \tag{11.7a}$$

$$b'_{43}(t) = 0 \tag{11.7b}$$

$$b'_{53}(t) = -\frac{\mu_f \omega_0}{a^3 J_{22}} \sin(i_m)\sin(\omega_0 t) \tag{11.7c}$$

$$b'_{51}(t) = -\frac{2\mu_f \omega_0}{a^3 J_{22}} \sin(i_m)\cos(\omega_0 t) = -b'_{42}\frac{J_{11}}{J_{22}} \tag{11.7d}$$

$$b'_{61}(t) = 0 \tag{11.7e}$$

$$b'_{62}(t) = \frac{\mu_f \omega_0}{a^3 J_{33}} \sin(i_m)\sin(\omega_0 t) = -b_{53}\frac{J_{22}}{J_{33}} \tag{11.7f}$$

and

$$b''_{42}(t) = -\frac{2\mu_f \omega_0^2}{a^3 J_{11}} \sin(i_m)\sin(\omega_0 t) \tag{11.8a}$$

$$b''_{43}(t) = 0 \tag{11.8b}$$

$$b''_{53}(t) = -\frac{\mu_f \omega_0^2}{a^3 J_{22}} \sin(i_m)\cos(\omega_0 t) \tag{11.8c}$$

$$b''_{51}(t) = \frac{2\mu_f \omega_0^2}{a^3 J_{22}} \sin(i_m)\sin(\omega_0 t) = -b''_{42}\frac{J_{11}}{J_{22}} \tag{11.8d}$$

$$b''_{61}(t) = 0 \tag{11.8e}$$

$$b''_{62}(t) = \frac{\mu_f \omega_0^2}{a^3 J_{33}} \sin(i_m)\cos(\omega_0 t) = -b''_{53}\frac{J_{22}}{J_{33}} \tag{11.8f}$$

In matrix format, we have

$$\mathbf{B}'_2(t) = \begin{bmatrix} 0 & b'_{42} & 0 \\ b'_{51} & 0 & b'_{53} \\ 0 & b'_{62} & 0 \end{bmatrix} \tag{11.9}$$

and

$$\mathbf{B}''_2(t) = \begin{bmatrix} 0 & b''_{42} & 0 \\ b''_{51} & 0 & b''_{53} \\ 0 & b''_{62} & 0 \end{bmatrix} \tag{11.10}$$

A special case is when $i_m = 0$, i.e., the spacecraft orbit is on the equator plane of the Earth's magnetic field. In this case, $\mathbf{b}(t) = [0, -\frac{\mu_f}{a^3}, 0]^{\mathrm{T}}$ is a constant vector. The linear time-varying system of this special case is reduced to a linear time-invariant system whose model is given by

$$
\begin{aligned}
\begin{bmatrix} \dot{q}_1 \\ \dot{q}_2 \\ \dot{q}_3 \\ \dot{\omega}_1 \\ \dot{\omega}_2 \\ \dot{\omega}_3 \end{bmatrix}
&= \begin{bmatrix} 0 & 0 & 0 & .5 & 0 & 0 \\ 0 & 0 & 0 & 0 & .5 & 0 \\ 0 & 0 & 0 & 0 & 0 & .5 \\ f_{41} & 0 & 0 & 0 & 0 & f_{46} \\ 0 & f_{52} & 0 & 0 & 0 & 0 \\ 0 & 0 & f_{63} & f_{64} & 0 & 0 \end{bmatrix}
\begin{bmatrix} q_1 \\ q_2 \\ q_3 \\ \omega_1 \\ \omega_2 \\ \omega_3 \end{bmatrix} \\
&+ \begin{bmatrix} 0 & 0 & 0 \\ 0 & 0 & 0 \\ 0 & 0 & 0 \\ 0 & 0 & -b_2/J_{11} \\ 0 & 0 & 0 \\ b_2/J_{33} & 0 & 0 \end{bmatrix}
\begin{bmatrix} m_1 \\ m_2 \\ m_3 \end{bmatrix} \\
&= \mathbf{A}\mathbf{x} + \mathbf{B}\mathbf{m}
\end{aligned}
\tag{11.11}
$$

11.2 Spacecraft Controllability Using Magnetic Torques

The definition of controllability of linear time-varying systems can be found in [169, page 124].

Definition 11.1 The linear state equation (11.4) is called controllable on $[t_0, t_f]$ if given any $\mathbf{x}_0$, there exists a continuous input signal $\mathbf{m}(t)$ defined on $[t_0, t_f]$ such that the corresponding solution of (11.4) satisfies $\mathbf{x}(t_f) = 0$.

A main theorem used to prove the controllability of (11.4) is also given in [169, page 127].

Theorem 11.1
Let the state transition matrix $\Phi(t, \tau) = e^{\mathbf{A}(t-\tau)}$. Denote

$$
\mathbf{K}_j(t) = \frac{\partial^j}{\partial \tau^j} [\Phi(t, \tau)\mathbf{B}(\tau)]\bigg|_{\tau=t}, \quad j = 1, 2, \ldots \tag{11.12}
$$

if p is a positive integer such that, for $t \in [t_0, t_f]$, $\mathbf{B}(t)$ is p time continuously differentiable. Then, the linear time-varying equation (11.4) is controllable on $[t_0, t_f]$ if for some $t_c \in [t_0, t_f]$

$$
rank\,[\mathbf{K}_0(t_c), \mathbf{K}_1(t_c), \ldots, \mathbf{K}_p(t_c)] = n. \tag{11.13}
$$

Remark 11.1 If $\mathbf{A}$ and $\mathbf{B}$ are constant matrices, the rank condition of (11.13) for the linear time-varying system is reduced to the rank condition for the linear time-invariant system [169, page 128], i.e., if

$$\text{rank}\left[\mathbf{B}, \mathbf{AB}, \ldots, \mathbf{A}^{n-1}\mathbf{B}\right] = n. \tag{11.14}$$

then the linear time-invariant system $(\mathbf{A}, \mathbf{B})$ is controllable. ■

First, we consider the special case of (11.11), the time-invariant system when the spacecraft orbit is on the equator plane of the Earth's magnetic field ($i_m = 0$). Let Σ denote any 3×3 anti-diagonal and Π be any diagonal matrix with the second row composed of zeros

$$\Sigma := \left\{ \begin{bmatrix} 0 & 0 & \times \\ 0 & 0 & 0 \\ \times & 0 & 0 \end{bmatrix} \right\} \quad \text{and} \quad \Pi := \left\{ \begin{bmatrix} \times & 0 & 0 \\ 0 & 0 & 0 \\ 0 & 0 & \times \end{bmatrix} \right\}$$

and Λ denote any 3×3 diagonal matrix with the form

$$\Lambda := \left\{ \begin{bmatrix} \times & 0 & 0 \\ 0 & \times & 0 \\ 0 & 0 & \times \end{bmatrix} \right\}$$

It is easy to verify that if $\Sigma_i \in \Sigma$, $\Sigma_j \in \Sigma$, and $\Lambda_k \in \Lambda$, then $\Sigma_i\Sigma_j \in \Sigma$, $\Sigma_i + \Sigma_j \in \Sigma$, and $\Lambda_k\Sigma_i \in \Sigma$. A similar claim is true for Π. Using this fact to expand the matrix $[\mathbf{B}, \mathbf{AB}, \mathbf{A}^2\mathbf{B}, \mathbf{A}^3\mathbf{B}, \mathbf{A}^4\mathbf{B}, \mathbf{A}^5\mathbf{B}]$, where $\mathbf{A}$ and $\mathbf{B}$ are defined in (11.11), shows that the second row of the controllability matrix in (11.14) is composed of all zeros. This proves that if the spacecraft orbit is on the equator plane of the Earth's magnetic field, the spacecraft attitude cannot be stabilized by using only magnetic torques.

Now we show that under some simple conditions, the linear time-varying system (11.4) is controllable for any orbit which is not on the equator plane of the Earth's magnetic field, i.e., $i_m \neq 0$. From (11.12), we have

$$\mathbf{K}_0(t) = \Phi(t,t)\mathbf{B}(t) = e^{\mathbf{A}(t-t)}\mathbf{B}(t) = \mathbf{B}(t)$$

$$\begin{aligned} \mathbf{K}_1(t) &= \frac{\partial}{\partial \tau}\left[\Phi(t,\tau)\mathbf{B}(\tau)\right]\Big|_{\tau=t} = \frac{\partial}{\partial \tau}\left[e^{\mathbf{A}(t-\tau)}\mathbf{B}(\tau)\right]\Big|_{\tau=t} \\ &= \left[-\mathbf{A}e^{\mathbf{A}(t-\tau)}\mathbf{B}(\tau) + e^{\mathbf{A}(t-\tau)}\mathbf{B}'(\tau)\right]\Big|_{\tau=t} \\ &= -\mathbf{AB}(t) + \mathbf{B}'(t) \end{aligned} \tag{11.15}$$

$$\mathbf{K}_2(t) = \frac{\partial^2}{\partial \tau^2}\left[\Phi(t,\tau)\mathbf{B}(\tau)\right]\Big|_{\tau=t}$$

$$
\begin{aligned}
&= \left.\left[\mathbf{A}^2 e^{\mathbf{A}(t-\tau)}\mathbf{B}(\tau) - 2\mathbf{A}e^{\mathbf{A}(t-\tau)}\mathbf{B}'(\tau) + e^{\mathbf{A}(t-\tau)}\mathbf{B}''(\tau)\right]\right|_{\tau=t} \\
&= \mathbf{A}^2\mathbf{B}(t) - 2\mathbf{A}\mathbf{B}'(t) + \mathbf{B}''(t) \qquad (11.16)
\end{aligned}
$$

Using the notation of (11.4), we can rewrite equation (11.15) as

$$
\mathbf{K}_1(t) = -\begin{bmatrix} \mathbf{0}_3 & \frac{1}{2}\mathbf{I}_3 \\ \Lambda_1 & \Sigma_1 \end{bmatrix}\begin{bmatrix} \mathbf{0}_3 \\ \mathbf{B}_2 \end{bmatrix} + \begin{bmatrix} \mathbf{0}_3 \\ \mathbf{B}_2' \end{bmatrix} = \begin{bmatrix} -\frac{1}{2}\mathbf{B}_2 \\ -\Sigma_1\mathbf{B}_2 + \mathbf{B}_2' \end{bmatrix}
$$

Since

$$
\mathbf{A}^2\mathbf{B} = \mathbf{A}\begin{bmatrix} \mathbf{0}_3 & \frac{1}{2}\mathbf{I}_3 \\ \Lambda_1 & \Sigma_1 \end{bmatrix}\begin{bmatrix} \mathbf{0}_3 \\ \mathbf{B}_2 \end{bmatrix} = \begin{bmatrix} \mathbf{0}_3 & \frac{1}{2}\mathbf{I}_3 \\ \Lambda_1 & \Sigma_1 \end{bmatrix}\begin{bmatrix} \frac{1}{2}\mathbf{B}_2 \\ \Sigma_1\mathbf{B}_2 \end{bmatrix} = \begin{bmatrix} \frac{1}{2}\Sigma_1\mathbf{B}_2 \\ \frac{1}{2}\Lambda_1\mathbf{B}_2 + \Sigma_1^2\mathbf{B}_2 \end{bmatrix}
$$

and

$$
-2\mathbf{A}\mathbf{B}' = -2\begin{bmatrix} \mathbf{0}_3 & \frac{1}{2}\mathbf{I}_3 \\ \Lambda_1 & \Sigma_1 \end{bmatrix}\begin{bmatrix} \mathbf{0}_3 \\ \mathbf{B}_2' \end{bmatrix} = \begin{bmatrix} -\mathbf{B}_2' \\ -2\Sigma_1\mathbf{B}_2' \end{bmatrix}
$$

Equation (11.16) is reduced to

$$
\mathbf{K}_2(t) = \mathbf{A}^2\mathbf{B} - 2\mathbf{A}\mathbf{B}' + \mathbf{B}'' = \begin{bmatrix} \frac{1}{2}\Sigma_1\mathbf{B}_2 - \mathbf{B}_2' \\ \frac{1}{2}\Lambda_1\mathbf{B}_2 + \Sigma_1^2\mathbf{B}_2 - 2\Sigma_1\mathbf{B}_2' + \mathbf{B}_2'' \end{bmatrix}
$$

Hence,

$$
\begin{aligned}
&[\mathbf{K}_0(t), \mathbf{K}_1(t), \mathbf{K}_2(t)] \\
&= [\mathbf{B}(t) \mid -\mathbf{A}\mathbf{B}(t) + \mathbf{B}'(t) \mid \mathbf{A}^2\mathbf{B}(t) - 2\mathbf{A}\mathbf{B}'(t) + \mathbf{B}''(t)] \\
&= \left[\begin{array}{c|c|c} \mathbf{0}_3 & -\frac{1}{2}\mathbf{B}_2 & \frac{1}{2}\Sigma_1\mathbf{B}_2 - \mathbf{B}_2'(t) \\ \mathbf{B}_2 & -\Sigma_1\mathbf{B}_2 + \mathbf{B}_2' & \frac{1}{2}\Lambda_1\mathbf{B}_2 + \Sigma_1^2\mathbf{B}_2 - 2\Sigma_1\mathbf{B}_2' + \mathbf{B}_2'' \end{array}\right] \qquad (11.17)
\end{aligned}
$$

Notice that

$$
\begin{aligned}
&\text{rank}[\mathbf{K}_0(t), \mathbf{K}_1(t), \mathbf{K}_2(t)] \\
&= \text{rank}\left(\begin{bmatrix} \mathbf{I}_3 & \mathbf{0}_3 \\ -2\Sigma_1 & \mathbf{I}_3 \end{bmatrix}\left[\begin{array}{c|c|c} \mathbf{0}_3 & -\frac{1}{2}\mathbf{B}_2 & \frac{1}{2}\Sigma_1\mathbf{B}_2 - \mathbf{B}_2'(t) \\ \mathbf{B}_2 & -\Sigma_1\mathbf{B}_2 + \mathbf{B}_2' & \frac{1}{2}\Lambda_1\mathbf{B}_2 + \Sigma_1^2\mathbf{B}_2 - 2\Sigma_1\mathbf{B}_2' + \mathbf{B}_2'' \end{array}\right]\right) \\
&= \text{rank}\left[\begin{array}{c|c|c} \mathbf{0}_3 & -\frac{1}{2}\mathbf{B}_2 & \frac{1}{2}\Sigma_1\mathbf{B}_2 - \mathbf{B}_2'(t) \\ \mathbf{B}_2 & \mathbf{B}_2' & \frac{1}{2}\Lambda_1\mathbf{B}_2 + \mathbf{B}_2'' \end{array}\right] \\
&= \text{rank}\left[\begin{array}{c|c|c} \mathbf{0}_3 & -\mathbf{B}_2 & \Sigma_1\mathbf{B}_2 - 2\mathbf{B}_2'(t) \\ \mathbf{B}_2 & \mathbf{B}_2' & \frac{1}{2}\Lambda_1\mathbf{B}_2 + \mathbf{B}_2'' \end{array}\right] \qquad (11.18)
\end{aligned}
$$

$$
\begin{aligned}
&\Sigma_1\mathbf{B}_2 - 2\mathbf{B}_2'(t) \\
&= \begin{bmatrix} 0 & 0 & f_{46} \\ 0 & 0 & 0 \\ f_{64} & 0 & 0 \end{bmatrix}\begin{bmatrix} 0 & b_{42}(t) & b_{43}(t) \\ b_{51}(t) & 0 & b_{53}(t) \\ b_{61}(t) & b_{62}(t) & 0 \end{bmatrix} - 2\begin{bmatrix} 0 & b_{42}' & 0 \\ b_{51}' & 0 & b_{53}' \\ 0 & b_{62}' & 0 \end{bmatrix} \\
&= \begin{bmatrix} f_{46}b_{61}(t) & f_{46}b_{62}(t) - 2b_{42}' & 0 \\ -2b_{51}' & 0 & 2b_{53}' \\ 0 & f_{64}b_{42}(t) - 2b_{62}' & f_{64}b_{43}(t) \end{bmatrix} \qquad (11.19)
\end{aligned}
$$

and

$$\frac{1}{2}\Lambda_1\mathbf{B}_2+\mathbf{B}_2''(t)$$
$$=\frac{1}{2}\begin{bmatrix} f_{41} & 0 & 0 \\ 0 & f_{52} & 0 \\ 0 & 0 & f_{63}\end{bmatrix}\begin{bmatrix} 0 & b_{42}(t) & b_{43}(t) \\ b_{51}(t) & 0 & b_{53}(t) \\ b_{61}(t) & b_{62}(t) & 0\end{bmatrix}+\begin{bmatrix} 0 & b_{42}'' & 0 \\ b_{51}'' & 0 & b_{53}'' \\ 0 & b_{62}'' & 0\end{bmatrix}$$
$$=\begin{bmatrix} 0 & \frac{1}{2}f_{41}b_{42}(t)+b_{42}'' & \frac{1}{2}f_{41}b_{43}(t) \\ \frac{1}{2}f_{52}b_{51}(t)+b_{51}'' & 0 & \frac{1}{2}f_{52}b_{53}(t)+b_{53}'' \\ \frac{1}{2}f_{63}b_{61}(t) & \frac{1}{2}f_{63}b_{62}(t)+b_{62}'' & 0\end{bmatrix} \quad (11.20)$$

we have

$$\left[\begin{array}{c|c|c} \mathbf{0}_3 & -\mathbf{B}_2 & \Sigma_1\mathbf{B}_2-2\mathbf{B}_2'(t) \\ \mathbf{B}_2 & \mathbf{B}_2' & \frac{1}{2}\Lambda_1\mathbf{B}_2+\mathbf{B}_2'' \end{array}\right]$$
$$=\begin{bmatrix} 0 & 0 & 0 & 0 & a_{15} & a_{16} & a_{17} & a_{18} & 0 \\ 0 & 0 & 0 & a_{24} & 0 & a_{26} & a_{27} & 0 & a_{29} \\ 0 & 0 & 0 & a_{34} & a_{35} & 0 & 0 & a_{38} & a_{39} \\ 0 & a_{42} & a_{43} & 0 & a_{45} & 0 & 0 & a_{48} & a_{49} \\ a_{51} & 0 & a_{53} & a_{54} & 0 & a_{56} & a_{57} & 0 & a_{59} \\ a_{61} & a_{62} & 0 & 0 & a_{65} & 0 & a_{67} & a_{68} & 0 \end{bmatrix}$$

where

$$a_{15}=-b_{42}(t),\ a_{16}=-b_{43}(t),\ a_{17}=f_{46}b_{61}(t),\ a_{18}=f_{46}b_{62}(t)-2b_{42}',$$
$$a_{24}=-b_{51}(t),\ a_{26}=-b_{53}(t),\ a_{27}=-2b_{51}',\ a_{29}=2b_{53}',$$
$$a_{34}=-b_{61}(t),\ a_{35}=-b_{62}(t),\ a_{38}=f_{64}b_{42}(t)-2b_{62}',\ a_{39}=f_{64}b_{43}(t),$$
$$a_{42}=b_{42}(t),\ a_{43}=b_{43}(t),\ a_{45}=b_{42}',\ a_{48}=\frac{1}{2}f_{41}b_{42}(t)+b_{42}'',\ a_{49}=\frac{1}{2}f_{41}b_{43}(t),$$
$$a_{51}=b_{51}(t),\ a_{53}=b_{53}(t),\ a_{54}=b_{51}',\ a_{56}=b_{53}'(t),$$
$$a_{57}=\frac{1}{2}f_{52}b_{51}(t)+b_{51}'',\ a_{59}=\frac{1}{2}f_{52}b_{53}(t)+b_{53}'',$$
$$a_{61}=b_{61}(t),\ a_{62}=b_{62}(t),\ a_{65}=b_{62}',\ a_{67}=\frac{1}{2}f_{63}b_{61}(t),\ a_{68}=\frac{1}{2}f_{63}b_{62}(t)+b_{62}''$$

To show that this matrix is full rank for some t_c, we show that there is a 6×6 sub-matrix whose determinant is not zero for $\omega_0 t_c=\frac{\pi}{2}$. In view of (11.6), (11.7), and (11.8), for this t_c, we have

$$b_{53}(t_c)=b_{62}(t_c)=b_{51}'(t_c)=b_{42}'(t_c)=b_{53}''(t_c)=b_{62}''(t_c)=0 \quad (11.21)$$

Considering the sub-matrix composed of the 1st, 2nd, 4th, 5th, 7th, 8th columns, and using (11.21), we have

$$
\begin{aligned}
&\det \begin{bmatrix} 0 & 0 & 0 & a_{15} & a_{17} & a_{18} \\ 0 & 0 & a_{24} & 0 & a_{27} & 0 \\ 0 & 0 & a_{34} & a_{35} & 0 & a_{38} \\ 0 & a_{42} & 0 & a_{45} & 0 & a_{48} \\ a_{51} & 0 & a_{54} & 0 & a_{57} & 0 \\ a_{61} & a_{62} & 0 & a_{65} & a_{67} & a_{68} \end{bmatrix} \\
= &\det \begin{bmatrix} 0 & 0 & 0 & a_{15} & a_{17} & 0 \\ 0 & 0 & a_{24} & 0 & 0 & 0 \\ 0 & 0 & a_{34} & 0 & 0 & a_{38} \\ 0 & a_{42} & 0 & 0 & 0 & a_{48} \\ a_{51} & 0 & 0 & 0 & a_{57} & 0 \\ a_{61} & 0 & 0 & a_{65} & a_{67} & 0 \end{bmatrix} \\
= &-a_{24} \det \begin{bmatrix} 0 & 0 & a_{15} & a_{17} & 0 \\ 0 & 0 & 0 & 0 & a_{38} \\ 0 & a_{42} & 0 & 0 & a_{48} \\ a_{51} & 0 & 0 & a_{57} & 0 \\ a_{61} & 0 & a_{65} & a_{67} & 0 \end{bmatrix} \\
= &\, a_{38}a_{24} \det \begin{bmatrix} 0 & 0 & a_{15} & a_{17} \\ 0 & a_{42} & 0 & 0 \\ a_{51} & 0 & 0 & a_{57} \\ a_{61} & 0 & a_{65} & a_{67} \end{bmatrix} \\
= &\, a_{42}a_{38}a_{24} \det \begin{bmatrix} 0 & a_{15} & a_{17} \\ a_{51} & 0 & a_{57} \\ a_{61} & a_{65} & a_{67} \end{bmatrix} \\
= &\, a_{42}a_{38}a_{24}\left(a_{15}a_{57}a_{61} + a_{51}a_{65}a_{17} - a_{15}a_{51}a_{67}\right) \\
= &-b_{42}(t_c)\left(f_{64}b_{42}(t_c) - 2b'_{62}\right)b_{51}(t_c) \\
&\left[b_{51}b'_{62}f_{46}b_{61} - b_{42}\left(\frac{1}{2}f_{52}b_{51} + b''_{51}\right)b_{61} + \frac{1}{2}f_{63}b_{61}b_{42}b_{51}\right]
\end{aligned}
\tag{11.22}
$$

Therefore, in view of Theorem 11.1, the time-varying system is controllable if

$$
f_{64}b_{42}(t_c) - 2b'_{62} \neq 0 \tag{11.23}
$$

and

$$
b_{51}b'_{62}f_{46}b_{61} - b_{42}\left(\frac{1}{2}f_{52}b_{51} + b''_{51}\right)b_{61} + \frac{1}{2}f_{63}b_{61}b_{42}b_{51} \neq 0 \tag{11.24}
$$

Using (11.2), (11.6), (11.7), (11.8), and noticing that $\sin(\omega_0 t_c) = \sin(\frac{\pi}{2}) = 1$, we have

$$
\begin{aligned}
& f_{64}b_{42}(t_c) - 2b'_{62} \\
= & \frac{(-J_{11}+J_{22}-J_{33})\omega_0}{J_{33}} \frac{2\mu_f}{a^3J_{11}}\sin(i_m) - 2\frac{\mu_f\omega_0}{a^3J_{33}}\sin(i_m) \\
= & \frac{2\mu_f\omega_0\sin(i_m)}{a^3(J_{11}J_{33})}(-2J_{11}-J_{33}+J_{22})
\end{aligned}
$$

the first condition (11.23) is reduced to

$$2J_{11}+J_{33} \neq J_{22} \tag{11.25}$$

Repeatedly using the same relations, we have

$$
\begin{aligned}
& b_{51}b'_{62}f_{46}b_{61} \\
= & \left(-\frac{2\mu_f}{a^3J_{22}}\sin(i_m)\right)\left(\frac{\mu_f\omega_0}{a^3J_{33}}\sin(i_m)\right) \\
& \left(\frac{(J_{11}-J_{22}+J_{33})\omega_0}{J_{11}}\right)\left(-\frac{\mu_f}{a^3J_{33}}\cos(i_m)\right) \\
= & \frac{2\mu_f^3\omega_0^2(J_{11}-J_{22}+J_{33})}{a^9J_{11}J_{22}J_{33}^2}\sin^2(i_m)\cos(i_m)
\end{aligned} \tag{11.26}
$$

$$
\begin{aligned}
& -b_{42}\left(\frac{1}{2}f_{52}b_{51}+b''_{51}\right)b_{61} \\
= & -\left(\frac{2\mu_f}{a^3J_{11}}\sin(i_m)\right)\left(\frac{3(J_{33}-J_{11})\omega_0^2}{J_{22}}\left(-\frac{2\mu_f}{a^3J_{22}}\sin(i_m)\right)+\frac{2\mu_f\omega_0^2}{a^3J_{22}}\sin(i_m)\right) \\
& \left(-\frac{\mu_f}{a^3J_{33}}\cos(i_m)\right) \\
= & -\left(\frac{2\mu_f}{a^3J_{11}}\sin(i_m)\right)\left(\frac{2\mu_f\omega_0^2}{a^3J_{22}^2}\sin(i_m)(-3J_{33}+3J_{11}+J_{22})\right) \\
& \left(-\frac{\mu_f}{a^3J_{33}}\cos(i_m)\right) \\
= & \frac{4\mu_f^3\omega_0^2(-3J_{33}+3J_{11}+J_{22})}{a^9J_{11}J_{22}^2J_{33}}\sin^2(i_m)\cos(i_m)
\end{aligned} \tag{11.27}
$$

and

$$
\begin{aligned}
\frac{1}{2}f_{63}b_{61}b_{42}b_{51} = & \frac{(J_{11}-J_{22})\omega_0^2}{J_{33}}\left(-\frac{\mu_f}{a^3J_{33}}\cos(i_m)\right) \\
& \left(\frac{2\mu_f}{a^3J_{11}}\sin(i_m)\right)\left(-\frac{2\mu_f}{a^3J_{22}}\sin(i_m)\right)
\end{aligned}
$$

$$= \frac{4\mu_f^3\omega_0^2(J_{11}-J_{22})}{a^9 J_{11}J_{22}J_{33}^2}\sin^2(i_m)\cos(i_m) \tag{11.28}$$

Combining (11.26), (11.27), and (11.28), we can rewrite (11.24) as

$$\begin{aligned}
& b_{51}b'_{62}f_{46}b_{61} - b_{42}\left(\frac{1}{2}f_{52}b_{51}+b''_{51}\right)b_{61} + \frac{1}{2}f_{63}b_{61}b_{42}b_{51} \\
= & \frac{\mu_f^3\omega_0^2}{a^9 J_{11}J_{22}^2J_{33}^2}\sin^2(i_m)\cos(i_m) \\
& [2J_{22}(J_{11}-J_{22}+J_{33})+4J_{33}(-3J_{33}+3J_{11}+J_{22})+4J_{22}(J_{11}-J_{22})] \\
= & \frac{\mu_f^3\omega_0^2}{a^9 J_{11}J_{22}^2J_{33}^2}\sin^2(i_m)\cos(i_m) \\
& [2J_{11}J_{22}-2J_{22}^2+2J_{22}J_{33}-12J_{33}^2+12J_{11}J_{33}+4J_{22}J_{33}+4J_{11}J_{22}-4J_{22}^2] \\
= & \frac{\mu_f^3\omega_0^2}{a^9 J_{11}J_{22}^2J_{33}^2}\sin^2(i_m)\cos(i_m)[6J_{11}J_{22}-6J_{22}^2+6J_{22}J_{33}-12J_{33}^2+12J_{11}J_{33}] \\
= & \frac{6\mu_f^3\omega_0^2}{a^9 J_{11}J_{22}^2J_{33}^2}\sin^2(i_m)\cos(i_m)[J_{11}J_{22}-J_{22}^2+J_{22}J_{33}-2J_{33}^2+2J_{11}J_{33}]
\end{aligned} \tag{11.29}$$

Therefore, the second condition of (11.24) is reduced to

$$J_{22}(J_{11}-J_{22}+J_{33}) \neq 2J_{33}(J_{33}-J_{11}) \tag{11.30}$$

We summarize the above result as the main theorem of this section.

Theorem 11.2
For the linear time-varying spacecraft attitude control system (11.4) using only magnetic torques, if the orbit is on the equator plane of the Earth's magnetic field, then the spacecraft attitude is not fully controllable. If the orbit is not on the equator plane of the Earth's magnetic field, and the following two conditions hold:

$$2J_{11}+J_{33} \neq J_{22} \tag{11.31a}$$

$$J_{22}(J_{11}-J_{22}+J_{33}) \neq 2J_{33}(J_{33}-J_{11}) \tag{11.31b}$$

then the spacecraft attitude is fully controllable by magnetic coils.

Remark 11.2 The controllability conditions include only the spacecraft orbit plane and the spacecraft inertia matrix, which can easily be verified. ▮

The idea developed in this section is applied to attitude control of a 2U cubesat by magnetic and air drag torques [197].

11.3 LQR Design Based on Periodic Riccati Equation

This section discusses the attitude control system design using only magnetic torque. We consider linear quadratic regulator (LQR) design method for this problem. Riccati equation plays an important role in the LQR problem [108]. For continuous-time linear systems, the optimal solution of the LQR problem is associated with the differential Riccati equation. For discrete-time linear systems, the optimal solution of the LQR is associated with the algebraic Riccati equation. The numerical algorithms for these Riccati equations have been thoroughly studied since the work of Macfarlane [118], Kleinman [95], and Vaughan [208]. If the linear system is periodic, the optimal solution of the LQR is then associated with the *periodic* Riccati equation [19]. For continuous-time periodic linear system, algorithms and solutions of the differential periodic Riccati equation have been studied, for example, in [20, 21, 206]. For discrete-time periodic linear system, an efficient algorithm was proposed for the *algebraic periodic* Riccati equation in [70].

Because the spacecraft attitude control system using magnetic torques is a time-varying period system, using a periodic feedback control will improve the system performance [48]. However, many researches, for example [159, 113, 114], were still focused on time-invariant feedbacks. Others [237, 30] sought feedbacks that approximate the optimal solution even though the optimal feedback exists. Most optimal control designs for this problem [160, 265, 163, 138, 228] solved the continuous differential Riccati equation using some traditional backward integration, which is inefficient and needs large memory space. As a matter of fact, a more efficient algorithm [70] developed for general periodic time-varying optimal control system has been available since 1994, even though the algorithm in [70] is not designed to use the features of this specific problem.

In this section, we will explore the features of the problem of attitude control using only magnetic torques. By utilizing these features, we are able to propose an efficient algorithm to solve the *discrete-time periodic* Riccati equation. We show that the new algorithm is more efficient than the widely recognized algorithm developed in [70] for this problem.

Note that the orbital period in system (11.4) is given by (2.54) (see also [181])

$$T = \frac{2\pi}{\omega_0} = 2\pi\sqrt{\frac{a^3}{\mu}} \tag{11.32}$$

where a is the orbital radius (for circular orbit) and $\mu = 3.986005 * 10^{14} m^3/s^2$ is the standard gravitational parameter (see also [219]). Oftentimes, a spacecraft controller is implemented in a discrete computer system. Therefore, the following discrete model is used for the design in real implementation:

$$\mathbf{x}_{k+1} = \mathbf{A}_k \mathbf{x}_k + \mathbf{B}_k \mathbf{m}_k \tag{11.33}$$

The system matrices $(\mathbf{A}_k, \mathbf{B}_k)$ in the discrete model can be derived from (11.4) and (11.5) by different methods. Let t_s be the sample time, we use the following formulations:

$$\mathbf{A}_k = (\mathbf{I} + \mathbf{A}t_s), \quad \mathbf{B}_k = \mathbf{B}(kt_s)t_s \tag{11.34}$$

Note that

$$\det(\mathbf{I} + \mathbf{A}t_s) = \det \begin{bmatrix} \mathbf{I} & 0.5t_s\mathbf{I} \\ t_s\Lambda_1 & \mathbf{I} + t_s\Sigma_1 \end{bmatrix} = \det \begin{bmatrix} \mathbf{I} & 0.5t_s\mathbf{I} \\ \mathbf{0}_3 & \mathbf{I} + t_s\Sigma_1 - 0.5t_s^2\Lambda_1 \end{bmatrix}$$

is invertible as long as t_s is selected small enough. It is worthwhile to mention that in both continuous-time and discrete-time models, the time-varying feature is introduced by time-varying matrices $\mathbf{B}(t)$ or $\mathbf{B}_k$; the system matrices $\mathbf{A}$ and $\mathbf{A}_k$ are constants and invertible, which are important for us to derive an efficient computational algorithm.

The discussion about the computational algorithm is focused on the solution of the periodic discrete Riccati equation using the special properties of (11.33), i.e., $\mathbf{A}_k$ is constant and invertible for all k.

11.3.1 Preliminary Results

First, a matrix $\mathbf{M}$ is called a real quasi-upper-triangular if (a) $\mathbf{M}$ is a real block triangular matrix, (b) each diagonal block is either 1×1 or 2×2, (c) for each 2×2 block, it has the form of

$$\begin{bmatrix} c & -s \\ s & c \end{bmatrix}$$

and $c \pm js$ is a pair of complex conjugate eigenvalues of $\mathbf{M}$. We use $\sigma(\mathbf{M})$ to denote the set of all eigenvalues of $\mathbf{M}$. Let

$$\mathbf{L} = \begin{bmatrix} \mathbf{0} & \mathbf{I} \\ -\mathbf{I} & \mathbf{0} \end{bmatrix} \in \mathbf{R}^{2n \times 2n} \tag{11.35}$$

where n is the dimension of $\mathbf{A}$ or $\mathbf{A}_k$. Note that $\mathbf{L}^{\mathrm{T}} = \mathbf{L}^{-1} = -\mathbf{L}$. A matrix $\mathbf{M} \in \mathbf{R}^{2n \times 2n}$ is said to be *symplectic* if it meets the condition $\mathbf{L}^{-1}\mathbf{M}^{\mathrm{T}}\mathbf{L} = \mathbf{M}^{-1}$. The symplectic matrix plays a fundamental role in finding the solution of the Riccati equation [103]. An important property for the symplectic matrix is given as the following theorem which is shown in [102, 208].

Theorem 11.3

If $\mathbf{M}$ is symplectic, then $\lambda \in \sigma(\mathbf{M})$ implies $\frac{1}{\lambda} \in \sigma(\mathbf{M})$ with the same multiplicity.

Proof 11.1 Let $\lambda \in \sigma(\mathbf{M})$ be an eigenvalue of $\mathbf{M}$, $\mathbf{f}$ and $\mathbf{g}$ be n-dimensional vectors such that

$$\mathbf{M}\begin{bmatrix} \mathbf{f} \\ \mathbf{g} \end{bmatrix} = \begin{bmatrix} \mathbf{M}_{11} & \mathbf{M}_{12} \\ \mathbf{M}_{21} & \mathbf{M}_{22} \end{bmatrix}\begin{bmatrix} \mathbf{f} \\ \mathbf{g} \end{bmatrix} = \lambda \begin{bmatrix} \mathbf{f} \\ \mathbf{g} \end{bmatrix}$$

Then,

$$\begin{aligned} \mathbf{M}^{-1} &= \mathbf{L}^{-1}\mathbf{M}^{\mathrm{T}}\mathbf{L} = \begin{bmatrix} \mathbf{0} & -\mathbf{I} \\ \mathbf{I} & \mathbf{0} \end{bmatrix}\begin{bmatrix} \mathbf{M}_{11}^{\mathrm{T}} & \mathbf{M}_{21}^{\mathrm{T}} \\ \mathbf{M}_{12}^{\mathrm{T}} & \mathbf{M}_{22}^{\mathrm{T}} \end{bmatrix}\begin{bmatrix} \mathbf{0} & \mathbf{I} \\ -\mathbf{I} & \mathbf{0} \end{bmatrix} \\ &= \begin{bmatrix} \mathbf{M}_{22}^{\mathrm{T}} & -\mathbf{M}_{12}^{\mathrm{T}} \\ -\mathbf{M}_{21}^{\mathrm{T}} & \mathbf{M}_{11}^{\mathrm{T}} \end{bmatrix} \end{aligned}$$

Therefore,

$$\mathbf{M}^{-\mathrm{T}}\begin{bmatrix} \mathbf{g} \\ -\mathbf{f} \end{bmatrix} = \begin{bmatrix} \mathbf{M}_{22} & -\mathbf{M}_{21} \\ -\mathbf{M}_{12} & \mathbf{M}_{11} \end{bmatrix}\begin{bmatrix} \mathbf{g} \\ -\mathbf{f} \end{bmatrix} = \lambda \begin{bmatrix} \mathbf{g} \\ -\mathbf{f} \end{bmatrix}$$

which means that λ is an eigenvalue of $\mathbf{M}^{-1}$. Since λ is an eigenvalue of $\mathbf{M}^{-1}$, $1/\lambda$ is an eigenvalue of $\mathbf{M}$. ■

A stable numerical solution of the Riccati equation depends on the so-called real Schur decomposition [137]. The following Proposition is a natural extension of the real Schur decomposition for the symplectic matrix.

Proposition 11.1
Let $\mathbf{M} \in \mathbf{R}^{2n\times 2n}$ be symplectic. Then there exists an orthogonal similarity transformation $\mathbf{U}$ such that

$$\begin{bmatrix} \mathbf{U}_{11} & \mathbf{U}_{12} \\ \mathbf{U}_{21} & \mathbf{U}_{22} \end{bmatrix}^{\mathrm{T}} \mathbf{M} \begin{bmatrix} \mathbf{U}_{11} & \mathbf{U}_{12} \\ \mathbf{U}_{21} & \mathbf{U}_{22} \end{bmatrix} = \begin{bmatrix} \mathbf{S}_{11} & \mathbf{S}_{12} \\ \mathbf{0} & \mathbf{S}_{22} \end{bmatrix} \tag{11.36}$$

where $\mathbf{U}_{11}, \mathbf{U}_{12}, \mathbf{U}_{21}, \mathbf{U}_{22}, \mathbf{S}_{11}, \mathbf{S}_{12}, \mathbf{S}_{22} \in \mathbf{R}^{n\times n}$, and $\mathbf{S}_{11}$, $\mathbf{S}_{22}$ are quasi-upper-triangular. Moreover, $\sigma(\mathbf{S}_{11})$ lies inside (or outside) the unit circle and $\sigma(\mathbf{S}_{22})$ lies outside (or inside) the unit circle.

We will also use a simple result in our derivation of the main result.

Proposition 11.2
If $\mathbf{M}_1$ and $\mathbf{M}_2$ are symplectic, then $\mathbf{M}_1\mathbf{M}_2$ is symplectic.

Proof 11.2 Since $\mathbf{L}^{-1}\mathbf{M}_1^{\mathrm{T}}\mathbf{L} = \mathbf{M}_1^{-1}$ and $\mathbf{L}^{-1}\mathbf{M}_2^{\mathrm{T}}\mathbf{L} = \mathbf{M}_2^{-1}$, we have

$$\mathbf{L}^{-1}(\mathbf{M}_1\mathbf{M}_2)^{\mathrm{T}}\mathbf{L} = \mathbf{L}^{-1}\mathbf{M}_2^{\mathrm{T}}\mathbf{M}_1^{\mathrm{T}}\mathbf{L} = \mathbf{L}^{-1}\mathbf{M}_2^{\mathrm{T}}\mathbf{L}\mathbf{L}^{-1}\mathbf{M}_1^{\mathrm{T}}\mathbf{L} = \mathbf{M}_2^{-1}\mathbf{M}_1^{-1} = (\mathbf{M}_1\mathbf{M}_2)^{-1}.$$

This concludes the proof. ■

11.3.2 Solution of the Algebraic Riccati Equation

For a discrete linear time-varying system (11.33), the LQR state feedback control is to find the optimal solution $\mathbf{m}_k$ to minimize the following quadratic cost function

$$\min \frac{1}{2}\mathbf{x}_N^{\mathrm{T}}\mathbf{Q}_N\mathbf{x}_N + \frac{1}{2}\sum_{k=0}^{N-1}\mathbf{x}_k^{\mathrm{T}}\mathbf{Q}_k\mathbf{x}_k + \mathbf{m}_k^{\mathrm{T}}\mathbf{R}_k\mathbf{m}_k \tag{11.37}$$

where

$$\mathbf{Q}_k \geq 0 \tag{11.38}$$

$$\mathbf{R}_k > 0 \tag{11.39}$$

and the initial condition $\mathbf{x}_0$ is given. The existence of the solution implicitly depends on the controllability of spacecraft attitude control using only magnetic torques which is discussed in the previous section. Let the co-state vector of $\mathbf{x}_k$ be denoted by $\mathbf{y}_k$. A very important assumption in the so-called sweep method [24] to solve the optimization problem (11.37) under the state constraint of (11.33) is the relation between $\mathbf{y}_k$ and $\mathbf{x}_k$ which is given as follows:

$$\mathbf{y}_k = \mathbf{P}_k\mathbf{x}_k \tag{11.40}$$

If $(\mathbf{A}_k, \mathbf{Q}_k)$ is detectable or $\mathbf{Q}_k > 0$, the optimal feedback $\mathbf{m}_k$ is given in Appendix B (B.21) (see also [70, 108])

$$\mathbf{m}_k = -(\mathbf{R}_k + \mathbf{B}_k^{\mathrm{T}}\mathbf{P}_{k+1}\mathbf{B}_k)^{-1}\mathbf{B}_k^{\mathrm{T}}\mathbf{P}_{k+1}\mathbf{A}_k\mathbf{x}_k \tag{11.41}$$

where $\mathbf{P}_k$ defined in (11.40) is the unique positive semi-definite solution of the discrete Riccati equation (B.19) (see also [70, 103, 108])

$$\mathbf{P}_k = \mathbf{Q}_k + \mathbf{A}_k^{\mathrm{T}}\mathbf{P}_{k+1}\mathbf{A}_k - \mathbf{A}_k^{\mathrm{T}}\mathbf{P}_{k+1}\mathbf{B}_k(\mathbf{R}_k + \mathbf{B}_k^{\mathrm{T}}\mathbf{P}_{k+1}\mathbf{B}_k)^{-1}\mathbf{B}_k^{\mathrm{T}}\mathbf{P}_{k+1}\mathbf{A}_k \tag{11.42}$$

with the boundary condition $\mathbf{P}_N = \mathbf{Q}_N$. For this discrete Riccati equation (not necessarily periodic) given as (11.42), it can be solved using a symplectic system associated with (11.33) and (11.37) as follows:

Let $\mathbf{z}_k = [\mathbf{x}_k^{\mathrm{T}}, \mathbf{y}_k^{\mathrm{T}}]^{\mathrm{T}}$. Appendix B gives (B.26), which is repeated below.

$$\begin{bmatrix} \mathbf{x}_k \\ \mathbf{y}_k \end{bmatrix} = \begin{bmatrix} \mathbf{A}_k^{-1} & \mathbf{A}_k^{-1}\mathbf{B}_k\mathbf{R}_k^{-1}\mathbf{B}_k^{\mathrm{T}} \\ \mathbf{Q}_k\mathbf{A}_k^{-1} & \mathbf{A}_k^{\mathrm{T}} + \mathbf{Q}_k\mathbf{A}_k^{-1}\mathbf{B}_k\mathbf{R}_k^{-1}\mathbf{B}_k^{\mathrm{T}} \end{bmatrix} \begin{bmatrix} \mathbf{x}_{k+1} \\ \mathbf{y}_{k+1} \end{bmatrix} := \mathbf{H}_k \begin{bmatrix} \mathbf{x}_{k+1} \\ \mathbf{y}_{k+1} \end{bmatrix} \tag{11.43}$$

Let

$$\mathbf{E}_k = \begin{bmatrix} \mathbf{I} & \mathbf{B}_k\mathbf{R}_k^{-1}\mathbf{B}_k^{\mathrm{T}} \\ \mathbf{0} & \mathbf{A}_k^{\mathrm{T}} \end{bmatrix} \tag{11.44}$$

$$\mathbf{F}_k = \begin{bmatrix} \mathbf{A}_k & \mathbf{0} \\ -\mathbf{Q}_k & \mathbf{I} \end{bmatrix} \tag{11.45}$$

Assume that $\mathbf{E}_k$ is invertible, which is true for $\det(\mathbf{I}+T\Sigma_1-\frac{1}{2}T^2\Lambda_1)\neq 0$. It is easy to verify that

$$\begin{aligned}\mathbf{Z}_k &:= \mathbf{H}_k^{-1}=\begin{bmatrix}\mathbf{A}_k+\mathbf{B}_k\mathbf{R}_k^{-1}\mathbf{B}_k^{\mathrm{T}}\mathbf{A}_k^{-\mathrm{T}}\mathbf{Q}_k & -\mathbf{B}_k\mathbf{R}_k^{-1}\mathbf{B}_k^{\mathrm{T}}\mathbf{A}_k^{-\mathrm{T}}\\ -\mathbf{A}_k^{-\mathrm{T}}\mathbf{Q}_k & \mathbf{A}_k^{-\mathrm{T}}\end{bmatrix}\\ &= \mathbf{E}_k^{-1}\mathbf{F}_k=\begin{bmatrix}\mathbf{I} & -\mathbf{B}_k\mathbf{R}_k^{-1}\mathbf{B}_k^{\mathrm{T}}\mathbf{A}_k^{-\mathrm{T}}\\ \mathbf{0} & \mathbf{A}_k^{-\mathrm{T}}\end{bmatrix}\begin{bmatrix}\mathbf{A}_k & \mathbf{0}\\ -\mathbf{Q}_k & \mathbf{I}\end{bmatrix}\end{aligned}\tag{11.46}$$

Therefore, (11.43) can be rewritten as

$$\mathbf{E}_k\mathbf{z}_{k+1}=\mathbf{E}_k\begin{bmatrix}\mathbf{x}_{k+1}\\ \mathbf{y}_{k+1}\end{bmatrix}=\mathbf{F}_k\begin{bmatrix}\mathbf{x}_k\\ \mathbf{y}_k\end{bmatrix}=\mathbf{F}_k\mathbf{z}_k\tag{11.47}$$

It is straightforward to verify that $\mathbf{L}^{-1}\mathbf{Z}_k^{\mathrm{T}}\mathbf{L}=\mathbf{Z}_k^{-1}$, therefore, from Proposition 11.1, there exists an orthogonal matrix $\mathbf{U}$ such that

$$\begin{bmatrix}\mathbf{U}_{11} & \mathbf{U}_{12}\\ \mathbf{U}_{21} & \mathbf{U}_{22}\end{bmatrix}^{\mathrm{T}}\mathbf{Z}_k\begin{bmatrix}\mathbf{U}_{11} & \mathbf{U}_{12}\\ \mathbf{U}_{21} & \mathbf{U}_{22}\end{bmatrix}=\begin{bmatrix}\mathbf{S}_{11} & \mathbf{S}_{12}\\ \mathbf{0} & \mathbf{S}_{22}\end{bmatrix}\tag{11.48}$$

and all eigenvalues of $\mathbf{S}_{11}$ are inside unit circle. For *linear time-invariant system*, $\mathbf{A}_k=\mathbf{A}$, $\mathbf{B}_k=\mathbf{B}$, $\mathbf{Q}_k=\mathbf{Q}$, $\mathbf{R}_k=\mathbf{R}$, and $\mathbf{Z}_k=\mathbf{Z}$ are all constant matrices, the (steady state) solution of (11.42) is given as follows (see Appendix B.3 and [103, Theorem 6])

$$\mathbf{P}=\mathbf{U}_{21}\mathbf{U}_{11}^{-1}$$

11.3.3 Solution of the Periodic Riccati Algebraic Equation

Now, we consider the periodic time-varying system

$$\lim_{N\to\infty}\left[\min\frac{1}{2}\mathbf{x}_N^{\mathrm{T}}\mathbf{Q}_N\mathbf{x}_N+\frac{1}{2}\sum_{k=0}^{N-1}\mathbf{x}_k^{\mathrm{T}}\mathbf{Q}_k\mathbf{x}_k+\mathbf{m}_k^{\mathrm{T}}\mathbf{R}_k\mathbf{m}_k\right]\tag{11.49a}$$

$$\mathbf{x}_{k+1}=\mathbf{A}_k\mathbf{x}_k+\mathbf{B}_k\mathbf{m}_k\tag{11.49b}$$

where

$$\mathbf{A}_k=\mathbf{A}_{k+1}=\ldots=\mathbf{A}_{k+p}\tag{11.50}$$

$$\mathbf{B}_k=\mathbf{B}_{k+p}\tag{11.51}$$

$$\mathbf{Q}_k=\mathbf{Q}_{k+1}=\ldots=\mathbf{Q}_{k+p}\geq 0\tag{11.52}$$

$$\mathbf{R}_k=\mathbf{R}_{k+p}>0\tag{11.53}$$

only $\mathbf{B}_k$ (and possibly $\mathbf{R}_k$) are periodic with period $p=\frac{T}{t_s}$. It is worthwhile to mention that $\mathbf{A}_k$ and $\mathbf{Q}_k$ are actually constant matrices. The optimal feedback given by (11.42) is periodic with $\mathbf{P}_k=\mathbf{P}_{k+p}$, a unique periodic positive semi-definite solution of the periodic Riccati equation (cf. [19]). Therefore, using the similar process for general discrete Riccati equation and noticing that $\mathbf{F}_k = \mathbf{F}$ in (11.45) is a constant matrix because $\mathbf{A}_k$ and $\mathbf{Q}_k$ are constant matrices, we get

$$\mathbf{E}_k \mathbf{z}_{k+1} = \mathbf{F}\mathbf{z}_k \tag{11.54}$$

$$\mathbf{E}_{k+1} \mathbf{z}_{k+2} = \mathbf{F}\mathbf{z}_{k+1} \tag{11.55}$$

$$\vdots \tag{11.56}$$

$$\mathbf{E}_{k+p-1} \mathbf{z}_{k+p} = \mathbf{F}\mathbf{z}_{k+p-1} \tag{11.57}$$

This gives

$$\mathbf{z}_{k+p} = \Pi_k \mathbf{z}_k \tag{11.58}$$

with

$$\Pi_k = \mathbf{E}_{k+p-1}^{-1}\mathbf{F}\ldots\mathbf{E}_{k+1}^{-1}\mathbf{F}\mathbf{E}_k^{-1}\mathbf{F} \tag{11.59}$$

Using Proposition 11.2, we conclude that Π_k is a symplectic matrix. Therefore, from Proposition 11.1 there is an orthogonal matrix $\mathbf{T}_k$ such that

$$\begin{bmatrix} \mathbf{T}_{11k} & \mathbf{T}_{12k} \\ \mathbf{T}_{21k} & \mathbf{T}_{22k} \end{bmatrix}^{\mathrm{T}} \Pi_k \begin{bmatrix} \mathbf{T}_{11k} & \mathbf{T}_{12k} \\ \mathbf{T}_{21k} & \mathbf{T}_{22k} \end{bmatrix} = \begin{bmatrix} \mathbf{S}_{11k} & \mathbf{S}_{12k} \\ \mathbf{0} & \mathbf{S}_{22k} \end{bmatrix} \tag{11.60}$$

According to [70, pp. 1197–1198], the matrix $\mathbf{S}_{11k}$ has eigenvalues in the open unit disk, and for each sampling time $k \in \{0, 1, \ldots, p-1\}$ the steady state solution of the Riccati equation corresponding to (11.58) is given by

$$\mathbf{P}_k = \mathbf{T}_{21k}\mathbf{T}_{11k}^{-1} \tag{11.61}$$

Since $\mathbf{F}$ is invertible in the problem of spacecraft attitude control using only magnetic torques, this method is more efficient than the one in [151] because the latter is designed for singular $\mathbf{F}$. However, the method of calculating (11.59), (11.60), and (11.61) as described above (proposed in [70]) is still not the best way for the problem of spacecraft attitude control using only magnetic torques. As a matter of fact, equation (11.58) can be written as

$$\begin{bmatrix} \mathbf{x}_k \\ \mathbf{y}_k \end{bmatrix} = \mathbf{z}_k = \Gamma_k \mathbf{z}_{k+p} = \Gamma_k \begin{bmatrix} \mathbf{x}_{k+p} \\ \mathbf{y}_{k+p} \end{bmatrix} \tag{11.62}$$

with the initial state $\mathbf{x}_0$, the boundary condition [108]

$$\mathbf{y}_N = \mathbf{Q}_N \mathbf{x}_N \tag{11.63}$$

and

$$\Gamma_k = \mathbf{F}^{-1}\mathbf{E}_k\mathbf{F}^{-1}\mathbf{E}_{k+1}\ldots,\mathbf{F}^{-1}\mathbf{E}_{k+p-2}\mathbf{F}^{-1}\mathbf{E}_{k+p-1} \tag{11.64}$$

Remark 11.3 Since the same $\mathbf{F}^{-1}$ is a constant matrix and is used repeatedly in Γ_k, the computation of Γ_k avoids $p-1$ matrix inverse comparing to the computation of Π_k. For large p, the difference is tremendous. ■

We propose a better way to solve (11.49). The derivations are similar to the method proposed in [103]. Since

$$\mathbf{F}^{-1} = \begin{bmatrix} \mathbf{A}_k^{-1} & \mathbf{0} \\ \mathbf{Q}_k\mathbf{A}_k^{-1} & \mathbf{I} \end{bmatrix}$$

$$\begin{aligned} \mathbf{M} &= \mathbf{F}^{-1}\mathbf{E}_k = \begin{bmatrix} \mathbf{A}_k^{-1} & \mathbf{0} \\ \mathbf{Q}_k\mathbf{A}_k^{-1} & \mathbf{I} \end{bmatrix} \begin{bmatrix} \mathbf{I} & \mathbf{B}_k\mathbf{R}_k^{-1}\mathbf{B}_k^{\mathrm{T}} \\ \mathbf{0} & \mathbf{A}_k^{\mathrm{T}} \end{bmatrix} \\ &= \begin{bmatrix} \mathbf{A}_k^{-1} & \mathbf{A}_k^{-1}\mathbf{B}_k\mathbf{R}_k^{-1}\mathbf{B}_k^{\mathrm{T}} \\ \mathbf{Q}_k\mathbf{A}_k^{-1} & \mathbf{Q}_k\mathbf{A}_k^{-1}\mathbf{B}_k\mathbf{R}_k^{-1}\mathbf{B}_k^{\mathrm{T}} + \mathbf{A}_k^{\mathrm{T}} \end{bmatrix} \end{aligned} \tag{11.65}$$

which is a similar formula as given in [208]. It is straightforward to verify that $\mathbf{M}$ is symplectic.

$$\begin{aligned} \mathbf{L}^{-1}\mathbf{M}^{\mathrm{T}}\mathbf{L} &= \begin{bmatrix} \mathbf{0} & -\mathbf{I} \\ \mathbf{I} & \mathbf{0} \end{bmatrix} \begin{bmatrix} \mathbf{A}_k^{-\mathrm{T}} & \mathbf{A}_k^{-\mathrm{T}}\mathbf{Q}_k \\ \mathbf{B}_k\mathbf{R}_k^{-1}\mathbf{B}_k^{\mathrm{T}}\mathbf{A}_k^{-\mathrm{T}} & \mathbf{A}_k + \mathbf{B}_k\mathbf{R}_k^{-1}\mathbf{B}_k^{\mathrm{T}}\mathbf{A}_k^{-\mathrm{T}}\mathbf{Q}_k \end{bmatrix} \mathbf{L} \\ &= \begin{bmatrix} -\mathbf{B}_k\mathbf{R}_k^{-1}\mathbf{B}_k^{\mathrm{T}}\mathbf{A}_k^{-\mathrm{T}} & -\mathbf{A}_k - \mathbf{B}_k\mathbf{R}_k^{-1}\mathbf{B}_k^{\mathrm{T}}\mathbf{A}_k^{-\mathrm{T}}\mathbf{Q}_k \\ \mathbf{A}_k^{-\mathrm{T}} & \mathbf{A}_k^{-\mathrm{T}}\mathbf{Q}_k \end{bmatrix} \begin{bmatrix} \mathbf{0} & \mathbf{I} \\ -\mathbf{I} & \mathbf{0} \end{bmatrix} \\ &= \begin{bmatrix} \mathbf{A}_k + \mathbf{B}_k\mathbf{R}_k^{-1}\mathbf{B}_k^{\mathrm{T}}\mathbf{A}_k^{-\mathrm{T}}\mathbf{Q}_k & -\mathbf{B}_k\mathbf{R}_k^{-1}\mathbf{B}_k^{\mathrm{T}}\mathbf{A}_k^{-\mathrm{T}} \\ -\mathbf{A}_k^{-\mathrm{T}}\mathbf{Q}_k & \mathbf{A}_k^{-\mathrm{T}} \end{bmatrix} \\ &= \mathbf{M}^{-1} \end{aligned} \tag{11.66}$$

Since $\mathbf{M}$ is symplectic, using Proposition 11.2 again, Γ_k is symplectic. Let

$$\mathbf{V}_k = \begin{bmatrix} \mathbf{V}_{11k} & \mathbf{V}_{12k} \\ \mathbf{V}_{21k} & \mathbf{V}_{22k} \end{bmatrix}$$

be a matrix that transform Γ_k into a Jordon form, we have

$$\Gamma_k\mathbf{V}_k = \mathbf{V}_k \begin{bmatrix} \Delta_k & \mathbf{0} \\ \mathbf{0} & \Delta_k^{-1} \end{bmatrix} \tag{11.67}$$

where Δ_k is the Jordan block matrix of the n eigenvalues outside of the unit circle. One of the main results of this section is the following theorem.

Theorem 11.4

The solution of the Riccati equation corresponding to (11.62) is given by

$$\mathbf{P}_k = \mathbf{V}_{21k}\mathbf{V}_{11k}^{-1}, \quad k = 0, \ldots, p-1 \tag{11.68}$$

Proof 11.3 The proof uses similar ideas to [208, 108]. Since the system is periodic, the Riccati equation corresponding to (11.62) represents any one of $k \in$

$\{0,1,\ldots,p-1\}$ equations, which has a sample period increasing by p with the patent $k, k+p, k+2p, \ldots, k+\ell p, \ldots$. In the following discussion, we consider one Riccati equation and drop the subscript k to simplify the notation to $0, p, 2p, \ldots, \ell p, \ldots$. To make the notation simpler, we will drop p and use ℓ for this step increment. Assume that the solution has the form

$$\mathbf{y}_\ell = \mathbf{P}\mathbf{x}_\ell \tag{11.69}$$

Using the method described in Appendix B, one can show that $\mathbf{P}$ satisfies the discrete-time periodic Riccati equation

$$\mathbf{0} = \mathbf{Q} + \mathbf{A}^{\mathrm{T}}\mathbf{P}\mathbf{A} - \mathbf{P} - \mathbf{A}^{\mathrm{T}}\mathbf{P}\mathbf{B}(\mathbf{R} + \mathbf{B}^{\mathrm{T}}\mathbf{P}\mathbf{B})\mathbf{B}^{\mathrm{T}}\mathbf{P}\mathbf{A}$$

Further, we assume for simplicity that the eigenvalues of Γ are distinct; therefore, Δ is diagonal. For any integer $\ell \geq 0$, let

$$\begin{bmatrix} \mathbf{x}_\ell \\ \mathbf{y}_\ell \end{bmatrix} = \begin{bmatrix} \mathbf{V}_{11} & \mathbf{V}_{12} \\ \mathbf{V}_{21} & \mathbf{V}_{22} \end{bmatrix} \begin{bmatrix} \mathbf{t}_\ell \\ \mathbf{s}_\ell \end{bmatrix} \tag{11.70}$$

from (11.62), (11.67) and (11.70), we have

$$\mathbf{V}\begin{bmatrix} \mathbf{t}_\ell \\ \mathbf{s}_\ell \end{bmatrix} = \begin{bmatrix} \mathbf{x}_\ell \\ \mathbf{y}_\ell \end{bmatrix} = \Gamma \begin{bmatrix} \mathbf{x}_{\ell+1} \\ \mathbf{y}_{\ell+1} \end{bmatrix} = \Gamma\mathbf{V}\begin{bmatrix} \mathbf{t}_{\ell+1} \\ \mathbf{s}_{\ell+1} \end{bmatrix} = \mathbf{V}\begin{bmatrix} \Delta & \mathbf{0} \\ \mathbf{0} & \Delta^{-1} \end{bmatrix}\begin{bmatrix} \mathbf{t}_{\ell+1} \\ \mathbf{s}_{\ell+1} \end{bmatrix}$$

which is equivalent to

$$\begin{bmatrix} \mathbf{t}_\ell \\ \mathbf{s}_\ell \end{bmatrix} = \begin{bmatrix} \Delta & \mathbf{0} \\ \mathbf{0} & \Delta^{-1} \end{bmatrix}\begin{bmatrix} \mathbf{t}_{\ell+1} \\ \mathbf{s}_{\ell+1} \end{bmatrix}$$

Hence,

$$\begin{bmatrix} \mathbf{t}_\ell \\ \mathbf{s}_\ell \end{bmatrix} = \begin{bmatrix} \Delta^{N-\ell} & \mathbf{0} \\ \mathbf{0} & \Delta^{-(N-\ell)} \end{bmatrix}\begin{bmatrix} \mathbf{t}_N \\ \mathbf{s}_N \end{bmatrix} \tag{11.71}$$

Using the boundary condition (11.63) and (11.70), we have

$$\mathbf{Q}_N(\mathbf{V}_{11}\mathbf{t}_N + \mathbf{V}_{12}\mathbf{s}_N) = \mathbf{Q}_N\mathbf{x}_N = \mathbf{y}_N = \mathbf{V}_{21}\mathbf{t}_N + \mathbf{V}_{22}\mathbf{s}_N$$

this gives

$$-(\mathbf{V}_{21} - \mathbf{Q}_N\mathbf{V}_{11})\mathbf{t}_N = (\mathbf{V}_{22} - \mathbf{Q}_N\mathbf{V}_{12})\mathbf{s}_N$$

or equivalently

$$\mathbf{s}_N = -(\mathbf{V}_{22} - \mathbf{Q}_N\mathbf{V}_{12})^{-1}(\mathbf{V}_{21} - \mathbf{Q}_N\mathbf{V}_{11})\mathbf{t}_N := \mathbf{H}\mathbf{t}_N \tag{11.72}$$

Combining (11.71) and (11.72) yields

$$\mathbf{s}_\ell = \Delta^{-(N-\ell)}\mathbf{s}_N = \Delta^{-(N-\ell)}\mathbf{H}\mathbf{t}_N = \Delta^{-(N-\ell)}\mathbf{H}\Delta^{-(N-\ell)}\mathbf{t}_\ell := \mathbf{G}\mathbf{t}_\ell$$

with $\mathbf{G} = \Delta^{-(N-\ell)}\mathbf{H}\Delta^{-(N-\ell)}$. Finally, using this relation, equations (6.22) and (11.69), we conclude that

$$\mathbf{y}_\ell = \mathbf{V}_{21}\mathbf{t}_\ell + \mathbf{V}_{22}\mathbf{s}_\ell = (\mathbf{V}_{21} + \mathbf{V}_{22}\mathbf{G})\mathbf{t}_\ell = \mathbf{P}\mathbf{x}_\ell = \mathbf{P}(\mathbf{V}_{11}\mathbf{t}_\ell + \mathbf{V}_{12}\mathbf{s}_\ell) = \mathbf{P}(\mathbf{V}_{11} + \mathbf{V}_{12}\mathbf{G})\mathbf{t}_\ell$$

holds for all $\mathbf{t}_\ell$; therefore

$$(\mathbf{V}_{21}+\mathbf{V}_{22}\mathbf{G}) = \mathbf{P}(\mathbf{V}_{11}+\mathbf{V}_{12}\mathbf{G})$$

or

$$\mathbf{P} = (\mathbf{V}_{21}+\mathbf{V}_{22}\mathbf{G})(\mathbf{V}_{11}+\mathbf{V}_{12}\mathbf{G})^{-1} \tag{11.73}$$

Note that $\mathbf{G} \to 0$ as $N \to \infty$. This finishes the proof. ■

Since the eigen-decomposition is not numerically stable, we suggest using the Schur decomposition instead. Since Γ_k is symplectic, Proposition 11.1 claims that there is an orthogonal matrix $\mathbf{W}_k$ such that

$$\begin{bmatrix} \mathbf{W}_{11k} & \mathbf{W}_{12k} \\ \mathbf{W}_{21k} & \mathbf{W}_{22k} \end{bmatrix}^T \Gamma_k \begin{bmatrix} \mathbf{W}_{11k} & \mathbf{W}_{12k} \\ \mathbf{W}_{21k} & \mathbf{W}_{22k} \end{bmatrix} = \begin{bmatrix} \mathbf{S}_{11k} & \mathbf{S}_{12k} \\ \mathbf{0} & \mathbf{S}_{22k} \end{bmatrix} \tag{11.74}$$

where $\mathbf{S}_{11k}$ is upper-triangular and has all of its eigenvalues outside the unit circle. We have the main result of the section as follows.

Theorem 11.5
Let the Schur decomposition of Γ_k be given by (11.74). The solution of the Riccati equation corresponding to (11.62) is given by

$$\mathbf{P}_k = \mathbf{W}_{21k}\mathbf{W}_{11k}^{-1} \tag{11.75}$$

Proof 11.4 The proof follows the same argument of [103, Remark 1]. From (11.67), we have

$$\Gamma_k \begin{bmatrix} \mathbf{V}_{11k} \\ \mathbf{V}_{21k} \end{bmatrix} = \begin{bmatrix} \mathbf{V}_{11k} \\ \mathbf{V}_{21k} \end{bmatrix} \Delta_k \tag{11.76}$$

From (11.74), we have

$$\Gamma_k \begin{bmatrix} \mathbf{W}_{11k} \\ \mathbf{W}_{21k} \end{bmatrix} = \begin{bmatrix} \mathbf{W}_{11k} \\ \mathbf{W}_{21k} \end{bmatrix} \mathbf{S}_{11k}$$

Let $\mathbf{T}$ be an invertible transformation matrix such that

$$\mathbf{T}^{-1}\mathbf{S}_{11k}\mathbf{T} = \Delta_k$$

then we have

$$\Gamma_k \begin{bmatrix} \mathbf{W}_{11k} \\ \mathbf{W}_{21k} \end{bmatrix} \mathbf{T} = \begin{bmatrix} \mathbf{W}_{11k} \\ \mathbf{W}_{21k} \end{bmatrix} \mathbf{T}\mathbf{T}^{-1}\mathbf{S}_{11k}\mathbf{T} = \begin{bmatrix} \mathbf{W}_{11k} \\ \mathbf{W}_{21k} \end{bmatrix} \mathbf{T}\Delta_k \tag{11.77}$$

Comparing (11.76) and (11.77) we must have

$$\begin{bmatrix} \mathbf{W}_{11k} \\ \mathbf{W}_{21k} \end{bmatrix} \mathbf{T} = \begin{bmatrix} \mathbf{V}_{11k} \\ \mathbf{V}_{21k} \end{bmatrix} \mathbf{D}$$

where **D** is a diagonal and invertible matrix. Thus,

$$\mathbf{W}_{21k}\mathbf{W}_{11k}^{-1} = \mathbf{V}_{21k}\mathbf{D}\mathbf{T}^{-1}\mathbf{T}\mathbf{D}^{-1}\mathbf{V}_{11k}^{-1} = \mathbf{V}_{21k}\mathbf{V}_{11k}^{-1}$$

This finishes the proof. ■

We can apply the algorithm to the problem described in (11.49).

Algorithm 11.1

Data: $\mathbf{J}$, i_m, $\mathbf{Q}$, $\mathbf{R}$, *altitude of the spacecraft, and selected sample period* t_s.

Step 1: Calculate $\mathbf{A}_k$ *and* $\mathbf{B}_k$ *using (11.33-11.34).*

Step 2: Calculate $\mathbf{E}_k$ *and* $\mathbf{F}_k$ *using (11.44-11.45).*

Step 3: Calculate Γ_k *using (11.64).*

Step 4: Use Schur decomposition (11.74) to get $\mathbf{W}_k$.

Step 5: Calculate $\mathbf{P}_k$ *using (11.75).*

Remark 11.4 This algorithm makes full use of the fact that **A** is a constant matrix in (11.45). Therefore, **F** is a constant matrix and the inverse of **F** in (11.64) does not need to be repeated many times which is the main difference between the method discussed in this section and the method in [70]. ■

11.3.4 Simulation Test

The following problem is used to demonstrate the effectiveness of proposed design algorithm. Let the spacecraft inertia matrix be $\mathbf{J} = \text{diag}(250, 150, 100)\, kg \cdot m^2$. The orbital inclination $i_m = 57^o$, the orbit is circular with an altitude of 657 km. In view of equation (11.32), the orbital period is 5863 seconds, and the orbital rate is $\omega_0 = 0.0011$ rad/second. Assuming that the total number of samples taken in one orbit is $p = 100$, then, each sample period is 58.6352 seconds. Select $\mathbf{Q} = \text{diag}(1.5 * 10^{-9}, 1.5 * 10^{-9}, 1.5 * 10^{-9}, 0.001, 0.001, 0.001)$ and $\mathbf{R} = \text{diag}(2 * 10^{-3}, 2 * 10^{-3}, 2 * 10^{-3})$. The Riccati equation solutions $\mathbf{P}_k$ for $k = 0, 1, 2, \ldots, 99$ are calculated using Algorithm 11.1 and are stored. Assuming that the initial quaternion error is $(0.01, 0.01, 0.01)$ and the initial body rate is $(0.00001, 0.00001, 0.00001)$ radians per second, applying the feedback (11.41) to the system (11.33), the simulated spacecraft attitude response is given in Figures 11.1–11.6.

The designed controller stabilizes the spacecraft using only magnetic torques. This shows the effectiveness of the design method. Since this time-varying system has a long period 5863 seconds and the number of samples in each period is 100, this means that using Γ_k in (11.64) instead of Π_k in (11.59) saves about 100 matrix inverses, a significant improvement in the computation compared to the well-known algorithm of [70]. For more detailed discussion of the computational comparison, readers are referred to [254].

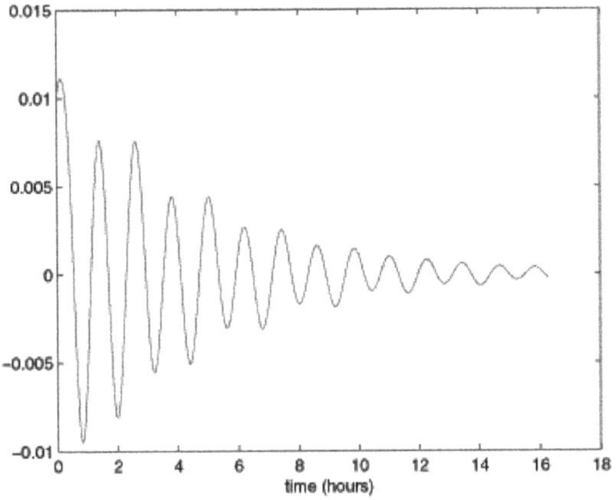

Figure 11.1: Attitude response q_1.

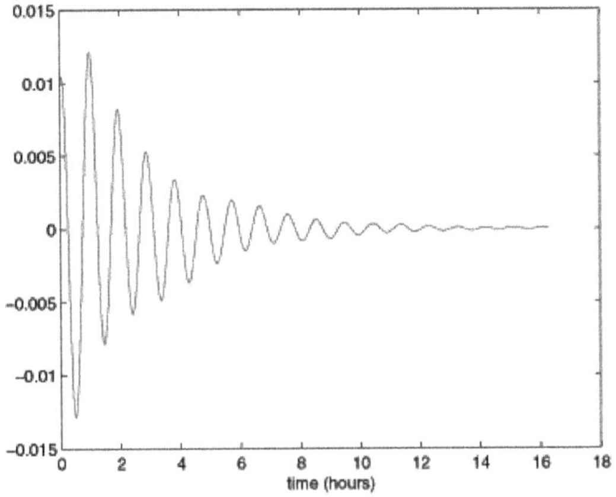

Figure 11.2: Attitude response q_2.

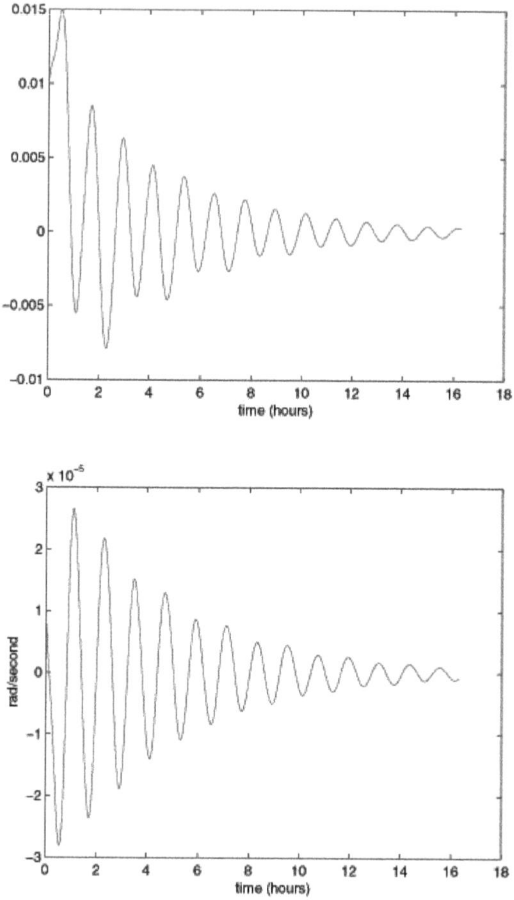

Figure 11.4: Body rate response ω_1.

11.4 Attitude and Desaturation Combined Control

Spacecraft attitude control and reaction wheel desaturation are normally regarded as two different control system design problems and are discussed in separate chapters in textbooks, such as [181, 219]. While spacecraft attitude control using magnetic torques has been one of the main research areas (see, for example, [166, 182] and extensive references therein), there are many research papers that address reaction wheel momentum management, see for example, [29, 44, 56] and references therein. In [44], Dzielsk et al. formulated the problem as an optimization problem and a nonlinear programming method was proposed to find the solution. His method can be very expensive and there is no guarantee to find the global optimal solution. Chen et al. [29] discussed optimal desaturation

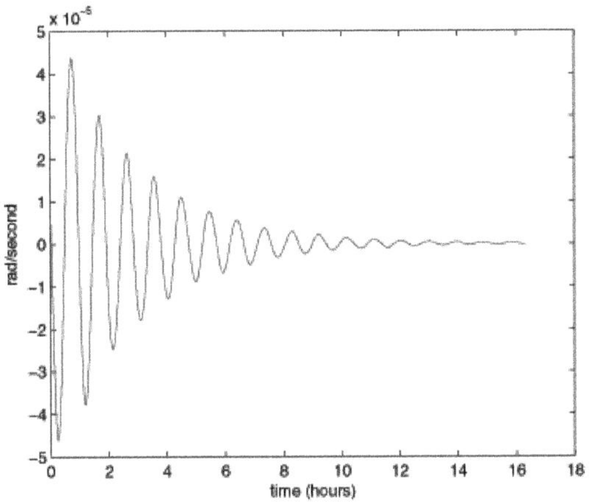

Figure 11.5: Body rate response ω_2.

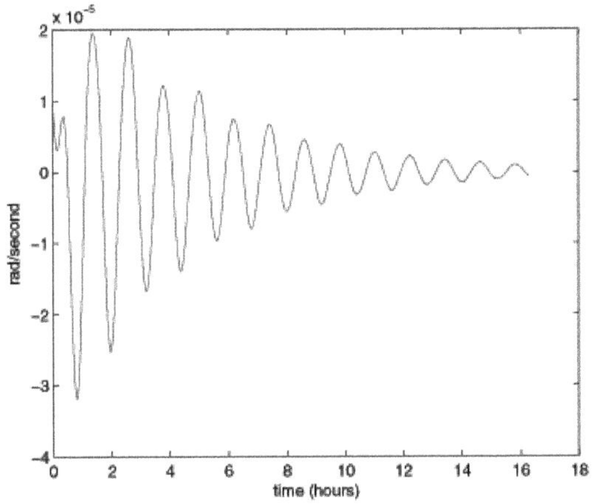

Figure 11.6: Body rate response ω_3.

controllers using magnetic torques and thrusters. Their methods find the optimal torques which, however, may not be able to achievable by magnetic torque coils because given the desired torques in a three-dimensional space, magnetic torque coils can only generate torques in a two dimensional plane [181]. Like most publications on this problem, the above two papers do not consider the time-varying effect of the geomagnetic field in body frame, which arises when a spacecraft flies around the Earth. Giulietti et al. [56] considered the same problem with more details on the geomagnetic field, but the periodic feature of the magnetic field along the orbit was not used in their proposed design. In addition, all these proposed designs considered only momentum management but not attitude control.

Since both attitude control and reaction wheel desaturation are performed at the same time using the same magnetic torque coils, the control system design should consider these two design objectives at the same time. Some recent research papers tackled the problem in this direction, for example, [5, 203]. In [203], Tregouet et al. studied the problem of the spacecraft stabilization and reaction wheel desaturation at the same time. They considered time-variation of the magnetic field in body frame, and their reference frame was the inertial frame. However, for a Low Earth Orbit (LEO) spacecraft that uses Earth's magnetic field, the reference frame for the spacecraft is most likely Local Vertical Local Horizontal (LVLH) frame. In addition, their design method depends on some assumption which is not easy to verify and their proposed design does not use the periodic feature of the magnetic field. Moreover, their design is composed of two loops, which is essentially an idea of dealing with attitude control and wheel momentum management in separate considerations. In [5], a heuristic proportional controller was proposed and a Lyapunov function was used to prove that the controller can simultaneously stabilize the spacecraft with respect to the LVLH frame and achieve reaction wheel management. But this design method does not consider the the time-varying effect of the geomagnetic field in body frame. Although these two designs are impressive, as we have seen, they do not consider some factors in reality and their solutions are not optimal.

This section proposes a more attractive design method which considers as many factors as practical. The controlled attitude is aligned with LVLH frame. A general reduced quaternion model, including (a) reaction wheels, (b) magnetic torque coils, (c) the gravity gradient torque, and (d) the periodic time-varying effects of the geomagnetic field along the orbit and its interaction with magnetic torque coils, is proposed. The model is an extension of the one discussed in Chapter 4 (see also [243]). A single objective function, which considers the performance of both attitude control and reaction wheel management at the same time, is suggested. Since a well-designed periodic controller for a period system is better than constant controllers as pointed out in [48, 94], this objective function is optimized using the solution of a matrix periodic Riccati equation described earlier in this chapter, which leads to a periodic time-varying optimal control. It is shown that the design can be calculated in an efficient way and the designed controller is optimal for both the spacecraft attitude control and for the reaction wheel momentum management at the same time. A simulation test is then provided to demonstrate that the designed system achieves more accurate attitude than the optimal control system that uses only magnetic torques. Moreover, it will be shown that the designed controller based on LQR method works on the nonlinear spacecraft system.

11.4.1 Spacecraft Model for Attitude and Reaction Wheel Desaturation Control

Throughout the rest of this section, it is assumed that the inertia matrix of a spacecraft $\mathbf{J} = \mathrm{diag}(J_1, J_2, J_3)$ is a diagonal matrix. This assumption is reasonable because in practical spacecraft design, spacecraft inertia matrix J is always designed as close to a diagonal matrix as possible [246]. (It is actually very close to a diagonal matrix.) For spacecraft using Earth's magnetic torques, the nadir pointing model is probably the mostly desired one by the missions. Therefore, the attitude of the spacecraft is represented by the rotation of the spacecraft body frame relative to the local vertical and local horizontal frame. This means that the quaternion and spacecraft body rate should be represented in terms of the rotation of the spacecraft body frame relative to the LVLH frame.

Let $\omega = [\omega_1, \omega_2, \omega_3]^{\mathrm{T}}$ be the body rate with respect to the LVLH frame represented in the body frame, $\omega_{lvlh} = [0, -\omega_0, 0]^{\mathrm{T}}$ the orbit rate (the rotation of LVLH frame) with respect to the inertial frame represented in the LVLH frame[3], and $\omega_I = [\omega_{I1}, \omega_{I2}, \omega_{I3}]^{\mathrm{T}}$ be the angular velocity vector of the spacecraft body with respect to the inertial frame, represented in the spacecraft body frame. Let A_l^b represent the rotational transformation matrix from the LVLH frame to the spacecraft body frame. Then, ω_I is expressed as in (4.17)

$$\omega_I = \omega + \mathbf{A}_l^b \omega_{lvlh} = \omega + \omega_{lvlh}^b \tag{11.78}$$

where ω_{lvlh}^b is the rotational rate of LVLH frame relative to the inertial frame represented in the spacecraft body frame. Assuming that the orbit is circular, i.e., $\dot{\omega}_{lvlh} = 0$, using the fact of (3.15)

$$\dot{\mathbf{A}}_l^b = -\omega \times \mathbf{A}_l^b \tag{11.79}$$

and taking the derivative of (11.78) give

$$\begin{aligned} \dot{\omega}_I &= \dot{\omega} + \dot{\mathbf{A}}_l^b \omega_{lvlh} + \mathbf{A}_l^b \dot{\omega}_{lvlh} \\ &= \dot{\omega} - \omega \times \mathbf{A}_l^b \omega_{lvlh} = \dot{\omega} - \omega \times \omega_{lvlh}^b \end{aligned} \tag{11.80}$$

Assuming that the three reaction wheels are aligned with the body frame axes, the total angular momentum of the spacecraft $\mathbf{h}_T$ in the body frame comprises the angular momentum of the spacecraft $\mathbf{J}\omega_I$ and the angular momentum of the reaction wheels $\mathbf{h}_w = [h_{w1}, h_{w2}, h_{w3}]^{\mathrm{T}}$, which is given by:

$$\mathbf{h}_T = \mathbf{J}\omega_I + \mathbf{h}_w \tag{11.81}$$

where

$$\mathbf{h}_w = \mathbf{J}_w \Omega \tag{11.82}$$

[3]For a circular orbit, given the spacecraft orbital period around the Earth P, $\omega_0 = \frac{2\pi}{P}$ is a known constant.

$\mathbf{J}_w = \text{diag}(\mathbf{J}_{w_1}, \mathbf{J}_{w_2}, \mathbf{J}_{w_3})$ is the inertia matrix of the three reaction wheels aligned with the spacecraft body axes, and $\Omega = [\Omega_1, \Omega_2, \Omega_3]^{\mathrm{T}}$ is the angular rate vector of the three reaction wheels. Let $\mathbf{h}'_T$ be the same vector of $\mathbf{h}_T$ represented in inertial frame. Let $\mathbf{t}_T$ be the total external torques acting on the spacecraft, then it must have (see [176])

$$\mathbf{t}_T = \left.\frac{d\mathbf{h}'_T}{dt}\right|_b$$

Taking derivative of (11.81) and using the above equation and (3.16) lead to the dynamic equations of the spacecraft as follows

$$\begin{aligned} \mathbf{J}\dot{\omega}_I + \dot{\mathbf{h}}_w &= \left.\left(\frac{d\mathbf{h}_T}{dt}\right)\right|_b = -\omega_I \times \mathbf{h}_T + \left.\left(\frac{d\mathbf{h}'_T}{dt}\right)\right|_b \\ &= -\omega_I \times (\mathbf{J}\omega_I + \mathbf{h}_w) + \mathbf{t}_T \end{aligned} \tag{11.83}$$

where $\mathbf{t}_T$ includes the gravity gradient torque $\mathbf{t}_g$, magnetic control torque $\mathbf{t}_m$, and internal and external disturbance torque $\mathbf{t}_d$ (including residual magnetic moment induced torque, atmosphere induced torque, solar radiation torque, etc). The torques generated by the reaction wheels $\mathbf{t}_w$ are given by

$$\mathbf{t}_w = -\dot{\mathbf{h}}_w = -\mathbf{J}_w\dot{\Omega}$$

Substituting these relations into (11.83) gives

$$\mathbf{J}\dot{\omega}_I = -\omega_I \times (\mathbf{J}\omega_I + \mathbf{J}_w\Omega) + \mathbf{t}_w + \mathbf{t}_g + \mathbf{t}_m + \mathbf{t}_d \tag{11.84}$$

Substituting (11.78) and (11.80) into (11.84) yields

$$\begin{aligned} \mathbf{J}\dot{\omega} &= \mathbf{J}\omega \times \omega^b_{lvlh} - (\omega + \omega^b_{lvlh}) \times [\mathbf{J}(\omega + \omega^b_{lvlh}) + \mathbf{J}_w\Omega] \\ &\quad + \mathbf{t}_w + \mathbf{t}_g + \mathbf{t}_m + \mathbf{t}_d \end{aligned} \tag{11.85}$$

Let

$$\bar{\mathbf{q}} = [q_0, q_1, q_2, q_3]^{\mathrm{T}} = [q_0, \mathbf{q}^{\mathrm{T}}]^{\mathrm{T}} = \left[\cos(\frac{\alpha}{2}), \hat{\mathbf{e}}^{\mathrm{T}}\sin(\frac{\alpha}{2})\right]^{\mathrm{T}} \tag{11.86}$$

be the quaternion representing the rotation of the body frame relative to the LVLH frame, where $\hat{\mathbf{e}}$ is the unit length rotational axis and α is the rotation angle about $\hat{\mathbf{e}}$. Therefore, from the derivation of (4.9), the reduced kinematics equation becomes (see also [243])

$$\begin{aligned} \begin{bmatrix} \dot{q}_1 \\ \dot{q}_2 \\ \dot{q}_3 \end{bmatrix} &= \frac{1}{2}\begin{bmatrix} q_0 & -q_3 & q_2 \\ q_3 & q_0 & -q_1 \\ -q_2 & q_1 & q_0 \end{bmatrix}\begin{bmatrix} \omega_1 \\ \omega_2 \\ \omega_3 \end{bmatrix} \\ &= \mathbf{g}(q_1, q_2, q_3, \omega) \end{aligned} \tag{11.87}$$

since $q_0 = \sqrt{1 - q_1^2 - q_2^2 - q_3^2}$. It can be rewritten simply as

$$\dot{\mathbf{q}} = \mathbf{g}(\mathbf{q}, \omega) \tag{11.88}$$

From (3.60), (see also [243, 246]),

$$\mathbf{A}_l^b = \begin{bmatrix} 2q_0^2-1+2q_1^2 & 2q_1q_2+2q_0q_3 & 2q_1q_3-2q_0q_2 \\ 2q_1q_2-2q_0q_3 & 2q_0^2-1+2q_2^2 & 2q_2q_3+2q_0q_1 \\ 2q_1q_3+2q_0q_2 & 2q_2q_3-2q_0q_1 & 2q_0^2-1+2q_3^2 \end{bmatrix}$$

we have

$$\omega_{lvlh}^b = \mathbf{A}_l^b \omega_{lvlh} = \begin{bmatrix} 2q_1q_2+2q_0q_3 \\ 2q_0^2-1+2q_2^2 \\ 2q_2q_3-2q_0q_1 \end{bmatrix} (-\omega_0) \tag{11.89}$$

which is a function of $\mathbf{q}$. Interestingly, given spacecraft inertia matrix $\mathbf{J}$, $\mathbf{t}_g$ is also a function of $\mathbf{q}$. Using the facts (a) the spacecraft mass is negligible compared to the Earth mass, and (b) the size of the spacecraft is negligible compared to the magnitude of the vector from the center of the Earth to the center of the mass of the spacecraft $\mathbf{R}$, the gravitational torque is given by (5.5) (see also [220, page 367]):

$$\mathbf{t}_g = \frac{3\mu}{|\mathbf{R}|^5} \mathbf{R} \times \mathbf{J}\mathbf{R} \tag{11.90}$$

where $\mu = GM, G = 6.669 * 10^{-11} m^3/kg - s^2$ is the universal constant of gravitation, and M is the mass of the Earth. Noticing that in local vertical local horizontal frame, $\mathbf{R}_l = [0, 0, -|\mathbf{R}|]^{\mathrm{T}}$, we can represent $\mathbf{R}$ in body frame as

$$\mathbf{R} = \mathbf{A}_l^b \mathbf{R}_l = \begin{bmatrix} 2q_0^2-1+2q_1^2 & 2q_1q_2+2q_0q_3 & 2q_1q_3-2q_0q_2 \\ 2q_1q_2-2q_0q_3 & 2q_0^2-1+2q_2^2 & 2q_2q_3+2q_0q_1 \\ 2q_1q_3+2q_0q_2 & 2q_2q_3-2q_0q_1 & 2q_0^2-1+2q_3^2 \end{bmatrix} \begin{bmatrix} 0 \\ 0 \\ -|\mathbf{R}| \end{bmatrix} \tag{11.91}$$

Denote the last column of $\mathbf{A}_l^b$ as $\mathbf{A}_l^b(:,3)$. Using the relation (2.55) (see also [181, page 109])

$$\omega_0 = \sqrt{\frac{\mu}{|\mathbf{R}|^3}} \tag{11.92}$$

and (11.91), we can rewrite (11.90) as

$$\mathbf{t}_g = 3\omega_0^2 \mathbf{A}_l^b(:,3) \times \mathbf{J}\mathbf{A}_l^b(:,3) \tag{11.93}$$

Let $\mathbf{b}(t) = [b_1(t), b_2(t), b_3(t)]^{\mathrm{T}}$ be the Earth's magnetic field in the spacecraft coordinates, computed using the spacecraft position, the spacecraft attitude, and a spherical harmonic model of the Earth's magnetic field [219]. Let $\mathbf{m} = [m_1, m_2, m_3]^{\mathrm{T}}$ be the spacecraft magnetic torque coils' induced magnetic moment in the spacecraft coordinates. The desired magnetic control torque $\mathbf{t}_m$ may not be achievable because

$$\mathbf{t}_m = \mathbf{m} \times \mathbf{b} = -\mathbf{b} \times \mathbf{m} \tag{11.94}$$

provides only a torque in a two-dimensional plane but not in the three dimensional space [181]. However, the spacecraft magnetic torque coils' induced magnetic moment $\mathbf{m}$ is an achievable engineering variable. Therefore, equation (11.85) should be rewritten as

$$\mathbf{J}\dot{\omega} = \mathbf{f}(\omega, \Omega, \mathbf{q}) + \mathbf{t}_w + \mathbf{t}_g - \mathbf{b} \times \mathbf{m} + \mathbf{t}_d \tag{11.95}$$

where

$$\mathbf{f}(\omega, \Omega, \mathbf{q}) = \mathbf{J}\omega \times \omega^b_{lvlh} - (\omega + \omega^b_{lvlh}) \times [\mathbf{J}(\omega + \omega^b_{lvlh}) + \mathbf{J}_w\Omega] \tag{11.96}$$

Notice that the cross product of $\mathbf{b} \times \mathbf{m}$ can be expressed as a product of an asymmetric matrix $\mathbf{b}^{\times}$ and the vector $\mathbf{m}$ with

$$\mathbf{b}^{\times} = \begin{bmatrix} 0 & -b_3(t) & b_2(t) \\ b_3(t) & 0 & -b_1(t) \\ -b_2(t) & b_1(t) & 0 \end{bmatrix} \tag{11.97}$$

Denote the system states $\mathbf{x} = [\omega^{\mathrm{T}}, \Omega^{\mathrm{T}}, \mathbf{q}^{\mathrm{T}}]^{\mathrm{T}}$ and control inputs $\mathbf{u} = [\mathbf{t}_w^{\mathrm{T}}, \mathbf{m}^{\mathrm{T}}]^{\mathrm{T}}$. The spacecraft control system model can be written as follows:

$$\mathbf{J}\dot{\omega} = \mathbf{f}(\omega, \Omega, \mathbf{q}) + \mathbf{t}_g + [\mathbf{I}, -\mathbf{b}^{\times}]\mathbf{u} + \mathbf{t}_d \tag{11.98a}$$

$$\mathbf{J}_w\dot{\Omega} = -\mathbf{t}_w \tag{11.98b}$$

$$\dot{\mathbf{q}} = \mathbf{g}(\mathbf{q}, \omega) \tag{11.98c}$$

Remark 11.5 The reduced quaternion, instead of the full quaternion, is proposed in this model because of many merits discussed in Chapter 9 (see also [243, 245, 250]). ▮

11.4.2 Linearized Model for Attitude and Reaction Wheel Desaturation Control

The nonlinear model of (11.98) can be used to design control systems. One popular design method for nonlinear model involves Lyapunov stability theorem, which is actually used in [5, 203]. A design based on this method focuses on stability but not on performance. Another widely known method is nonlinear optimal control design [44], it normally produces an open-loop controller which is not robust [108] and its computational cost is high. Therefore, it is proposed to use Linear Quadratic Regulator (LQR) which achieves the optimal performance for the linearized system and is a closed-loop feedback control. The task in this section is to derive the linearized model for the nonlinear system (11.98).

In view of (4.21), ω_{lvlh}^{b} in (11.89) can be expressed approximately as a linear function of $\mathbf{q}$ as follows

$$\omega_{lvlh}^{b} \approx \begin{bmatrix} 0 & 0 & -2\omega_0 \\ 0 & 0 & 0 \\ 2\omega_0 & 0 & 0 \end{bmatrix} \mathbf{q} - \begin{bmatrix} 0 \\ \omega_0 \\ 0 \end{bmatrix} \tag{11.99}$$

Similarly, $\mathbf{t}_g$ in (11.93) can be expressed approximately as a linear function of $\mathbf{q}$ as in (5.9):

$$\mathbf{t}_g \approx \begin{bmatrix} 6\omega_0^2(J_3-J_2) & 0 & 0 \\ 0 & 6\omega_0^2(J_3-J_1) & 0 \\ 0 & 0 & 0 \end{bmatrix} \mathbf{q} := \mathbf{T}\mathbf{q} \tag{11.100}$$

Since $\mathbf{t}_g$ and ω_{lvlh}^{b} are functions of $\mathbf{q}$, the linearized spacecraft model can be expressed as follows:

$$\begin{bmatrix} \mathbf{J} & \mathbf{0} & \mathbf{0} \\ \mathbf{0} & \mathbf{J}_w & \mathbf{0} \\ \mathbf{0} & \mathbf{0} & \mathbf{I} \end{bmatrix} \begin{bmatrix} \dot{\omega} \\ \dot{\Omega} \\ \dot{\mathbf{q}} \end{bmatrix} = \begin{bmatrix} \frac{\partial \mathbf{f}}{\partial \omega} & \frac{\partial \mathbf{f}}{\partial \Omega} & \frac{\partial \mathbf{f}}{\partial \mathbf{q}} + \mathbf{T} \\ \mathbf{0} & \mathbf{0} & \mathbf{0} \\ \frac{\partial \mathbf{g}}{\partial \omega} & \mathbf{0} & \frac{\partial \mathbf{g}}{\partial \mathbf{q}} \end{bmatrix} \begin{bmatrix} \omega \\ \Omega \\ \mathbf{q} \end{bmatrix} + \begin{bmatrix} \mathbf{I} & -\mathbf{b}^{\times} \\ -\mathbf{I} & \mathbf{0} \\ \mathbf{0} & \mathbf{0} \end{bmatrix} \begin{bmatrix} \mathbf{t}_w \\ \mathbf{m} \end{bmatrix} + \begin{bmatrix} \mathbf{t}_d \\ \mathbf{0} \\ \mathbf{0} \end{bmatrix} \tag{11.101}$$

where $\frac{\partial \mathbf{f}}{\partial \omega}$, $\frac{\partial \mathbf{f}}{\partial \Omega}$, $\frac{\partial \mathbf{f}}{\partial \mathbf{q}}$, $\frac{\partial \mathbf{g}}{\partial \omega}$, and $\frac{\partial \mathbf{g}}{\partial \mathbf{q}}$ are evaluated at the desired equilibrium point $\omega = 0$, $\Omega = 0$, and $\mathbf{q} = 0$. Using the definition of (11.97), (11.99), (11.100), and (11.96), we have

$$\begin{aligned} \left.\frac{\partial \mathbf{f}}{\partial \omega}\right|_{\substack{\omega\approx 0\\ \Omega\approx 0\\ \mathbf{q}\approx 0}} &\approx \left. -\mathbf{J}(\omega_{lvlh}^{b})^{\times} + (\mathbf{J}\omega_{lvlh}^{b})^{\times} - (\omega_{lvlh}^{b})^{\times}\mathbf{J} \right|_{\substack{\omega\approx 0\\ \Omega\approx 0\\ \mathbf{q}\approx 0}} \\ &= -\mathbf{J}\begin{bmatrix} 0 & 0 & -\omega_0 \\ 0 & 0 & 0 \\ \omega_0 & 0 & 0 \end{bmatrix} + \begin{bmatrix} 0 & 0 & -J_2\omega_0 \\ 0 & 0 & 0 \\ J_2\omega_0 & 0 & 0 \end{bmatrix} - \begin{bmatrix} 0 & 0 & -\omega_0 \\ 0 & 0 & 0 \\ \omega_0 & 0 & 0 \end{bmatrix}\mathbf{J} \\ &= \begin{bmatrix} 0 & 0 & \omega_0(J_1-J_2+J_3) \\ 0 & 0 & 0 \\ \omega_0(-J_1+J_2-J_3) & 0 & 0 \end{bmatrix} \end{aligned} \tag{11.102}$$

$$\left.\frac{\partial \mathbf{f}}{\partial \Omega}\right|_{\substack{\omega\approx 0\\ \Omega\approx 0\\ \mathbf{q}\approx 0}} \approx \left. -(\omega)^{\times}\mathbf{J}_w - (\omega_{lvlh}^{b})^{\times}\mathbf{J}_w \right|_{\substack{\omega\approx 0\\ \Omega\approx 0\\ \mathbf{q}\approx 0}}$$

$$
\begin{aligned}
&= \begin{bmatrix} 0 & 0 & \omega_0 \\ 0 & 0 & 0 \\ -\omega_0 & 0 & 0 \end{bmatrix} \mathbf{J}_w \\
&= \begin{bmatrix} 0 & 0 & \omega_0 J_{w_3} \\ 0 & 0 & 0 \\ -\omega_0 J_{w_1} & 0 & 0 \end{bmatrix}
\end{aligned} \tag{11.103}
$$

and

$$
\begin{aligned}
\left.\frac{\partial \mathbf{f}}{\partial \mathbf{q}}\right|_{\substack{\omega\approx 0\\ \Omega\approx 0\\ \mathbf{q}\approx 0}} &\approx -\frac{\partial}{\partial \mathbf{q}}\left(\omega_{lvlh}^b \times \mathbf{J}\omega_{lvlh}^b\right)\Big|_{\substack{\omega\approx 0\\ \Omega\approx 0\\ \mathbf{q}\approx 0}} \\
&= (\mathbf{J}\omega_{lvlh}^b)^\times \frac{\partial \omega_{lvlh}^b}{\partial \mathbf{q}} - \omega_{lvlh}^b \times \mathbf{J}\frac{\partial \omega_{lvlh}^b}{\partial \mathbf{q}}\Big|_{\substack{\omega\approx 0\\ \Omega\approx 0\\ \mathbf{q}\approx 0}} \\
&\approx (\mathbf{J}\omega_{lvlh}^b)^\times \begin{bmatrix} 0 & 0 & -2\omega_0 \\ 0 & 0 & 0 \\ 2\omega_0 & 0 & 0 \end{bmatrix} - (\omega_{lvlh}^b)^\times \mathbf{J} \begin{bmatrix} 0 & 0 & -2\omega_0 \\ 0 & 0 & 0 \\ 2\omega_0 & 0 & 0 \end{bmatrix} \\
&\approx \left(\begin{bmatrix} 0 \\ -\omega_0 J_2 \\ 0 \end{bmatrix}^\times - \begin{bmatrix} 0 & 0 & -\omega_0 \\ 0 & 0 & 0 \\ \omega_0 & 0 & 0 \end{bmatrix} \mathbf{J} \right) \begin{bmatrix} 0 & 0 & -2\omega_0 \\ 0 & 0 & 0 \\ 2\omega_0 & 0 & 0 \end{bmatrix} \\
&\approx \begin{bmatrix} 0 & 0 & -\omega_0(J_2 - J_3) \\ 0 & 0 & 0 \\ -\omega_0(J_1 - J_2) & 0 & 0 \end{bmatrix} \begin{bmatrix} 0 & 0 & -2\omega_0 \\ 0 & 0 & 0 \\ 2\omega_0 & 0 & 0 \end{bmatrix} \\
&= \begin{bmatrix} 2\omega_0^2(J_3 - J_2) & 0 & 0 \\ 0 & 0 & 0 \\ 0 & 0 & 2\omega_0^2(J_1 - J_2) \end{bmatrix}
\end{aligned} \tag{11.104}
$$

From (11.88), we have

$$
\left.\frac{\partial \mathbf{g}}{\partial \omega}\right|_{\substack{\omega\approx 0\\ \mathbf{q}\approx 0}} \approx \frac{1}{2}\mathbf{I} \tag{11.105}
$$

$$
\left.\frac{\partial \mathbf{g}}{\partial \mathbf{q}}\right|_{\substack{\omega\approx 0\\ \mathbf{q}\approx 0}} \approx \mathbf{0} \tag{11.106}
$$

Substituting (11.100), (11.97), (11.102), (11.103), (11.104), (11.105), and (11.106) into (11.101) yields

$$
\begin{aligned}
\dot{\mathbf{x}} &:= \begin{bmatrix} \dot{\omega} \\ \dot{\Omega} \\ \dot{\mathbf{q}} \end{bmatrix} \\
&= \begin{bmatrix} \mathbf{J}^{-1}\frac{\partial \mathbf{f}}{\partial \omega} & \mathbf{J}^{-1}\frac{\partial \mathbf{f}}{\partial \Omega} & \mathbf{J}^{-1}\left(\frac{\partial \mathbf{f}}{\partial \mathbf{q}} + \mathbf{T}\right) \\ \mathbf{0} & \mathbf{0} & \mathbf{0} \\ \frac{\partial \mathbf{g}}{\partial \omega} & \mathbf{0} & \frac{\partial \mathbf{g}}{\partial \mathbf{q}} \end{bmatrix} \begin{bmatrix} \omega \\ \Omega \\ \mathbf{q} \end{bmatrix} \\
&+ \begin{bmatrix} \mathbf{J}^{-1} & -\mathbf{J}^{-1}\mathbf{b}^{\times} \\ -\mathbf{J}_w^{-1} & \mathbf{0} \\ \mathbf{0} & \mathbf{0} \end{bmatrix} \begin{bmatrix} \mathbf{t}_w \\ \mathbf{m} \end{bmatrix} + \begin{bmatrix} \mathbf{J}^{-1} \\ \mathbf{0} \\ \mathbf{0} \end{bmatrix} \mathbf{t}_d \\
&= \begin{bmatrix}
0 & 0 & a_{13} & 0 & 0 & a_{16} & a_{17} & 0 & 0 \\
0 & 0 & 0 & 0 & 0 & 0 & 0 & a_{28} & 0 \\
a_{31} & 0 & 0 & a_{34} & 0 & 0 & 0 & 0 & a_{39} \\
0 & 0 & 0 & 0 & 0 & 0 & 0 & 0 & 0 \\
0 & 0 & 0 & 0 & 0 & 0 & 0 & 0 & 0 \\
0 & 0 & 0 & 0 & 0 & 0 & 0 & 0 & 0 \\
0.5 & 0 & 0 & 0 & 0 & 0 & 0 & 0 & 0 \\
0 & 0.5 & 0 & 0 & 0 & 0 & 0 & 0 & 0 \\
0 & 0 & 0.5 & 0 & 0 & 0 & 0 & 0 & 0
\end{bmatrix}
\begin{bmatrix} \omega_1 \\ \omega_2 \\ \omega_3 \\ \Omega_1 \\ \Omega_2 \\ \Omega_3 \\ q_1 \\ q_2 \\ q_3 \end{bmatrix} \\
&+ \begin{bmatrix}
J_1^{-1} & 0 & 0 & 0 & \frac{b_3(t)}{J_1} & -\frac{b_2(t)}{J_1} \\
0 & J_2^{-1} & 0 & -\frac{b_3(t)}{J_2} & 0 & \frac{b_1(t)}{J_2} \\
0 & 0 & J_3^{-1} & \frac{b_2(t)}{J_3} & -\frac{b_1(t)}{J_3} & 0 \\
-J_{w_1}^{-1} & 0 & 0 & 0 & 0 & 0 \\
0 & -J_{w_2}^{-1} & 0 & 0 & 0 & 0 \\
0 & 0 & -J_{w_3}^{-1} & 0 & 0 & 0 \\
0 & 0 & 0 & 0 & 0 & 0 \\
0 & 0 & 0 & 0 & 0 & 0 \\
0 & 0 & 0 & 0 & 0 & 0
\end{bmatrix}
\begin{bmatrix} t_{w_1} \\ t_{w_1} \\ t_{w_1} \\ m_1 \\ m_2 \\ m_3 \end{bmatrix}
+ \begin{bmatrix} \frac{t_{d_1}}{J_1} \\ \frac{t_{d_2}}{J_2} \\ \frac{t_{d_3}}{J_3} \\ 0 \\ 0 \\ 0 \\ 0 \\ 0 \\ 0 \end{bmatrix} \\
&:= \mathbf{A}\mathbf{x} + \mathbf{B}\mathbf{u} + \mathbf{d}
\end{aligned}
\tag{11.107}
$$

where $a_{13} = \omega_0 \frac{J_1 - J_2 + J_3}{J_1}$, $a_{16} = \frac{\omega_0 J_{w3}}{J_1}$, $a_{17} = 8\omega_0^2 \frac{J_3 - J_2}{J_1}$, $a_{28} = 6\omega_0^2 \frac{J_3 - J_1}{J_2}$, $a_{31} = \omega_0 \frac{J_1 - J_2 + J_3}{-J_3}$, $a_{34} = \frac{\omega_0 J_{w1}}{-J_3}$, $a_{39} = 2\omega_0^2 \frac{J_1 - J_2}{J_3}$. It is worthwhile to notice that (11.107) is in general a time-varying system. The time-variation of the system arises from an approximately periodic function of $\mathbf{b}(t) = \mathbf{b}(t+T)$, where T is the orbital period given in (11.32). This magnetic field $\mathbf{b}(t)$ is given in (11.3). The time $t = 0$ is measured at the ascending-node crossing of the magnetic equator. Therefore, the periodic time-varying matrix $\mathbf{B}$ in (11.107) can be written as

$$\mathbf{B}=\begin{bmatrix} J_1^{-1} & 0 & 0 & 0 & b_{15} & b_{16} \\ 0 & J_2^{-1} & 0 & b_{24} & 0 & b_{26} \\ 0 & 0 & J_3^{-1} & b_{34} & b_{35} & 0 \\ -J_{w_1}^{-1} & 0 & 0 & 0 & 0 & 0 \\ 0 & -J_{w_2}^{-1} & 0 & 0 & 0 & 0 \\ 0 & 0 & -J_{w_3}^{-1} & 0 & 0 & 0 \\ 0 & 0 & 0 & 0 & 0 & 0 \\ 0 & 0 & 0 & 0 & 0 & 0 \\ 0 & 0 & 0 & 0 & 0 & 0 \end{bmatrix} \tag{11.108}$$

where

$$\begin{aligned} b_{15} &= \tfrac{2\mu_f}{a^3 J_1}\sin(i_m)\sin(\omega_0 t) \\ b_{16} &= \tfrac{\mu_f}{a^3 J_1}\cos(i_m) \\ b_{24} &= -\tfrac{2\mu_f}{a^3 J_2}\sin(i_m)\sin(\omega_0 t) \\ b_{26} &= \tfrac{\mu_f}{a^3 J_2}\sin(i_m)\cos(\omega_0 t) \\ b_{34} &= -\tfrac{\mu_f}{a^3 J_3}\cos(i_m) \\ b_{35} &= -\tfrac{\mu_f}{a^3 J_3}\sin(i_m)\cos(\omega_0 t) \end{aligned}$$

A special case is when $i_m = 0$, i.e., the spacecraft orbit is on the equator plane of the Earth's magnetic field. In this case, $\mathbf{b}(t) = [0, -\frac{\mu_f}{a^3}, 0]^{\mathrm{T}}$ is a constant vector and $\mathbf{B}$ is reduced to a constant matrix given as follows:

$$\mathbf{B}=\begin{bmatrix} J_1^{-1} & 0 & 0 & 0 & 0 & \frac{\mu_f}{a^3 J_1} \\ 0 & J_2^{-1} & 0 & 0 & 0 & 0 \\ 0 & 0 & J_3^{-1} & -\frac{\mu_f}{a^3 J_3} & 0 & 0 \\ -J_{w_1}^{-1} & 0 & 0 & 0 & 0 & 0 \\ 0 & -J_{w_2}^{-1} & 0 & 0 & 0 & 0 \\ 0 & 0 & -J_{w_3}^{-1} & 0 & 0 & 0 \\ 0 & 0 & 0 & 0 & 0 & 0 \\ 0 & 0 & 0 & 0 & 0 & 0 \\ 0 & 0 & 0 & 0 & 0 & 0 \end{bmatrix} \tag{11.109}$$

In the remainder of the discussion, we will consider the discrete time system of (11.107) because it is more suitable for computer controlled system implementations. The discrete time system is given as follows:

$$\mathbf{x}_{k+1} = \mathbf{A}\mathbf{x}_k + \mathbf{B}_k\mathbf{u}_k + \mathbf{d}_k \tag{11.110}$$

Assuming that the sampling time is t_s, the simplest but less accurate discretization formulas to get $\mathbf{A}_k$ and $\mathbf{B}_k$ are given as (11.34). A slightly more complex but more accurate discretization formulas to get $\mathbf{A}_k$ and $\mathbf{B}_k$ are given as follows [108, page 53]:

$$\mathbf{A}_k = e^{\mathbf{A}t_s}, \quad \mathbf{B}_k = \int_0^{t_s} e^{\mathbf{A}\tau}B(\tau)d\tau \tag{11.111}$$

11.4.3 The LQR Design

Given the linearized spacecraft model (11.107) which has the state variables composed of spacecraft quaternion $\mathbf{q}$, the spacecraft rotational rate with respect to the LVLH frame ω, and the reaction wheel rotational speed Ω, one can see that controlling the spacecraft attitude and managing the reaction wheel momentum are equivalent to minimizing the following objective function

$$\int_0^{\infty} (\mathbf{x}^{\mathrm{T}}\mathbf{Q}\mathbf{x} + \mathbf{u}^{\mathrm{T}}\mathbf{R}\mathbf{u})dt \tag{11.112}$$

under the constraints of (11.107). The corresponding discrete time system is given as follows:

$$\lim_{N\to\infty} \left(\min \frac{1}{2}\mathbf{x}_N^{\mathrm{T}}\mathbf{Q}_N\mathbf{x}_N + \frac{1}{2}\sum_{k=0}^{N-1} \mathbf{x}_k^{\mathrm{T}}\mathbf{Q}_k\mathbf{x}_k + \mathbf{u}_k^{\mathrm{T}}\mathbf{R}_k\mathbf{u}_k \right)$$
$$\text{s.t.} \qquad \mathbf{x}_{k+1} = \mathbf{A}\mathbf{x}_k + \mathbf{B}_k\mathbf{u}_k + \mathbf{d}_k \tag{11.113}$$

This is clearly a LQR design problem which has known efficient methods to solve. However, in each special case, this system has some special properties which should be fully utilized to select the most efficient and effective method for each of these cases.

11.4.3.1 Case 1: $\mathbf{i}_m = 0$

It was shown earlier in this chapter that a spacecraft in this orbit is not controllable if only magnetic torque bars are used. But for a spacecraft with three reaction wheels as discussed in this section, the system is fully controllable. The controllability condition can be checked straightforward but the check is tedious and is omitted in this section (also the controllability check is not the focus of this section). In this case, as we have seen from (11.107), (11.109), and (11.34) that the linear system is time-invariant. Therefore, a method for time-varying system is not appropriate for this simple problem. For this *linear time-invariant* system (LTI), the optimal solution of (11.113) is given by (B.50) (see also [108, page 69])

$$\mathbf{u}_k = -(\mathbf{R} + \mathbf{B}^{\mathrm{T}}\mathbf{P}\mathbf{B})^{-1}\mathbf{B}^{\mathrm{T}}\mathbf{P}\mathbf{A}\mathbf{x}_k = -\mathbf{K}\mathbf{x}_k \tag{11.114}$$

where $\mathbf{P}$ is a constant positive semi-definite solution of the following discrete-time algebraic Riccati equation (DARE) (B.29)

$$\mathbf{P} = \mathbf{Q} + \mathbf{A}^{\mathrm{T}}\mathbf{P}\mathbf{A} - \mathbf{A}^{\mathrm{T}}\mathbf{P}\mathbf{B}(\mathbf{R} + \mathbf{B}^{\mathrm{T}}\mathbf{P}\mathbf{B})^{-1}\mathbf{B}^{\mathrm{T}}\mathbf{P}\mathbf{A} \tag{11.115}$$

The solution of (11.115) is discussed in Appendix B.3. There is an efficient algorithms [6] for this DARE system and an MATLAB® function `dare` implements this algorithm.

11.4.3.2 Case 2: $\mathbf{i}_m \neq 0$

It was shown earlier in this chapter that a spacecraft without any reaction wheel in any orbit of this case is controllable if the spacecraft design satisfies some additional conditions imposed on **J** matrix. By intuition, the system is also controllable by adding reaction wheels. As a matter of fact, adding reaction wheels will achieve better performance of spacecraft attitude as will be seen later in this section (which is also pointed out in [219, page 19]). A better algorithm for this case is the one developed earlier in this chapter because **B** is a time-varying matrix but **A** is a constant matrix. The optimal solution of (11.113) is discussed in the previous section, which is given by

$$\mathbf{u}_k = -(\mathbf{R}_k + \mathbf{B}_k^{\mathrm{T}}\mathbf{P}_k\mathbf{B}_k)^{-1}\mathbf{B}_k^{\mathrm{T}}\mathbf{P}_k\mathbf{A}_k\mathbf{x}_k = -\mathbf{K}_k\mathbf{x}_k \tag{11.116}$$

where $\mathbf{P}_k$ is a periodic positive semi-definite solution of the periodic time-varying Riccati (PTVR) equation (B.19) which is rewritten here

$$\begin{aligned}\mathbf{P}_k &= \mathbf{Q}_k + \mathbf{A}_k^{\mathrm{T}}\mathbf{P}_k\mathbf{A}_k \\ &\quad -\mathbf{A}_k^{\mathrm{T}}\mathbf{P}_k\mathbf{B}_k(\mathbf{R}_k + \mathbf{B}_k^{\mathrm{T}}\mathbf{P}_k\mathbf{B}_k)^{-1}\mathbf{B}_k^{\mathrm{T}}\mathbf{P}_k\mathbf{A}_k\end{aligned} \tag{11.117}$$

The periodic Riccati equation for this case is discussed in the previous section and the algorithm is presented below:

Algorithm 11.2

Data: i_m, **J**, $\mathbf{J}_w$, **Q**, **R**, *the altitude of the spacecraft (for the calculation of a in (11.3)),* t_s *(the selected sample time period), and* p *(the total samples in one period* $T = \frac{2\pi}{\omega_0}$*).*

Step 1: For $k = 1, \ldots, p$*, calculate* $\mathbf{A}_k$ *and* $\mathbf{B}_k$ *using (11.34) or (11.111).*

Step 2: Calculate $\mathbf{E}_k$ *and* $\mathbf{F}_k$ *using*

$$\mathbf{E}_k = \begin{bmatrix} \mathbf{I} & \mathbf{B}_k\mathbf{R}^{-1}\mathbf{B}_k^{\mathrm{T}} \\ \mathbf{0} & \mathbf{A}^{\mathrm{T}} \end{bmatrix} \tag{11.118}$$

$$\mathbf{F}_k = \begin{bmatrix} \mathbf{A} & \mathbf{0} \\ -\mathbf{Q} & \mathbf{I} \end{bmatrix} = \mathbf{F} \tag{11.119}$$

Step 3: Calculate Γ_k*, for* $k = 1, \ldots, p$*, using*

$$\Gamma_k = \mathbf{F}^{-1}\mathbf{E}_k\mathbf{F}^{-1}\mathbf{E}_{k+1}\ldots,\mathbf{F}^{-1}\mathbf{E}_{k+p-2}\mathbf{F}^{-1}\mathbf{E}_{k+p-1} \tag{11.120}$$

Step 4: Use Schur decomposition

$$\begin{bmatrix} \mathbf{W}_{11k} & \mathbf{W}_{12k} \\ \mathbf{W}_{21k} & \mathbf{W}_{22k} \end{bmatrix}^{\mathrm{T}} \Gamma_k \begin{bmatrix} \mathbf{W}_{11k} & \mathbf{W}_{12k} \\ \mathbf{W}_{21k} & \mathbf{W}_{22k} \end{bmatrix} = \begin{bmatrix} \mathbf{S}_{11k} & \mathbf{S}_{12k} \\ \mathbf{0} & \mathbf{S}_{22k} \end{bmatrix} \tag{11.121}$$

Step 5: Calculate $\mathbf{P}_k$ *using*

$$\mathbf{P}_k = \mathbf{W}_{21k}\mathbf{W}_{11k}^{-1} \tag{11.122}$$

11.4.4 Simulation Test and Implementation Consideration

This section has several goals. First, it shows, by using a design example, that the proposed design achieves both attitude control and reaction wheel momentum management. Second, it compares with the design in the previous section which does not use reaction wheels for the purpose to show that using reaction wheels achieves better attitude pointing accuracy. More importantly, it demonstrates that the LQR design works very well for both attitude and desaturation control for the nonlinear spacecraft in the environment close to the reality. Finally, it discusses the strategy in real spacecraft control system implementation.

11.4.4.1 Comparison with the Design without Reaction Wheels

The proposed design algorithm has been tested using the same spacecraft model and orbit parameters as in the previous section with the spacecraft inertia matrix given by

$$\mathbf{J} = \mathrm{diag}\,(250, 150, 100)\,kg \cdot m^2$$

The orbital inclination $i_m = 57^o$ and the orbit is assumed to be circular with the altitude 657 km. In view of equation (11.32), the orbital period is 5863 seconds and the orbital rate is $\omega_0 = 0.0011$ rad/second. Assuming that the total number of samples taken in one orbit is 100, then, each sample period is 58.6352 second. It is easy to see that all parameters are selected the same as the simulation example in the previous section so that the two different designs can be compared. Select $\mathbf{Q} = \mathrm{diag}([0.001, 0.001, 0.001, 0.001, 0.001, 0.001, 0.02, 0.02, 0.02])$ and $\mathbf{R} = \mathrm{diag}([10^3, 10^3, 10^3, 10^2, 10^2, 10^2])$. The solution of the periodic Riccati equations $\mathbf{P}_k$ for $k = 0, 1, 2, \ldots, 99$ have been calculated and stored using Algorithm 11.2. Assuming that the initial quaternion error is $(0.01, 0.01, 0.01)$, initial body rate vector is $(0.00001, 0.00001, 0.00001)$ radians/second, and the initial wheel speed vector is $(0.00001, 0.00001, 0.00001)$ radians/second, applying the feedback (11.116) to the linearized system (11.107) and (11.108), the linearized spacecraft rotational rate response is obtained and given in Figure 11.7, the reaction wheel response is given in Figure 11.8, and the spacecraft attitude response is given in Figure 11.9.

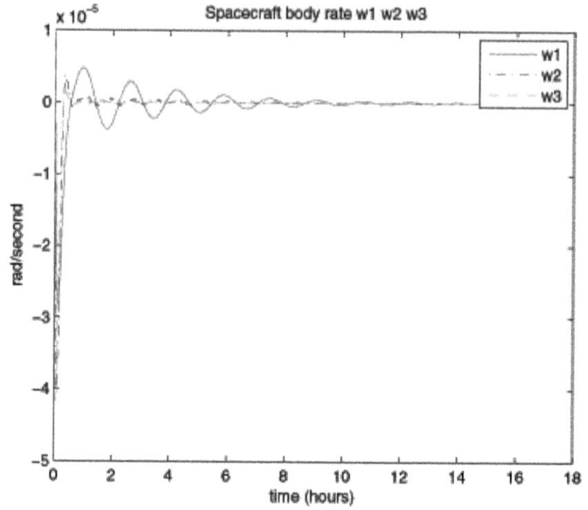

Figure 11.7: Body rate response ω_1, ω_2, and ω_3.

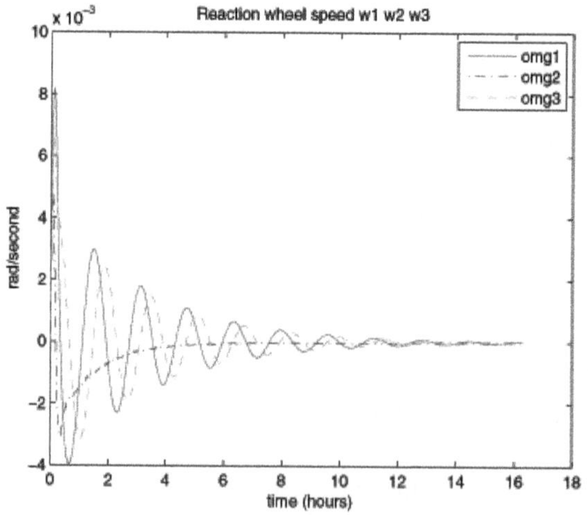

Figure 11.8: Reaction wheel response Ω_1, Ω_2, and Ω_3.

Comparing the response obtained here using both reaction wheels and magnetic torque coils and the response obtained in the previous section that uses magnetic torques only, one can see that both control methods stabilize the spacecraft, but using reaction wheels achieve much accurate nadir pointing. Also reaction wheel speeds approach to zero as t goes to infinity. Therefore, the second design goal for reaction wheel desaturation is achieved nicely.

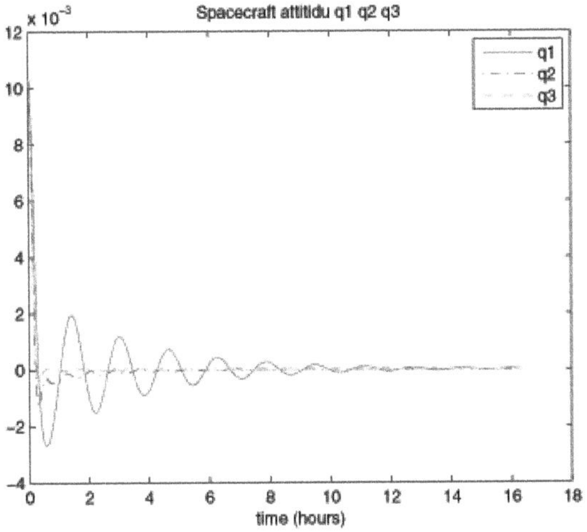

Figure 11.9: Attitude response q_1, q_2, and q_3.

11.4.4.2 Control of the Nonlinear System

It is nature to ask the following question: can the designed controller (11.116), which is based on the linearized model, stabilize the original nonlinear spacecraft system (11.98) with satisfied performance? This question is answered by applying the designed controller to the original nonlinear spacecraft system (11.98). More specifically, the LVLH frame rotational rate ω^b_{lvlh} is calculated using the accurate nonlinear formula (11.89) rather than the approximated linear model (11.99). The gravity gradient torque $\mathbf{t}_g$ is calculated using the accurate nonlinear formula (11.93) rather than the approximated linear model (11.100). The Earth's magnetic field is calculated using the much accurate International Geomagnetic Reference Field (IGRF) model [47] rather than the simplified model (11.3). This is done as follows. First, combining (2.32) and (2.55) gives the lateral speed of the spacecraft $v = R\omega_0$. Given the altitude of the spacecraft (657 km) and the orbital radius R is 7028 kilometers, the lateral speed of the spacecraft is obtained. Assuming that the ascending node at $t = 0$ ("now") is the $\mathbf{X}$ axis of the ECEF frame, the velocity vector $\mathbf{v} = [0, v\cos(i_m), v\sin(i_m)]^{\mathrm{T}}$. Using Algorithm 3.4 of [37, page 142], one can get the spacecraft coordinate in ECI frame at any time after $t = 0$. Converting ECI coordinate to ECEF coordinate, one can calculate a much accurate Earth magnetic field vector $\mathbf{b}$ using IGRF model [47], which has been implemented in MATLAB. Applying this Earth's magnetic field vector $\mathbf{b}$ and feedback control $\mathbf{u}_k = -\mathbf{K}_k\mathbf{x}_k$ designed by the LQR method to (11.98), the nonlinear spacecraft system is controlled by using the LQR controller. Also, larger initial errors in 100 test cases (possibly 10 times larger than they were used in the previous simulation test) are randomly generated.

The nonlinear spacecraft system response to the LQR controller is given in Figures 11.10, 11.11, and 11.12. These figures show that the proposed design does achieve the design goals. Moreover, the difference between the linear (approximate) system response and nonlinear (true) system response for the LQR design is very small.

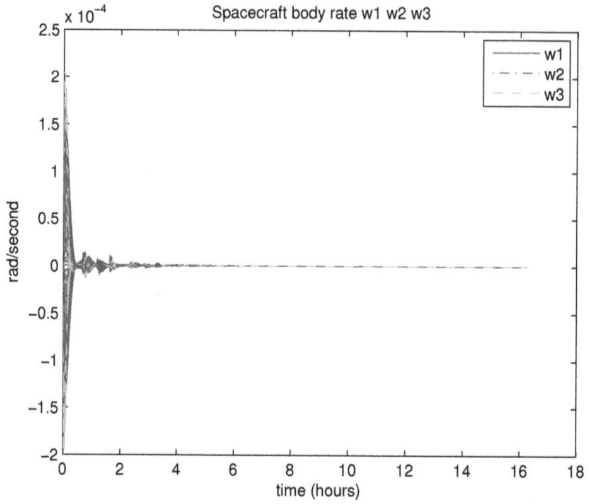

Figure 11.10: Body rate response ω_1, ω_2, and ω_3.

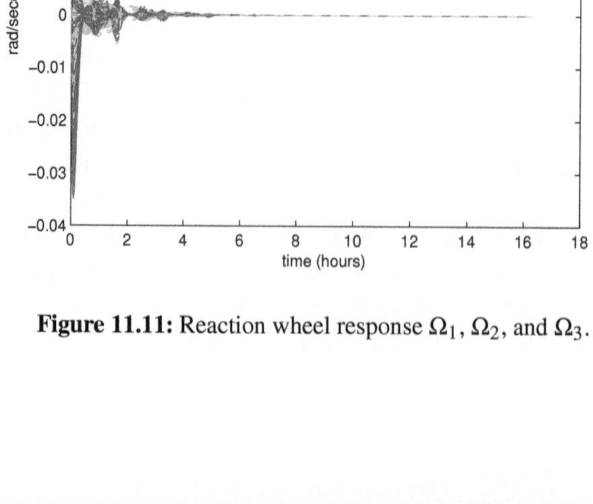

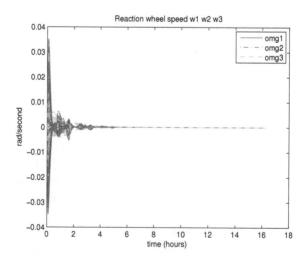

Figure 11.11: Reaction wheel response Ω_1, Ω_2, and Ω_3.

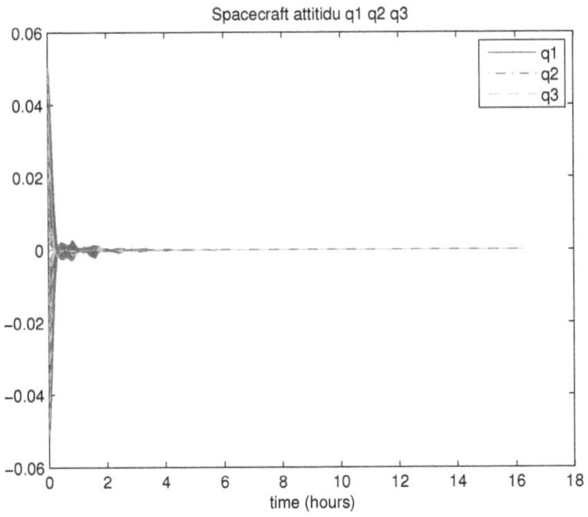

Figure 11.12: Attitude response q_1, q_2, and q_3.

11.4.4.3 Implementation to Real System

In real space environment, even the magnetic field vector obtained from the high fidelity IGRF model may not be identical to the real magnetic field vector which can be measured by magnetometer installed on spacecraft. Therefore, it is suggested to use the measured magnetic field vector $\mathbf{b}$ to form $\mathbf{B}_k$ in the state feedback (11.116). Because of the interaction between the magnetic torque coils and the magnetometer, it is a common practice that measurement and control are not taken at the same time (some time slot in a sample period is allocated to the measurement and the rest time in the sample period is allocated for control). Therefore, a scaling for the control gain should be taken to compensate for the time loss in the sample period when measurement is taken. For example, if the magnetic field measurement uses half time of the sample period, the control gain should be doubled because only half sample period is used for control. This is similar to the method used in [250], which will be discussed in the Chapter 12.

11.5 LQR Design Based on a Novel Lifting Method

It has been known for about six decades that linear periodic time-varying system can be converted to some equivalent linear time-invariant systems [85, 86]. The most popular and widely used methods that convert the linear periodic time-varying model into linear time-invariant models are the so-called *lifting methods* proposed in [60, 130]. But the LQR design for linear periodic system has been focused on the periodic system not on the equivalent linear time-invariant systems proposed in [60, 130]. This strategy leads to extensive research on the solutions of the periodic Riccati equations (see [19, 20, 21, 206, 207] and references therein). For the discrete-time *linear periodic* system, two efficient algorithms have been discussed in this chapter for Discrete-time Periodic Algebraic Riccati Equation (DPARE).

This section considers a novel lifting method that converts the linear periodic system to an augmented Linear Time-Invariant (LTI) system. It shows that the LQR design method can be directly applied to this LTI system. Moreover, by making full use of the structure of the augmented LTI system, one can derive a very efficient algorithm. The new algorithm is compared to the ones discussed in the previous sections of this chapter. In addition to some simple analysis, the efficiency and effectiveness of the new algorithm is demonstrated by the simulation test for the design problems of spacecraft attitude control using magnetic torques.

11.5.1 Periodic LQR Design Based on Linear Periodic System

First, the two efficient algorithms for solving DPARE discussed in the previous sections are briefly reviewed. This will be beneficial later in the comparison of the proposed method to the existing methods.

Let p be an integer representing the total number of samples in one period of a periodic discrete-time system. The following discrete-time linear periodic system is considered:

$$\mathbf{x}_{k+1} = \mathbf{A}_k \mathbf{x}_k + \mathbf{B}_k \mathbf{u}_k \tag{11.123}$$

where $\mathbf{A}_k = \mathbf{A}_{k+p} \in \mathbf{R}^{n \times n}$ and $\mathbf{B}_k = \mathbf{B}_{k+p} \in \mathbf{R}^{n \times m}$ are periodic time-varying matrices. For this discrete-time linear periodic system (11.33), the LQR state feedback control is to find the optimal $\mathbf{u}_k$ to minimize the following quadratic cost function

$$\lim_{N \to \infty} \left(\min \frac{1}{2} \mathbf{x}_N^{\mathrm{T}} \mathbf{Q}_N \mathbf{x}_N + \frac{1}{2} \sum_{k=0}^{N-1} \mathbf{x}_k^{\mathrm{T}} \mathbf{Q}_k \mathbf{x}_k + \mathbf{u}_k^{\mathrm{T}} \mathbf{R}_k \mathbf{u}_k \right) \tag{11.124}$$

where

$$\mathbf{Q}_k = \mathbf{Q}_{k+p} \geq 0 \tag{11.125}$$

$$\mathbf{R}_k = \mathbf{R}_{k+p} > 0 \tag{11.126}$$

and the initial condition $\mathbf{x}_0$ is given. It is now known that the LQR design for problem (11.33–11.37) can be solved by using the periodic solution of the discrete-time periodic algebraic Riccati equation described in the previous sections. These two algorithms solve p n-dimensional matrix Riccati equations to find p positive semidefinite matrices $\mathbf{P}_k$, $k = 1, \ldots, p$. Given $\mathbf{P}_k$, the periodic feedback controllers are given by the following equations:

$$\mathbf{u}_k = -(\mathbf{R}_k + \mathbf{B}_k^T \mathbf{P}_k \mathbf{B}_k)^{-1} \mathbf{B}_k^T \mathbf{P}_k \mathbf{A}_k \mathbf{x}_k \tag{11.127}$$

These two algorithms are summarized as follows: Let

$$\mathbf{E}_k = \begin{bmatrix} \mathbf{I} & \mathbf{B}_k \mathbf{R}_k^{-1} \mathbf{B}_k^T \\ \mathbf{0} & \mathbf{A}_k^T \end{bmatrix} = \mathbf{E}_{k+p} \tag{11.128}$$

$$\mathbf{F}_k = \begin{bmatrix} \mathbf{A}_k & \mathbf{0} \\ -\mathbf{Q}_k & \mathbf{I} \end{bmatrix} = \mathbf{F}_{k+p} \tag{11.129}$$

If $\mathbf{A}_k$ is invertible, then $\mathbf{E}_k$ and $\mathbf{F}_k$ are invertible, and

$$\mathbf{E}_k^{-1} = \begin{bmatrix} \mathbf{I} & -\mathbf{B}_k \mathbf{R}_k^{-1} \mathbf{B}_k^T \mathbf{A}_k^{-T} \\ \mathbf{0} & \mathbf{A}_k^{-T} \end{bmatrix} = \mathbf{E}_{k+p}^{-1}$$

and

$$\mathbf{F}_k^{-1} = \begin{bmatrix} \mathbf{A}_k^{-1} & \mathbf{0} \\ \mathbf{Q}_k \mathbf{A}_k^{-1} & \mathbf{I} \end{bmatrix} = \mathbf{F}_{k+p}^{-1}$$

Let $\mathbf{y}_k$ be the costate of $\mathbf{x}_k$, $\mathbf{z}_k = [\mathbf{x}_k^T, \mathbf{y}_k^T]^T$, and

$$\Pi_k = \mathbf{E}_{k+p-1}^{-1} \mathbf{F}_{k+p-1} \mathbf{E}_{k+p-2}^{-1} \mathbf{F}_{k+p-2} \ldots \mathbf{E}_{k+1}^{-1} \mathbf{F}_{k+1} \mathbf{E}_k^{-1} \mathbf{F}_k = \Pi_{k+p} \tag{11.130}$$

$$\Gamma_k = \mathbf{F}_k^{-1} \mathbf{E}_k \mathbf{F}_{k+1}^{-1} \mathbf{E}_{k+1} \ldots, \mathbf{F}_{k+p-2}^{-1} \mathbf{E}_{k+p-2} \mathbf{F}_{k+p-1}^{-1} \mathbf{E}_{k+p-1} = \Gamma_{k+p} \tag{11.131}$$

The solutions of p discrete-time periodic algebraic Riccati equations are symmetric positive semi-definite matrices, $\mathbf{P}_k$, $k = 1, \ldots, p$, which are related to the solutions of either one of the two linear systems of equations:

$$\mathbf{z}_{k+p} = \Pi_k \mathbf{z}_k \tag{11.132}$$

$$\mathbf{z}_k = \Gamma_k \mathbf{z}_{k+p} \tag{11.133}$$

Therefore, $\mathbf{P}_k$, $k = 1, \ldots, p$, can be obtained by two methods. The first method uses Schur decomposition:

$$\begin{bmatrix} \mathbf{T}_{11k} & \mathbf{T}_{12k} \\ \mathbf{T}_{21k} & \mathbf{T}_{22k} \end{bmatrix}^T \Pi_k \begin{bmatrix} \mathbf{T}_{11k} & \mathbf{T}_{12k} \\ \mathbf{T}_{21k} & \mathbf{T}_{22k} \end{bmatrix} = \begin{bmatrix} \mathbf{S}_{11k} & \mathbf{S}_{12k} \\ \mathbf{0} & \mathbf{S}_{22k} \end{bmatrix} \tag{11.134}$$

where $\mathbf{S}_{11k}$ is upper-triangular and has all of its eigenvalues inside the unit circle. The periodic solution $\mathbf{P}_k$, $k = 1, \ldots, p$, is given by

$$\mathbf{P}_k = \mathbf{T}_{21k} \mathbf{T}_{11k}^{-1} \tag{11.135}$$

The second method uses Schur decomposition:

$$\begin{bmatrix} \mathbf{W}_{11k} & \mathbf{W}_{12k} \\ \mathbf{W}_{21k} & \mathbf{W}_{22k} \end{bmatrix}^T \Gamma_k \begin{bmatrix} \mathbf{W}_{11k} & \mathbf{W}_{12k} \\ \mathbf{W}_{21k} & \mathbf{W}_{22k} \end{bmatrix} = \begin{bmatrix} \mathbf{U}_{11k} & \mathbf{U}_{12k} \\ \mathbf{0} & \mathbf{U}_{22k} \end{bmatrix} \tag{11.136}$$

where $\mathbf{U}_{11k}$ is upper-triangular and has all of its eigenvalues outside the unit circle. The periodic solution $\mathbf{P}_k$, $k=1,\ldots,p$, is given by

$$\mathbf{P}_k = \mathbf{W}_{21k}\mathbf{W}_{11k}^{-1} \tag{11.137}$$

Remark 11.6 When $\mathbf{A}_k$ and $\mathbf{Q}_k$ are constant matrices, the second method is much efficient because $\mathbf{F}_k$ becomes a constant matrix and $\mathbf{F}_k^{-1} = \cdots = \mathbf{F}_{k+p-1}^{-1} = \mathbf{F}^{-1}$, which makes the computation of (11.131) much more efficient than the computation of (11.130). ▮

11.5.2 Periodic LQR Design Based on Linear Time-invariant System

This section discusses a lifting method that converts the discrete-time linear periodic system into an augmented linear time-invariant system. Thereby, the periodic LQR design is reduced to the LQR design for the *augmented linear time-invariant* system.

To simplify the discussion, assume that the number of samples in a period is $p=3$. In this section, the small case k is used for the discrete-time in the periodic system and the capital K is used for the discrete-time in the augmented system.

$$\begin{aligned}
\mathbf{x}_1 &= \mathbf{A}_0\mathbf{x}_0 + \mathbf{B}_0\mathbf{u}_0 \\
\mathbf{x}_2 &= \mathbf{A}_1\mathbf{x}_1 + \mathbf{B}_1\mathbf{u}_1 \\
\mathbf{x}_3 &= \mathbf{A}_2\mathbf{x}_2 + \mathbf{B}_2\mathbf{u}_2 \\
\mathbf{x}_4 &= \mathbf{A}_0\mathbf{x}_3 + \mathbf{B}_0\mathbf{u}_3 \\
\mathbf{x}_5 &= \mathbf{A}_1\mathbf{x}_4 + \mathbf{B}_1\mathbf{u}_4 \\
\mathbf{x}_6 &= \mathbf{A}_2\mathbf{x}_5 + \mathbf{B}_2\mathbf{u}_5 \\
\mathbf{x}_7 &= \mathbf{A}_0\mathbf{x}_6 + \mathbf{B}_0\mathbf{u}_6 \\
&\ \vdots
\end{aligned}$$

It is easy to regroup the periodic system and to rewrite it in the following form:

$$\begin{aligned}
\bar{\mathbf{x}}_1 &= \begin{bmatrix} \mathbf{x}_1 \\ \mathbf{x}_2 \\ \mathbf{x}_3 \end{bmatrix} = \begin{bmatrix} \mathbf{0} & \mathbf{0} & \mathbf{A}_0 \\ \mathbf{0} & \mathbf{0} & \mathbf{A}_1\mathbf{A}_0 \\ \mathbf{0} & \mathbf{0} & \mathbf{A}_2\mathbf{A}_1\mathbf{A}_0 \end{bmatrix} \begin{bmatrix} \mathbf{0} \\ \mathbf{0} \\ \mathbf{x}_0 \end{bmatrix} \\
&+ \begin{bmatrix} \mathbf{B}_0 & \mathbf{0} & \mathbf{0} \\ \mathbf{A}_1\mathbf{B}_0 & \mathbf{B}_1 & \mathbf{0} \\ \mathbf{A}_2\mathbf{A}_1\mathbf{B}_0 & \mathbf{A}_2\mathbf{B}_1 & \mathbf{B}_2 \end{bmatrix} \begin{bmatrix} \mathbf{u}_0 \\ \mathbf{u}_1 \\ \mathbf{u}_2 \end{bmatrix} \\
&= \bar{\mathbf{A}}\bar{\mathbf{x}}_0 + \bar{\mathbf{B}}\bar{\mathbf{u}}_0
\end{aligned}$$

$$\begin{aligned}
\bar{\mathbf{x}}_2 &= \begin{bmatrix} \mathbf{x}_4 \\ \mathbf{x}_5 \\ \mathbf{x}_6 \end{bmatrix} = \begin{bmatrix} \mathbf{0} & \mathbf{0} & \mathbf{A}_0 \\ \mathbf{0} & \mathbf{0} & \mathbf{A}_1\mathbf{A}_0 \\ \mathbf{0} & \mathbf{0} & \mathbf{A}_2\mathbf{A}_1\mathbf{A}_0 \end{bmatrix} \begin{bmatrix} \mathbf{x}_1 \\ \mathbf{x}_2 \\ \mathbf{x}_3 \end{bmatrix} \\
&+ \begin{bmatrix} \mathbf{B}_0 & \mathbf{0} & \mathbf{0} \\ \mathbf{A}_1\mathbf{B}_0 & \mathbf{B}_1 & \mathbf{0} \\ \mathbf{A}_2\mathbf{A}_1\mathbf{B}_0 & \mathbf{A}_2\mathbf{B}_1 & \mathbf{B}_2 \end{bmatrix} \begin{bmatrix} \mathbf{u}_3 \\ \mathbf{u}_4 \\ \mathbf{u}_5 \end{bmatrix} \\
&= \bar{\mathbf{A}}\bar{\mathbf{x}}_1 + \bar{\mathbf{B}}\bar{\mathbf{u}}_1
\end{aligned}$$

in general, for $k \geq 0$ ($K \geq 0$), we have

$$\begin{aligned}
\bar{\mathbf{x}}_{K+1} &:= \begin{bmatrix} \mathbf{x}_{pk+1} \\ \mathbf{x}_{pk+2} \\ \mathbf{x}_{pk+3} \end{bmatrix} \\
&= \begin{bmatrix} \mathbf{0} & \mathbf{0} & \mathbf{A}_0 \\ \mathbf{0} & \mathbf{0} & \mathbf{A}_1\mathbf{A}_0 \\ \mathbf{0} & \mathbf{0} & \mathbf{A}_2\mathbf{A}_1\mathbf{A}_0 \end{bmatrix} \begin{bmatrix} \mathbf{x}_{p(k-1)+1} \\ \mathbf{x}_{p(k-1)+2} \\ \mathbf{x}_{p(k-1)+3} \end{bmatrix} \\
&+ \begin{bmatrix} \mathbf{B}_0 & \mathbf{0} & \mathbf{0} \\ \mathbf{A}_1\mathbf{B}_0 & \mathbf{B}_1 & \mathbf{0} \\ \mathbf{A}_2\mathbf{A}_1\mathbf{B}_0 & \mathbf{A}_2\mathbf{B}_1 & \mathbf{B}_2 \end{bmatrix} \begin{bmatrix} \mathbf{u}_{pk} \\ \mathbf{u}_{pk+1} \\ \mathbf{u}_{pk+2} \end{bmatrix} \\
&:= \bar{\mathbf{A}}\bar{\mathbf{x}}_K + \bar{\mathbf{B}}\bar{\mathbf{u}}_K
\end{aligned} \tag{11.139}$$

where

$$\bar{\mathbf{x}}_0 = \begin{bmatrix} \mathbf{x}_{-2} \\ \mathbf{x}_{-1} \\ \mathbf{x}_0 \end{bmatrix} := \begin{bmatrix} \mathbf{0} \\ \mathbf{0} \\ \mathbf{x}_0 \end{bmatrix}, \quad \bar{\mathbf{u}}_0 = \begin{bmatrix} \mathbf{u}_0 \\ \mathbf{u}_1 \\ \mathbf{u}_2 \end{bmatrix}$$

It is worthwhile to note that (11.139) is a *linear time-invariant* system. It is easy to extend the result to the general case. Let

$$\bar{\mathbf{x}}_K = \begin{bmatrix} \mathbf{x}_{p(k-1)+1} \\ \mathbf{x}_{p(k-1)+2} \\ \vdots \\ \mathbf{x}_{p(k-1)+p} \end{bmatrix}, \quad \bar{\mathbf{x}}_0 := \begin{bmatrix} \mathbf{0} \\ \vdots \\ \mathbf{0} \\ \mathbf{x}_0 \end{bmatrix}, \quad \bar{\mathbf{u}}_K = \begin{bmatrix} \mathbf{u}_{pk} \\ \mathbf{u}_{pk+1} \\ \cdots \\ \mathbf{u}_{pk+p-1} \end{bmatrix}$$

and

$$\bar{\mathbf{x}}_{K+1} = \begin{bmatrix} \mathbf{x}_{pk+1} \\ \mathbf{x}_{pk+2} \\ \vdots \\ \mathbf{x}_{pk+p} \end{bmatrix}, \quad \bar{\mathbf{u}}_{K+1} = \begin{bmatrix} \mathbf{u}_{p(k+1)} \\ \mathbf{u}_{p(k+1)+1} \\ \cdots \\ \mathbf{u}_{p(k+1)+p-1} \end{bmatrix}$$

Theorem 11.6

Given a linear periodic discrete-time system with period of p as follows:

$$\mathbf{x}_{pk+1} = \mathbf{A}_0\mathbf{x}_{pk} + \mathbf{B}_0\mathbf{u}_{pk}$$

$$
\begin{aligned}
\mathbf{x}_{pk+2} &= \mathbf{A}_1\mathbf{x}_{pk+1}+\mathbf{B}_1\mathbf{u}_{pk+1} \\
&\vdots \\
\mathbf{x}_{pk+p} &= \mathbf{A}_{p-1}\mathbf{x}_{pk+p-1}+\mathbf{B}_{p-1}\mathbf{u}_{pk+p-1}
\end{aligned} \tag{11.140}
$$

Then, this discrete-time periodic system is equivalent to the linear time-invariant system given as follows:

$$
\begin{aligned}
\bar{\mathbf{x}}_{K+1} &:= \begin{bmatrix} \mathbf{x}_{pk+1} \\ \mathbf{x}_{pk+2} \\ \vdots \\ \mathbf{x}_{pk+p} \end{bmatrix} = \begin{bmatrix} \mathbf{0} & \dots & \mathbf{0} & \mathbf{A}_0 \\ \mathbf{0} & \dots & \mathbf{0} & \mathbf{A}_1\mathbf{A}_0 \\ \vdots & \vdots & \vdots & \vdots \\ \mathbf{0} & \dots & \mathbf{0} & \mathbf{A}_{p-1}\dots\mathbf{A}_2\mathbf{A}_1\mathbf{A}_0 \end{bmatrix} \begin{bmatrix} \mathbf{x}_{p(k-1)+1} \\ \mathbf{x}_{p(k-1)+2} \\ \vdots \\ \mathbf{x}_{p(k-1)+p} \end{bmatrix} \\
&+ \begin{bmatrix} \mathbf{B}_0 & \mathbf{0} & \dots & \mathbf{0} \\ \mathbf{A}_1\mathbf{B}_0 & \mathbf{B}_1 & \dots & \mathbf{0} \\ \vdots & \vdots & \vdots & \vdots \\ \mathbf{A}_{p-1}\dots\mathbf{A}_1\mathbf{B}_0 & \mathbf{A}_{p-1}\dots\mathbf{A}_2\mathbf{B}_1 & \dots & \mathbf{B}_{p-1} \end{bmatrix} \begin{bmatrix} \mathbf{u}_{pk} \\ \mathbf{u}_{pk+1} \\ \dots \\ \mathbf{u}_{pk+p-1} \end{bmatrix} \\
&:= \bar{\mathbf{A}}\bar{\mathbf{x}}_K + \bar{\mathbf{B}}\bar{\mathbf{u}}_K
\end{aligned} \tag{11.141}
$$

where $\bar{\mathbf{A}} \in \mathbf{R}^{pn\times pn}$ and $\bar{\mathbf{B}} \in \mathbf{R}^{pn\times pm}$. Moreover, the structure of $\bar{\mathbf{B}}$ matrix guarantees the causality of the system (11.141). ■

It is worthwhile to emphasize that there is no overlap between $\bar{\mathbf{x}}_{K+1}$ and $\bar{\mathbf{x}}_K$; in addition, there is no overlap between $\bar{\mathbf{u}}_{K+1}$ and $\bar{\mathbf{u}}_K$. This is the major difference between the proposed lifting method and the existing lifting methods in [60, 130] (see also [207]). This feature makes it possible to apply existing design methods to the linear time-invariant system (11.141) which is equivalent to the linear periodic system (11.140). The remainder of this section discusses the LQR design for the system (11.141). The LQR state feedback control is to find the optimal $\bar{\mathbf{u}}_K$ to minimize the following quadratic cost function

$$
\lim_{N\to\infty}\left(\min \frac{1}{2}\bar{\mathbf{x}}_N^{\mathrm{T}}\bar{\mathbf{Q}}_N\bar{\mathbf{x}}_N + \frac{1}{2}\sum_{K=0}^{N-1}\bar{\mathbf{x}}_K^{\mathrm{T}}\bar{\mathbf{Q}}_K\bar{\mathbf{x}}_K + \bar{\mathbf{u}}_K^{\mathrm{T}}\bar{\mathbf{R}}_K\bar{\mathbf{u}}_K \right) \tag{11.142}
$$

where

$$
\bar{\mathbf{Q}}_K := \mathrm{diag}(\mathbf{Q}_0,\dots,\mathbf{Q}_{p-1}) \geq 0, \quad \bar{\mathbf{R}}_K := \mathrm{diag}(\mathbf{R}_0,\dots,\mathbf{R}_{p-1}) > 0 \tag{11.143}
$$

are constant matrices and the initial condition $\bar{\mathbf{x}}_0$ is given. It is straightforward to see that the optimal control problem described by (11.141) and (11.142) is *time-invariant but equivalent to the time-varying periodic* system described by (11.33) and (11.37). Moreover, the optimal feedback matrix of the system (11.141 - 11.142) is given in (B.21) as follows:

$$
\bar{\mathbf{u}}_K = -(\bar{\mathbf{R}}+\bar{\mathbf{B}}^{\mathrm{T}}\bar{\mathbf{P}}\bar{\mathbf{B}})^{-1}\bar{\mathbf{B}}^{\mathrm{T}}\bar{\mathbf{P}}\bar{\mathbf{A}}\bar{\mathbf{x}}_K \tag{11.144}
$$

where $\bar{\mathbf{P}}$ is the solution of the following *time-invariant algebraic Riccati equation* (see (B.29) and (B.50)):

$$\bar{\mathbf{A}}^{\mathrm{T}}\bar{\mathbf{P}}\bar{\mathbf{A}} - \bar{\mathbf{P}} - \bar{\mathbf{A}}^{\mathrm{T}}\bar{\mathbf{P}}\bar{\mathbf{B}}(\bar{\mathbf{R}} + \bar{\mathbf{B}}^{\mathrm{T}}\bar{\mathbf{P}}\bar{\mathbf{B}})^{-1}\bar{\mathbf{B}}^{\mathrm{T}}\bar{\mathbf{P}}\bar{\mathbf{A}} + \bar{\mathbf{Q}} = \mathbf{0}_S \tag{11.145}$$

Notice that $\bar{\mathbf{A}}$ is not invertible, this time-invariant algebraic Riccati equation cannot be directly solved by using the algorithms either described in Appendix B.3 or proposed in [103, 208], but it can be solved by using the algorithm proposed in [151]. However, because of the structure of $\bar{\mathbf{A}}$, there is a more efficient algorithm than the one of [151]. The new algorithm makes full use of the specific structure of $\bar{\mathbf{A}}$ in which the first $(p-1)n$ columns are zeros. Denote

$$\bar{\mathbf{Q}} := \bar{\mathbf{Q}}_K = \left[\begin{array}{c|c} \mathrm{diag}(\mathbf{Q}_0,\ldots,\mathbf{Q}_{p-2}) & \mathbf{0} \\ \hline \mathbf{0} & \mathbf{Q}_{p-1} \end{array}\right] = \mathrm{diag}(\bar{\mathbf{Q}}_1, \bar{\mathbf{Q}}_2) \tag{11.146}$$

where $\bar{\mathbf{Q}}_1 = \mathrm{diag}(\mathbf{Q}_0,\ldots,\mathbf{Q}_{p-2}) \in \mathbf{R}^{(p-1)n\times(p-1)n}$ and $\bar{\mathbf{Q}}_2 = \mathbf{Q}_{p-1} \in \mathbf{R}^{n\times n}$

$$\bar{\mathbf{R}} := \bar{\mathbf{R}}_K = \left[\begin{array}{c|c} \mathrm{diag}(\mathbf{R}_0,\ldots,\mathbf{R}_{p-2}) & \mathbf{0} \\ \hline \mathbf{0} & \mathbf{R}_{p-1} \end{array}\right] = \mathrm{diag}(\bar{\mathbf{R}}_1, \bar{\mathbf{R}}_2) \tag{11.147}$$

where $\bar{\mathbf{R}}_1 = \mathrm{diag}(\mathbf{R}_0,\ldots,\mathbf{R}_{p-2}) \in \mathbf{R}^{(p-1)m\times(p-1)m}$ and $\bar{\mathbf{R}}_2 = \mathbf{R}_{p-1} \in \mathbf{R}^{m\times m}$. Let

$$\begin{aligned}
\bar{\mathbf{A}} &= \left[\begin{array}{ccc|c} \mathbf{0} & \ldots & \mathbf{0} & \mathbf{A}_0 \\ \mathbf{0} & \ldots & \mathbf{0} & \mathbf{A}_1\mathbf{A}_0 \\ \vdots & \vdots & \vdots & \vdots \\ \mathbf{0} & \ldots & \mathbf{0} & \mathbf{A}_{p-2}\ldots\mathbf{A}_1\mathbf{A}_0 \\ \hline \mathbf{0} & \ldots & \mathbf{0} & \mathbf{A}_{p-1}\ldots\mathbf{A}_2\mathbf{A}_1\mathbf{A}_0 \end{array}\right] = \left[\begin{array}{ccc|c} \mathbf{0} & \ldots & \mathbf{0} & \mathbf{A}_0 \\ \mathbf{0} & \ldots & \mathbf{0} & \mathbf{A}_1\mathbf{A}_0 \\ \vdots & \vdots & \vdots & \vdots \\ \mathbf{0} & \ldots & \mathbf{0} & \mathbf{A}_{p-2}\ldots\mathbf{A}_1\mathbf{A}_0 \\ \hline \mathbf{0} & \ldots & \mathbf{0} & \mathbf{A}_{p-1}\ldots\mathbf{A}_2\mathbf{A}_1\mathbf{A}_0 \end{array}\right] \\
&= \left[\begin{array}{c|c} \underbrace{0}_{(p-1)n \text{ columns}} & \underbrace{\begin{array}{c}\bar{\mathbf{A}}_1 \\ \hline \bar{\mathbf{A}}_2\end{array}}_{n \text{ columns}} \end{array}\right] = \left[\mathbf{0} \,|\, \bar{\mathbf{F}}\right]
\end{aligned} \tag{11.148}$$

where $\bar{\mathbf{A}}_1 \in \mathbf{R}^{(p-1)n\times n}$, $\bar{\mathbf{A}}_2 \in \mathbf{R}^{n\times n}$, and $\bar{\mathbf{F}} = \left[\bar{\mathbf{A}}_1^{\mathrm{T}}, \bar{\mathbf{A}}_2^{\mathrm{T}}\right]^{\mathrm{T}} \in \mathbf{R}^{pn\times n}$

$$\bar{\mathbf{B}} = \left[\begin{array}{ccccc} \mathbf{B}_0 & \mathbf{0} & \ldots & \mathbf{0} & \mathbf{0} \\ \mathbf{A}_1\mathbf{B}_0 & \mathbf{B}_1 & \ldots & \mathbf{0} & \mathbf{0} \\ \vdots & \vdots & \vdots & \vdots & \vdots \\ \mathbf{A}_{p-2}\ldots\mathbf{A}_1\mathbf{B}_0 & \mathbf{A}_{p-2}\ldots\mathbf{A}_2\mathbf{B}_1 & \ldots & \mathbf{B}_{p-2} & \mathbf{0} \\ \hline \mathbf{A}_{p-1}\ldots\mathbf{A}_1\mathbf{B}_0 & \mathbf{A}_{p-1}\ldots\mathbf{A}_2\mathbf{B}_1 & \ldots & \mathbf{A}_{p-1}\mathbf{B}_{p-2} & \mathbf{B}_{p-1} \end{array}\right] = \left[\begin{array}{c} \bar{\mathbf{B}}_1 \\ \hline \bar{\mathbf{B}}_2 \end{array}\right] \tag{11.149}$$

where $\bar{\mathbf{B}}_1 \in \mathbf{R}^{(p-1)n\times pm}$ and $\bar{\mathbf{B}}_2 \in \mathbf{R}^{n\times pm}$; and

$$\bar{\mathbf{P}} = \begin{bmatrix} \bar{\mathbf{P}}_{11} & \bar{\mathbf{P}}_{12} \\ \bar{\mathbf{P}}_{21} & \bar{\mathbf{P}}_{22} \end{bmatrix} \tag{11.150}$$

where $\bar{\mathbf{P}}_{11} \in \mathbf{R}^{(p-1)n\times(p-1)n}$, $\bar{\mathbf{P}}_{12} \in \mathbf{R}^{(p-1)n\times n}$, $\bar{\mathbf{P}}_{21} \in \mathbf{R}^{n\times(p-1)n}$, and $\bar{\mathbf{P}}_{22} \in \mathbf{R}^{n\times n}$. Let

$$\mathbf{Y} = \bar{\mathbf{P}}\bar{\mathbf{B}}(\bar{\mathbf{R}}+\bar{\mathbf{B}}^{\mathrm{T}}\bar{\mathbf{P}}\bar{\mathbf{B}})^{-1}\bar{\mathbf{B}}^{\mathrm{T}}\bar{\mathbf{P}} \tag{11.151}$$

Substituting (11.146), (11.147), (11.148), (11.149), (11.150), and (11.151) into (11.145) yields

$$\begin{bmatrix} \mathbf{0} \\ \bar{\mathbf{F}}^{\mathrm{T}} \end{bmatrix} \bar{\mathbf{P}} \begin{bmatrix} \mathbf{0} & \bar{\mathbf{F}} \end{bmatrix} - \begin{bmatrix} \bar{\mathbf{P}}_{11} & \bar{\mathbf{P}}_{12} \\ \bar{\mathbf{P}}_{21} & \bar{\mathbf{P}}_{22} \end{bmatrix} - \begin{bmatrix} \mathbf{0} \\ \bar{\mathbf{F}}^{\mathrm{T}} \end{bmatrix} \bar{\mathbf{Y}} \begin{bmatrix} \mathbf{0} & \bar{\mathbf{F}} \end{bmatrix} + \begin{bmatrix} \bar{\mathbf{Q}}_1 & \mathbf{0} \\ \mathbf{0} & \bar{\mathbf{Q}}_2 \end{bmatrix} = \mathbf{0} \tag{11.152}$$

or equivalently

$$\begin{bmatrix} \mathbf{0} & \mathbf{0} \\ \mathbf{0} & \bar{\mathbf{F}}^{\mathrm{T}}\bar{\mathbf{P}}\bar{\mathbf{F}} \end{bmatrix} - \begin{bmatrix} \bar{\mathbf{P}}_{11} & \bar{\mathbf{P}}_{12} \\ \bar{\mathbf{P}}_{21} & \bar{\mathbf{P}}_{22} \end{bmatrix} - \begin{bmatrix} \mathbf{0} & \mathbf{0} \\ \mathbf{0} & \bar{\mathbf{F}}^{\mathrm{T}}\bar{\mathbf{Y}}\bar{\mathbf{F}} \end{bmatrix} + \begin{bmatrix} \bar{\mathbf{Q}}_1 & \mathbf{0} \\ \mathbf{0} & \bar{\mathbf{Q}}_2 \end{bmatrix} = \mathbf{0} \tag{11.153}$$

This proves that $\bar{\mathbf{P}}_{12} = \bar{\mathbf{P}}_{21}^{\mathrm{T}} = \mathbf{0}$ and $\bar{\mathbf{P}}_{11} = \bar{\mathbf{P}}_{11}^{\mathrm{T}} = \bar{\mathbf{Q}}_1$. By examining the lower right block of (11.153), it is easy to see

$$\bar{\mathbf{F}}^{\mathrm{T}}\bar{\mathbf{P}}\bar{\mathbf{F}} = \bar{\mathbf{A}}_1^{\mathrm{T}}\bar{\mathbf{Q}}_1\bar{\mathbf{A}}_1 + \bar{\mathbf{A}}_2^{\mathrm{T}}\bar{\mathbf{P}}_{22}\bar{\mathbf{A}}_2 \in \mathbf{R}^{n\times n} \tag{11.154}$$

and

$$\begin{aligned} &\bar{\mathbf{F}}^{\mathrm{T}}\bar{\mathbf{Y}}\bar{\mathbf{F}} \\ = \; & \begin{bmatrix} \bar{\mathbf{A}}_1^{\mathrm{T}}, \bar{\mathbf{A}}_2^{\mathrm{T}} \end{bmatrix} \begin{bmatrix} \bar{\mathbf{Q}}_1\bar{\mathbf{B}}_1 \\ \bar{\mathbf{P}}_{22}\bar{\mathbf{B}}_2 \end{bmatrix} \begin{bmatrix} \bar{\mathbf{R}}+\bar{\mathbf{B}}_1^{\mathrm{T}}\bar{\mathbf{Q}}_1\bar{\mathbf{B}}_1+\bar{\mathbf{B}}_2^{\mathrm{T}}\bar{\mathbf{P}}_{22}^{\mathrm{T}}\bar{\mathbf{B}}_2 \end{bmatrix}^{-1} \begin{bmatrix} \bar{\mathbf{B}}_1^{\mathrm{T}}\bar{\mathbf{Q}}_1 & \bar{\mathbf{B}}_2^{\mathrm{T}}\bar{\mathbf{P}}_{22} \end{bmatrix} \begin{bmatrix} \bar{\mathbf{A}}_1 \\ \bar{\mathbf{A}}_2 \end{bmatrix} \\ = \; & \underbrace{\begin{bmatrix} \bar{\mathbf{A}}_1^{\mathrm{T}}\bar{\mathbf{Q}}_1\bar{\mathbf{B}}_1+\bar{\mathbf{A}}_2^{\mathrm{T}}\bar{\mathbf{P}}_{22}\bar{\mathbf{B}}_2 \end{bmatrix}}_{n\times pm} \underbrace{\begin{bmatrix} \bar{\mathbf{R}}+\bar{\mathbf{B}}_1^{\mathrm{T}}\bar{\mathbf{Q}}_1\bar{\mathbf{B}}_1+\bar{\mathbf{B}}_2^{\mathrm{T}}\bar{\mathbf{P}}_{22}\bar{\mathbf{B}}_2 \end{bmatrix}^{-1}}_{pm\times pm} \underbrace{\begin{bmatrix} \bar{\mathbf{B}}_1^{\mathrm{T}}\bar{\mathbf{Q}}_1\bar{\mathbf{A}}_1+\bar{\mathbf{B}}_2^{\mathrm{T}}\bar{\mathbf{P}}_{22}\bar{\mathbf{A}}_2 \end{bmatrix}}_{pm\times n} \end{aligned} \tag{11.155}$$

Let

$$\hat{\mathbf{A}} = \bar{\mathbf{A}}_2 \in \mathbf{R}^{n\times n} \tag{11.156a}$$

$$\hat{\mathbf{B}} = \bar{\mathbf{B}}_2 \in \mathbf{R}^{n\times pm} \tag{11.156b}$$

$$\hat{\mathbf{Q}} = \bar{\mathbf{Q}}_2 + \bar{\mathbf{A}}_1^{\mathrm{T}}\bar{\mathbf{Q}}_1\bar{\mathbf{A}}_1 \in \mathbf{R}^{n\times n} \tag{11.156c}$$

$$\hat{\mathbf{R}} = \bar{\mathbf{R}} + \bar{\mathbf{B}}_1^{\mathrm{T}}\bar{\mathbf{Q}}_1\bar{\mathbf{B}}_1 \in \mathbf{R}^{pm\times pm} \tag{11.156d}$$

$$\hat{\mathbf{S}} = \bar{\mathbf{A}}_1^{\mathrm{T}}\bar{\mathbf{Q}}_1\bar{\mathbf{B}}_1 \in \mathbf{R}^{n\times pm} \tag{11.156e}$$

$$\hat{\mathbf{P}} = \bar{\mathbf{P}}_{22} \in \mathbf{R}^{n\times n} \tag{11.156f}$$

The lower right block of (11.153) can be rewritten as follows:

$$\hat{\mathbf{A}}^{\mathrm{T}}\hat{\mathbf{P}}\hat{\mathbf{A}} - \hat{\mathbf{P}} - \underbrace{(\hat{\mathbf{A}}^{\mathrm{T}}\hat{\mathbf{P}}\hat{\mathbf{B}}+\hat{\mathbf{S}})}_{n\times pm} \underbrace{(\hat{\mathbf{B}}^{\mathrm{T}}\hat{\mathbf{P}}\hat{\mathbf{B}}+\hat{\mathbf{R}})^{-1}}_{pm\times pm} \underbrace{(\hat{\mathbf{B}}^{\mathrm{T}}\hat{\mathbf{P}}\hat{\mathbf{A}}+\hat{\mathbf{S}}^{\mathrm{T}})}_{pm\times n} + \hat{\mathbf{Q}} = \mathbf{0} \tag{11.157}$$

The Riccati equation (11.157) is a special case discussed in [6, Eq. (6)]. An efficient MATLAB function `dare` that implements an algorithm of [6] is available to solve (11.157).

Remark 11.7 Comparing to the methods described in the previous section which need to solve p n-dimensional discrete-time Riccati equations, one needs only to solve one n-dimensional discrete-time Riccati equation using the method proposed in this section. ∎

To compare the efficiency of the method to the ones discussed in Section 11.5.1, The MATLAB function `dare` is not used directly because `dare` calculates more information than the solution of the Riccati equation (11.157). Let $\tilde{\mathbf{B}} = \hat{\mathbf{B}}$, $\tilde{\mathbf{R}} = \hat{\mathbf{R}}$,

$$\tilde{\mathbf{A}} = \hat{\mathbf{A}} - \hat{\mathbf{B}}\hat{\mathbf{R}}^{-1}\hat{\mathbf{S}}^{\mathrm{T}} \tag{11.158}$$

and

$$\tilde{\mathbf{Q}} = \hat{\mathbf{Q}} - \hat{\mathbf{S}}\hat{\mathbf{R}}^{-1}\hat{\mathbf{S}}^{\mathrm{T}} \tag{11.159}$$

Riccati equation (11.157) can be solved by either eigen-decomposition or Schur decomposition for the following generalized eigenvalue problem [6, page 1748, equation (8)]:

$$\lambda \begin{bmatrix} \mathbf{I} & \tilde{\mathbf{B}}\tilde{\mathbf{R}}^{-1}\tilde{\mathbf{B}}^{\mathrm{T}} \\ \mathbf{0} & \tilde{\mathbf{A}}^{\mathrm{T}} \end{bmatrix} - \begin{bmatrix} \tilde{\mathbf{A}} & \mathbf{0} \\ -\tilde{\mathbf{Q}} & \mathbf{I} \end{bmatrix} := \lambda \mathbf{E} - \mathbf{F} \tag{11.160}$$

If $\tilde{\mathbf{A}}$ is invertible, then $\det(\mathbf{E}) \neq 0$ and $0 = \det(\lambda\mathbf{E} - \mathbf{F}) = \det(\lambda\mathbf{I} - \mathbf{E}^{-1}\mathbf{F})$, the problem is reduced to solve the eigenvalue for problem (11.46):

$$\mathbf{Z} = \mathbf{E}^{-1}\mathbf{F} = \begin{bmatrix} \mathbf{I} & -\tilde{\mathbf{B}}\tilde{\mathbf{R}}^{-1}\tilde{\mathbf{B}}^{\mathrm{T}}\tilde{\mathbf{A}}^{-\mathrm{T}} \\ \mathbf{0} & \tilde{\mathbf{A}}^{-\mathrm{T}} \end{bmatrix} \begin{bmatrix} \tilde{\mathbf{A}} & \mathbf{0} \\ -\tilde{\mathbf{Q}} & \mathbf{I} \end{bmatrix} \tag{11.161}$$

Using Schur decomposition for (11.161), the following equation holds:

$$\begin{bmatrix} \mathbf{W}_{11} & \mathbf{W}_{12} \\ \mathbf{W}_{21} & \mathbf{W}_{22} \end{bmatrix}^{\mathrm{T}} \mathbf{Z} \begin{bmatrix} \mathbf{W}_{11} & \mathbf{W}_{12} \\ \mathbf{W}_{21} & \mathbf{W}_{22} \end{bmatrix} = \begin{bmatrix} \mathbf{S}_{11} & \mathbf{S}_{12} \\ \mathbf{0} & \mathbf{S}_{22} \end{bmatrix} \tag{11.162}$$

where $\mathbf{S}_{11}$ is upper-triangular and has all of its eigenvalues inside the unit circle. The solution of the discrete algebraic Riccati equation (11.157) is given by

$$\hat{\mathbf{P}} = \mathbf{W}_{21}\mathbf{W}_{11}^{-1} \tag{11.163}$$

The proposed algorithm is as follows:

Algorithm 11.3

Data: $\mathbf{A}_0, \ldots, \mathbf{A}_{p-1}$, $\mathbf{B}_0, \ldots, \mathbf{B}_{p-1}$, $\mathbf{Q}_0, \ldots, \mathbf{Q}_{p-1}$, $\mathbf{R}_0, \ldots, \mathbf{R}_{p-1}$.

Step 1: Form

$$\bar{\mathbf{A}}_1 = \begin{bmatrix} \mathbf{A}_0 \\ \mathbf{A}_1\mathbf{A}_0 \\ \vdots \\ \mathbf{A}_{p-2}\ldots\mathbf{A}_2\mathbf{A}_1\mathbf{A}_0 \end{bmatrix} \tag{11.164a}$$

$$\bar{\mathbf{B}}_1 = \begin{bmatrix} \mathbf{B}_0 & \mathbf{0} & \cdots & \mathbf{0} & \mathbf{0} \\ \mathbf{A}_1\mathbf{B}_0 & \mathbf{B}_1 & \cdots & \mathbf{0} & \mathbf{0} \\ \vdots & \vdots & \vdots & \vdots & \vdots \\ \mathbf{A}_{p-2}\ldots\mathbf{A}_1\mathbf{B}_0, & \mathbf{A}_{p-2}\ldots\mathbf{A}_2\mathbf{B}_1, & \cdots & \mathbf{B}_{p-2}, & \mathbf{0} \end{bmatrix} \tag{11.164b}$$

$$\bar{\mathbf{A}}_2 = \mathbf{A}_{p-1}\ldots\mathbf{A}_2\mathbf{A}_1\mathbf{A}_0 \tag{11.164c}$$

$$\bar{\mathbf{B}}_2 = \begin{bmatrix} \mathbf{A}_{p-1}\ldots\mathbf{A}_1\mathbf{B}_0, & \mathbf{A}_{p-1}\ldots\mathbf{A}_2\mathbf{B}_1, & \cdots & \mathbf{B}_{p-1} \end{bmatrix} \tag{11.164d}$$

$$\bar{\mathbf{Q}}_1 = \text{diag}(\mathbf{Q}_0,\ldots,\mathbf{Q}_{p-2}), \quad \bar{\mathbf{Q}}_2 = \mathbf{Q}_{p-1} \tag{11.164e}$$

$$\bar{\mathbf{R}}_1 = \text{diag}(\mathbf{R}_0,\ldots,\mathbf{R}_{p-2}), \quad \bar{\mathbf{R}}_2 = \mathbf{R}_{p-1} \tag{11.164f}$$

Step 2: Form $\hat{\mathbf{A}}$, $\hat{\mathbf{B}}$, $\hat{\mathbf{Q}}$, $\hat{\mathbf{R}}$, and $\hat{\mathbf{S}}$ using (11.156).

Step 3: Find the solution $\hat{\mathbf{P}}$ of the discrete-time algebraic Riccati equation (11.157) using the algorithm of [6] implemented as `dare` *or using the algorithm described in (11.162) and (11.163).*

Step 4: The solution of the discrete-time algebraic Riccati equation (11.145) is given by

$$\bar{\mathbf{P}} = \text{diag}(\bar{\mathbf{Q}}_1, \hat{\mathbf{P}}) \tag{11.165}$$

Given $\bar{\mathbf{x}}_K$, the feedback control can be calculated by (11.144). Applying this feedback control to (11.141) yields the next state $\bar{\mathbf{x}}_{K+1}$.

11.5.3 Implementation and Numerical Simulation

In this section, some details of implementation, which will reduce some computation time compared to the direct implementation described in the previous section, have been discussed. The test result of the proposed algorithm for the problems discussed in 11.3.4 and 11.4.4 is reported. The comparison of the test results obtained from the method discussed here and the ones obtained in 11.3.4 and 11.4.4 is performed.

11.5.3.1 Implementation Consideration

The most expensive calculations in Algorithm 11.3 are the calculation of $\hat{\mathbf{Q}}$, $\hat{\mathbf{R}}$, and $\hat{\mathbf{S}}$ in Step 2, and the calculation of $\hat{\mathbf{R}}^{-1} = \tilde{\mathbf{R}}^{-1}$ in Step 3. It is easy to check (cf. [58]):

(1) direct calculation of $\hat{\mathbf{Q}}$ requires

$$\mathcal{O}(2(p-1)^2n^3) + \mathcal{O}(2(p-1)n^3) + \mathcal{O}(n^2) \text{ flops,}$$

(2) direct calculation of $\hat{\mathbf{R}}$ requires

$$\mathcal{O}(2p(p-1)^2n^2m) + \mathcal{O}(2p^2(p-1)nm^2) + \mathcal{O}(p^2m^2) \text{ flops,}$$

(3) direct calculation of $\hat{\mathbf{S}}$ requires

$$\mathcal{O}(2(p-1)^2n^3)+\mathcal{O}(2(p-1)n^3) \text{ flops},$$

(4) directly calculation of $\hat{\mathbf{R}}^{-1}$ requires

$$\mathcal{O}(p^3m^3) \text{ flops}.$$

For extremely large p, i.e., very long period of the system, the majority of the computation is the computation of $\hat{\mathbf{R}}$ and $\hat{\mathbf{R}}^{-1}$.

Let $\mathbf{Q}_A = \mathbf{Q}_1^{\frac{1}{2}}\bar{\mathbf{A}}_1 \in \mathbf{R}^{(p-1)n\times n}$ and $\mathbf{Q}_B = \mathbf{Q}_1^{\frac{1}{2}}\bar{\mathbf{B}}_1 \in \mathbf{R}^{(p-1)n\times pm}$. We use MATLAB notation for sub-matrices. Since $\bar{\mathbf{Q}}_1$, $\bar{\mathbf{Q}}_2$, and $\bar{\mathbf{R}}$ are positive diagonal matrices, $\hat{\mathbf{Q}}_1$, $\hat{\mathbf{S}}_2$, and $\hat{\mathbf{R}}$ in (11.156) can be calculated more efficiently as follows:

for $i = 1:(p-1)n$

$$\mathbf{Q}_A(i,:) = \mathbf{Q}_1^{\frac{1}{2}}(i,i)\bar{\mathbf{A}}_1(i,:);$$

end

$\hat{\mathbf{Q}} = \mathbf{Q}_A^{\mathrm{T}}\mathbf{Q}_A$

for $i = 1:n$

$$\hat{\mathbf{Q}}(i,i) = \hat{\mathbf{Q}}(i,i) + \bar{\mathbf{Q}}_2(i,i);$$

end

for $i = 1:(p-1)n$

$$\mathbf{Q}_B(i,:) = \mathbf{Q}_1^{\frac{1}{2}}(i,i)\bar{\mathbf{B}}_1(i,:);$$

end

$\hat{\mathbf{R}} = \mathbf{Q}_B^{\mathrm{T}}\mathbf{Q}_B$

for $i = 1:pm$

$$\hat{\mathbf{R}}(i,i) = \hat{\mathbf{R}}(i,i) + \bar{\mathbf{R}}(i,i);$$

end

$\hat{\mathbf{S}} = \mathbf{Q}_A^{\mathrm{T}}\mathbf{Q}_B$

It is easy to check (cf. [58]) the flops for the following calculations:

(1) the calculation of $\hat{\mathbf{Q}}$ requires

$$\mathcal{O}((p-1)n)+\mathcal{O}((p-1)n^2)+\mathcal{O}(2(p-1)^2n^3)+\mathcal{O}(n) \text{ flops},$$

(2) the calculation of $\hat{\mathbf{R}}$ requires

$$\mathcal{O}(p(p-1)nm) + \mathcal{O}(2p^2(p-1)nm^2) + \mathcal{O}(pm) \text{ flops},$$

(3) the calculation of $\hat{\mathbf{S}}$ requires

$$\mathcal{O}(2(p-1)pn^2m) \text{ flops},$$

(4) this does not reduce the computation of $\hat{\mathbf{R}}^{-1}$.

11.5.3.2 Simulation Test for the Problem in Section 11.3

The first simulation test problem is the spacecraft attitude control design using only magnetic torques discussed in Section 11.3. The number of states of this system is $n = 6$. The number of control inputs of this system is $m = 3$. The controllability of this problem is established in Section 11.2. In this simulation test, the same discrete-time linear periodic model as in Section 11.3 with the same parameters, such as the spacecraft inertia matrix, orbital inclination, orbital altitude, weight matrices $\mathbf{Q}$ and $\mathbf{R}$, and the same initial conditions, is used.

Using $p = 100$, $p = 500$, and $p = 1000$, all three algorithms discussed in this chapter are used for this design and the CPU times for all three algorithms are recorded. The result is presented in Table 11.1.

Table 11.1: CPU time comparison for problem in [254].

Samples per period	Algorithm 11.3	Algorithm 11.1	Algorithm [70]
100	0.0097 (s)	0.0757 (s)	0.2711 (s)
500	0.2528 (s)	1.6042 (s)	6.5435 (s)
1000	4.2821 (s)	6.3155 (s)	25.8996 (s)

Clearly, the proposed Algorithm 11.3 is significantly cheaper than the algorithms 11.1 and the algorithm proposed in [70].

11.5.3.3 Simulation Test for the Problem in Section 11.4

The second simulation test problem is a combined method for the spacecraft attitude and desaturation control design using both reaction wheels and magnetic torques discussed in Section 11.4. The number of states of this system is $n = 9$. The number of control inputs of this system is $m = 6$. The controllability of problem is guaranteed because three reaction wheels are assumed to be available.

Using the parameters provided in Section 11.4, for $p = 100$, $p = 500$, and $p = 1000$, the solutions for the corresponding algebraic Riccati equations is obtained and the CPU times for all three algorithms are recorded. The result is presented in Table 11.2.

Table 11.2: CPU time comparison for problem in [252].

Samples per period	Algorithm 11.3	Algorithm 11.2	Algorithm [70]
100	0.0284 (s)	0.1120 (s)	0.3807 (s)
500	3.6376 (s)	2.5629 (s)	9.0144 (s)
1000	38.4912 (s)	10.0629 (s)	36.0690 (s)

For this problem, $m = 6$ is twice as large as the previous problem, the algorithm 11.3 is faster than the algorithm 11.2 and the algorithm developed in [70] when the total number of samples in one period is moderate ($p = 100$ samples per period), but when the total number of samples in one period increases (to $p = 500$ or $p = 1000$ samples per period), the advantage of the proposed algorithm will be lost because the computation of the inverse of $\tilde{\mathbf{R}} \in \mathbf{R}^{6000\times 6000}$ is $\mathcal{O}(p^3m^3)$ which is very expensive.

Chapter 12

Attitude Maneuver and Orbit-Raising

During its life span, spacecraft normally needs to change the attitude from one orientation to another one to accommodate different mission requirements. One example is *orbital-raising* described in [190]. This chapter discusses the same problem but with a design based on reduced quaternion model.

The coordinate system of Orbview-2 satellite is provided in [190], which is defined in Figure 12.1. The satellite is sent to the *parking orbit* about 300 kilometers (km) above the Earth by the launch vehicle. The spacecraft thrust control system is designed to transfer the satellite from the parking orbit to a *sun-synchronous* orbit. The attitude of the satellite before orbit-raising is stabilized in the nadir-pointing orientation as in Figure 12.1. To perform the orbit raising task, the spacecraft needs to rotate $90°$ around y-axis so that the thrusters, which are mounted on the anti-nadir face, are aligned parallel to the velocity vector as described in Figure 12.2. This is a typical example of *spacecraft attitude maneuver*.

12.1 Attitude Maneuver

Attitude maneuver has been discussed in most popular textbooks, such as [219], [181], and [220]. The controller design is normally very simple, and it can use either Euler angle error, the direction cosine error matrix, Euler axis command, or the quaternion error vector. Among these different methods, Euler angle method and quaternion method are the most widely used because they have fewer param-

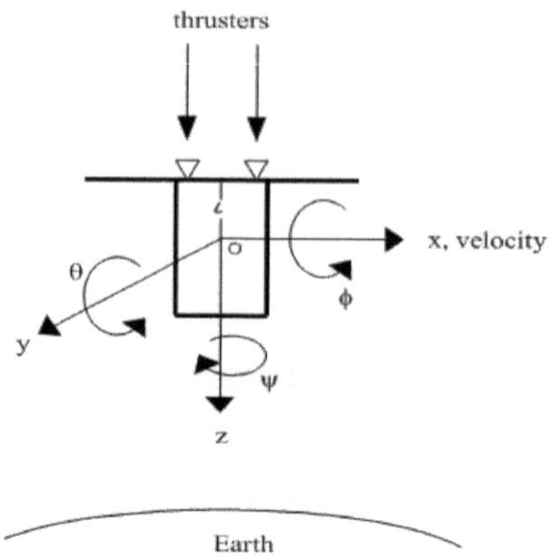

Figure 12.1: Coordinate definition.

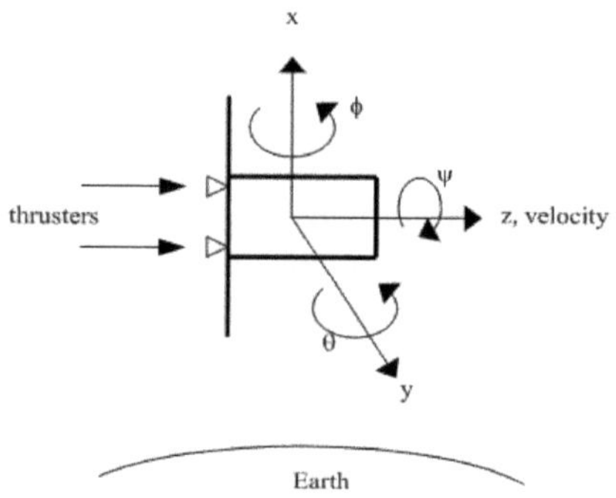

Figure 12.2: Coordinate definition.

eters and these parameters are measured directly in all spacecraft. Sidi [181] has shown, by numerical simulations, that the quaternion-based maneuver control law is clearly superior to the Euler angle-based maneuver control law.

Let the current attitude quaternions be $\bar{\mathbf{q}} = (q_0, q_1, q_2, q_3) = (q_0, \mathbf{q})$ and the desired (or target) attitude quaternion be $\bar{\mathbf{p}} = (p_0, p_1, p_2, p_3) = (p_0, \mathbf{p})$. Then the error quaternion is defined by $\bar{\mathbf{r}} = (r_0, r_1, r_2, r_3) = (r_0, \mathbf{r})$ which is given by

$$\bar{\mathbf{r}} = \bar{\mathbf{p}}^{-1} \otimes \bar{\mathbf{q}} = \bar{\mathbf{p}}^* \otimes \bar{\mathbf{q}} = (p_0 - \mathbf{p}) \otimes (q_0 + \mathbf{q})$$

In view of (3.63), $\bar{\mathbf{r}}$ can be written as

$$\begin{bmatrix} r_0 \\ r_1 \\ r_2 \\ r_3 \end{bmatrix} = \begin{bmatrix} p_0 & p_1 & p_2 & p_3 \\ -p_1 & p_0 & p_3 & -p_2 \\ -p_2 & -p_3 & p_0 & p_1 \\ -p_3 & p_2 & -p_1 & p_0 \end{bmatrix} \begin{bmatrix} q_0 \\ q_1 \\ q_2 \\ q_3 \end{bmatrix} \tag{12.1}$$

The obvious *PD* controller is therefore given by

$$\mathbf{u} = -\mathbf{Kr} - \mathbf{D}\omega \tag{12.2}$$

where **K** and **D** are positive gain matrix. This control law can be verified intuitively using the example of OrbView-2 satellite where to perform the orbit raising task, the spacecraft needs to rotate 90° around y-axis so that the thrusters are aligned parallel to the velocity vector (see Figures 12.1 and 12.2). Assume that the initial attitude is perfectly aligned with local vertical local horizontal frame, i.e., $\bar{\mathbf{q}} = (q_0, q_1, q_2, q_3) = (1,0,0,0)$. The target quaternion is $\bar{\mathbf{p}} = (p_0, p_1, p_2, p_3) = \left(\cos\left(\frac{\pi}{4}\right), 0, \sin\left(\frac{\pi}{4}\right), 0\right)$ which require the spacecraft to rotate around y-axis 90°. Substituting $\bar{\mathbf{q}}$ and $\bar{\mathbf{p}}$ into (12.1) yields

$$\begin{bmatrix} r_0 \\ r_1 \\ r_2 \\ r_3 \end{bmatrix} = \begin{bmatrix} \frac{\sqrt{2}}{2} & 0 & \frac{\sqrt{2}}{2} & 0 \\ 0 & \frac{\sqrt{2}}{2} & 0 & -\frac{\sqrt{2}}{2} \\ -\frac{\sqrt{2}}{2} & 0 & \frac{\sqrt{2}}{2} & 0 \\ 0 & \frac{\sqrt{2}}{2} & 0 & \frac{\sqrt{2}}{2} \end{bmatrix} \begin{bmatrix} 1 \\ 0 \\ 0 \\ 0 \end{bmatrix} = \begin{bmatrix} \frac{\sqrt{2}}{2} \\ 0 \\ -\frac{\sqrt{2}}{2} \\ 0 \end{bmatrix} \tag{12.3}$$

Therefore, $-\mathbf{r}^{\mathrm{T}} = \left(0, \frac{\sqrt{2}}{2}, 0\right)$ is a vector that a torque should be applied around y-axis. If the spacecraft is rotated 90° around y-axis, the attitude quaternion is given by $\bar{\mathbf{q}} = \left(\frac{\sqrt{2}}{2}, 0, \frac{\sqrt{2}}{2}, 0\right)$. From (12.1), the error quaternion $\bar{\mathbf{r}}$ is given by

$$\begin{bmatrix} r_0 \\ r_1 \\ r_2 \\ r_3 \end{bmatrix} = \begin{bmatrix} \frac{\sqrt{2}}{2} & 0 & \frac{\sqrt{2}}{2} & 0 \\ 0 & \frac{\sqrt{2}}{2} & 0 & -\frac{\sqrt{2}}{2} \\ -\frac{\sqrt{2}}{2} & 0 & \frac{\sqrt{2}}{2} & 0 \\ 0 & \frac{\sqrt{2}}{2} & 0 & \frac{\sqrt{2}}{2} \end{bmatrix} \begin{bmatrix} \frac{\sqrt{2}}{2} \\ 0 \\ \frac{\sqrt{2}}{2} \\ 0 \end{bmatrix} = \begin{bmatrix} 1 \\ 0 \\ 0 \\ 0 \end{bmatrix} \tag{12.4}$$

Therefore, $-\mathbf{r}^{\mathrm{T}} = (0,0,0)$ requires no torque as the spacecraft has been reached the required attitude.

12.2 Orbit-raising

The quaternion model for orbit raising depends on the spacecraft design. This section uses OrbView-2 spacecraft [190] as an example to describe the modeling process. Most materials in this section are directly from [250].

OrbView-2 has a momentum wheel with the angular momentum vector aligned parallel to the orbit-normal (–y axis), the spacecraft attitude control is performed by this wheel and three magnetic torque bars. The parking-orbit of OrbView-2 is about 300 km above the Earth's surface, and the working-orbit is about 705 km. Orbit-raising is performed by four fixed thrusters (1 Newton) with on/off switches which are mounted on the anti-nadir face of the spacecraft in each corner of a square with a side length of $2d$ as shown in Figure 12.3.

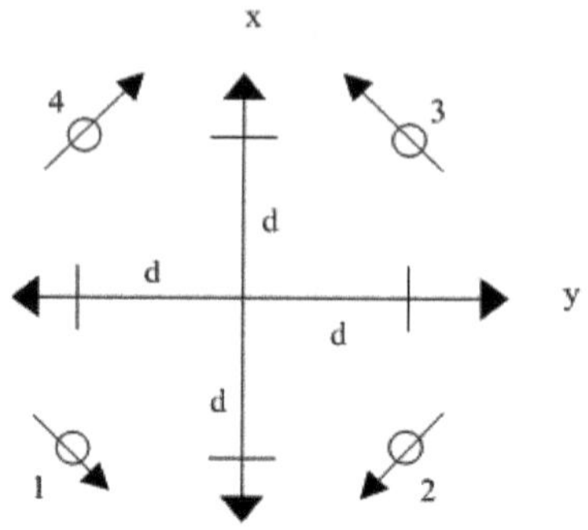

Figure 12.3: Thrusters coordinate definition.

The thrusters point to $+z$ direction (into the page) and are canted 5° from z-axis to produce moments to maintain the spacecraft attitude during the burns. they are mounted a distance l along $-z$ axis from the spacecraft center of mass (based on the coordinate system origin). To conduct *Hohmann transfers* [181] to raise the orbit, the momentum wheel provides the torque to rotate the spacecraft $\pm 90°$ to align the thrusters along with or anti-parallel to the velocity vector (see Figure 12.2).

At this orientation, the thruster burns will raise the spacecraft orbit. Let h_w be the angular momentum produced by the *momentum wheel*, $\bar{\mathbf{q}} = [q_0, q_1, q_2, q_3]^{\mathrm{T}} = [q_0, \mathbf{q}^{\mathrm{T}}]^{\mathrm{T}}$ be the quaternion that represents the rotation of the spacecraft body frame relative to the frame described by Figure 12.2 (with x-axis aligned with anti-nadir direction) represented in the body frame, $\omega = [\omega_x, \omega_y, \omega_z]^{\mathrm{T}}$ be the angular rate of the rotation represented in the body frame,

$$\mathbf{J} = \begin{bmatrix} J_x & 0 & 0 \\ 0 & J_y & 0 \\ 0 & 0 & J_z \end{bmatrix} \tag{12.5}$$

be the diagonal inertia matrix of the spacecraft, $\mathbf{m} = [m_x, m_y, m_z]^{\mathrm{T}}$ be the *control torques* generated by the thrusters, $\mathbf{h} = [J_x\omega_x, J_y\omega_y + h_w, J_z\omega_z]^{\mathrm{T}}$ be the inertial angular momentum vector of the spacecraft, then the spacecraft dynamics equation is given by (4.2)

$$\dot{\mathbf{h}} = \mathbf{J}\dot{\omega} = -\omega \times \mathbf{h} + \mathbf{m} = \mathbf{h} \times \omega + \mathbf{m} \tag{12.6}$$

or equivalently

$$\begin{aligned} &\begin{bmatrix} J_x & 0 & 0 \\ 0 & J_y & 0 \\ 0 & 0 & J_z \end{bmatrix} \begin{bmatrix} \dot{\omega}_x \\ \dot{\omega}_y \\ \dot{\omega}_z \end{bmatrix} \\ = &\begin{bmatrix} 0 & -J_z\omega_z & J_y\omega_y + h_w \\ J_z\omega_z & 0 & -J_x\omega_x \\ -J_y\omega_y - h_w & J_x\omega_x & 0 \end{bmatrix} \begin{bmatrix} \omega_x \\ \omega_y \\ \omega_z \end{bmatrix} + \begin{bmatrix} m_x \\ m_y \\ m_z \end{bmatrix} \end{aligned} \tag{12.7}$$

From Figure 12.3, the matrices of thruster force directions **F** and moment arms **R** in the body frame are given as

$$\mathbf{F} = [\mathbf{f}_1, \mathbf{f}_2, \mathbf{f}_3, \mathbf{f}_4] = \begin{bmatrix} -a & -a & a & a \\ a & -a & -a & a \\ 1 & 1 & 1 & 1 \end{bmatrix} \tag{12.8}$$

and

$$\mathbf{R} = [\mathbf{r}_1, \mathbf{r}_2, \mathbf{r}_3, \mathbf{r}_4] = \begin{bmatrix} -d & -d & d & d \\ -d & d & d & -d \\ -l & -l & -l & -l \end{bmatrix} \tag{12.9}$$

where $a = \frac{\sqrt{2}}{2}\sin(5 \times \frac{\pi}{180}) \approx 0.707 \times 5 \times (\frac{\pi}{180})$ Newton, columns 1, 2, 3, and 4 represent the thruster 1, 2, 3, and 4. Denote T_1, T_2, T_3, and T_4 the *thruster levels of thrusters* 1, 2, 3, 4, and $\mathbf{u} = [T_1, T_2, T_3, T_4]^{\mathrm{T}}$, then the control torque $\mathbf{m}$ can be expressed as

$$\mathbf{m} = \begin{bmatrix} \mathbf{r}_1 \times \mathbf{f}_1, \mathbf{r}_2 \times \mathbf{f}_2, \mathbf{r}_3 \times \mathbf{f}_3, \mathbf{r}_4 \times \mathbf{f}_4 \end{bmatrix} \begin{bmatrix} T_1 \\ T_2 \\ T_3 \\ T_4 \end{bmatrix} \tag{12.10}$$

Combining (12.7) and (12.10) gives

$$\begin{aligned} \begin{bmatrix} \dot{\omega}_x \\ \dot{\omega}_y \\ \dot{\omega}_z \end{bmatrix} &= \begin{bmatrix} 0 & -\frac{J_z\omega_z}{J_x} & \frac{J_y\omega_y+h_w}{J_x} \\ \frac{J_z\omega_z}{J_y} & 0 & -\frac{J_x\omega_x}{J_y} \\ -\frac{J_y\omega_y+h_w}{J_z} & \frac{J_x\omega_x}{J_z} & 0 \end{bmatrix} \begin{bmatrix} \omega_x \\ \omega_y \\ \omega_z \end{bmatrix} \\ &+ \begin{bmatrix} \frac{1}{J_x} & 0 & 0 \\ 0 & \frac{1}{J_y} & 0 \\ 0 & 0 & \frac{1}{J_z} \end{bmatrix} \begin{bmatrix} \mathbf{r}_1 \times \mathbf{f}_1 \\ \mathbf{r}_2 \times \mathbf{f}_2 \\ \mathbf{r}_3 \times \mathbf{f}_3 \\ \mathbf{r}_4 \times \mathbf{f}_4 \end{bmatrix}^{\mathrm{T}} \begin{bmatrix} T_1 \\ T_2 \\ T_3 \\ T_4 \end{bmatrix} \end{aligned} \tag{12.11}$$

From [243], the vector part of the quaternion $\bar{\mathbf{q}}$ meets the following relation

$$\begin{bmatrix} \dot{q}_1 \\ \dot{q}_2 \\ \dot{q}_3 \end{bmatrix} = \frac{1}{2} \begin{bmatrix} f & -q_3 & q_2 \\ q_3 & f & -q_1 \\ -q_2 & q_1 & f \end{bmatrix} \begin{bmatrix} \omega_x \\ \omega_y \\ \omega_z \end{bmatrix} \tag{12.12}$$

where $f = \sqrt{1 - q_1^2 - q_2^2 - q_3^2}$. The linearized form of (12.11) is given as

$$\begin{bmatrix} \dot{\omega}_x \\ \dot{\omega}_y \\ \dot{\omega}_z \end{bmatrix} = \begin{bmatrix} 0 & 0 & \frac{h_w}{J_x} \\ 0 & 0 & 0 \\ -\frac{h_w}{J_z} & 0 & 0 \end{bmatrix} \begin{bmatrix} \omega_x \\ \omega_y \\ \omega_z \end{bmatrix}$$

$$
+\begin{bmatrix} J_x^{-1} & 0 & 0 \\ 0 & J_y^{-1} & 0 \\ 0 & 0 & J_z^{-1} \end{bmatrix} \begin{bmatrix} \mathbf{r}_1 \times \mathbf{f}_1 \\ \mathbf{r}_2 \times \mathbf{f}_2 \\ \mathbf{r}_3 \times \mathbf{f}_3 \\ \mathbf{r}_4 \times \mathbf{f}_4 \end{bmatrix}^{\mathrm{T}} \begin{bmatrix} T_1 \\ T_2 \\ T_3 \\ T_4 \end{bmatrix} \tag{12.13}
$$

The linearized form of (12.12) is given as

$$
\dot{\mathbf{q}} = \frac{1}{2}\mathbf{I}_3\omega \tag{12.14}
$$

Combining (12.13) and (12.14) gives the linearized quaternion-based thruster control system equation as follows

$$
\begin{aligned}
\begin{bmatrix} \dot{\omega}_x \\ \dot{\omega}_y \\ \dot{\omega}_z \\ \dot{q}_1 \\ \dot{q}_2 \\ \dot{q}_3 \end{bmatrix} &= \begin{bmatrix} 0 & 0 & \frac{h_w}{J_x} & 0 & 0 & 0 \\ 0 & 0 & 0 & 0 & 0 & 0 \\ -\frac{h_w}{J_z} & 0 & 0 & 0 & 0 & 0 \\ \frac{1}{2} & 0 & 0 & 0 & 0 & 0 \\ 0 & \frac{1}{2} & 0 & 0 & 0 & 0 \\ 0 & 0 & \frac{1}{2} & 0 & 0 & 0 \end{bmatrix} \begin{bmatrix} \omega_x \\ \omega_y \\ \omega_z \\ q_1 \\ q_2 \\ q_3 \end{bmatrix} \\
&+ \begin{bmatrix} J_x^{-1} & 0 & 0 \\ 0 & J_y^{-1} & 0 \\ 0 & 0 & J_z^{-1} \\ 0 & 0 & 0 \\ 0 & 0 & 0 \\ 0 & 0 & 0 \end{bmatrix} \begin{bmatrix} \mathbf{r}_1 \times \mathbf{f}_1 \\ \mathbf{r}_2 \times \mathbf{f}_2 \\ \mathbf{r}_3 \times \mathbf{f}_3 \\ \mathbf{r}_4 \times \mathbf{f}_4 \end{bmatrix}^{\mathrm{T}} \begin{bmatrix} T_1 \\ T_2 \\ T_3 \\ T_4 \end{bmatrix} \\
&:= \mathbf{Ax} + \mathbf{Bu}
\end{aligned} \tag{12.15}
$$

where

$$
\mathbf{A} = \begin{bmatrix} 0 & 0 & \frac{h_w}{J_x} & 0 & 0 & 0 \\ 0 & 0 & 0 & 0 & 0 & 0 \\ -\frac{h_w}{J_z} & 0 & 0 & 0 & 0 & 0 \\ \frac{1}{2} & 0 & 0 & 0 & 0 & 0 \\ 0 & \frac{1}{2} & 0 & 0 & 0 & 0 \\ 0 & 0 & \frac{1}{2} & 0 & 0 & 0 \end{bmatrix}
$$

and

$$
\mathbf{B} = \begin{bmatrix} J_x^{-1} & 0 & 0 \\ 0 & J_y^{-1} & 0 \\ 0 & 0 & J_z^{-1} \\ 0 & 0 & 0 \\ 0 & 0 & 0 \\ 0 & 0 & 0 \end{bmatrix} \begin{bmatrix} \mathbf{r}_1 \times \mathbf{f}_1 \\ \mathbf{r}_2 \times \mathbf{f}_2 \\ \mathbf{r}_3 \times \mathbf{f}_3 \\ \mathbf{r}_4 \times \mathbf{f}_4 \end{bmatrix}^{\mathrm{T}}
$$

For the convenience of computer control system design, following the same steps performed in [190], the continuous system is converted to discrete form given by

$$\mathbf{x}_6(n+1) = \Phi_6 \mathbf{x}_6(n) + \Gamma_{6\times 4}\mathbf{u}(n) \tag{12.16}$$

where $\mathbf{x}_6 = [\omega_x, \omega_y, \omega_z, q_1, q_2, q_3]^{\mathrm{T}}$, $\Phi_6 = e^{\mathbf{A}t_s}$, $\Gamma_{6\times 4} = \int_0^{t_s} e^{\mathbf{A}(t-\tau)}\mathbf{B}d\tau$, and t_s is the *sample period.*

In [190], it is shown that a PID control design is very successful for orbit-raising. To incorporate the integral terms, the discrete integrators defined by $\mathbf{iq} = [iq_1, iq_2, iq_3]^{\mathrm{T}} = \left[\int_0^{t_s} q_1, \int_0^{t_s} q_2, \int_0^{t_s} q_3\right]^{\mathrm{T}}$ are added simply as

$$\mathbf{iq}(n+1) = \mathbf{iq}(n) + t_s * \mathbf{q}(n) \tag{12.17}$$

where $\mathbf{q}(n)$ is the vector value of the quaternion at n-sampling time. Combining (12.16) and (12.17) gives

$$\begin{aligned} \mathbf{x}_9(n+1) &= \begin{bmatrix} \mathbf{x}_6(n+1) \\ \mathbf{iq}(n+1) \end{bmatrix} \\ &= \begin{bmatrix} \Phi_6 & \mathbf{0}_{6\times 3} \\ t_s[\mathbf{0}_{3\times 3} \;\; \mathbf{I}_{3\times 3}] & \mathbf{I}_{3\times 3} \end{bmatrix} \begin{bmatrix} \mathbf{x}_6(n) \\ \mathbf{iq}(n) \end{bmatrix} + \begin{bmatrix} \Gamma_{6\times 4} \\ \mathbf{0}_{3\times 4} \end{bmatrix} \mathbf{u}(n) \end{aligned} \tag{12.18}$$

The thrust control design is to select $\mathbf{u}(n)$ to maintain the attitude in the orbit-raising operation: this can be represented as a LQR design which minimizes a quadratic cost function (B.23) under the constraints of (12.18). Using MATLAB® control toolbox [59], the discrete state feedback control can be directly obtained by function `dare` as

$$\mathbf{u}(n) = -\mathbf{K}\mathbf{x}_9(n) \tag{12.19}$$

where $\mathbf{K}$ is the 4×9 state feedback matrix.

12.3 Comparing Quaternion and Euler Angle Designs

This section compares two different orbit-raising designs, the design based on the reduced quaternion model established in the previous section and the design based on the Euler angle model given in [190]. Both designs use the standard LQR method. The same spacecraft parameters as reported in [190] are used in both the designs. In particular, the sampling interval is 4 second; the diagonal elements of the inertia matrix are $J_x = 189(kg\cdot m^2)$, $J_y = 159(kg\cdot m^2)$, $J_z = 114(kg\cdot m^2)$; the momentum wheel moment is $-2.8(N\cdot m\cdot sec)$; the diagonal elements of the $\mathbf{Q}$ matrix are $Q_1 = Q_2 = Q_3 = 1/(2.5rad/sec)^2$ and $Q_4 = Q_5 = Q_6 = 1/(9rad)^2$, $Q_7 = Q_8 = Q_9 = 1/(182^2 rad^2 sec^2)$; the diagonal elements of the $\mathbf{R}$ matrix are $R_1 = R_2 = R_3 = R_4 = 1N^2$. It is assumed further that the same thrusters are installed and the same alignments are used as in Figure 12.3 where $d = 0.248m$ and $l = 0.815m$.

The LQR design based on Euler angle model has been successfully used for OrbView-2 orbit-raising and the results have been reported in [190]. Using the parameters listed above and the design model described in [190] and applying `dlqr` command in MATLAB toolbox [59] yields the feedback matrix

$$\mathbf{K}_e = \begin{bmatrix} -23.3459 & 12.0068 & -40.7442 & 0.0473 & 0.4753 & -1.1156 & -0.0002 & 0.0023 & -0.0034 \\ 23.3459 & 12.0068 & 40.7442 & -0.0473 & 0.4753 & 1.1156 & 0.0002 & 0.0023 & 0.0034 \\ 17.4922 & -12.0068 & -25.5628 & 1.0759 & -0.4753 & -0.2488 & 0.0035 & -0.0023 & -0.0004 \\ -17.4922 & -12.0068 & 25.5628 & -1.0759 & -0.4753 & 0.2488 & -0.0035 & -0.0023 & 0.0004 \end{bmatrix}$$

For the reduced quaternion model (12.18) with the same set of parameters listed above, applying `dlqr` command in MATLAB toolbox gives the feedback matrix of the LQR design

$$\mathbf{K}_q = \begin{bmatrix} -19.0183 & 9.6756 & -30.6404 & 0.2488 & 0.6127 & -1.3928 & 0.0003 & 0.0024 & -0.0035 \\ 19.0183 & 9.6756 & 30.6404 & -0.2488 & 0.6127 & 1.3928 & -0.0003 & 0.0024 & 0.0035 \\ 14.1902 & -9.6756 & -17.8344 & 1.4606 & -0.6127 & -0.1312 & 0.0036 & -0.0024 & 0.0000 \\ -14.1902 & -9.6756 & 17.8344 & -1.4606 & -0.6127 & 0.1312 & -0.0036 & -0.0024 & -0.0000 \end{bmatrix}$$

These feedback matrices ($\mathbf{K}_e$ and $\mathbf{K}_q$) are applied to the orginal nonlinear system (12.11) and (12.12) in their discretized form as follows:

$$\begin{aligned} \begin{bmatrix} \omega_x(n+1) \\ \omega_y(n+1) \\ \omega_z(n+1) \end{bmatrix} &= t_s \begin{bmatrix} 1 & -\frac{J_z\omega_z(n)}{J_x} & \frac{J_y\omega_y(n)+h_w}{J_x} \\ \frac{J_z\omega_z(n)}{J_y} & 1 & -\frac{J_x\omega_x(n)}{J_y} \\ -\frac{J_y\omega_y(n)+h_w}{J_z} & \frac{J_x\omega_x(n)}{J_z} & 1 \end{bmatrix} \begin{bmatrix} \omega_x(n) \\ \omega_y(n) \\ \omega_z(n) \end{bmatrix} \\ &+ t_s \begin{bmatrix} \frac{1}{J_x} & 0 & 0 \\ 0 & \frac{1}{J_y} & 0 \\ 0 & 0 & \frac{1}{J_z} \end{bmatrix} \begin{bmatrix} \mathbf{r}_1\times\mathbf{f}_1, \mathbf{r}_2\times\mathbf{f}_2, \mathbf{r}_3\times\mathbf{f}_3, \mathbf{r}_4\times\mathbf{f}_4 \end{bmatrix} \mathbf{K}\mathbf{x}_9(n) \end{aligned} \tag{12.20}$$

where $\mathbf{K}$ is either $\mathbf{K}_e$ or $\mathbf{K}_q$. For Euler angle model, the nonlinear kinematics equation of motion is given as follows [99]:

$$\begin{bmatrix} \dot{\phi} \\ \dot{\theta} \\ \dot{\psi} \end{bmatrix} = \begin{bmatrix} 1 & \sin(\phi)\tan(\theta) & \cos(\phi)\tan(\theta) \\ 0 & \cos(\phi) & -\sin(\phi) \\ 0 & \sin(\phi)\sec(\theta) & \cos(\phi)\sec(\theta) \end{bmatrix} \begin{bmatrix} \omega_x \\ \omega_y \\ \omega_z \end{bmatrix} \tag{12.21}$$

which has its discretized form as follows:

$$\begin{bmatrix} \phi(n+1) \\ \theta(n+1) \\ \psi(n+1) \end{bmatrix} - \begin{bmatrix} \phi(n) \\ \theta(n) \\ \psi(n) \end{bmatrix}$$

$$= t_s \begin{bmatrix} 1 & \sin(\phi(n))\tan(\theta(n)) & \cos(\phi(n))\tan(\theta(n)) \\ 0 & \cos(\phi(n)) & -\sin(\phi(n)) \\ 0 & \sin(\phi(n))\sec(\theta(n)) & \cos(\phi(n))\sec(\theta(n)) \end{bmatrix} \begin{bmatrix} \omega_x(n) \\ \omega_y(n) \\ \omega_z(n) \end{bmatrix} \tag{12.22}$$

For reduced quaternion model, the nonlinear kinematics equation of motion has its discretized form as follows:

$$\begin{bmatrix} q_1(n+1) \\ q_2(n+1) \\ q_3(n+1) \end{bmatrix} - \begin{bmatrix} q_1(n) \\ q_2(n) \\ q_3(n) \end{bmatrix}$$
$$= \frac{t_s}{2} \begin{bmatrix} \sqrt{1-q_1^2(n)-q_2^2(n)-q_3^2(n)}\,\omega_x(n) - q_3(n)\omega_y(n) + q_2(n)\omega_z(n) \\ q_3(n)\omega_x(n) + \sqrt{1-q_1^2(n)-q_2^2(n)-q_3^2(n)}\,\omega_y(n) - q_1(n)\omega_z(n) \\ -q_2(n)\omega_x(n) + q_1(n)\omega_y(n) + \sqrt{1-q_1^2(n)-q_2^2(n)-q_3^2(n)}\,\omega_z(n) \end{bmatrix} \tag{12.23}$$

It is worthwhile to note that (12.17) is used to propagate for the last three integral states for the feedback control. For the Euler angle feedback control, the discrete Euler angle integrators

$$\mathbf{ie} = [ie_1,\ ie_2,\ ie_3]^{\mathrm{T}} = \left[\ \int_0^{t_s} \phi,\ \ \int_0^{t_s} \theta,\ \ \int_0^{t_s} \psi\ \right]^{\mathrm{T}}$$

is given by

$$\mathbf{ie}(n+1) = \mathbf{ie}(n) + t_s * [\phi(n), \theta(n), \psi(n)]^{\mathrm{T}}$$

to propagate the last three integral states.

In the simulation test, it is assumed that the initial quaternion rates are zeros; the initial Euler angles are 2 degrees in roll, pitch, and yaw which is about $2\pi/180$ radians; the initial Euler angle is converted to initial quaternion and used as the initial feedback in quaternion model based design; the initial integral terms for quaternion and for Euler angles are all set to zeros. At the end of every iteration for quaternion-based design simulation, the quaternion is converted back to the Euler angles and saved so that the responses of the two different designs can be compared using the same error measurement. The simulation results are provided in Figures 12.4–12.9. In these figures, the solid lines are the response of the closed-loop system of quaternion based design; the dashed lines are the response of the closed-loop system of Euler angle based design. Clearly, the system based on quaternion model design has slightly better responses than the system based on Euler angle model design in terms of widely used metrics, such as percentage of overshoot, settling time, etc [41].

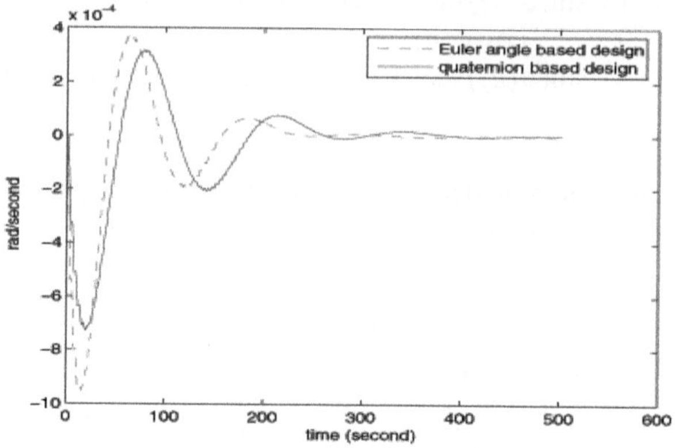

Figure 12.4: Design comparison for quaternion rate ω_x.

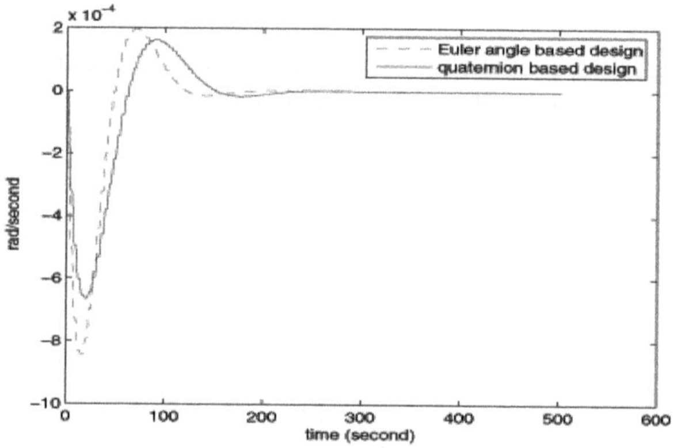

Figure 12.5: Design comparison for quaternion rate ω_y.

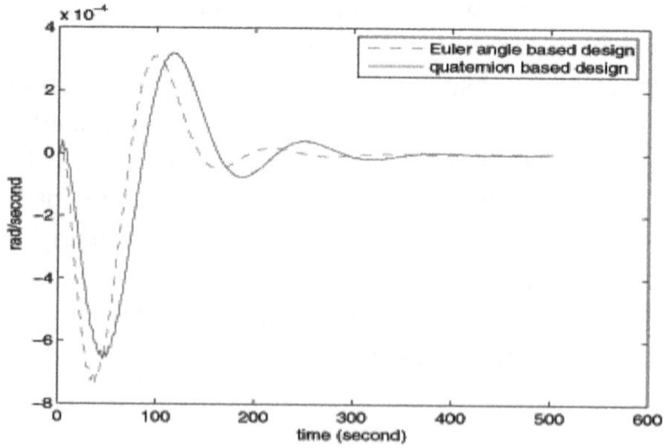

Figure 12.6: Design comparison for quaternion rate ω_z.

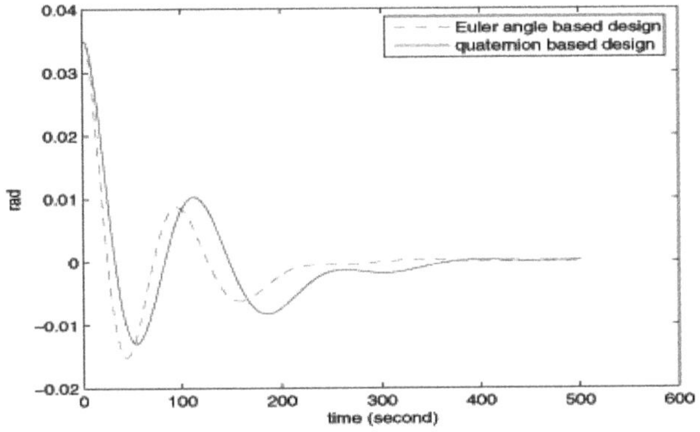

Figure 12.7: Design comparison for quaternion q_1.

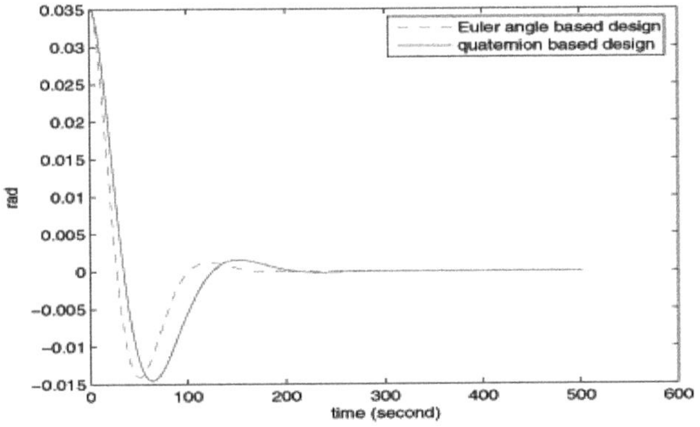

Figure 12.8: Design comparison for quaternion q_2.

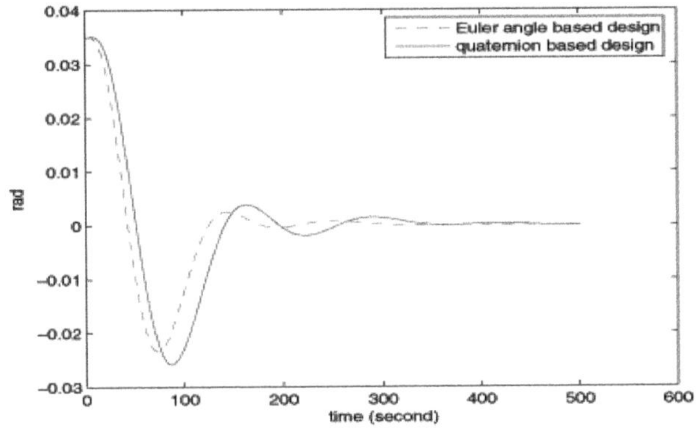

Figure 12.9: Design comparison for quaternion q_3.

Chapter 13

Attitude MPC Control

Model predictive control (MPC) design [12, 129, 161, 2] has been a major research area and many successful applications have been reported [161]. The main idea of the model predictive control is to repeatedly solve a continuously updated control problem based on the latest information and apply the control action to the system based on the latest solution of the updated control problem. This requires significantly more on-line computational effort than most other control strategies. Therefore, model predictive control was not immediately adopted in spacecraft attitude control system designs when on-board computational power was limited. But as computers become more and more powerful, research of model predictive control designs for spacecraft application becomes very active, for example, Hegrenæs et al. in [69, 231] discussed model predictive control in different scenarios for spacecraft attitude control, Hartley et al. in [67] considered model predictive control design for spacecraft rendezvous problem, Di Cairano et al. in [25] investigated spacecraft rendezvous and proximity maneuvering, and Morgan et al. in [135] proposed model predictive control design for swarms of spacecraft using sequential convex programming.

One of the most attractive and popular methods in model predictive control is to repeatedly solve a Constrained Linear Quadratic Regulator (CLQR) design problem because (a) the problem is relatively easy to solve on-line and (b) most nonlinear systems may be appropriately simplified as a linear system. To ensure that the model predictive control will be workable for the purpose of on-line application, researchers have been working on efficient and effective algorithms for the CLQR even though some algorithms were available as early as 1982 (see [15]). Since the CLQR problem can be reduced to a quadratic programming (QP) problem, most efficient algorithms up to the date are focused on the efficient solutions of QP. For example, Rao et al. in [162] proposed an *interior-point* algorithm with desirable theoretical properties (polynomial complexity). Bemporad et al. in

[13, 202] proposed a multi-parametric program method which was aimed at reducing on-line computational burden and use off-line computation as much as possible. Wang et al. in [216] suggested some fast algorithm specially designed for on-line convex QP for the model predictive control. Though these methods proposed innovative ideas to enhance on-line optimization efficiency, there are rooms and needs to further improve these methods. For example, the most efficient interior-point algorithm for general QP problems is infeasible interior-point algorithms (finding a feasible starting point for general QP is expensive) which is a defect for MPC application, i.e., if early termination has to be enforced because of the on-line application requirement, the solution may not be feasible (because the intermediate iterates of infeasible interior-point algorithm are very likely infeasible). The multi-parametric QP proposed in [13, 202] would generate a look-up table growing exponentially with the horizon, state, and input dimensions, as noted in [216]; therefore, multi-parametric QP can be used for some very small problems (state dimensions is no more than 5). The convex QP algorithm proposed in [216] also uses infeasible interior-point; therefore, its intermediate iterates are likely infeasible. Moreover, like the method in [162], the size of the QP problem obtained by [216] is big (for a system of $n = 20$, $m = 3$, and a horizon $N = 30$, the corresponding QP has 450 variables and 1260 constraints).

In this chapter, constrained MPC design problem subject to actuators saturation is considered. This problem is slightly simpler than the problems considered in [162, 13, 202, 216] but is still general enough for most real world problems. Several significant improvements over the aforementioned methods are proposed. First, the numbers of the variables and constraints of the corresponding QP problem can be reduced significantly and all equality constraints can be removed. This means that the corresponding QP problem is not only much smaller but also has a special structure, i.e., the problem is reduced to a convex quadratic programming *subject to box constraints*, for which we can easily find a feasible starting point. Therefore, the second improvement over [162, 216] is to solve the reduced problem using a *feasible interior-point* algorithm which has several advantages over infeasible interior-point algorithms: (a) in general, the feasible interior-point algorithms have lower polynomial bound (more efficient) and (b) all intermediate iterates are feasible; therefore, early termination will give a feasible near-optimal solution. To further reduce the on-line computational cost, the third improvement is to devise a new algorithm that improves the efficiency of existing algorithms. By using the special structures of the problem, one can show that the algorithm proposed in this section enhances the general QP algorithm proposed in [247] in two aspects: (a) search in a larger neighborhood (the algorithm is more efficient) and (b) use an explicit initial feasible interior point (the algorithm does not need a phase-one process to find a feasible point). It is also shown that this algorithm has the best polynomial complexity bound, a very desirable theoretical property. By using the MATLAB® code to a spacecraft orbit-raising MPC design example, it is then verified that the proposed constrained MPC design has superior performance in computation because of the above mentioned improvements. The content of this chapter is based on [248]. Throughout the chapter, $\mathbf{e}$ denotes a vector whose elements are all ones.

13.1 Some Technical Lemmas

Some technical lemmas, which are independent of the problem, are introduced in this section. The first two simple lemmas are given in [247, 248].

Lemma 13.1
Let $p > 0$, $q > 0$, and $r > 0$ be some constants. If $p+q \le r$, then $pq \le \frac{r^2}{4}$.

Lemma 13.2
For $\alpha \in [0, \frac{\pi}{2}]$,

$$\sin(\alpha) \ge \sin^2(\alpha) = 1 - \cos^2(\alpha) \ge 1 - \cos(\alpha).$$

The next Lemma is proved in [134].

Lemma 13.3
Let $\mathbf{u}$, $\mathbf{v}$, and $\mathbf{w}$ be real vectors of same size satisfying $\mathbf{u}+\mathbf{v}=\mathbf{w}$ and $\mathbf{u}^{\mathrm{T}}\mathbf{v} \ge \mathbf{0}$. Then,

$$2\|\mathbf{u}\| \cdot \|\mathbf{v}\| \le \|\mathbf{u}\|^2 + \|\mathbf{v}\|^2 \le \|\mathbf{u}\|^2 + \|\mathbf{v}\|^2 + 2\mathbf{u}^{\mathrm{T}}\mathbf{v} = \|\mathbf{u}+\mathbf{v}\|^2 = \|\mathbf{w}\|^2 \tag{13.1}$$

Let $\circ$ denote the Hadamard (entrywise) product operator. The following technical lemma is from [233, page 88].

Lemma 13.4
Let $\mathbf{u}$ and $\mathbf{v}$ be the vectors of the same dimension, and $\mathbf{u}^{\mathrm{T}}\mathbf{v} \ge \mathbf{0}$. Then

$$\|\mathbf{u} \circ \mathbf{v}\| \le 2^{-\frac{3}{2}} \|\mathbf{u}+\mathbf{v}\|^2$$

The famous *Cardano's formula* can be found in [157].

Lemma 13.5
Let p and q be the real numbers that are related to the following cubic algebra equation

$$x^3 + px + q = 0$$

If

$$\Delta = \left(\frac{q}{2}\right)^2 + \left(\frac{p}{3}\right)^3 > 0$$

then the cubic equation has one real root that is given by

$$x = \sqrt[3]{-\frac{q}{2} + \sqrt{\left(\frac{q}{2}\right)^2 + \left(\frac{p}{3}\right)^3}} + \sqrt[3]{-\frac{q}{2} - \sqrt{\left(\frac{q}{2}\right)^2 + \left(\frac{p}{3}\right)^3}}$$

For quartic polynomials, the roots can be represented by several different formulas, which are not discussed here but are referred to [178] and references therein. The last technical lemma in this section is as follows.

Lemma 13.6

Let **u** *and* **v** *be any n-dimensional vectors. Then*

$$\left\| \mathbf{u} \circ \mathbf{v} - \frac{1}{n} \left(\mathbf{u}^{\mathrm{T}} \mathbf{v}\right) \mathbf{e} \right\| \le \left\| \mathbf{u} \circ \mathbf{v} \right\|$$

Proof 13.1 Simple calculation gives

$$\begin{aligned}
& \left\| \mathbf{u} \circ \mathbf{v} - \frac{1}{n} \left(\mathbf{u}^{\mathrm{T}} \mathbf{v}\right) \mathbf{e} \right\|^2 \\
= & \sum_{i=1}^{n} \left(u_i v_i - \frac{1}{n} \sum_{i=1}^{n} u_i v_i \right)^2 \\
= & \sum_{i=1}^{n} \left(u_i^2 v_i^2 - \frac{2 u_i v_i}{n} \sum_{i=1}^{n} u_i v_i + \frac{1}{n^2} \left(\sum_{i=1}^{n} u_i v_i \right)^2 \right) \\
= & \sum_{i=1}^{n} \left(u_i^2 v_i^2 \right) - \frac{2}{n} \left(\sum_{i=1}^{n} u_i v_i \right)^2 + \frac{1}{n} \left(\sum_{i=1}^{n} u_i v_i \right)^2 \\
= & \sum_{i=1}^{n} \left(u_i^2 v_i^2 \right) - \frac{1}{n} \left(\sum_{i=1}^{n} u_i v_i \right)^2 \le \left\| \mathbf{u} \circ \mathbf{v} \right\|^2
\end{aligned}$$

This finishes the proof. ■

13.2 Constrained MPC and Convex QP with Box Constraints

Constrained MPC design under consideration repeatedly solves the following CLQR design problem. Let $\mathbf{x} \in \mathbf{R}^r$ be the system state, $\mathbf{u} \in \mathbf{R}^m$ be the control vector, $\mathbf{A} \in \mathbf{R}^{r \times r}$ and $\mathbf{B} \in \mathbf{R}^{r \times m}$ be system matrices. The discrete linear time-invariant system is given by

$$\mathbf{x}_{s+1} = \mathbf{A}\mathbf{x}_s + \mathbf{B}\mathbf{u}_s \tag{13.2}$$

while fulfilling the constraints

$$-\mathbf{e} \leq \mathbf{u}_s \leq \mathbf{e} \tag{13.3}$$

where $s = t, \ldots, t+N-1$. Let $\mathbf{P} \in \mathbf{R}^{r\times r}$, $\mathbf{Q} \in \mathbf{R}^{r\times r}$, and $\mathbf{R} \in \mathbf{R}^{m\times m}$ be positive definite matrices. The design is to optimize the following cost function

$$J = \min_{\mathbf{u}_t, \mathbf{u}_{t+1}, \cdots, \mathbf{u}_{t+N-1}} \frac{1}{2}\mathbf{x}_{t+N}^{\mathrm{T}}\mathbf{P}\mathbf{x}_{t+N} + \frac{1}{2}\sum_{k=0}^{N-1}\left[\mathbf{x}_{t+k}^{\mathbf{T}}\mathbf{Q}\mathbf{x}_{t+k} + \mathbf{u}_{t+k}^{\mathrm{T}}\mathbf{R}\mathbf{u}_{t+k}\right] \tag{13.4}$$

under the system dynamics equality constraints (13.2) and control saturation inequality constraints (13.3). Given current state $\mathbf{x}_t$, this CLQR (or MPC) design problem is a typical *convex quadratic programming* problems with $Nr+Nm$ variables $\mathbf{x}_{t+1}, \cdots, \mathbf{x}_{t+N}$, $\mathbf{u}_t, \cdots, \mathbf{u}_{t+N-1}$. Though this problem can be directly solved as suggested by [15], it can be significantly reduced to an equivalent but much smaller convex quadratic programming problem subject only to box constraints. Denote

$$\mathbf{A}^k = \underbrace{\mathbf{A}\cdots\mathbf{A}}_{\text{product of k } \mathbf{A}} := \mathbf{A}_k \in \mathbf{R}^{r\times r}$$

with $\mathbf{A}_0 = \mathbf{I}$. Since

$$\begin{aligned}
\mathbf{x}_{t+k} &= \mathbf{A}\mathbf{x}_{t+k-1} + \mathbf{B}\mathbf{u}_{t+k-1} = \mathbf{A}^k\mathbf{x}_t + \sum_{j=0}^{k-1}\mathbf{A}^j\mathbf{B}\mathbf{u}_{t+k-j-1} \\
&= \mathbf{A}_k\mathbf{x}_t + \sum_{j=0}^{k-1}\mathbf{A}_j\mathbf{B}\mathbf{u}_{t+k-j-1}
\end{aligned} \tag{13.5}$$

Equation (13.4) can be rewritten as

$$\begin{aligned}
J &= \min_{\mathbf{u}_t, \mathbf{u}_{t+1}, \cdots, \mathbf{u}_{t+N-1}} \frac{1}{2}\left(\mathbf{A}_N\mathbf{x}_t + \sum_{j=0}^{N-1}\mathbf{A}_j\mathbf{B}\mathbf{u}_{t+N-j-1}\right)^{\mathrm{T}}\mathbf{P}\left(\mathbf{A}_N\mathbf{x}_t + \sum_{j=0}^{N-1}\mathbf{A}_j\mathbf{B}\mathbf{u}_{t+N-j-1}\right) \\
&+\frac{1}{2}\sum_{k=1}^{N-1}\left(\mathbf{A}_k\mathbf{x}_t + \sum_{j=0}^{k-1}\mathbf{A}_j\mathbf{B}\mathbf{u}_{t+k-j-1}\right)^{\mathrm{T}}\mathbf{Q}\left(\mathbf{A}_k\mathbf{x}_t + \sum_{j=0}^{k-1}\mathbf{A}_j\mathbf{B}\mathbf{u}_{t+k-j-1}\right) \\
&+\frac{1}{2}\sum_{k=0}^{N-1}\left(\mathbf{u}_{t+k}^{\mathrm{T}}\mathbf{R}\mathbf{u}_{t+k}\right)
\end{aligned} \tag{13.6}$$

Notice that $\mathbf{x}_t$ is a constant vector, $\mathbf{A}_j$, $\mathbf{P}$, $\mathbf{Q}$, and $\mathbf{R}$ are constant matrices, the (13.6) can be reduced to

$$J_0 = \min_{\mathbf{u}_t, \mathbf{u}_{t+1}, \cdots, \mathbf{u}_{t+N-1}} \frac{1}{2}\left(\sum_{j=0}^{N-1}\mathbf{A}_j\mathbf{B}\mathbf{u}_{t+N-j-1}\right)^{\mathrm{T}}\mathbf{P}\left(\sum_{j=0}^{N-1}\mathbf{A}_j\mathbf{B}\mathbf{u}_{t+N-j-1}\right)$$

$$
\begin{aligned}
&+ \quad (\mathbf{A}_N\mathbf{x}_t)^{\mathrm{T}}\mathbf{P}\left(\sum_{j=0}^{N-1}\mathbf{A}_j\mathbf{B}\mathbf{u}_{t+N-j-1}\right) \\
&+ \quad \frac{1}{2}\sum_{k=1}^{N-1}\left(\sum_{j=0}^{k-1}\mathbf{A}_j\mathbf{B}\mathbf{u}_{t+k-j-1}\right)^{\mathrm{T}}\mathbf{Q}\left(\sum_{j=0}^{k-1}\mathbf{A}_j\mathbf{B}\mathbf{u}_{t+k-j-1}\right) \\
&+ \quad \sum_{k=1}^{N-1}\left((\mathbf{A}_k\mathbf{x}_t)^{\mathrm{T}}\mathbf{Q}\left(\sum_{j=0}^{k-1}\mathbf{A}_j\mathbf{B}\mathbf{u}_{t+k-j-1}\right)\right) \\
&+ \quad \frac{1}{2}\sum_{k=0}^{N-1}\left(\mathbf{u}_{t+k}^{\mathrm{T}}\mathbf{R}\mathbf{u}_{t+k}\right)
\end{aligned} \tag{13.7}
$$

Denote

$$
\begin{aligned}
\sum_{j=0}^{k-1}\mathbf{A}_j\mathbf{B}\mathbf{u}_{t+k-j-1} &= \underbrace{[\mathbf{A}_{k-1}\mathbf{B},\mathbf{A}_{k-2}\mathbf{B},\cdots,\mathbf{B}]}_{\phi_k\in\mathbf{R}^{r\times(km)}}\underbrace{\begin{bmatrix}\mathbf{u}_t\\ \vdots \\ \mathbf{u}_{t+k-1}\end{bmatrix}}_{\mathbf{v}_k\in\mathbf{R}^{km}} \\
&= \phi_k\mathbf{v}_k, \quad k\in\{1,2,\ldots,N\}
\end{aligned} \tag{13.8}
$$

$$
\mathbf{Q}_k = \begin{bmatrix}\phi_k^{\mathrm{T}}\mathbf{Q}\phi_k & \mathbf{0}\\ \mathbf{0} & \mathbf{0}\end{bmatrix}\in\mathbf{R}^{(Nm)\times(Nm)}, \quad \phi_k^{\mathrm{T}}\mathbf{Q}\phi_k\in\mathbf{R}^{(km)\times(km)}, \quad k\in\{1,2,\ldots,N-1\} \tag{13.9}
$$

$$
\mathbf{R}_N = \underbrace{\begin{bmatrix}\mathbf{R} & \cdots & \mathbf{0}\\ \vdots & \ddots & \vdots\\ \mathbf{0} & \cdots & \mathbf{R}\end{bmatrix}}_{\text{N diagonal matrices}} \in\mathbf{R}^{(Nm)\times(Nm)} \tag{13.10}
$$

and

$$
\begin{aligned}
\mathbf{S}_k &= \begin{bmatrix}\mathbf{A}_k^{\mathrm{T}}\mathbf{Q}\phi_k & \mathbf{0}\end{bmatrix}\in\mathbf{R}^{r\times(Nm)}, \quad \mathbf{A}_k^{\mathrm{T}}\mathbf{Q}\phi_k\in\mathbf{R}^{r\times(km)}, \\
& k\in\{1,2,\ldots,N-1\}
\end{aligned} \tag{13.11}
$$

where $\mathbf{0}$s are zero matrices with appropriate dimensions. The CLQR (or MPC) design is reduced further to

$$
\begin{aligned}
J_0 &= \min_{\mathbf{u}_t,\mathbf{u}_{t+1},\cdots,\mathbf{u}_{t+N-1}} \frac{1}{2}\mathbf{v}_N^{\mathrm{T}}\left(\phi_N^{\mathrm{T}}\mathbf{P}\phi_N+\sum_{k=1}^{N-1}\mathbf{Q}_k+\mathbf{R}_N\right)\mathbf{v}_N+\mathbf{x}_t^{\mathrm{T}}\left(\mathbf{A}_N^{\mathrm{T}}\mathbf{P}\phi_N+\sum_{k=1}^{N-1}\mathbf{S}_k\right)\mathbf{v}_N \\
& s.t. \quad -\mathbf{e}\le\mathbf{v}_N\le\mathbf{e}
\end{aligned} \tag{13.12}
$$

Let $n = Nm$,

$$\mathbf{x} = \mathbf{v}_N \tag{13.13}$$

$$\mathbf{H} = \left(\phi_N^{\mathrm{T}} \mathbf{P} \phi_N + \sum_{k=1}^{N-1} \mathbf{Q}_k + \mathbf{R}_N \right) \tag{13.14}$$

$$\mathbf{c}^{\mathrm{T}} = \mathbf{x}_t^{\mathrm{T}} \left(\mathbf{A}_N^{\mathrm{T}} \mathbf{P} \phi_N + \sum_{k=1}^{N-1} \mathbf{S}_k \right) \tag{13.15}$$

The CLQR (or MPC) design problem can be written in a standard form of convex quadratic problem with *box constraints*:

$$(QP) \quad \min \ \tfrac{1}{2}\mathbf{x}^{\mathrm{T}}\mathbf{H}\mathbf{x} + \mathbf{c}^{\mathrm{T}}\mathbf{x}, \quad \text{subject to} \quad -\mathbf{e} \leq \mathbf{x} \leq \mathbf{e} \tag{13.16}$$

where $\mathbf{0} < \mathbf{H} \in \mathbf{R}^{n \times n}$ is a positive definite matrix, $\mathbf{c} \in \mathbf{R}^n$ is given, and $\mathbf{x} \in \mathbf{R}^n$ is the control vector to be optimized. This convex quadratic programming problem has Nm variables and $2Nm$ box constraints, and its size is independent of the system dimension r, a much smaller and simpler problem than the original one. A quick comparison of the MPC problem sizes and reduced QP sizes using the method of this section and methods mentioned in [216] (cf. [216, Table 1]) is given in Table 13.1.

Table 13.1: Comparison of reduced QP sizes of the proposed method and other methods.

System state r	Control input m	Horizon N	QP size of this section		QP size of [216] and other papers	
			# of variables	# of constraints	# of variables	# of constraints
4	2	20	40	80	100	320
10	3	30	90	180	360	1080
16	4	30	120	240	570	1680
30	8	30	240	480	1110	3180

The bigger the linear system is, the more advantage of the proposed method will be. A bigger advantage of the proposed method is that the constraints in (13.16) is very simple that admits an feasible initial interior point (see Section 13.6), and the proposed method allows users to use more efficient feasible (initial) interior-point algorithms rather than an infeasible (initial) interior-point algorithm as used in [216]. Moreover, if a premature termination is enforced due to the on-line computational requirement, the solution is feasible.

All the simplifications described in this section is off-line. But it greatly reduces the on-line problem size and simplifies the problem constraints. However, it is not wise to use an interior-point algorithm designed for general problems for this very special convex quadratic programming problem which has `only box constraints`. In the remainder of this chapter, the structure of the box constraints will be fully investigated and a very efficient algorithm for the problem (13.16) will be devised.

13.3 Central Path of Convex QP with Box Constraints

In view of the KKT conditions (see Appendix A or [142]), since $\mathbf{H}$ is a positive definite matrix, $\mathbf{x}$ is an optimal solution of (13.16) if and only if $\mathbf{x}$, λ, and γ satisfy

$$-\lambda+\gamma-\mathbf{Hx}=\mathbf{c} \tag{13.17a}$$

$$-\mathbf{e}\leq\mathbf{x}\leq\mathbf{e} \tag{13.17b}$$

$$(\lambda,\gamma)\geq\mathbf{0} \tag{13.17c}$$

$$\lambda_i(e_i-x_i)=0,\quad \gamma_i(e_i+x_i)=0,\quad i=1,\dots,n \tag{13.17d}$$

Denote $\mathbf{y}=\mathbf{e}-\mathbf{x}\geq\mathbf{0}$, $\mathbf{z}=\mathbf{e}+\mathbf{x}\geq\mathbf{0}$. The KKT conditions can be rewritten as

$$\mathbf{Hx}+\mathbf{c}+\lambda-\gamma=\mathbf{0} \tag{13.18a}$$

$$\mathbf{x}+\mathbf{y}=\mathbf{e},\quad \mathbf{x}-\mathbf{z}=-\mathbf{e} \tag{13.18b}$$

$$(\mathbf{y},\mathbf{z},\lambda,\gamma)\geq\mathbf{0} \tag{13.18c}$$

$$\lambda_i y_i=0,\quad \gamma_i z_i=0,\quad i=1,\dots,n \tag{13.18d}$$

For the convex (QP) problem, the KKT conditions are also sufficient for $\mathbf{x}$ to be a global optimal solution (see Appendix A). Denote the *feasible set* $\mathcal{F}$ as a collection of all points that meet the constraints (13.18a), (13.18b), (13.18c)

$$\mathcal{F}=\{(\mathbf{x},\mathbf{y},\mathbf{z},\lambda,\gamma):\mathbf{Hx}+\mathbf{c}+\lambda-\gamma=\mathbf{0},(\mathbf{y},\mathbf{z},\lambda,\gamma)\geq\mathbf{0},\mathbf{x}+\mathbf{y}=\mathbf{e},\mathbf{x}-\mathbf{z}=-\mathbf{e}\} \tag{13.19}$$

and the *strictly feasible* set $\mathcal{F}^o$ as a collection of all points that meet the constraints (13.18a), (13.18b), and are strictly positive in (13.18c)

$$\mathcal{F}^o=\{(\mathbf{x},\mathbf{y},\mathbf{z},\lambda,\gamma):\mathbf{Hx}+\mathbf{c}+\lambda-\gamma=\mathbf{0},(\mathbf{y},\mathbf{z},\lambda,\gamma)>\mathbf{0},\mathbf{x}+\mathbf{y}=\mathbf{e},\mathbf{x}-\mathbf{z}=-\mathbf{e}\} \tag{13.20}$$

Similar to the linear programming, the *central path* $\mathcal{C}\in\mathcal{F}^o\subset\mathcal{F}$ is defined as a curve in finite dimensional space parameterized by a scalar $\tau>0$ as follows. For each interior point $(\mathbf{x},\mathbf{y},\mathbf{z},\lambda,\gamma)\in\mathcal{F}^o$ on the central path, there is a $\tau>0$ such that

$$\mathbf{Hx}+\mathbf{c}+\lambda-\gamma=\mathbf{0} \tag{13.21a}$$

$$\mathbf{x}+\mathbf{y}=\mathbf{e},\quad \mathbf{x}-\mathbf{z}=-\mathbf{e} \tag{13.21b}$$

$$(\mathbf{y},\mathbf{z},\lambda,\gamma)>\mathbf{0} \tag{13.21c}$$

$$\lambda_i y_i=\tau,\quad \gamma_i z_i=\tau,\quad i=1,\dots,n \tag{13.21d}$$

Therefore, the central path is an arc that is parameterized as a function of τ and is denoted as

$$\mathcal{C}=\{(\mathbf{x}(\tau),\mathbf{y}(\tau),\mathbf{z}(\tau),\lambda(\tau),\gamma(\tau)):\tau>0\} \tag{13.22}$$

As $\tau \to 0$, the moving point $(\mathbf{x}(\tau), \mathbf{y}(\tau), \mathbf{z}(\tau), \lambda(\tau), \gamma(\tau))$ on the central path represented by (13.21) approaches the solution of (QP) represented by (13.16). Throughout the rest of this chapter, the following assumption is made.

Assumption:

1. $\mathcal{F}^o$ is not empty.

Assumption 1 implies the existence of a central path. This assumption is always true for the CLQR problem. An explicit initial interior point will be provided later in this chapter.

Let $1 > \theta > 0$, denote $\mathbf{p} = (\mathbf{y}, \mathbf{z})$, $\omega = (\lambda, \gamma)$, and the *duality measure*

$$\mu = \frac{\lambda^{\mathrm{T}}\mathbf{y} + \gamma^{\mathrm{T}}\mathbf{z}}{2n} = \frac{\mathbf{p}^{\mathrm{T}}\omega}{2n} \tag{13.23}$$

A set of neighborhood of the central path is defined as

$$\mathcal{N}_2(\theta) = \{(\mathbf{x}, \mathbf{y}, \mathbf{z}, \lambda, \gamma) \in \mathcal{F}^o : \|\mathbf{p} \circ \omega - \mu \mathbf{e}\| \leq \theta \mu\} \subset \mathcal{F}^o \tag{13.24}$$

As the duality measure is reduced to zero, the neighborhood of $\mathcal{N}_2(\theta)$ will be a neighborhood of the central path that approaches the optimizer(s) of the QP problem, therefore, all points inside $\mathcal{N}_2(\theta)$ will approach the optimizer(s) of the QP problem. For $(\mathbf{x}, \mathbf{y}, \mathbf{z}, \lambda, \gamma) \in \mathcal{N}_2(\theta)$, since $(1-\theta)\mu \leq \omega_i p_i \leq (1+\theta)\mu$, where ω_i are either λ_i or γ_i, and p_i are either y_i or z_i, it must have

$$\frac{\omega_i p_i}{1+\theta} \leq \frac{\max_i \omega_i p_i}{1+\theta} \leq \mu \leq \frac{\min_i \omega_i p_i}{1-\theta} \leq \frac{\omega_i p_i}{1-\theta} \tag{13.25}$$

13.4 An Algorithm for Convex QP with Box Constraints

The idea of *arc-search* proposed in this section is very simple. The algorithm starts from a feasible point in $\mathcal{N}_2(\theta)$ close to the central path, constructs an arc that passes through the point and approximates the central path, searches along the arc to a new point in a larger area $\mathcal{N}_2(2\theta)$ that reduces the duality measure $\mathbf{p}^{\mathrm{T}}\omega$ and meets (13.21a), (13.21b), and (13.21c). The process is repeated by finding a better point close to the central path or on the central path in $\mathcal{N}_2(\theta)$ that simultaneously meets (13.21a), (13.21b), and (13.21c).

Following the idea used in [247], an *ellipse* $\mathcal{E}$ [28] in an appropriate dimensional space will be used to approximate the central path C described by (13.21), where

$$\mathcal{E} \quad = \quad \{(\mathbf{x}(\alpha), \mathbf{y}(\alpha), \mathbf{z}(\alpha), \lambda(\alpha), \gamma(\alpha)) :$$

$$(\mathbf{x}(\alpha),\mathbf{y}(\alpha),\mathbf{z}(\alpha),\lambda(\alpha),\gamma(\alpha)) = \vec{\mathbf{a}}\cos(\alpha)+\vec{\mathbf{b}}\sin(\alpha)+\vec{\mathbf{c}}\} \tag{13.26}$$

$\vec{\mathbf{a}} \in \mathbf{R}^{5n}$ and $\vec{\mathbf{b}} \in \mathbf{R}^{5n}$ are the axes of the ellipse, $\vec{\mathbf{c}} \in \mathbf{R}^{5n}$ is the center of the ellipse. Given a point $(\mathbf{x},\mathbf{y},\mathbf{z},\lambda,\gamma) = (\mathbf{x}(\alpha_0),\mathbf{y}(\alpha_0),\mathbf{z}(\alpha_0),\lambda(\alpha_0),\gamma(\alpha_0)) \in \mathcal{E}$ which is close to or on the central path, $\vec{\mathbf{a}}$, $\vec{\mathbf{b}}$, $\vec{\mathbf{c}}$ are functions of α, $(\mathbf{x},\lambda,\gamma,\mathbf{y},\mathbf{z})$, $(\dot{\mathbf{x}},\dot{\mathbf{y}},\dot{\mathbf{z}},\dot{\lambda},\dot{\gamma})$, and $(\ddot{\mathbf{x}},\ddot{\mathbf{y}},\ddot{\mathbf{z}},\ddot{\lambda},\ddot{\gamma})$, where $(\dot{\mathbf{x}},\dot{\mathbf{y}},\dot{\mathbf{z}},\dot{\lambda},\dot{\gamma})$ and $(\ddot{\mathbf{x}},\ddot{\mathbf{y}},\ddot{\mathbf{z}},\ddot{\lambda},\ddot{\gamma})$ are defined as

$$\begin{bmatrix} \mathbf{H} & \mathbf{0} & \mathbf{0} & \mathbf{I} & -\mathbf{I} \\ \mathbf{I} & \mathbf{I} & \mathbf{0} & \mathbf{0} & \mathbf{0} \\ \mathbf{I} & \mathbf{0} & -\mathbf{I} & \mathbf{0} & \mathbf{0} \\ \mathbf{0} & \Lambda & \mathbf{0} & \mathbf{Y} & \mathbf{0} \\ \mathbf{0} & \mathbf{0} & \Gamma & \mathbf{0} & \mathbf{Z} \end{bmatrix} \begin{bmatrix} \dot{\mathbf{x}} \\ \dot{\mathbf{y}} \\ \dot{\mathbf{z}} \\ \dot{\lambda} \\ \dot{\gamma} \end{bmatrix} = \begin{bmatrix} \mathbf{0} \\ \mathbf{0} \\ \mathbf{0} \\ \lambda \circ \mathbf{y} \\ \gamma \circ \mathbf{z} \end{bmatrix} \tag{13.27}$$

$$\begin{bmatrix} \mathbf{H} & \mathbf{0} & \mathbf{0} & \mathbf{I} & -\mathbf{I} \\ \mathbf{I} & \mathbf{I} & \mathbf{0} & \mathbf{0} & \mathbf{0} \\ \mathbf{I} & \mathbf{0} & -\mathbf{I} & \mathbf{0} & \mathbf{0} \\ \mathbf{0} & \Lambda & \mathbf{0} & \mathbf{Y} & \mathbf{0} \\ \mathbf{0} & \mathbf{0} & \Gamma & \mathbf{0} & \mathbf{Z} \end{bmatrix} \begin{bmatrix} \ddot{\mathbf{x}} \\ \ddot{\mathbf{y}} \\ \ddot{\mathbf{z}} \\ \ddot{\lambda} \\ \ddot{\gamma} \end{bmatrix} = \begin{bmatrix} \mathbf{0} \\ \mathbf{0} \\ \mathbf{0} \\ -2\dot{\lambda} \circ \dot{\mathbf{y}} \\ -2\dot{\gamma} \circ \dot{\mathbf{z}} \end{bmatrix} \tag{13.28}$$

where $\Lambda = \text{diag}(\lambda)$, $\Gamma = \text{diag}(\gamma)$, $\mathbf{Y} = \text{diag}(\mathbf{y})$, and $\mathbf{Z} = \text{diag}(\mathbf{z})$. The first rows of (13.27) and (13.28) are equivalent to

$$\mathbf{H}\dot{\mathbf{x}} = \dot{\gamma} - \dot{\lambda}, \quad \mathbf{H}\ddot{\mathbf{x}} = \ddot{\gamma} - \ddot{\lambda} \tag{13.29}$$

The next 2 rows of (13.27) and (13.28) are equivalent to

$$\dot{\mathbf{x}} = -\dot{\mathbf{y}}, \quad \dot{\mathbf{x}} = \dot{\mathbf{z}}, \quad \ddot{\mathbf{x}} = -\ddot{\mathbf{y}}, \quad \ddot{\mathbf{x}} = \ddot{\mathbf{z}} \tag{13.30}$$

The last 2 rows of (13.27) and (13.28) are equivalent to

$$\mathbf{p} \circ \dot{\omega} + \dot{\mathbf{p}} \circ \omega = \mathbf{p} \circ \omega \tag{13.31}$$

$$\mathbf{p} \circ \ddot{\omega} + \ddot{\mathbf{p}} \circ \omega = -2\dot{\mathbf{p}} \circ \dot{\omega} \tag{13.32}$$

where $\circ$ denotes the Hadamard product which will be used in the remainder of this chapter.

It has been shown in [245, 247] that one can avoid the calculation of $\vec{\mathbf{a}}$, $\vec{\mathbf{b}}$, and $\vec{\mathbf{c}}$ in the expression of the ellipse. The following formulas are used instead.

Theorem 13.1
Let $(\mathbf{x}(\alpha),\mathbf{y}(\alpha),\mathbf{z}(\alpha),\lambda(\alpha),\gamma(\alpha))$ be an arc defined by (13.26) passing through a point $(\mathbf{x},\mathbf{y},\mathbf{z},\lambda,\gamma) \in \mathcal{E}$, and its first and second derivatives at $(\mathbf{x},\mathbf{y},\mathbf{z},\lambda,\gamma)$ be $(\dot{\mathbf{x}},\dot{\mathbf{y}},\dot{\mathbf{z}},\dot{\lambda},\dot{\gamma})$ and $(\ddot{\mathbf{x}},\ddot{\mathbf{y}},\ddot{\mathbf{z}},\ddot{\lambda},\ddot{\gamma})$ which are defined by (13.27) and (13.28). Then an ellipse approximation of the central path is given by

$$\mathbf{x}(\alpha) = \mathbf{x} - \dot{\mathbf{x}}\sin(\alpha) + \ddot{\mathbf{x}}(1-\cos(\alpha)) \tag{13.33}$$

$$\mathbf{y}(\alpha) = \mathbf{y} - \dot{\mathbf{y}}\sin(\alpha) + \ddot{\mathbf{y}}(1 - \cos(\alpha)) \tag{13.34}$$

$$\mathbf{z}(\alpha) = \mathbf{z} - \dot{\mathbf{z}}\sin(\alpha) + \ddot{\mathbf{z}}(1 - \cos(\alpha)) \tag{13.35}$$

$$\lambda(\alpha) = \lambda - \dot{\lambda}\sin(\alpha) + \ddot{\lambda}(1 - \cos(\alpha)) \tag{13.36}$$

$$\gamma(\alpha) = \gamma - \dot{\gamma}\sin(\alpha) + \ddot{\gamma}(1 - \cos(\alpha)) \tag{13.37}$$

■

Two compact representations for $\mathbf{p}(\alpha) = (\mathbf{y}(\alpha), \mathbf{z}(\alpha))$ and $\omega(\alpha) = (\lambda(\alpha), \gamma(\alpha))$ are given below:

$$\mathbf{p}(\alpha) = \mathbf{p} - \dot{\mathbf{p}}\sin(\alpha) + \ddot{\mathbf{p}}(1 - \cos(\alpha)) \tag{13.38}$$

$$\omega(\alpha) = \omega - \dot{\omega}\sin(\alpha) + \ddot{\omega}(1 - \cos(\alpha)) \tag{13.39}$$

The duality measure at point $(\mathbf{x}(\alpha), \mathbf{p}(\alpha), \omega(\alpha))$ is defined as:

$$\mu(\alpha) = \frac{\lambda(\alpha)^{\mathrm{T}}\mathbf{y}(\alpha) + \gamma(\alpha)^{\mathrm{T}}\mathbf{z}(\alpha)}{2n} = \frac{\mathbf{p}(\alpha)^{\mathrm{T}}\omega(\alpha)}{2n} \tag{13.40}$$

Assuming $(\mathbf{y}, \mathbf{z}, \lambda, \gamma) > 0$, one can easily see that if $\frac{\dot{\mathbf{y}}}{\mathbf{y}}, \frac{\dot{\mathbf{z}}}{\mathbf{z}}, \frac{\dot{\lambda}}{\lambda}, \frac{\dot{\gamma}}{\gamma}, \frac{\ddot{\mathbf{y}}}{\mathbf{y}}, \frac{\ddot{\mathbf{z}}}{\mathbf{z}}, \frac{\ddot{\lambda}}{\lambda}, \frac{\ddot{\gamma}}{\gamma}$ are bounded (this will be shown to be true), and if α is small enough, then, $\mathbf{y}(\alpha) > 0$, $\mathbf{z}(\alpha) > 0$, $\lambda(\alpha) > 0$, and $\gamma(\alpha) > 0$. It will also be shown that searching along this ellipse will reduce the duality measure, i.e., $\mu(\alpha) < \mu$.

Lemma 13.7
Let $(\mathbf{x}, \mathbf{y}, \mathbf{z}, \lambda, \gamma)$ be a strictly feasible point of (QP), $(\dot{\mathbf{x}}, \dot{\mathbf{y}}, \dot{\mathbf{z}}, \dot{\lambda}, \dot{\gamma})$ and $(\ddot{\mathbf{x}}, \ddot{\mathbf{y}}, \ddot{\mathbf{z}}, \ddot{\lambda}, \ddot{\gamma})$ meet (13.27) and (13.28), $(\mathbf{x}(\alpha), \mathbf{y}(\alpha), \mathbf{z}(\alpha), \lambda(\alpha), \gamma(\alpha))$ be calculated using (13.33), (13.34), (13.35), (13.36), and (13.37), then the following conditions hold.

$$\mathbf{x}(\alpha) + \mathbf{y}(\alpha) = \mathbf{e}, \quad \mathbf{x}(\alpha) - \mathbf{z}(\alpha) = -\mathbf{e}, \quad \mathbf{H}\mathbf{x}(\alpha) + \mathbf{c} + \lambda(\alpha) + \gamma(\alpha) = \mathbf{0}$$

Proof 13.2 Since $(\mathbf{x}, \mathbf{y}, \mathbf{z}, \lambda, \gamma)$ is a strictly feasible point, the result follows from direct calculation by using (13.20), (13.27), (13.28), and Theorem 13.1. ■

Lemma 13.8
Let $(\dot{\mathbf{x}}, \dot{\mathbf{p}}, \dot{\omega})$ be defined by (13.27), $(\ddot{\mathbf{x}}, \ddot{\mathbf{p}}, \ddot{\omega})$ be defined by (13.28), and **H** *be positive definite matrix. Then the following relations hold:*

$$\dot{\mathbf{p}}^{\mathrm{T}}\dot{\omega} = \dot{\mathbf{x}}^{\mathrm{T}}(\dot{\gamma} - \dot{\lambda}) = \dot{\mathbf{x}}^{\mathrm{T}}\mathbf{H}\dot{\mathbf{x}} \geq \mathbf{0} \tag{13.41}$$

the equality holds if and only if $\|\dot{\mathbf{x}}\| = \mathbf{0}$

$$\ddot{\mathbf{p}}^{\mathrm{T}}\ddot{\omega} = \ddot{\mathbf{x}}^{\mathrm{T}}(\ddot{\gamma} - \ddot{\lambda}) = \ddot{\mathbf{x}}^{\mathrm{T}}\mathbf{H}\ddot{\mathbf{x}} \geq \mathbf{0} \tag{13.42}$$

the equality holds if and only if $\|\ddot{\mathbf{x}}\| = \mathbf{0}$

$$\ddot{\mathbf{p}}^{\mathrm{T}}\dot{\omega} = \ddot{\mathbf{x}}^{\mathrm{T}}(\dot{\gamma}-\dot{\lambda}) = \dot{\mathbf{x}}^{\mathrm{T}}(\ddot{\gamma}-\ddot{\lambda}) = \dot{\mathbf{p}}^{\mathrm{T}}\ddot{\omega} = \dot{\mathbf{x}}^{\mathrm{T}}\mathbf{H}\ddot{\mathbf{x}} \tag{13.43}$$

$$\begin{aligned} & -(\dot{\mathbf{x}}^{\mathrm{T}}\mathbf{H}\dot{\mathbf{x}})(1-\cos(\alpha))^2 - (\ddot{\mathbf{x}}^{\mathrm{T}}\mathbf{H}\ddot{\mathbf{x}})\sin^2(\alpha) \\ \leq\ & (\ddot{\mathbf{x}}^{\mathrm{T}}(\dot{\gamma}-\dot{\lambda}) + \dot{\mathbf{x}}^{\mathrm{T}}(\ddot{\gamma}-\ddot{\lambda}))\sin(\alpha)(1-\cos(\alpha)) \\ \leq\ & (\dot{\mathbf{x}}^{\mathrm{T}}\mathbf{H}\dot{\mathbf{x}})(1-\cos(\alpha))^2 + (\ddot{\mathbf{x}}^{\mathrm{T}}\mathbf{H}\ddot{\mathbf{x}})\sin^2(\alpha) \end{aligned} \tag{13.44}$$

and

$$\begin{aligned} & -(\dot{\mathbf{x}}^{\mathrm{T}}\mathbf{H}\dot{\mathbf{x}})\sin^2(\alpha) - (\ddot{\mathbf{x}}^{\mathrm{T}}\mathbf{H}\ddot{\mathbf{x}})(1-\cos(\alpha))^2 \\ \leq\ & (\ddot{\mathbf{x}}^{\mathrm{T}}(\dot{\gamma}-\dot{\lambda}) + \dot{\mathbf{x}}^{\mathrm{T}}(\ddot{\gamma}-\ddot{\lambda}))\sin(\alpha)(1-\cos(\alpha)) \\ \leq\ & (\dot{\mathbf{x}}^{\mathrm{T}}\mathbf{H}\dot{\mathbf{x}})\sin^2(\alpha) + (\ddot{\mathbf{x}}^{\mathrm{T}}\mathbf{H}\ddot{\mathbf{x}})(1-\cos(\alpha))^2 \end{aligned} \tag{13.45}$$

For $\alpha = \frac{\pi}{2}$*, (13.44) and (13.45) reduce to*

$$-\left(\dot{\mathbf{x}}^{\mathrm{T}}\mathbf{H}\dot{\mathbf{x}} + \ddot{\mathbf{x}}^{\mathrm{T}}\mathbf{H}\ddot{\mathbf{x}}\right) \leq (\ddot{\mathbf{x}}^{\mathrm{T}}\mathbf{H}\dot{\mathbf{x}} + \dot{\mathbf{x}}^{\mathrm{T}}\mathbf{H}\ddot{\mathbf{x}}) \leq \dot{\mathbf{x}}^{\mathrm{T}}\mathbf{H}\dot{\mathbf{x}} + \ddot{\mathbf{x}}^{\mathrm{T}}\mathbf{H}\ddot{\mathbf{x}} \tag{13.46}$$

The proof of this lemma is given in the last section.

From Lemmas 13.8, 13.1, and 13.3, it can be shown that $\frac{\dot{\mathbf{p}}}{\mathbf{p}} := \left(\frac{\dot{\mathbf{y}}}{\mathbf{y}}, \frac{\dot{\mathbf{z}}}{\mathbf{z}}\right)$, $\frac{\dot{\omega}}{\omega} := \left(\frac{\dot{\lambda}}{\lambda}, \frac{\dot{\gamma}}{\gamma}\right)$, $\frac{\ddot{\mathbf{p}}}{\mathbf{p}} := \left(\frac{\ddot{\mathbf{y}}}{\mathbf{y}}, \frac{\ddot{\mathbf{z}}}{\mathbf{z}}\right)$, and $\frac{\ddot{\omega}}{\omega} := \left(\frac{\ddot{\lambda}}{\lambda}, \frac{\ddot{\gamma}}{\gamma}\right)$ are all bounded as claimed in the following two Lemmas.

Lemma 13.9
Let $(\mathbf{x},\mathbf{p},\omega) = (\mathbf{x},\mathbf{y},\mathbf{z},\lambda,\gamma) \in \mathcal{N}_2(\theta)$ *and* $(\dot{\mathbf{x}},\dot{\mathbf{p}},\dot{\omega}) = (\dot{\mathbf{x}},\dot{\mathbf{y}},\dot{\mathbf{z}},\dot{\lambda},\dot{\gamma})$ *meet equation (13.27). Then,*

$$\left\|\frac{\dot{\mathbf{p}}}{\mathbf{p}}\right\|^2 + \left\|\frac{\dot{\omega}}{\omega}\right\|^2 \leq \frac{2n}{1-\theta} \tag{13.47}$$

$$\left\|\frac{\dot{\mathbf{p}}}{\mathbf{p}}\right\|^2 \left\|\frac{\dot{\omega}}{\omega}\right\|^2 \leq \left(\frac{n}{1-\theta}\right)^2 \tag{13.48}$$

$$0 \leq \frac{\dot{\mathbf{p}}^{\mathrm{T}}\dot{\omega}}{\mu} \leq \frac{1+\theta}{1-\theta}n := \delta_1 n \tag{13.49}$$

The proof of this lemma is given in the last section.

Lemma 13.10
Let $(\mathbf{x},\mathbf{p},\omega) = (\mathbf{x},\mathbf{y},\mathbf{z},\lambda,\gamma) \in \mathcal{N}_2(\theta)$*,* $(\dot{\mathbf{x}},\dot{\mathbf{y}},\dot{\mathbf{z}},\dot{\lambda},\dot{\gamma})$ *and* $(\ddot{\mathbf{x}},\ddot{\mathbf{y}},\ddot{\mathbf{z}},\ddot{\lambda},\ddot{\gamma})$ *meet equations (13.27) and (13.28). Then,*

$$\left\|\frac{\ddot{\mathbf{p}}}{\mathbf{p}}\right\|^2 + \left\|\frac{\ddot{\omega}}{\omega}\right\|^2 \leq \frac{4(1+\theta)n^2}{(1-\theta)^3} \tag{13.50}$$

$$\left\|\frac{\ddot{\mathbf{p}}}{\mathbf{p}}\right\|^2\left\|\frac{\ddot{\omega}}{\omega}\right\|^2 \leq \left(\frac{2(1+\theta)n^2}{(1-\theta)^3}\right)^2 \tag{13.51}$$

$$0 \leq \frac{\ddot{\mathbf{p}}^{\mathrm{T}}\ddot{\omega}}{\mu} \leq \frac{2(1+\theta)^2}{(1-\theta)^3}n^2 := \delta_2 n^2 \tag{13.52}$$

$$\left|\frac{\dot{\mathbf{p}}^{\mathrm{T}}\ddot{\omega}}{\mu}\right| \leq \frac{(2n(1+\theta))^{\frac{3}{2}}}{(1-\theta)^2} := \delta_3 n^{\frac{3}{2}} \quad \left|\frac{\ddot{\mathbf{p}}^{\mathrm{T}}\dot{\omega}}{\mu}\right| \leq \frac{(2n(1+\theta))^{\frac{3}{2}}}{(1-\theta)^2} := \delta_3 n^{\frac{3}{2}} \tag{13.53}$$

The proof of this lemma is given in the last section.

From the bounds established in Lemmas 13.8, 13.9, 13.10, and 13.2, the lower bound and upper bound for $\mu(\alpha)$ can be obtained.

Lemma 13.11

Let $(\mathbf{x},\mathbf{p},\omega) = (\mathbf{x},\mathbf{y},\mathbf{z},\lambda,\gamma) \in \mathcal{N}_2(\theta)$, $(\dot{\mathbf{x}},\dot{\mathbf{y}},\dot{\mathbf{z}},\dot{\lambda},\dot{\gamma})$ *and* $(\ddot{\mathbf{x}},\ddot{\mathbf{y}},\ddot{\mathbf{z}},\ddot{\lambda},\ddot{\gamma})$ *meet equations (13.27) and (13.28). Let* $\mathbf{x}(\alpha)$, $\mathbf{y}(\alpha)$, $\mathbf{z}(\alpha)$, $\lambda(\alpha)$, *and* $\gamma(\alpha)$ *be defined by (13.33), (13.34), (13.35), (13.36), and (13.37). Then,*

$$\begin{aligned}
&\mu(1-\sin(\alpha)) - \frac{1}{2n}\ddot{\mathbf{x}}^{\mathrm{T}}\mathbf{H}\ddot{\mathbf{x}}\left((1-\cos(\alpha))^2+\sin^2(\alpha)\right)\\
\leq&\mu(\alpha) = \mu(1-\sin(\alpha)) + \frac{1}{2n}\left(\ddot{\mathbf{x}}^{\mathrm{T}}(\ddot{\gamma}-\ddot{\lambda}) - \dot{\mathbf{x}}^{\mathrm{T}}(\dot{\gamma}-\dot{\lambda})\right)(1-\cos(\alpha))^2\\
&-\frac{1}{2n}\left(\dot{\mathbf{x}}^{\mathrm{T}}(\ddot{\gamma}-\ddot{\lambda}) + \ddot{\mathbf{x}}^{\mathrm{T}}(\dot{\gamma}-\dot{\lambda})\right)\sin(\alpha)(1-\cos(\alpha))\\
\leq&\mu(1-\sin(\alpha)) + \frac{1}{2n}\ddot{\mathbf{x}}^{\mathrm{T}}\mathbf{H}\ddot{\mathbf{x}}\left((1-\cos(\alpha))^2+\sin^2(\alpha)\right)
\end{aligned} \tag{13.54}$$

The proof of this lemma is given in the last section.

To keep all the iterates of the algorithm inside the strictly feasible set, $(p(\alpha),\omega(\alpha)) > 0$ for all iterations is required. This is guaranteed when $\mu(\alpha) > 0$ holds. The following corollary states the condition for $\mu(\alpha) > 0$ to hold.

Corollary 13.1

If $\mu > 0$, *then for any fixed* $\theta \in (0,1)$, *there is an* $\bar{\alpha} > 0$ *depending on* θ, *such that for any* $\sin(\alpha) \leq \sin(\bar{\alpha})$, $\mu(\alpha) > 0$. *In particular, if* $\theta = 0.19$, $\sin(\bar{\alpha}) \geq 0.6158$.

Proof 13.3 From Lemmas 13.8 and 13.2, it is easy to see that $\dot{\mathbf{x}}^{\mathrm{T}}\mathbf{H}\dot{\mathbf{x}}^{\mathrm{T}} = \dot{\mathbf{x}}^{\mathrm{T}}(\dot{\gamma} - \dot{\lambda}) = \dot{\mathbf{p}}^{\mathrm{T}}\dot{\omega}$ and $((1-\cos(\alpha))^2 \leq \sin^4(\alpha)$. Therefore, from Lemmas 13.11 and 13.9, it must have

$$\mu(\alpha) \geq \mu\left(1-\sin(\alpha) - \frac{1}{2n\mu}\dot{\mathbf{p}}^{\mathrm{T}}\dot{\omega}\left(\sin^4(\alpha)+\sin^2(\alpha)\right)\right)$$

$$\geq \mu\left(1-\sin(\alpha)-\frac{(1+\theta)}{2(1-\theta)}\left(\sin^4(\alpha)+\sin^2(\alpha)\right)\right) := \mu r(\alpha).$$

Since $\mu > 0$, and $r(\alpha)$ is a monotonic decreasing function in $[0, \frac{\pi}{2}]$ with $r(0) > 0$ and $r(\frac{\pi}{2}) < 0$, there is a unique real solution $\sin(\bar{\alpha}) \in (0,1)$ of $r(\alpha) = 0$ such that for all $\sin(\alpha) < \sin(\bar{\alpha})$, $r(\alpha) > 0$, or $\mu(\alpha) > 0$. It is easy to check that if $\theta = 0.19$, $\sin(\bar{\alpha}) = 0.6158$ is the solution of $r(\alpha) = 0$. ∎

Remark 13.1 Corollary 13.1 indicates that for any $\theta \in (0,1)$, there is a positive $\bar{\alpha}$ such that for $\alpha \leq \bar{\alpha}$, $\mu(\alpha) > 0$. Intuitively, to search in a wider region will generate a longer step. Therefore, the larger the θ is, the better. But to derive the convergence result, $\theta \leq 0.22$ is imposed in Lemma 13.15 and $\theta \leq 0.19$ is imposed in Lemma 13.19. ∎

To reduce the duality measure in an iteration, it must have $\mu(\alpha) \leq \mu$. For linear programming, it is known [247] that $\mu(\alpha) \leq \mu$ for $\alpha \in [0, \hat{\alpha}]$ with $\hat{\alpha} = \frac{\pi}{2}$, and the larger the α in the interval is, the smaller the $\mu(\alpha)$ will be. This claim is not true for the convex quadratic programming with box constraints and it needs to be modified as follows.

Lemma 13.12
Let $(\mathbf{x},\mathbf{p},\omega) = (\mathbf{x},\mathbf{y},\mathbf{z},\lambda,\gamma) \in \mathcal{N}_2(\theta)$, $(\dot{\mathbf{x}},\dot{\mathbf{y}},\dot{\mathbf{z}},\dot{\lambda},\dot{\gamma})$ *and* $(\ddot{\mathbf{x}},\ddot{\mathbf{y}},\ddot{\mathbf{z}},\ddot{\lambda},\ddot{\gamma})$ *meet equations (13.27) and (13.28). Let* $\mathbf{x}(\alpha)$, $\mathbf{y}(\alpha)$, $\mathbf{z}(\alpha)$, $\lambda(\alpha)$, *and* $\gamma(\alpha)$ *be defined by (13.33), (13.34), (13.35), (13.36), and (13.37). Then, there exists*

$$\hat{\alpha} = \begin{cases} \frac{\pi}{2}, & \text{if } \frac{\ddot{\mathbf{x}}^{\mathrm{T}}\mathbf{H}\ddot{\mathbf{x}}}{n\mu} \leq 1 \\ \sin^{-1}(g), & \text{if } \frac{\ddot{\mathbf{x}}^{\mathrm{T}}\mathbf{H}\ddot{\mathbf{x}}}{n\mu} > 1 \end{cases} \tag{13.55}$$

where

$$g = \sqrt[3]{\frac{n\mu}{\ddot{\mathbf{x}}^{\mathrm{T}}\mathbf{H}\ddot{\mathbf{x}}} + \sqrt{\left(\frac{n\mu}{\ddot{\mathbf{x}}^{\mathrm{T}}\mathbf{H}\ddot{\mathbf{x}}}\right)^2 + \left(\frac{1}{3}\right)^3}} + \sqrt[3]{\frac{n\mu}{\ddot{\mathbf{x}}^{\mathrm{T}}\mathbf{H}\ddot{\mathbf{x}}} - \sqrt{\left(\frac{n\mu}{\ddot{\mathbf{x}}^{\mathrm{T}}\mathbf{H}\ddot{\mathbf{x}}}\right)^2 + \left(\frac{1}{3}\right)^3}}$$

such that for every $\alpha \in [0, \hat{\alpha}]$, $\mu(\alpha) \leq \mu$.

The proof of this lemma is given in the last section.

According to Theorem 13.1, Lemmas 13.7, 13.9, 13.10, and 13.12, if α is small enough, then $(\mathbf{p}(\alpha), \omega(\alpha)) > 0$, and $\mu(\alpha) < \mu$, i.e., the search along the ellipse defined by Theorem 13.1 will generate a strictly feasible point with a smaller duality measure. Since $(\mathbf{p}, \omega) > 0$ holds in all iterations, reducing the duality measure to zero means approaching the solution of the convex quadratic programming. This can be achieved by applying a similar idea used in [133], i.e., starting with an iterate in $\mathcal{N}_2(\theta)$, searching along the approximated central path to reduce the duality measure and to keep the iterate in $\mathcal{N}_2(2\theta)$, and then making a correction to move the iterate back to $\mathcal{N}_2(\theta)$. The following notations will be used.

$$a_0 = -\theta\mu < 0$$

$$a_1 = \theta\mu > 0$$

$$a_2 = 2\theta\frac{\dot{\mathbf{p}}^{\mathrm{T}}\dot{\omega}}{2n} = 2\theta\frac{\dot{\mathbf{x}}^{\mathrm{T}}(\dot{\gamma}-\dot{\lambda})}{2n} = 2\theta\frac{\dot{\mathbf{x}}^{\mathrm{T}}\mathbf{H}\dot{\mathbf{x}}}{2n} \geq 0$$

$$a_3 = \left\|\dot{\mathbf{p}}\circ\ddot{\omega}+\dot{\omega}\circ\ddot{\mathbf{p}}-\frac{1}{2n}(\dot{\mathbf{p}}^{\mathrm{T}}\ddot{\omega}+\dot{\omega}^{\mathrm{T}}\ddot{\mathbf{p}})\mathbf{e}\right\| \geq 0$$

and

$$\begin{aligned} a_4 &= \left\|\ddot{\mathbf{p}}\circ\ddot{\omega}-\dot{\omega}\circ\dot{\mathbf{p}}-\frac{1}{2n}(\ddot{\mathbf{p}}^{\mathrm{T}}\ddot{\omega}-\dot{\omega}^{\mathrm{T}}\dot{\mathbf{p}})\mathbf{e}\right\| + 2\theta\frac{\dot{\mathbf{p}}^{\mathrm{T}}\dot{\omega}}{2n} \\ &= \left\|\ddot{\mathbf{p}}\circ\ddot{\omega}-\dot{\omega}\circ\dot{\mathbf{p}}-\frac{1}{2n}(\ddot{\mathbf{p}}^{\mathrm{T}}\ddot{\omega}-\dot{\omega}^{\mathrm{T}}\dot{\mathbf{p}})\mathbf{e}\right\| + 2\theta\frac{\dot{\mathbf{x}}^{\mathrm{T}}\mathbf{H}\dot{\mathbf{x}}}{2n} \geq 0 \end{aligned}$$

Denote a quartic polynomial in terms of $\sin(\alpha)$ as follows:

$$q(\alpha) = a_4\sin^4(\alpha)+a_3\sin^3(\alpha)+a_2\sin^2(\alpha)+a_1\sin(\alpha)+a_0 = 0 \qquad (13.56)$$

Since $q(\alpha)$ is a monotonic increasing function of $\alpha \in [0, \frac{\pi}{2}]$, $q(0) = -\theta\mu < 0$ and $q(\frac{\pi}{2}) = a_2+a_3+a_4 > 0$ if $\dot{x} \neq 0$, the polynomial has exactly one positive root in $[0, \frac{\pi}{2}]$. Moreover, since (13.56) is a quartic equation, all the solutions are analytical and the computational cost is independent of the size of $\mathbf{H}$ and negligible [157].

Lemma 13.13
Let $(\mathbf{x},\mathbf{p},\omega) = (\mathbf{x},\mathbf{y},\mathbf{z},\lambda,\omega) \in \mathcal{N}_2(\theta)$, $(\dot{\mathbf{x}},\dot{\mathbf{y}},\dot{\mathbf{z}},\dot{\lambda},\dot{\omega})$ *and* $(\ddot{\mathbf{x}},\ddot{\mathbf{y}},\ddot{\mathbf{z}},\ddot{\lambda},\ddot{\omega})$ *be calculated from (13.27) and (13.28). Denote* $\sin(\tilde{\alpha})$ *the only positive real solution of (13.56) in* $[0,1]$. *Assume* $\sin(\alpha) \leq \min\{\sin(\tilde{\alpha}),\sin(\bar{\alpha})\}$, *let* $(\mathbf{x}(\alpha),\mathbf{y}(\alpha),\mathbf{z}(\alpha),\lambda(\alpha),\gamma(\alpha))$ *and* $\mu(\alpha)$ *be updated as follows:*

$$\begin{aligned} &(\mathbf{x}(\alpha),\mathbf{y}(\alpha),\mathbf{z}(\alpha),\lambda(\alpha),\gamma(\alpha)) \\ =\ &(\mathbf{x},\mathbf{y},\mathbf{z},\lambda,\gamma)-(\dot{\mathbf{x}},\dot{\mathbf{y}},\dot{\mathbf{z}},\dot{\lambda},\dot{\gamma})\sin(\alpha)+(\ddot{\mathbf{x}},\ddot{\mathbf{y}},\ddot{\mathbf{z}},\ddot{\lambda},\ddot{\gamma})(1-\cos(\alpha)) \end{aligned} \qquad (13.57)$$

$$\begin{aligned} \mu(\alpha) &= \mu(1-\sin(\alpha)) \\ &+ \frac{1}{2n}\Big((\ddot{\mathbf{p}}^{\mathrm{T}}\ddot{\omega}-\dot{\mathbf{p}}^{\mathrm{T}}\dot{\omega})(1-\cos(\alpha))^2-(\dot{\mathbf{p}}^{\mathrm{T}}\ddot{\omega}+\ddot{\mathbf{p}}^{\mathrm{T}}\dot{\omega})\sin(\alpha)(1-\cos(\alpha))\Big) \end{aligned} \qquad (13.58)$$

Then $(\mathbf{x}(\alpha),\mathbf{y}(\alpha),\mathbf{z}(\alpha),\lambda(\alpha),\gamma(\alpha)) \in \mathcal{N}_2(2\theta)$.

The proof of this lemma is given in the last section.

The lower bound of $\sin(\bar{\alpha})$ is estimated in Corollary 13.1. To estimate the lower bound of $\sin(\tilde{\alpha})$, the following lemma is needed.

Lemma 13.14

Let $(\mathbf{x},\mathbf{p},\omega)\in\mathcal{N}_2(\theta)$, $(\dot{\mathbf{x}},\dot{\mathbf{p}},\dot{\omega})$ *and* $(\ddot{\mathbf{x}},\ddot{\mathbf{p}},\ddot{\omega})$ *meet equations (13.27) and (13.28). Then,*

$$\left\|\dot{\mathbf{p}}\circ\dot{\omega}\right\| \le \frac{(1+\theta)}{(1-\theta)}n\mu \tag{13.59}$$

$$\left\|\ddot{\mathbf{p}}\circ\ddot{\omega}\right\| \le \frac{2(1+\theta)^2}{(1-\theta)^3}n^2\mu \tag{13.60}$$

$$\left\|\ddot{\mathbf{p}}\circ\dot{\omega}\right\| \le \frac{2\sqrt{2}(1+\theta)^{\frac{3}{2}}}{(1-\theta)^2}n^{\frac{3}{2}}\mu \tag{13.61}$$

$$\left\|\dot{\mathbf{p}}\circ\ddot{\omega}\right\| \le \frac{2\sqrt{2}(1+\theta)^{\frac{3}{2}}}{(1-\theta)^2}n^{\frac{3}{2}}\mu \tag{13.62}$$

The proof of this lemma is given in the last section.

Lemma 13.15

Let $\theta\le 0.22$. *Then* $\sin(\tilde{\alpha})\ge\frac{\theta}{\sqrt{n}}$

The proof of this lemma is given in the last section.

Corollary 13.1, Lemmas 13.13, and 13.15 prove the feasibility of searching optimizer along the ellipse. To move the iterate back to $\mathcal{N}_2(\theta)$, one can use the direction $(\Delta\mathbf{x},\Delta\mathbf{y},\Delta\mathbf{z},\Delta\lambda,\Delta\gamma)$ defined by

$$\begin{bmatrix} \mathbf{H} & \mathbf{0} & \mathbf{0} & \mathbf{I} & -\mathbf{I} \\ \mathbf{I} & \mathbf{I} & \mathbf{0} & \mathbf{0} & \mathbf{0} \\ \mathbf{I} & \mathbf{0} & -\mathbf{I} & \mathbf{0} & \mathbf{0} \\ \mathbf{0} & \Lambda(\alpha) & \mathbf{0} & \mathbf{Y}(\alpha) & \mathbf{0} \\ \mathbf{0} & \mathbf{0} & \Gamma(\alpha) & \mathbf{0} & \mathbf{Z}(\alpha) \end{bmatrix} \begin{bmatrix} \Delta\mathbf{x} \\ \Delta\mathbf{y} \\ \Delta\mathbf{z} \\ \Delta\lambda \\ \Delta\gamma \end{bmatrix} = \begin{bmatrix} \mathbf{0} \\ \mathbf{0} \\ \mathbf{0} \\ \mu(\alpha)\mathbf{e}-\lambda(\alpha)\circ\mathbf{y}(\alpha) \\ \mu(\alpha)\mathbf{e}-\gamma(\alpha)\circ\mathbf{z}(\alpha) \end{bmatrix} \tag{13.63}$$

and update $(\mathbf{x}^{k+1},\mathbf{p}^{k+1},\omega^{k+1})$ and μ^{k+1} by

$$(\mathbf{x}^{k+1},\mathbf{p}^{k+1},\omega^{k+1}) = (\mathbf{x}(\alpha),\mathbf{p}(\alpha),\omega(\alpha))+(\Delta\mathbf{x},\Delta\mathbf{p},\Delta\omega) \tag{13.64}$$

$$\mu^{k+1} = \frac{\mathbf{p}^{k+1^{\mathrm{T}}}\omega^{k+1}}{2n} \tag{13.65}$$

where $\Delta\mathbf{p}=(\Delta\mathbf{y},\Delta\mathbf{z})$ and $\Delta\omega=(\Delta\lambda,\Delta\gamma)$. Denote $\mathbf{P}(\alpha)=\begin{bmatrix}\mathbf{Y}(\alpha) & \mathbf{0}\\ \mathbf{0} & \mathbf{Z}(\alpha)\end{bmatrix}$,

$\Omega(\alpha)=\begin{bmatrix}\Lambda(\alpha) & \mathbf{0}\\ \mathbf{0} & \Gamma(\alpha)\end{bmatrix}$, and $\mathbf{D}=\mathbf{P}^{\frac{1}{2}}(\alpha)\Omega^{-\frac{1}{2}}(\alpha)$. Then, the last 2 rows of (13.63) can be rewritten as

$$\mathbf{P}\Delta\omega+\Omega\Delta\mathbf{p}=\mu(\alpha)\mathbf{e}-\mathbf{P}(\alpha)\Omega(\alpha)\mathbf{e} \tag{13.66}$$

Now, it is ready to show that the correction step brings the iterate from $\mathcal{N}_2(2\theta)$ back to $\mathcal{N}_2(\theta)$.

Lemma 13.16
Let $(\mathbf{x}(\alpha),\mathbf{p}(\alpha),\omega(\alpha))\in\mathcal{N}_2(2\theta)$ and $(\Delta\mathbf{x},\Delta\mathbf{p},\Delta\omega)$ be defined as in (13.63). Let $(\mathbf{x}^{k+1},\mathbf{p}^{k+1},\omega^{k+1})$ be updated by using (13.64). Then, for $\theta\le 0.29$ and $\sin(\alpha)\le\sin(\bar{\alpha})$, $(\mathbf{x}^{k+1},\mathbf{p}^{k+1},\omega^{k+1})\in\mathcal{N}_2(\theta)$.

The proof of this lemma is given in the last section.

The next step is to show that the combined step (searching along the arc in $\mathcal{N}_2(2\theta)$ and moving back to $\mathcal{N}_2(\theta)$) will reduce the duality measure of the iterate, i.e., $\mu^{k+1}<\mu^k$, if some appropriate θ and α are selected. The following two Lemmas are introduced for this purpose.

Lemma 13.17
Let $(\mathbf{x}(\alpha),\mathbf{p}(\alpha),\omega(\alpha))\in\mathcal{N}_2(2\theta)$ and $(\Delta\mathbf{x},\Delta\mathbf{p},\Delta\omega)$ be defined as in (13.63). Then,

$$0\le\frac{\Delta\mathbf{p}^{\mathrm{T}}\Delta\omega}{2n}\le\frac{\theta^2(1+2\theta)}{n(1-2\theta)^2}\mu(\alpha):=\frac{\delta_0}{n}\mu(\alpha) \tag{13.67}$$

The proof of this lemma is given in the last section.

Lemma 13.18
Let $(\mathbf{x}(\alpha),\mathbf{p}(\alpha),\omega(\alpha))\in\mathcal{N}_2(2\theta)$ and $(\Delta\mathbf{x},\Delta\mathbf{p},\Delta\omega)$ be defined as in (13.63). Let $(\mathbf{x}^{k+1},\mathbf{p}^{k+1},\omega^{k+1})$ be defined as in (13.64). Then,

$$\mu(\alpha)\le\mu^{k+1}:=\frac{\mathbf{p}^{k+1^{\mathrm{T}}}\omega^{k+1}}{2n}\le\mu(\alpha)\left(1+\frac{\theta^2(1+2\theta)}{n(1-2\theta)^2}\right)=\mu(\alpha)\left(1+\frac{\delta_0}{n}\right)$$

Proof 13.4 Using the fact that $\mathbf{p}(\alpha)^{\mathrm{T}}\Delta\omega+\omega(\alpha)^{\mathrm{T}}\Delta\mathbf{p}=0$ established in (13.114) in the proof of Lemma 13.16, and Lemma 13.17, it is straightforward to obtain

$$\mu(\alpha)\ \le\ \frac{\mathbf{p}(\alpha)^{\mathrm{T}}\omega(\alpha)}{2n}+\frac{1}{2n}\Delta\mathbf{p}^{\mathrm{T}}\Delta\omega$$

$$\begin{aligned}
&= \frac{(\mathbf{p}(\alpha)+\Delta\mathbf{p})^{\mathrm{T}}(\omega(\alpha)+\Delta\omega)}{2n} = \mu^{k+1} \\
&\le \mu(\alpha) + \frac{\theta^2(1+2\theta)}{n(1-2\theta)^2}\mu(\alpha)
\end{aligned} \tag{13.68}$$

This proves the lemma. ■

For linear programming, it is known [133, 247] that $\mu^{k+1} = \mu(\alpha)$. This claim is not always true for the convex quadratic programming as is pointed out in Lemma 13.18. Therefore, some extra work is needed to make sure that the μ^k will be reduced in every iteration.

Lemma 13.19
For $\theta \le 0.19$, if

$$\sin(\alpha) = \frac{\theta}{\sqrt{n}} \tag{13.69}$$

then $\mu^{k+1} < \mu^k$. Moreover, for $\sin(\alpha) = \frac{\theta}{\sqrt{n}} = \frac{0.19}{\sqrt{n}}$,

$$\mu^{k+1} \le \mu^k \left(1 - \frac{0.0185}{\sqrt{n}}\right) \tag{13.70}$$

The proof of this lemma is given in the last section.

Remark 13.2 As one has seen in this section that starting with $(\mathbf{x}^0, \mathbf{p}^0, \omega^0)$, the interior-point algorithm proceeds with finding $(\mathbf{x}(\alpha), \mathbf{p}(\alpha), \omega(\alpha)) \in \mathcal{N}_2(2\theta)$ and $(\mathbf{x}^{k+1}, \mathbf{p}^{k+1}, \omega^{k+1}) \in \mathcal{N}_2(\theta)$ such that $\mu^{k+1} < \mu^k$. In view of the proofs of Lemmas 13.13, 13.16, and 13.19, the positivity conditions of $(\mathbf{x}(\alpha), \mathbf{p}(\alpha), \omega(\alpha)) > 0$ and $(\mathbf{x}^{k+1}, \mathbf{p}^{k+1}, \omega^{k+1}) > 0$ relies on $\mu(\alpha) > 0$ which, according to Corollary 13.1, is achievable for any θ and is given by a bound in terms of $\bar{\alpha}$. The proximity condition for $(\mathbf{x}(\alpha), \mathbf{p}(\alpha), \omega(\alpha))$ relies on the real positive root of $q(\sin(\alpha)) = 0$, denoted by $\sin(\tilde{\alpha})$, which is conservatively estimated in Lemma 13.15 under the condition that $\theta \le 0.22$; the proximity condition for $(\mathbf{x}^{k+1}, \mathbf{p}^{k+1}, \omega^{k+1})$ is established in Lemma 13.16 under the condition that $\theta \le 0.29$. Finally, duality measure reduction $\mu^{k+1} < \mu^k$ is established in Lemma 13.19 under the condition that $\theta \le 0.19$. For all these results to hold, it just needs to take the smallest bound $\theta = 0.19$. ■

Summarizing all the results in this section leads to the following theorem.

Theorem 13.2
Let $\theta = 0.19$ and $(\mathbf{x}^k, \mathbf{p}^k, \omega^k) \in \mathcal{N}_2(\theta)$. Then, $(\mathbf{x}(\alpha), \mathbf{p}(\alpha), \omega(\alpha)) \in \mathcal{N}_2(2\theta)$; $(\mathbf{x}^{k+1}, \mathbf{p}^{k+1}, \omega^{k+1}) \in \mathcal{N}_2(\theta)$; and $\mu^{k+1} \le \mu^k \left(1 - \frac{0.0185}{\sqrt{n}}\right)$.

Proof 13.5 From Corollary 13.1 and Lemma 13.15, one can select $\sin(\alpha) \leq \min\{\sin(\tilde{\alpha}), \sin(\bar{\alpha})\}$. Therefore, Lemma 13.13 holds, i.e., $(\mathbf{x}(\alpha), \mathbf{p}(\alpha), \omega(\alpha)) \in \mathcal{N}_2(2\theta)$. Since $\sin(\alpha) \leq \sin(\bar{\alpha})$ and $(\mathbf{x}(\alpha), \mathbf{p}(\alpha), \omega(\alpha)) \in \mathcal{N}_2(2\theta)$, Lemma 13.16 states $(\mathbf{x}^{k+1}, \mathbf{p}^{k+1}, \omega^{k+1}) \in \mathcal{N}_2(\theta)$. For $\theta = 0.19$ and $\sin(\alpha) = \frac{\theta}{\sqrt{n}}$, Lemma 13.19 states $\mu^{k+1} \leq \mu^k \left(1 - \frac{0.0185}{\sqrt{n}}\right)$. This finishes the proof. ■

Remark 13.3 It is worthwhile to point out that $\theta = 0.19$ for the box constrained quadratic optimization problem is larger than the $\theta = 0.148$ for linearly constrained quadratic optimization problem. This makes the searching neighborhood larger and the following algorithm more efficient than the algorithm in [247]. ■

The proposed method can be presented as the following algorithm.

Algorithm 13.1
(Arc-search path-following)
Data: $\mathbf{H} \geq 0$, $\mathbf{c}$, n, $\theta = 0.19$, $\varepsilon > 0$.
Initial point $(\mathbf{x}^0, \mathbf{p}^0, \omega^0) \in \mathcal{N}_2(\theta)$, *and* $\mu^0 = \frac{\mathbf{p}^{0^{\mathrm{T}}} \omega^0}{2n}$.
for *iteration* $k = 1, 2, \ldots$

Step 1: Solve the linear systems of equations (13.27) and (13.28) to get $(\dot{\mathbf{x}}, \dot{\mathbf{p}}, \dot{\omega})$ *and* $(\ddot{\mathbf{x}}, \ddot{\mathbf{p}}, \ddot{\omega})$.

Step 2: Let $\sin(\alpha) = \frac{\theta}{\sqrt{n}}$. *Update* $(\mathbf{x}(\alpha), \mathbf{p}(\alpha), \omega(\alpha))$ *and* $\mu(\alpha)$ *by (13.57) and (13.58).*

Step 3: Solve (13.63) to get $(\Delta\mathbf{x}, \Delta\mathbf{p}, \Delta\omega)$, *update* $(\mathbf{x}^{k+1}, \mathbf{p}^{k+1}, \omega^{k+1})$ *and* μ^{k+1} *by using (13.64) and (13.65).*

Step 4: Set $k+1 \to k$. *Go back to Step 1.*

end (for)

13.5 Convergence Analysis

The first result in this section extends a result of linear programming (c.f. [233]) to convex quadratic programming subject to box constraints.

Lemma 13.20
Suppose $\mathcal{F}^o \neq \emptyset$. *Then for each* $K \geq 0$, *the set*

$$\{(\mathbf{x}, \mathbf{p}, \omega) \mid (\mathbf{x}, \mathbf{p}, \omega) \in \mathcal{F}, \quad \mathbf{p}^{\mathrm{T}} \omega \leq K\}$$

is bounded.

Proof 13.6 The proof is similar to the proof in [233]. It is given here for completeness. First, $\mathbf{x}$ is bounded because $-\mathbf{e} \leq \mathbf{x} \leq \mathbf{e}$. Since $\mathbf{x}+\mathbf{y}=\mathbf{e}$ and $-\mathbf{e} \leq \mathbf{x} \leq \mathbf{e}$, it is easy to see $\mathbf{0} \leq \mathbf{y}=\mathbf{e}-\mathbf{x} \leq 2\mathbf{e}$. Since $\mathbf{x}-\mathbf{z}=-\mathbf{e}$, it is easy to see $\mathbf{0} \leq \mathbf{z}=\mathbf{x}+\mathbf{e} \leq 2\mathbf{e}$. Therefore, $\mathbf{y}$ and $\mathbf{z}$ are also bounded. Let $(\bar{\mathbf{x}}, \bar{\mathbf{y}}, \bar{\mathbf{z}}, \bar{\lambda}, \bar{\gamma})$ be any fixed point in $\mathcal{F}^o$, and $(\mathbf{x}, \mathbf{y}, \mathbf{z}, \lambda, \gamma)$ be any point in $\mathcal{F}$ with $\mathbf{y}^{\mathrm{T}}\lambda+\mathbf{z}^{\mathrm{T}}\gamma \leq K$. Using the definition of $\mathcal{F}^o$ and $\mathcal{F}$ yields

$$\mathbf{H}(\bar{\mathbf{x}}-\mathbf{x})+(\bar{\lambda}-\lambda)-(\bar{\gamma}-\gamma)=0$$

Therefore,

$$(\bar{\mathbf{x}}-\mathbf{x})^{\mathrm{T}}\mathbf{H}(\bar{\mathbf{x}}-\mathbf{x})+(\bar{\mathbf{x}}-\mathbf{x})^{\mathrm{T}}(\bar{\lambda}-\lambda)-(\bar{\mathbf{x}}-\mathbf{x})^{\mathrm{T}}(\bar{\gamma}-\gamma)=0$$

or equivalently

$$(\bar{\mathbf{x}}-\mathbf{x})^{\mathrm{T}}(\bar{\gamma}-\gamma)-(\bar{\mathbf{x}}-\mathbf{x})^{\mathrm{T}}(\bar{\lambda}-\lambda)=(\bar{\mathbf{x}}-\mathbf{x})^{\mathrm{T}}\mathbf{H}(\bar{\mathbf{x}}-\mathbf{x}) \geq 0$$

This gives

$$((\bar{\mathbf{x}}+\mathbf{e})-(\mathbf{x}+\mathbf{e}))^{\mathrm{T}}(\bar{\gamma}-\gamma)-((\bar{\mathbf{x}}-\mathbf{e})-(\mathbf{x}-\mathbf{e}))^{\mathrm{T}}(\bar{\lambda}-\lambda) \geq 0$$

Substituting $\mathbf{x}-\mathbf{e}=-\mathbf{y}$ and $\mathbf{x}+\mathbf{e}=\mathbf{z}$ yields

$$(\bar{\mathbf{z}}-\mathbf{z})^{\mathrm{T}}(\bar{\gamma}-\gamma)+(\bar{\mathbf{y}}-\mathbf{y})^{\mathrm{T}}(\bar{\lambda}-\lambda) \geq 0$$

This leads to

$$\bar{\mathbf{z}}^{\mathrm{T}}\bar{\gamma}+\mathbf{z}^{\mathrm{T}}\gamma-\mathbf{z}^{\mathrm{T}}\bar{\gamma}-\bar{\mathbf{z}}^{\mathrm{T}}\gamma+\bar{\mathbf{y}}^{\mathrm{T}}\bar{\lambda}+\mathbf{y}^{\mathrm{T}}\lambda-\mathbf{y}^{\mathrm{T}}\bar{\lambda}-\bar{\mathbf{y}}^{\mathrm{T}}\lambda \geq 0$$

or in a compact form

$$\bar{\mathbf{p}}^{\mathrm{T}}\bar{\omega}+\mathbf{p}^{\mathrm{T}}\omega-\mathbf{p}^{\mathrm{T}}\bar{\omega}-\bar{\mathbf{p}}^{\mathrm{T}}\omega \geq 0$$

Sine $(\bar{\mathbf{p}}, \bar{\omega})>0$ is fixed, let

$$\xi=\min_{i=1,\cdots,n} \min\{\bar{p}_i, \bar{\omega}_i\}$$

then, using $\mathbf{p}^{\mathrm{T}}\omega \leq K$

$$\bar{\mathbf{p}}^{\mathrm{T}}\bar{\omega}+K \geq \xi\mathbf{e}^{\mathrm{T}}(\mathbf{p}+\omega) \geq \max_{i=1,\cdots,n} \max\{\xi p_i, \xi\omega_i\}$$

i.e., for $i \in \{1,\cdots,n\}$

$$0 \leq p_i \leq \frac{1}{\xi}(K+\bar{\mathbf{p}}^{\mathrm{T}}\bar{\omega}), \quad 0 \leq \omega_i \leq \frac{1}{\xi}(K+\bar{\mathbf{p}}^{\mathrm{T}}\bar{\omega})$$

This proves the lemma. ■

The following theorem is a direct result of Lemmas 13.20, 13.7, Theorem 13.2, KKT conditions, Theorem A.2 in [233].

Theorem 13.3
Suppose that Assumption 1 holds, then the sequence generated by Algorithm 13.1 converges to a set of accumulation points, and all these accumulation points are global optimal solutions of the convex quadratic programming subject to box constraints.

Let $(\mathbf{x}^*, \mathbf{p}^*, \omega^*)$ be any solution of (13.17), following the notation of [14], denote index sets $\mathcal{B}$, $\mathcal{S}$, and $\mathcal{T}$ as

$$\mathcal{B} = \{j \in \{1,\ldots,2n\} \mid p_j^* \neq 0\} \tag{13.71}$$

$$\mathcal{S} = \{j \in \{1,\ldots,2n\} \mid \omega_j^* \neq 0\} \tag{13.72}$$

$$\mathcal{T} = \{j \in \{1,\ldots,2n\} \mid p_j^* = \omega_j^* = 0\} \tag{13.73}$$

According to Goldman-Tucker theorem [57], for the linear programming, $\mathcal{B} \cap \mathcal{S} = \emptyset = \mathcal{T}$ and $\mathcal{B} \cup \mathcal{S} = \{1,\ldots,2n\}$. A solution with this property is called strictly complementary (see Appendix A). This property has been used in many papers to prove the locally super-linear convergence of interior-point algorithms in linear programming. However, it is pointed out in [63] that this partition does not hold for general quadratic programming problems. But a convex quadratic programming subject to box constraints has strictly complementary solution(s), an interior-point algorithm will generate a sequence to approach strict complementary solution(s). As a matter of fact, from Lemma 13.20, the result of [233, Lemma 5.13] can be extended to the case of convex quadratic programming subject to box constraints, and the following lemma, which is independent of any algorithm, holds.

Lemma 13.21
Let $\mu^0 > 0$, and $\rho \in (0,1)$. Assume that the convex QP (13.16) has strictly complementary solution(s). Then for all points $(\mathbf{x}, \mathbf{p}, \omega)$ with $(\mathbf{x}, \mathbf{p}, \omega) \in \mathcal{F}^o$, $p_i\omega_i > \rho\mu$, and $\mu < \mu^0$, there are constants M, C_1, and C_2 such that

$$\|(\mathbf{p}, \omega)\| \le M \tag{13.74}$$

$$0 < p_i \le \mu / C_1 \quad (i \in \mathcal{S}), \qquad 0 < \omega_i \le \mu / C_1 \quad (i \in \mathcal{B}) \tag{13.75}$$

$$\omega_i \ge C_2 \rho \quad (i \in \mathcal{S}), \qquad p_i \ge C_2 \rho \quad (i \in \mathcal{B}) \tag{13.76}$$

Proof 13.7 The proof mimics the one in [233, Lemma 5.13]. It is presented here for completeness. The first result (13.74) follows immediately from Lemma 13.20 by setting $K = 2n\mu^0$. Let $(\mathbf{x}^*, \mathbf{p}^*, \omega^*)$ be any strictly complementary solution. Since $(\mathbf{x}^*, \mathbf{p}^*, \omega^*)$ and $(\mathbf{x}, \mathbf{p}, \omega)$ are both feasible, it must have

$$(\mathbf{y} - \mathbf{y}^*) = -(\mathbf{x} - \mathbf{x}^*) = -(\mathbf{z} - \mathbf{z}^*), \qquad \mathbf{H}(\mathbf{x} - \mathbf{x}^*) + (\lambda - \lambda^*) - (\gamma - \gamma^*) = 0$$

Therefore,

$$(\mathbf{y}-\mathbf{y}^*)^{\mathrm{T}}(\lambda-\lambda^*)+(\mathbf{z}-\mathbf{z}^*)^{\mathrm{T}}(\gamma-\gamma^*)=(\mathbf{x}-\mathbf{x}^*)^{\mathrm{T}}\mathbf{H}(\mathbf{x}-\mathbf{x}^*)\geq 0 \tag{13.77}$$

Since $(\mathbf{x}^*,\mathbf{y}^*,\mathbf{z}^*,\lambda^*,\gamma^*)=(\mathbf{x}^*,\mathbf{p}^*,\omega^*)$ is a strictly complementary solution, it must have $\mathcal{T}=\emptyset$, $p_i^*=0$ for $i\in\mathcal{S}$, and $\omega_i^*=0$ for $i\in\mathcal{B}$. Since $\mathbf{p}^{\mathrm{T}}\omega=2n\mu$, $(\mathbf{p}^*)^{\mathrm{T}}\omega^*=0$, from (13.77), it must have

$$\begin{aligned}
&\mathbf{p}^{\mathrm{T}}\omega=\mathbf{y}^{\mathrm{T}}\lambda+\mathbf{z}^{\mathrm{T}}\gamma+\left((\mathbf{y}^*)^{\mathrm{T}}\lambda^*+(\mathbf{z}^*)^{\mathrm{T}}\gamma^*\right)\\
&\geq \mathbf{y}^{\mathrm{T}}\lambda^*+\mathbf{z}^{\mathrm{T}}\gamma^*+\left((\mathbf{y}^*)^{\mathrm{T}}\lambda+(\mathbf{z}^*)^{\mathrm{T}}\gamma\right)=\mathbf{p}^{\mathrm{T}}\omega^*+\omega^{\mathrm{T}}\mathbf{p}^*\\
\iff\quad &2n\mu\geq \mathbf{p}^{\mathrm{T}}\omega^*+\omega^{\mathbf{T}}\mathbf{p}^*=\textstyle\sum_{i\in\mathcal{S}}p_i\omega_i^*+\sum_{i\in\mathcal{B}}p_i^*\omega_i
\end{aligned} \tag{13.78}$$

Since each term in the summations is positive and bounded above by $2n\mu$, it must have $\omega_i^*>0$ for any $i\in\mathcal{S}$; therefore,

$$0<p_i\leq\frac{2n\mu}{\omega_i^*}$$

Denote $\Omega_D=\{(\mathbf{p}^*,\omega^*)|\omega_i^*>0\}$ and $\Omega_P=\{(\mathbf{p}^*,\omega^*)|p_i^*>0\}$, it must have

$$0<p_i\leq\frac{2n\mu}{\sup_{(\mathbf{p}^*,\omega^*)\in\Omega_D}\omega_i^*}$$

This leads to

$$\max_{i\in\mathcal{S}}p_i\leq\frac{2n\mu}{\min_{i\in\mathcal{S}}\sup_{(\mathbf{p}^*,\omega^*)\in\Omega_D}\omega_i^*}$$

Similarly,

$$\max_{i\in\mathcal{B}}\omega_i\leq\frac{2n\mu}{\min_{i\in\mathcal{B}}\sup_{(\mathbf{p}^*,\omega^*)\in\Omega_P}p_i^*}$$

Combining these two inequalities gives

$$\begin{aligned}
&\max\{\max_{i\in\mathcal{S}}p_i,\max_{i\in\mathcal{B}}\omega_i\}\\
\leq\;&\frac{2n\mu}{\min\{\min_{i\in\mathcal{S}}\sup_{(\mathbf{p}^*,\omega^*)\in\Omega_D}\omega_i^*,\min_{i\in\mathcal{B}}\sup_{(\mathbf{p}^*,\omega^*)\in\Omega_P}p_i^*\}}\\
=\;&\frac{\mu}{C_1}
\end{aligned} \tag{13.79}$$

This proves (13.75). Finally, since $p_i\omega_i\geq\rho\mu$, we have, for any $i\in\mathcal{S}$,

$$\omega_i\geq\frac{\rho\mu}{p_i}\geq\frac{\rho\mu}{\mu/C_1}=C_2\rho$$

Similarly, for any $i\in\mathcal{B}$,

$$p_i\geq\frac{\rho\mu}{\omega_i}\geq\frac{\rho\mu}{\mu/C_1}=C_2\rho$$

■

Lemma 13.21 leads to the following:

Theorem 13.4
Let $(\mathbf{x}^k, \mathbf{p}^k, \omega^k) \in \mathcal{N}_2(\theta)$ *be generated by Algorithms 13.1. Assume that the convex QP with box constraints has strictly complementary solution(s). Then every limit point of the sequence is a strictly complementary solution of the convex quadratic programming with box constraints, i.e.,*

$$\omega_i^* \geq C_2\rho \quad (i \in \mathcal{S}), \qquad p_i^* \geq C_2\rho \quad (i \in \mathcal{B}) \tag{13.80}$$

Proof 13.8 From Lemma 13.21, $(\mathbf{p}^k, \omega^k)$ is bounded; therefore there is at least one limit point $(\mathbf{p}^*, \omega^*)$. Since (p_i^k, ω_i^k) is in the neighborhood of the central path, i.e., $p_i^k \omega_i^k > \rho\mu^k := (1-3\theta)\mu^k$,

$$\omega_i^k \geq C_2\rho \quad (i \in \mathcal{S}), \qquad p_i^k \geq C_2\rho \quad (i \in \mathcal{B})$$

every limit point will meet (13.80) due to the fact that $C_2\rho$ is a constant. ■

It is now ready to show that the complexity bound of Algorithm 13.1 is $O(\sqrt{n}\log(1/\varepsilon))$. The following theorem from [233] is needed for this purpose.

Theorem 13.5
Let $\varepsilon \in (0,1)$ *be given. Suppose that an algorithm for solving (13.17) generates a sequence of iterations that satisfies*

$$\mu^{k+1} \leq \left(1 - \frac{\delta}{n^\chi}\right)\mu^k, \quad k = 0, 1, 2, \ldots, \tag{13.81}$$

for some positive constants δ *and* χ*. Suppose that the starting point* $(\mathbf{x}^0, \mathbf{p}^0, \omega^0)$ *satisfies* $\mu^0 \leq 1/\varepsilon$*. Then there exists an index K with*

$$K = O(n^\chi \log(1/\varepsilon))$$

such that

$$\mu^k \leq \varepsilon \quad \text{for} \quad \forall k \geq K$$

Combining Lemma 13.19 and Theorem 13.5 gives

Theorem 13.6
The complexity of Algorithm 13.1 is bounded by $O(\sqrt{n}\log(1/\varepsilon))$*.*

13.6 Implementation Issues

Algorithm 13.1 is presented in a form that is convenient for the convergence analysis. Some implementation details that make the algorithm more efficient are discussed in this section.

13.6.1 Termination Criterion

Algorithm 13.1 needs a termination criterion in real implementation. One can use

$$\mu^k \leq \varepsilon \tag{13.82a}$$
$$\|\mathbf{r}_X\| = \|\mathbf{H}\mathbf{x}^k + \lambda^k - \gamma^k + \mathbf{c}\| \leq \varepsilon \tag{13.82b}$$
$$\|\mathbf{r}_Y\| = \|\mathbf{x}^k + \mathbf{y}^k - \mathbf{e}\| \leq \varepsilon \tag{13.82c}$$
$$\|\mathbf{r}_Z\| = \|\mathbf{x}^k - \mathbf{z}^k + \mathbf{e}\| \leq \varepsilon \tag{13.82d}$$
$$\|\mathbf{r}_t\| = \|\mathbf{P}^k\Omega^k\mathbf{e} - \mu\mathbf{e}\| \leq \varepsilon \tag{13.82e}$$
$$(\mathbf{p}^k, \omega^k) > 0 \tag{13.82f}$$

An alternate criterion is similar to the one used in `linprog` [267]

$$\kappa := \frac{\|\mathbf{r}_Y\| + \|\mathbf{r}_Z\|}{2n} + \frac{\|\mathbf{r}_X\|}{\max\{1, \|\mathbf{c}\|\}} + \frac{\mu^k}{\max\{1, \|\mathbf{x}^{k\mathrm{T}}\mathbf{H}\mathbf{x}^k + \mathbf{c}^{\mathrm{T}}\mathbf{x}^k\|\}} \leq \varepsilon \tag{13.83}$$

13.6.2 Initial $(\mathbf{x}^0, \mathbf{y}^0, \mathbf{z}^0, \lambda^0, \gamma^0) \in \mathcal{N}_2(\theta)$

For feasible interior-point algorithms, an important prerequisite is to start with a feasible interior point. While finding an *initial feasible point* may not be a simple and trivial task for even linear programming with equality constraints [233], for quadratic programming subject to box constraints, finding the initial point is not an issue. As a matter of fact, the following initial point $(\mathbf{x}^0, \mathbf{y}^0, \mathbf{z}^0, \lambda^0, \gamma^0)$ is an interior point, moreover $(\mathbf{x}^0, \mathbf{y}^0, \mathbf{z}^0, \lambda^0, \gamma^0) \in \mathcal{N}_2(\theta)$.

$$\mathbf{x}^0 = 0, \quad \mathbf{y}^0 = \mathbf{z}^0 = \mathbf{e} > 0 \tag{13.84a}$$
$$\lambda_i^0 = 4(1 + \|\mathbf{c}\|^2) - \frac{c_i}{2} > 0 \tag{13.84b}$$
$$\gamma_i^0 = 4(1 + \|\mathbf{c}\|^2) + \frac{c_i}{2} > 0 \tag{13.84c}$$

It is easy to see that this selected point meets (13.20). Since

$$\mu^0 = \frac{\sum_{i=1}^n (\lambda_i^0 + \gamma_i^0)}{2n} = \frac{\sum_{i=1}^n (8(1 + \|\mathbf{c}\|^2))}{2n} = 4(1 + \|\mathbf{c}\|^2) \tag{13.85}$$

for $\theta = 0.19$, it must have

$$\begin{aligned}\left\|\mathbf{p}^0 \circ \omega^0 - \mu^0 \mathbf{e}\right\|^2 &= \sum_{i=1}^{n}(\lambda_i^0 - \mu^0)^2 + \sum_{i=1}^{n}(\gamma_i^0 - \mu^0)^2 \\ &= \frac{\|\mathbf{c}\|^2}{2} \leq 16\theta^2(1+\|\mathbf{c}\|^2)^2 = \theta^2(\mu^0)^2\end{aligned}$$

This shows that $(\mathbf{x}^0, \mathbf{y}^0, \mathbf{z}^0, \lambda^0, \gamma^0) \in \mathcal{N}_2(\theta)$.

13.6.3 Step Size

Directly using $\sin(\alpha) = \frac{\theta}{\sqrt{n}}$ in Algorithm 13.1 provides an effective formula to prove the polynomiality. However, this choice of $\sin(\alpha)$ is too conservative in practice because this search step in $\mathcal{N}_2(2\theta)$ is too small and the speed of duality measure reduction is slow. A better choice of $\sin(\alpha)$ should have a larger step in every iteration so that the polynomiality is reserved and fast convergence is achieved. In view of Remark 13.2, conditions that restrict step size are positivity conditions, proximity conditions, and duality reduction condition. This section examines how to enlarge the step size under these restrictions.

First, from (13.108) and (13.117), $\mu(\alpha) > 0$ is required for positivity conditions $(\mathbf{p}(\alpha), \omega(\alpha)) > 0$ and $(\mathbf{p}^{k+1}, \omega^{k+1}) > 0$ to hold. Since $\sin(\bar{\alpha})$ estimated in Corollary 13.1 is conservative, a better selection of $\bar{\alpha}$ is directly from (13.54), Lemmas 13.2 and 13.8:

$$\begin{aligned}\mu(\alpha) &\geq \mu(1-\sin(\alpha)) - \frac{1}{2n}\dot{\mathbf{x}}^{\mathrm{T}}\mathbf{H}\dot{\mathbf{x}}\left((1-\cos(\alpha))^2 + \sin^2(\alpha)\right) \\ &\geq \mu(1-\sin(\alpha)) - \frac{1}{2n}(\dot{\mathbf{p}}^{\mathrm{T}}\dot{\omega})\left(\sin^4(\alpha) + \sin^2(\alpha)\right) \\ &:= f(\sin(\alpha)) = \sigma \end{aligned} \tag{13.86}$$

where $\sigma > 0$ is a small number, and $f(\sin(\alpha))$ is a monotonic decreasing function of $\sin(\alpha)$ with $f(\sin(0)) = \mu$ and $f(\sin(\frac{\pi}{2})) < 0$. Therefore, (13.86) has a unique positive real solution for $\alpha \in [0, \frac{\pi}{2}]$. Since (13.86) is a quartic function of $\sin(\alpha)$, the cost of finding the smallest positive solution is negligible [157].

Second, in view of (13.116), the *proximity condition* for

$$(\mathbf{x}^{k+1}, \mathbf{y}^{k+1}, \mathbf{z}^{k+1}, \lambda^{k+1}, \gamma^{k+1}) \in \mathcal{N}_2(\theta)$$

holds for $\theta \leq 0.19$ without further restriction. The proximity condition (13.107) is met for $\sin(\alpha) \in [0, \sin(\tilde{\alpha})]$, where $\sin(\tilde{\alpha})$ is the smallest positive solution of (13.56) and it is estimated very conservatively in Lemma 13.15. An efficient implementation should use $\sin(\tilde{\alpha})$, the smallest positive solution of (13.56). Actually, there exist a $\acute{\alpha}$ which is normally larger than $\tilde{\alpha}$ such that the proximity condition (13.107) is met for $\sin(\alpha) \in [0, \sin(\acute{\alpha})]$. Let

$$b_0 = -\theta\mu < 0$$

$$b_1 = \theta\mu > 0$$

$$b_3 = \left\|\dot{\mathbf{p}} \circ \ddot{\omega} + \dot{\omega} \circ \ddot{\mathbf{p}} - \frac{1}{2n}(\dot{\mathbf{p}}^{\mathrm{T}}\ddot{\omega} + \dot{\omega}^{\mathrm{T}}\ddot{\mathbf{p}})\mathbf{e}\right\| + \frac{\theta}{n}\left(\dot{\mathbf{p}}^{\mathrm{T}}\ddot{\omega} + \ddot{\mathbf{p}}^{\mathrm{T}}\dot{\omega}\right)$$

$$b_4 = \left\|\ddot{\mathbf{p}} \circ \ddot{\omega} - \dot{\omega} \circ \dot{\mathbf{p}} - \frac{1}{2n}(\ddot{\mathbf{p}}^{\mathrm{T}}\ddot{\omega} - \dot{\omega}^{\mathrm{T}}\dot{\mathbf{p}})\mathbf{e}\right\| - \frac{\theta}{n}\left(\ddot{\mathbf{p}}^{\mathrm{T}}\ddot{\omega} - \dot{\mathbf{p}}^{\mathrm{T}}\dot{\omega}\right)$$

and

$$p(\alpha) := b_4(1-\cos(\alpha))^2 + b_3\sin(\alpha)(1-\cos(\alpha)) + b_1\sin(\alpha) + b_0 \qquad (13.87)$$

Applying the second inequality of (13.45) to $\frac{\theta}{n}\left(\dot{\mathbf{p}}^{\mathrm{T}}\ddot{\omega} + \ddot{\mathbf{p}}^{\mathrm{T}}\dot{\omega}\right)\sin(\alpha)(1-\cos(\alpha))$, one can easily show that

$$p(\alpha) \leq q(\alpha)$$

where $q(\alpha)$ is defined in (13.56). Therefore, the smallest positive solution $\grave{\alpha}$ of $p(\alpha)$ is larger than the smallest positive solution $\tilde{\alpha}$ of $q(\alpha)$. Hence, the goal is to show that for $\sin(\alpha) \in [0, \sin(\grave{\alpha})]$, the proximity condition (13.107) holds. Since for $\sin(\alpha) \in [0, \sin(\grave{\alpha})]$, $p(\alpha) \leq 0$, it must have

$$\begin{aligned}
& \left\|\ddot{\mathbf{p}} \circ \ddot{\omega} - \dot{\omega} \circ \dot{\mathbf{p}} - \frac{1}{2n}(\ddot{\mathbf{p}}^{\mathrm{T}}\ddot{\omega} - \dot{\omega}^{\mathrm{T}}\dot{\mathbf{p}})\mathbf{e}\right\|(1-\cos(\alpha))^2 \\
& + \left\|\dot{\mathbf{p}} \circ \ddot{\omega} + \dot{\omega} \circ \ddot{\mathbf{p}} - \frac{1}{2n}(\dot{\mathbf{p}}^{\mathrm{T}}\ddot{\omega} + \dot{\omega}^{\mathrm{T}}\ddot{\mathbf{p}})\mathbf{e}\right\|\sin(\alpha)(1-\cos(\alpha)) \\
\leq\ & (2\theta)\left(\frac{1}{2n}\left(\ddot{\mathbf{p}}^{\mathrm{T}}\ddot{\omega} - \dot{\mathbf{p}}^{\mathrm{T}}\dot{\omega}\right)(1-\cos(\alpha))^2 - \frac{1}{2n}\left(\dot{\mathbf{p}}^{\mathrm{T}}\ddot{\omega} + \ddot{\mathbf{p}}^{\mathrm{T}}\dot{\omega}\right)\sin(\alpha)(1-\cos(\alpha))\right) \\
& + \theta\mu(1-\sin(\alpha)) \qquad (13.88)
\end{aligned}$$

Substituting this inequality into (13.106) gives

$$\begin{aligned}
& \left\|\mathbf{p}(\alpha) \circ \omega(\alpha) - \mu(\alpha)\mathbf{e}\right\| \\
\leq\ & 2\theta\left[\mu(1-\sin(\alpha)) + \frac{1}{2n}\left(\ddot{\mathbf{x}}^{\mathrm{T}}(\ddot{\gamma} - \ddot{\lambda}) - \dot{\mathbf{x}}^{\mathrm{T}}(\dot{\gamma} - \dot{\lambda})\right)(1-\cos(\alpha))^2\right. \\
- & \left.\frac{1}{2n}\left(\dot{\mathbf{x}}^{\mathrm{T}}(\ddot{\gamma} - \ddot{\lambda}) + \ddot{\mathbf{x}}^{\mathrm{T}}(\dot{\gamma} - \dot{\lambda})\right)\sin(\alpha)(1-\cos(\alpha))\right] = 2\theta\mu(\alpha)
\end{aligned} \qquad (13.89)$$

This is the proximity condition for $(\mathbf{x}(\alpha), \mathbf{y}(\alpha), \mathbf{z}(\alpha), \lambda(\alpha), \gamma(\alpha))$. Denote $\hat{b}_0 = b_0$, $\hat{b}_1 = b_1$,

$$\hat{b}_3 = \begin{cases} b_3 & \text{if } b_3 \geq 0, \\ 0 & \text{if } b_3 < 0, \end{cases} \qquad \hat{b}_4 = \begin{cases} b_4 & \text{if } b_4 \geq 0, \\ 0 & \text{if } b_4 < 0, \end{cases}$$

and

$$\hat{p}(\alpha) := \hat{b}_4(1-\cos(\alpha))^2 + \hat{b}_3\sin(\alpha)(1-\cos(\alpha)) + \hat{b}_1\sin(\alpha) + \hat{b}_0 \qquad (13.90)$$

Since $\hat{p}(\alpha) \geq p(\alpha)$, the smallest positive solution $\acute{\alpha}$ of $\hat{p}(\alpha)$ is smaller than smallest positive solution $\grave{\alpha}$ of $p(\alpha)$. To estimate the smallest solution of $\acute{\alpha}$, by noticing that $\hat{p}(\alpha)$ is a monotonic increasing function of α and $\hat{p}(0) = -\theta\mu < 0$, one can simply use the bisection method. The computational cost is independent of the problem size n and is negligible. Since both estimated step sizes $\acute{\alpha}$ and $\tilde{\alpha}$ guarantee the proximity condition for $(\mathbf{x}(\alpha), \mathbf{y}(\alpha), \mathbf{z}(\alpha), \lambda(\alpha), \gamma(\alpha))$ to hold, one can select $\breve{\alpha} = \max\{\acute{\alpha}, \tilde{\alpha}\} \geq \tilde{\alpha}$ which guarantees the polynomiality claim to hold.

Third, from (C.76a) and Lemmas 13.11, 13.8, and 13.2, it must have

$$\begin{aligned} \mu^{k+1} \leq \; & \mu^k \left[1 + \frac{\theta^2(1+2\theta)}{n(1-2\theta)^2} - \left(1 + \frac{\theta^2(1+2\theta)}{n(1-2\theta)^2}\right) \sin(\alpha) \right. \\ & \left. + \left(1 + \frac{\theta^2(1+2\theta)}{n(1-2\theta)^2}\right) \frac{\ddot{\mathbf{p}}^{\mathrm{T}}\ddot{\omega}}{2n\mu} \left(\sin^2(\alpha) + \sin^4(\alpha)\right) \right] \end{aligned} \tag{13.91}$$

For $\mu^{k+1} \leq \mu^k$ to hold, one needs

$$\begin{aligned} & \frac{\theta^2(1+2\theta)}{n(1-2\theta)^2} - \left(1 + \frac{\theta^2(1+2\theta)}{n(1-2\theta)^2}\right) \sin(\alpha) \\ & + \left(1 + \frac{\theta^2(1+2\theta)}{n(1-2\theta)^2}\right) \frac{\ddot{\mathbf{p}}^{\mathrm{T}}\ddot{\omega}}{2n\mu} \left(\sin^2(\alpha) + \sin^4(\alpha)\right) \leq 0 \end{aligned}$$

For the sake of convenience in convergence analysis, a conservative estimate is used in Lemma 13.19. For efficient implementation, the following solution should be adopted. Denote $u = \frac{\theta^2(1+2\theta)}{n(1-2\theta)^2} > 0$, $v = \frac{\ddot{\mathbf{p}}^{\mathrm{T}}\ddot{\omega}}{2n\mu} > 0$, $z = \sin(\alpha) \in [0,1]$, and

$$F(z) = (1+u)vz^4 + (1+u)vz^2 - (1+u)z + u$$

For $z \in [0,1]$ and $v \leq \frac{1}{6}$, $F'(z) = (1+u)(4vz^3 + 2vz - 1) \leq 0$; therefore, the upper bound of the duality measure is a monotonic decreasing function of $\sin(\alpha)$ for $\alpha \in [0, \frac{\pi}{2}]$. The larger α is, the smaller the upper bound of the duality measure will be. For $v > \frac{1}{6}$, to minimize the upper bound of the duality measure, one can find the solution of $F'(z) = 0$. It is easy to check from discriminator [157] that the cubic polynomial $F'(z)$ has only one real solution which is given by (see Lemma 13.5)

$$\sin(\breve{\alpha}) = \sqrt[3]{\frac{n\mu}{4\ddot{\mathbf{p}}^{\mathrm{T}}\ddot{\omega}} + \sqrt{\left(\frac{n\mu}{4\ddot{\mathbf{p}}^{\mathrm{T}}\ddot{\omega}}\right)^2 + \left(\frac{1}{6}\right)^3}} + \sqrt[3]{\frac{n\mu}{4\ddot{\mathbf{p}}^{\mathrm{T}}\ddot{\omega}} - \sqrt{\left(\frac{n\mu}{4\ddot{\mathbf{p}}^{\mathrm{T}}\ddot{\omega}}\right)^2 + \left(\frac{1}{6}\right)^3}}$$

Since $F''(\sin(\breve{\alpha})) = (1+u)(12v\sin^2(\breve{\alpha}) + 2v) > 0$ at $\sin(\breve{\alpha}) \in [0,1)$, the upper bound of the duality measure is minimized. Therefore, one can define

$$\breve{\alpha} = \begin{cases} \frac{\pi}{2}, & \text{if } \frac{\ddot{\mathbf{p}}^{\mathrm{T}}\ddot{\omega}}{2n\mu} \le \frac{1}{6} \\ \sin^{-1}\left(\sqrt[3]{\frac{n\mu}{4\ddot{\mathbf{p}}^{\mathrm{T}}\ddot{\omega}} + \sqrt{\left(\frac{n\mu}{4\ddot{\mathbf{p}}^{\mathrm{T}}\ddot{\omega}}\right)^2 + \left(\frac{1}{6}\right)^3}} + \sqrt[3]{\frac{n\mu}{4\ddot{\mathbf{p}}^{\mathrm{T}}\ddot{\omega}} - \sqrt{\left(\frac{n\mu}{4\ddot{\mathbf{p}}^{\mathrm{T}}\ddot{\omega}}\right)^2 + \left(\frac{1}{6}\right)^3}}\right), & \text{if } \frac{\ddot{\mathbf{p}}^{\mathrm{T}}\ddot{\omega}}{2n\mu} > \frac{1}{6} \end{cases} \tag{13.92}$$

It is worthwhile to note that for $\alpha < \breve{\alpha}$, $F'(\sin(\alpha)) < 0$, i.e., $F(\sin(\alpha))$ is a monotonic decreasing function of $\alpha \in [0, \breve{\alpha}]$.

The step size selection process is therefore a simple algorithm as follows.

Algorithm 13.2
(Step Size Selection)
Data: $\sigma > 0$
Step 1: Find the positive real solution of (13.86) to get $\sin(\bar{\alpha})$.
Step 2: Find the smallest positive real solution of (13.90) to get $\sin(\acute{\alpha})$, *the smallest positive real solution of (13.56) to get* $\sin(\tilde{\alpha})$, *and set* $\sin(\check{\alpha}) = \max\{\sin(\tilde{\alpha}), \sin(\acute{\alpha})\}$.
Step 3: Calculate $\breve{\alpha}$ *given by (13.92).*
Step 4: The step size is obtained as $\sin(\alpha) = \min\{\sin(\bar{\alpha}), \sin(\check{\alpha}), \sin(\breve{\alpha})\}$.

13.6.4 The Practical Implementation

Therefore, Algorithm 13.1 can be implemented as follows:

Algorithm 13.3
(Arc-search path-following)
Data: $\mathbf{H} \geq \mathbf{0}$, $\mathbf{c}$, n, $\theta = 0.19$, $\varepsilon > \sigma > 0$
Step 0: Find initial point $(\mathbf{x}^0, \mathbf{p}^0, \omega^0) \in \mathcal{N}_2(\theta)$ *using (13.84),* κ *using (13.83), and* μ^0 *using (13.85).*
while $\kappa > \varepsilon$

Step 1: Compute $(\dot{\mathbf{x}}, \dot{\mathbf{p}}, \dot{\omega})$ *and* $(\ddot{\mathbf{x}}, \ddot{\mathbf{p}}, \ddot{\omega})$ *using (13.27) and (13.28).*

Step 2: Select $\sin(\alpha)$ *using Algorithm 13.2. Update* $(\mathbf{x}(\alpha), \mathbf{p}(\alpha), \omega(\alpha))$ *and* $\mu(\alpha)$ *using (13.57) and (13.58).*

Step 3: Compute $(\Delta\mathbf{x}, \Delta\mathbf{p}, \Delta\omega)$ *using (13.63), update* $(\mathbf{x}^{k+1}, \mathbf{p}^{k+1}, \omega^{k+1})$ *and* μ^{k+1} *using (13.64) and (13.65).*

Step 4: Computer κ *using (13.83).*

Step 5: Set $k+1 \rightarrow k$. *Go back to Step 1.*

end (while)

Remark 13.4 The condition $\mu > \sigma$ guarantees that the equation (13.86) has a positive solution before termination criterion is met. ∎

13.7 A Design Example

In this section, OrbView-2 spacecraft orbit-raising design example discussed in Chapter 12 is used to demonstrate the effectiveness and efficiency of the proposed algorithm. Let $\mathbf{w} = (w_x, w_y, w_z)$ be the spacecraft body rate with respect to the reference frame expressed in the body frame, $\bar{\mathbf{q}} = (q_0, q_1, q_2, q_3)$ be the quaternion of the spacecraft attitude with respect to the reference frame represented in the body frame and $\mathbf{q} = (q_1, q_2, q_3)$ be the reduced quaternion, $\mathbf{J} = \text{diag}(J_x, J_y, J_z)$ be the spacecraft inertia matrix, and h_w be the angular momentum produced by a momentum wheel. Orbit-raising is performed by four fixed thrusters (1 Newton) with on/off switches which are mounted on the anti-nadir face of the spacecraft in each corner of a square with a side length of $2d$ meter. The thrusters point to $+z$ direction and canted 5 degree from z-axis. (more details are provided in Chapter 12). The matrices of the thruster force direction $\mathbf{F}$ and moment arms $\mathbf{R}$ in the body frame are given as

$$\mathbf{F} = [\mathbf{f}_1, \mathbf{f}_2, \mathbf{f}_3, \mathbf{f}_4] = \begin{bmatrix} -a & -a & a & a \\ a & -a & -a & a \\ 1 & 1 & 1 & 1 \end{bmatrix}$$

$$\mathbf{R}_a = [\mathbf{r}_1, \mathbf{r}_2, \mathbf{r}_3, \mathbf{r}_4] = \begin{bmatrix} -d & -d & d & d \\ -d & d & d & -d \\ -\ell & -\ell & -\ell & -\ell \end{bmatrix}$$

Let $\mathbf{x} = (w_x, w_y, w_z, q_1, q_2, q_3)$ the states of the attitude and $\mathbf{u} = (T_1, T_2, T_3, T_4)$ be the control variable with T_1, T_2, T_3, T_4 the thrust level of the four thrusters. The linear time-invariant system under consideration is represented in a reduced quaternion model (see Chapter 12).

$$\begin{aligned} \dot{\mathbf{x}} &= \begin{bmatrix} 0 & 0 & \frac{h_w}{J_x} & 0 & 0 & 0 \\ 0 & 0 & 0 & 0 & 0 & 0 \\ \frac{h_w}{J_z} & 0 & 0 & 0 & 0 & 0 \\ 0.5 & 0 & 0 & 0 & 0 & 0 \\ 0 & 0.5 & 0 & 0 & 0 & 0 \\ 0 & 0 & 0.5 & 0 & 0 & 0 \end{bmatrix} \mathbf{x} \\ &+ \begin{bmatrix} \frac{1}{J_x} 0 & 0 \\ 0 & \frac{1}{J_y} & 0 \\ 0 & 0 & \frac{1}{J_z} \\ 0 & 0 & 0 \\ 0 & 0 & 0 \\ 0 & 0 & 0 \end{bmatrix} \begin{bmatrix} \mathbf{r}_1 \times \mathbf{f}_1 \\ \mathbf{r}_2 \times \mathbf{f}_2 \\ \mathbf{r}_3 \times \mathbf{f}_3 \\ \mathbf{r}_4 \times \mathbf{f}_4 \end{bmatrix}^{\mathbf{T}} \begin{bmatrix} T_1 \\ T_2 \\ T_3 \\ T_4 \end{bmatrix} \\ &= \mathbf{Ax} + \mathbf{Bu} \end{aligned} \tag{13.93}$$

with the control constraints

$$-\mathbf{e} \le \mathbf{u} = (T_1, T_2, T_3, T_4) \le \mathbf{e} \tag{13.94}$$

The problem is converted to discrete model using MATLAB function `c2d` with sampling time 1 second. The design is to minimize

$$J = \min_{\mathbf{u}_0, \mathbf{u}_1, \cdots, \mathbf{u}_{N-1}} \frac{1}{2}\mathbf{x}_N^{\mathrm{T}} \mathbf{P} \mathbf{x}_N + \frac{1}{2}\sum_{k=0}^{N-1} \left[\mathbf{x}_k^{\mathrm{T}} \mathbf{Q} \mathbf{x}_k + \mathbf{u}_k^{\mathrm{T}} \mathbf{R} \mathbf{u}_k\right] \tag{13.95}$$

where the horizon number $N = 30$, the matrices $\mathbf{P}$, $\mathbf{Q}$, and $\mathbf{R}$ are given by

$$\mathbf{P} = \mathbf{Q} = \begin{bmatrix} \frac{1}{2.5}\mathbf{I}_3 & \mathbf{0} \\ \mathbf{0} & 10000\mathbf{I}_3 \end{bmatrix}, \quad \mathbf{R} = \mathbf{I}_6$$

Other spacecraft parameters ($d = 0.248$m, $\ell = 0.815$m, $I_x = 189 kg.m^2$, $I_y = 159 kg.m^2$, and $I_z = 114 kg.m^2$, and $h_w = -2.8 N.m.s$) are the same as the ones of Chapter 12 and are taken from [190]. The algorithm is implemented in MATLAB. In our implementation of Algorithm 13.3, $\varepsilon = 10^{-6}$ and $\sigma = 10^{-10}$ are selected. Since MATLAB is an interpreted language (meaning that in the execution, every line has to be translated into machine language before the computer executes this line), MATLAB code is normally magnitudes slower than compiled languages, such as C, C++, and Fortran. But it turns out that even this MATLAB code is very fast. In 0.88 second, after 20 iterations, the algorithm converges (any intermediate result can be used in real time because they are all feasible). Using the optimal control inputs, we can calculate the state space response from (13.93). The control inputs and state space response are displayed in Figures 13.1, 13.2, and 13.3.

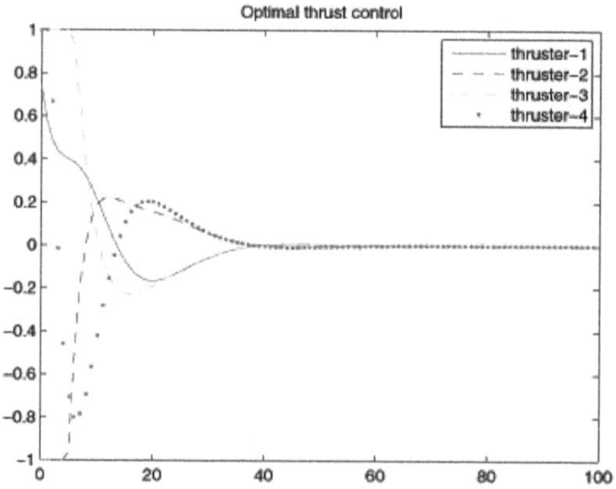

Figure 13.1: Optimal control with saturation constraint.

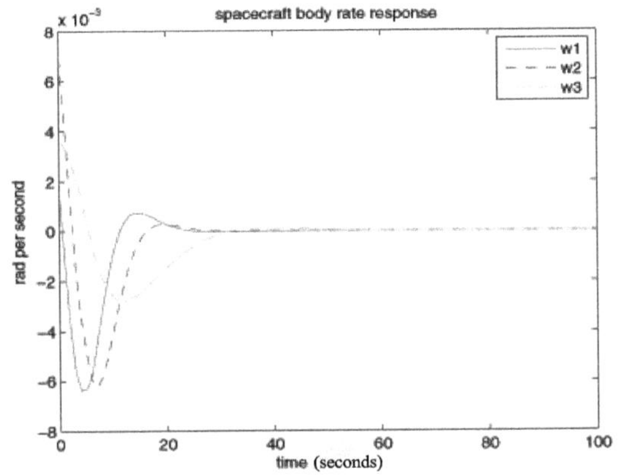

Figure 13.2: spacecraft body rate response.

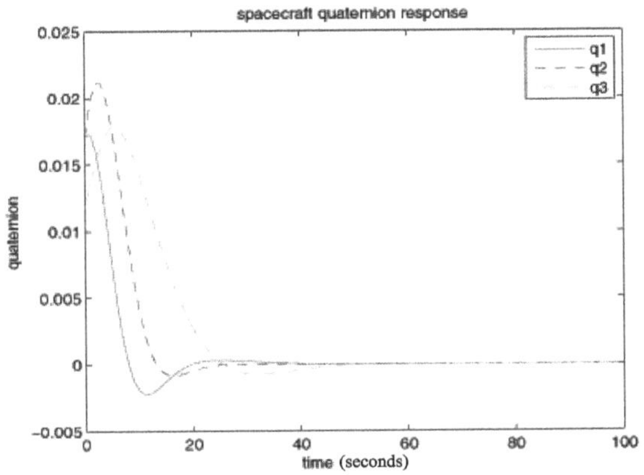

Figure 13.3: spacecraft quaternion response.

13.8 Proofs of Technical lemmas

Proof of Lemma 13.8:
From (13.30), we have

$$\dot{\mathbf{x}}^{\mathrm{T}}(\dot{\gamma}-\dot{\lambda}) = \dot{\mathbf{z}}^{\mathrm{T}}\dot{\gamma}+\dot{\mathbf{y}}^{\mathrm{T}}\dot{\lambda} = \dot{\mathbf{p}}^{\mathrm{T}}\dot{\omega}$$

$$\ddot{\mathbf{x}}^{\mathrm{T}}(\ddot{\gamma}-\ddot{\lambda}) = \ddot{\mathbf{z}}^{\mathrm{T}}\ddot{\gamma}+\ddot{\mathbf{y}}^{\mathrm{T}}\ddot{\lambda} = \ddot{\mathbf{p}}^{\mathrm{T}}\ddot{\omega}$$

$$\ddot{\mathbf{x}}^{\mathrm{T}}(\dot{\gamma}-\dot{\lambda}) = \ddot{\mathbf{p}}^{\mathrm{T}}\dot{\omega}$$

and

$$\dot{\mathbf{x}}^{\mathrm{T}}(\ddot{\gamma}-\ddot{\lambda}) = \dot{\mathbf{p}}^{\mathrm{T}}\ddot{\omega}$$

Pre-multiplying $\dot{\mathbf{x}}^{\mathrm{T}}$ and $\ddot{\mathbf{x}}^{\mathrm{T}}$ to (13.29) gives

$$\dot{\mathbf{x}}^{\mathrm{T}}(\dot{\gamma}-\dot{\lambda}) = \dot{\mathbf{x}}^{\mathrm{T}}\mathbf{H}\dot{\mathbf{x}}$$

$$\ddot{\mathbf{x}}^{\mathrm{T}}(\ddot{\gamma}-\ddot{\lambda}) = \ddot{\mathbf{x}}^{\mathrm{T}}\mathbf{H}\ddot{\mathbf{x}}$$
$$\ddot{\mathbf{x}}^{\mathrm{T}}(\dot{\gamma}-\dot{\lambda}) = \ddot{\mathbf{x}}^{\mathrm{T}}\mathbf{H}\dot{\mathbf{x}} = \dot{\mathbf{x}}^{\mathrm{T}}\mathbf{H}\ddot{\mathbf{x}} = \dot{\mathbf{x}}^{\mathrm{T}}(\ddot{\gamma}-\ddot{\lambda})$$

Equations (13.41) and (13.42) follow from the first two equations and the fact that $\mathbf{H}$ is a positive definite. The last equation is equivalent to (13.43). Using (13.41), (13.42), and (13.43) gives

$$\begin{aligned}
& (\dot{\mathbf{x}}(1-\cos(\alpha))+\ddot{\mathbf{x}}\sin(\alpha))^{\mathrm{T}}\mathbf{H}(\dot{\mathbf{x}}(1-\cos(\alpha))+\ddot{\mathbf{x}}\sin(\alpha)) \\
= \; & (\dot{\mathbf{x}}^{\mathrm{T}}\mathbf{H}\dot{\mathbf{x}})(1-\cos(\alpha))^2 + 2(\dot{\mathbf{x}}^{\mathrm{T}}\mathbf{H}\ddot{\mathbf{x}})\sin(\alpha)(1-\cos(\alpha)) + (\ddot{\mathbf{x}}^{\mathrm{T}}\mathbf{H}\ddot{\mathbf{x}})\sin^2(\alpha) \\
= \; & (\dot{\mathbf{x}}^{\mathrm{T}}\mathbf{H}\dot{\mathbf{x}})(1-\cos(\alpha))^2 + (\ddot{\mathbf{x}}^{\mathrm{T}}\mathbf{H}\ddot{\mathbf{x}})\sin^2(\alpha) \\
+ \; & (\ddot{\mathbf{x}}^{\mathrm{T}}(\dot{\gamma}-\dot{\lambda}) + \dot{\mathbf{x}}^{\mathrm{T}}(\ddot{\gamma}-\ddot{\lambda}))\sin(\alpha)(1-\cos(\alpha)) \geq 0
\end{aligned}$$

which is the first inequality of (13.44). Using (13.41), (13.42), and (13.43) also gives

$$\begin{aligned}
& (\dot{\mathbf{x}}(1-\cos(\alpha))-\ddot{\mathbf{x}}\sin(\alpha))^{\mathrm{T}}\mathbf{H}(\dot{\mathbf{x}}(1-\cos(\alpha))-\ddot{\mathbf{x}}\sin(\alpha)) \\
= \; & (\dot{\mathbf{x}}^{\mathrm{T}}\mathbf{H}\dot{\mathbf{x}})(1-\cos(\alpha))^2 - 2(\dot{\mathbf{x}}^{\mathrm{T}}\mathbf{H}\ddot{\mathbf{x}})\sin(\alpha)(1-\cos(\alpha)) + (\ddot{\mathbf{x}}^{\mathrm{T}}\mathbf{H}\ddot{\mathbf{x}})\sin^2(\alpha) \\
= \; & (\dot{\mathbf{x}}^{\mathrm{T}}\mathbf{H}\dot{\mathbf{x}})(1-\cos(\alpha))^2 + (\ddot{\mathbf{x}}^{\mathrm{T}}\mathbf{H}\ddot{\mathbf{x}})\sin^2(\alpha) \\
& -(\ddot{\mathbf{x}}^{\mathrm{T}}(\dot{\gamma}-\dot{\lambda}) + \dot{\mathbf{x}}^{\mathrm{T}}(\ddot{\gamma}-\ddot{\lambda}))\sin(\alpha)(1-\cos(\alpha)) \geq 0
\end{aligned}$$

which is the second inequality of (13.44). Replacing $\dot{\mathbf{x}}(1-\cos(\alpha))$ and $\ddot{\mathbf{x}}\sin(\alpha)$ by $\dot{\mathbf{x}}\sin(\alpha)$ and $\ddot{\mathbf{x}}(1-\cos(\alpha))$, and using the same method, one can obtain equation (13.45). ■

Proof of Lemma 13.9:
From the last two rows of (13.27) or equivalently (13.31), it must have

$$\Lambda\dot{\mathbf{y}} + \mathbf{Y}\dot{\lambda} = \Lambda\mathbf{Y}\mathbf{e}$$
$$\Gamma\dot{\mathbf{z}} + \mathbf{Z}\dot{\gamma} = \Gamma\mathbf{Z}\mathbf{e}$$

Pre-multiplying $\mathbf{Y}^{-\frac{1}{2}}\Lambda^{-\frac{1}{2}}$ on both sides of the first equality gives

$$\mathbf{Y}^{-\frac{1}{2}}\Lambda^{\frac{1}{2}}\dot{\mathbf{y}} + \mathbf{Y}^{\frac{1}{2}}\Lambda^{-\frac{1}{2}}\dot{\lambda} = \mathbf{Y}^{\frac{1}{2}}\Lambda^{\frac{1}{2}}\mathbf{e}$$

Pre-multiplying $\mathbf{Z}^{-\frac{1}{2}}\Gamma^{-\frac{1}{2}}$ on both sides of the second equality gives

$$\mathbf{Z}^{-\frac{1}{2}}\Gamma^{\frac{1}{2}}\dot{\mathbf{z}} + \mathbf{Z}^{\frac{1}{2}}\Gamma^{-\frac{1}{2}}\dot{\gamma} = \mathbf{Z}^{\frac{1}{2}}\Gamma^{\frac{1}{2}}\mathbf{e} \tag{13.96}$$

Let $\mathbf{u} = \begin{bmatrix} \mathbf{Y}^{-\frac{1}{2}}\Lambda^{\frac{1}{2}}\dot{\mathbf{y}} \\ \mathbf{Z}^{-\frac{1}{2}}\Gamma^{\frac{1}{2}}\dot{\mathbf{z}} \end{bmatrix}$, $\mathbf{v} = \begin{bmatrix} \mathbf{Y}^{\frac{1}{2}}\Lambda^{-\frac{1}{2}}\dot{\lambda} \\ \mathbf{Z}^{\frac{1}{2}}\Gamma^{-\frac{1}{2}}\dot{\gamma} \end{bmatrix}$, and $\mathbf{w} = \begin{bmatrix} \mathbf{Y}^{\frac{1}{2}}\Lambda^{\frac{1}{2}}\mathbf{e} \\ \mathbf{Z}^{\frac{1}{2}}\Gamma^{\frac{1}{2}}\mathbf{e} \end{bmatrix}$, using (13.30) and Lemma 13.8 yields $\mathbf{u}^{\mathrm{T}}\mathbf{v} = \dot{\mathbf{y}}^{\mathrm{T}}\dot{\lambda} + \dot{\mathbf{z}}^{\mathrm{T}}\dot{\gamma} = \dot{\mathbf{x}}^{\mathrm{T}}(\dot{\gamma}-\dot{\lambda}) \geq 0$. Using Lemma 13.3 and (13.23) yields

$$\|\mathbf{u}\|^2 + \|\mathbf{v}\|^2 = \sum_{i=1}^{n}\left(\frac{\dot{y}_i^2\lambda_i}{y_i} + \frac{\dot{z}_i^2\gamma_i}{z_i}\right) + \sum_{i=1}^{n}\left(\frac{\dot{\lambda}_i^2 y_i}{\lambda_i} + \frac{\dot{\gamma}_i^2 z_i}{\gamma_i}\right)$$

$$\leq \sum_{i=1}^{n}(y_i\lambda_i + z_i\gamma_i) = \sum_{i=1}^{2n} p_i\omega_i = 2n\mu \tag{13.97}$$

Since $p_i > 0$ and $\omega_i > 0$, dividing both sides of the inequality by $\min_j p_i\omega_i$ and using (13.25) gives

$$\sum_{i=1}^{n}\left(\frac{\dot{y}_i^2}{y_i^2} + \frac{\dot{z}_i^2}{z_i^2}\right) + \sum_{i=1}^{n}\left(\frac{\dot{\gamma}_i^2}{\gamma_i^2} + \frac{\dot{\lambda}_i^2}{\lambda_i^2}\right) = \left\|\frac{\dot{\mathbf{p}}}{\mathbf{p}}\right\|^2 + \left\|\frac{\dot{\omega}}{\omega}\right\|^2 \leq \frac{2n\mu}{\min_j p_i\omega_i} \leq \frac{2n}{1-\theta} \tag{13.98}$$

This proves (13.47). Combining (13.47) and Lemma 13.1 yields

$$\left\|\frac{\dot{\mathbf{p}}}{\mathbf{p}}\right\|^2 \left\|\frac{\dot{\omega}}{\omega}\right\|^2 \leq \left(\frac{n}{(1-\theta)}\right)^2$$

This leads to

$$\left\|\frac{\dot{\mathbf{p}}}{\mathbf{p}}\right\| \left\|\frac{\dot{\omega}}{\omega}\right\| \leq \frac{n}{(1-\theta)} \tag{13.99}$$

Therefore, using (13.25) and Cauchy-Schwarz inequality yields

$$\begin{aligned}\frac{\dot{\mathbf{p}}^{\mathrm{T}}\dot{\omega}}{\mu} &\leq \frac{|\dot{\mathbf{p}}|^{\mathrm{T}}|\dot{\omega}|}{\mu} \leq (1+\theta)\frac{|\dot{\mathbf{p}}|^{\mathrm{T}}|\dot{\omega}|}{\max_i p_i\omega_i} \leq (1+\theta)\left(\frac{|\dot{\mathbf{p}}|}{\mathbf{p}}\right)^{\mathrm{T}}\left(\frac{|\dot{\omega}|}{\omega}\right) \\ &\leq (1+\theta)\left\|\frac{\dot{\mathbf{p}}}{\mathbf{p}}\right\| \left\|\frac{\dot{\omega}}{\omega}\right\| \leq \frac{1+\theta}{1-\theta}n \end{aligned} \tag{13.100}$$

which is the second inequality of (13.49). From Lemma 13.8, $\dot{\mathbf{p}}^{\mathrm{T}}\dot{\omega} = \dot{\mathbf{x}}^{\mathrm{T}}(\dot{\gamma} - \dot{\lambda}) = \dot{\mathbf{x}}^{\mathrm{T}}\mathbf{H}\dot{\mathbf{x}} \geq 0$, the first inequality of (13.49) follows. ■

Proof of Lemma 13.10:
Similar to the proof of Lemma 13.9, from (13.32), it must have

$$\begin{aligned} &\Lambda\ddot{\mathbf{y}} + \mathbf{Y}\ddot{\lambda} = -2\left(\dot{\mathbf{y}}\circ\dot{\lambda}\right) \\ \iff\ &\mathbf{Y}^{-\frac{1}{2}}\Lambda^{\frac{1}{2}}\ddot{\mathbf{y}} + \mathbf{Y}^{\frac{1}{2}}\Lambda^{-\frac{1}{2}}\ddot{\lambda} = -2\mathbf{Y}^{-\frac{1}{2}}\Lambda^{-\frac{1}{2}}\left(\dot{\mathbf{y}}\circ\dot{\lambda}\right)\end{aligned}$$

and

$$\begin{aligned} &\Gamma\ddot{\mathbf{z}} + \mathbf{Z}\ddot{\gamma} = -2\left(\dot{\mathbf{z}}\circ\dot{\gamma}\right) \\ \iff\ &\mathbf{Z}^{-\frac{1}{2}}\Gamma^{\frac{1}{2}}\ddot{\mathbf{z}} + \mathbf{Z}^{\frac{1}{2}}\Gamma^{-\frac{1}{2}}\ddot{\gamma} = -2\mathbf{Z}^{-\frac{1}{2}}\Gamma^{-\frac{1}{2}}\left(\dot{\mathbf{z}}\circ\dot{\gamma}\right)\end{aligned}$$

Let $\mathbf{u} = \begin{bmatrix}\mathbf{Y}^{-\frac{1}{2}}\Lambda^{\frac{1}{2}}\ddot{\mathbf{y}} \\ \mathbf{Z}^{-\frac{1}{2}}\Gamma^{\frac{1}{2}}\ddot{\mathbf{z}}\end{bmatrix}$, $\mathbf{v} = \begin{bmatrix}\mathbf{Y}^{\frac{1}{2}}\Lambda^{-\frac{1}{2}}\ddot{\lambda} \\ \mathbf{Z}^{\frac{1}{2}}\Gamma^{-\frac{1}{2}}\ddot{\gamma}\end{bmatrix}$, and $\mathbf{w} = \begin{bmatrix}-2\mathbf{Y}^{-\frac{1}{2}}\Lambda^{-\frac{1}{2}}\left(\dot{\mathbf{y}}\circ\dot{\lambda}\right) \\ -2\mathbf{Z}^{-\frac{1}{2}}\Gamma^{-\frac{1}{2}}\left(\dot{\mathbf{z}}\circ\dot{\gamma}\right)\end{bmatrix}$

using (13.30) and Lemma 13.8 yields $\mathbf{u}^{\mathrm{T}}\mathbf{v} = \ddot{\mathbf{y}}^{\mathrm{T}}\ddot{\lambda} + \ddot{\mathbf{z}}^{\mathrm{T}}\ddot{\gamma} = \ddot{\mathbf{x}}^{\mathrm{T}}(\ddot{\gamma} - \ddot{\lambda}) \geq 0$. Using Lemma 13.3 yields

$$\begin{aligned}
\|\mathbf{u}\|^2 + \|\mathbf{v}\|^2 &= \sum_{i=1}^{n}\left(\frac{\ddot{y}_i^2\lambda_i}{y_i} + \frac{\ddot{z}_i^2\gamma_i}{z_i}\right) + \sum_{i=1}^{n}\left(\frac{\ddot{\lambda}_i^2 y_i}{\lambda_i} + \frac{\ddot{\gamma}_i^2 z_i}{\gamma_i}\right) \\
&\leq \left\|-2\mathbf{Y}^{-\frac{1}{2}}\Lambda^{-\frac{1}{2}}\left(\dot{\mathbf{y}} \circ \dot{\lambda}\right)\right\|^2 + \left\|-2\mathbf{Z}^{-\frac{1}{2}}\Gamma^{-\frac{1}{2}}\left(\dot{\mathbf{z}} \circ \dot{\gamma}\right)\right\|^2 \\
&= 4\sum_{i=1}^{n}\left(\frac{\dot{y}_i^2}{y_i}\frac{\dot{\lambda}_i^2}{\lambda_i} + \frac{\dot{z}_i^2}{z_i}\frac{\dot{\gamma}_i^2}{\gamma_i}\right)
\end{aligned}$$

Dividing both sides of the inequality by μ and using (13.25) gives

$$\begin{aligned}
&(1-\theta)\left(\sum_{i=1}^{n}\left(\frac{\ddot{y}_i^2}{y_i^2} + \frac{\ddot{z}_i^2}{z_i^2}\right) + \sum_{i=1}^{n}\left(\frac{\ddot{\lambda}_i^2}{\lambda_i^2} + \frac{\ddot{\gamma}_i^2}{\gamma_i^2}\right)\right) \\
= \; &(1-\theta)\left(\left\|\frac{\ddot{\mathbf{p}}}{\mathbf{p}}\right\|^2 + \left\|\frac{\ddot{\omega}}{\omega}\right\|^2\right) \\
\leq \; &4(1+\theta)\left(\sum_{i=1}^{n}\left(\frac{\dot{y}_i^2}{y_i^2}\frac{\dot{\lambda}_i^2}{\lambda_i^2} + \frac{\dot{z}_i^2}{z_i^2}\frac{\dot{\gamma}_i^2}{\gamma_i^2}\right)\right)
\end{aligned}$$

in view of Lemma 13.9, this leads to

$$\left\|\frac{\ddot{\mathbf{p}}}{\mathbf{p}}\right\|^2 + \left\|\frac{\ddot{\omega}}{\omega}\right\|^2 \leq 4\frac{1+\theta}{1-\theta}\left\|\frac{\dot{\mathbf{p}}}{\mathbf{p}} \circ \frac{\dot{\omega}}{\omega}\right\|^2 \leq 4\frac{1+\theta}{1-\theta}\left\|\frac{\dot{\mathbf{p}}}{\mathbf{p}}\right\|^2\left\|\frac{\dot{\omega}}{\omega}\right\|^2 \leq \frac{4(1+\theta)n^2}{(1-\theta)^3} \tag{13.101}$$

This proves (13.50). Combining (13.50) and Lemma 13.1 yields

$$\left\|\frac{\ddot{\mathbf{p}}}{\mathbf{p}}\right\|^2\left\|\frac{\ddot{\omega}}{\omega}\right\|^2 \leq \left(\frac{2(1+\theta)n^2}{(1-\theta)^3}\right)^2$$

Using (13.25) and Cauchy-Schwarz inequality yields

$$\begin{aligned}
\frac{\ddot{\mathbf{p}}^{\mathrm{T}}\ddot{\omega}}{\mu} &\leq \frac{|\ddot{\mathbf{p}}|^{\mathrm{T}}|\ddot{\omega}|}{\mu} \leq (1+\theta)\frac{|\ddot{\mathbf{p}}|^{\mathrm{T}}|\ddot{\omega}|}{\max_i p_i\omega_i} \leq (1+\theta)\left(\frac{|\ddot{\mathbf{p}}|}{\mathbf{p}}\right)^{\mathrm{T}}\left(\frac{|\ddot{\omega}|}{\omega}\right) \\
&\leq (1+\theta)\left\|\frac{\ddot{\mathbf{p}}}{\mathbf{p}}\right\|\left\|\frac{\ddot{\omega}}{\omega}\right\| \leq \frac{2n^2(1+\theta)^2}{(1-\theta)^3}
\end{aligned}$$

which is the second inequality of (13.52). Using (13.30) and Lemma 13.8, one must have $\ddot{\mathbf{p}}^{\mathrm{T}}\ddot{\omega} = \ddot{\mathbf{y}}^{\mathrm{T}}\ddot{\lambda} + \ddot{\mathbf{z}}^{\mathrm{T}}\ddot{\gamma} = \ddot{\mathbf{x}}^{\mathrm{T}}(\ddot{\gamma} - \ddot{\lambda}) = \ddot{\mathbf{x}}^{\mathrm{T}}\mathbf{H}\ddot{\mathbf{x}} \geq 0$. This proves the first inequality of (13.52). Finally, using (13.25), Cauchy-Schwarz inequality, (13.47), and (13.50) yields

$$\frac{|\dot{\mathbf{p}}^{\mathrm{T}}\ddot{\omega}|}{\mu} \leq \frac{|\dot{\mathbf{p}}|^{\mathrm{T}}|\ddot{\omega}|}{\mu} \leq (1+\theta)\frac{|\dot{\mathbf{p}}|^{\mathrm{T}}|\ddot{\omega}|}{\max_i p_i\omega_i} \leq (1+\theta)\left(\frac{|\dot{\mathbf{p}}|}{\mathbf{p}}\right)^{\mathrm{T}}\left(\frac{|\ddot{\omega}|}{\omega}\right)$$

$$\leq (1+\theta)\left\|\frac{\dot{\mathbf{p}}}{\mathbf{p}}\right\|\left\|\frac{\ddot{\omega}}{\omega}\right\| \leq (1+\theta)\left(\frac{2n}{1-\theta}\right)^{\frac{1}{2}}\left(\frac{4(1+\theta)n^2}{(1+\theta)^3}\right)^{\frac{1}{2}} \leq \frac{(2n(1+\theta))^{\frac{3}{2}}}{(1-\theta)^2}$$

This proves the first inequality of (13.53). Replacing $\dot{\mathbf{p}}$ by $\ddot{\mathbf{p}}$ and $\ddot{\omega}$ by $\dot{\omega}$, then using the same reasoning, one can prove the second inequality of (13.53). ■

Proof of Lemma 13.11:
Using (13.34), (13.36), (13.31), and (13.32), one must have

$$\begin{aligned}
&\mathbf{y}^{\mathrm{T}}(\alpha)\lambda(\alpha)\\
=&\left(\mathbf{y}^{\mathrm{T}}-\dot{\mathbf{y}}^{\mathrm{T}}\sin(\alpha)+\ddot{\mathbf{y}}^{\mathrm{T}}(1-\cos(\alpha))\right)\left(\lambda-\dot{\lambda}\sin(\alpha)+\ddot{\lambda}(1-\cos(\alpha))\right)\\
=&\mathbf{y}^{\mathrm{T}}\lambda-\mathbf{y}^{\mathrm{T}}\dot{\lambda}\sin(\alpha)+\mathbf{y}^{\mathrm{T}}\ddot{\lambda}(1-\cos(\alpha))\\
&-\dot{\mathbf{y}}^{\mathrm{T}}\lambda\sin(\alpha)+\dot{\mathbf{y}}^{\mathrm{T}}\dot{\lambda}\sin^2(\alpha)-\dot{\mathbf{y}}^{\mathrm{T}}\ddot{\lambda}\sin(\alpha)(1-\cos(\alpha))\\
&+\ddot{\mathbf{y}}^{\mathrm{T}}\lambda(1-\cos(\alpha))-\ddot{\mathbf{y}}^{\mathrm{T}}\dot{\lambda}\sin(\alpha)(1-\cos(\alpha))+\ddot{\mathbf{y}}^{\mathrm{T}}\ddot{\lambda}(1-\cos(\alpha))^2\\
=&\mathbf{y}^{\mathrm{T}}\lambda-(\mathbf{y}^{\mathrm{T}}\dot{\lambda}+\lambda^{\mathrm{T}}\dot{\mathbf{y}})\sin(\alpha)+(\mathbf{y}^{\mathrm{T}}\ddot{\lambda}+\lambda^{\mathrm{T}}\ddot{\mathbf{y}})(1-\cos(\alpha))\\
&-(\dot{\mathbf{y}}^{\mathrm{T}}\ddot{\lambda}+\dot{\lambda}^{\mathrm{T}}\ddot{\mathbf{y}})\sin(\alpha)(1-\cos(\alpha))+\dot{\mathbf{y}}^{\mathrm{T}}\dot{\lambda}\sin^2(\alpha)+\ddot{\mathbf{y}}^{\mathrm{T}}\ddot{\lambda}(1-\cos(\alpha))^2\\
=&\mathbf{y}^{\mathrm{T}}\lambda(1-\sin(\alpha))-2\dot{\mathbf{y}}^{\mathrm{T}}\dot{\lambda}(1-\cos(\alpha))\\
&-(\dot{\mathbf{y}}^{\mathrm{T}}\ddot{\lambda}+\dot{\lambda}^{\mathrm{T}}\ddot{\mathbf{y}})\sin(\alpha)(1-\cos(\alpha))\\
&+\dot{\mathbf{y}}^{\mathrm{T}}\dot{\lambda}(1-\cos^2(\alpha))+\ddot{\mathbf{y}}^{\mathrm{T}}\ddot{\lambda}(1-\cos(\alpha))^2\\
=&\mathbf{y}^{\mathrm{T}}\lambda(1-\sin(\alpha))+(\ddot{\mathbf{y}}^{\mathrm{T}}\ddot{\lambda}-\dot{\mathbf{y}}^{\mathrm{T}}\dot{\lambda})(1-\cos(\alpha))^2\\
&-(\dot{\mathbf{y}}^{\mathrm{T}}\ddot{\lambda}+\dot{\lambda}^{\mathrm{T}}\ddot{\mathbf{y}})\sin(\alpha)(1-\cos(\alpha))
\end{aligned}\tag{13.102}$$

Using (13.35), (13.37), (13.31), (13.32), and a similar derivation of (13.102), one gets

$$\begin{aligned}
\mathbf{z}^{\mathrm{T}}(\alpha)\gamma(\alpha) \;=\;& \mathbf{z}^{\mathrm{T}}\gamma(1-\sin(\alpha))+(\ddot{\mathbf{z}}^{\mathrm{T}}\ddot{\gamma}-\dot{\mathbf{z}}^{\mathrm{T}}\dot{\gamma})(1-\cos(\alpha))^2\\
&-(\dot{\mathbf{z}}^{\mathrm{T}}\ddot{\gamma}+\dot{\gamma}^{\mathrm{T}}\ddot{\mathbf{z}})\sin(\alpha)(1-\cos(\alpha))
\end{aligned}\tag{13.103}$$

Combining (13.102) and (13.103), then using (13.30) and (13.44) yield

$$\begin{aligned}
&2n\mu(\alpha)=\mathbf{p}^{\mathrm{T}}(\alpha)\omega(\alpha)\\
=&\mathbf{y}^{\mathrm{T}}(\alpha)\lambda(\alpha)+\mathbf{z}^{\mathrm{T}}(\alpha)\gamma(\alpha)\\
=&(\mathbf{y}^{\mathrm{T}}\lambda+\mathbf{z}^{\mathrm{T}}\gamma)(1-\sin(\alpha))+(\ddot{\mathbf{y}}^{\mathrm{T}}\ddot{\lambda}+\ddot{\mathbf{z}}^{\mathrm{T}}\ddot{\gamma}-\dot{\mathbf{y}}^{\mathrm{T}}\dot{\lambda}-\dot{\mathbf{z}}^{\mathrm{T}}\dot{\gamma})(1-\cos(\alpha))^2\\
&-(\dot{\mathbf{y}}^{\mathrm{T}}\ddot{\lambda}+\dot{\mathbf{z}}^{\mathrm{T}}\ddot{\gamma}+\ddot{\mathbf{y}}^{\mathrm{T}}\dot{\lambda}+\ddot{\mathbf{z}}^{\mathrm{T}}\dot{\gamma})\sin(\alpha)(1-\cos(\alpha))\\
=&(\mathbf{y}^{\mathrm{T}}\lambda+\mathbf{z}^{\mathrm{T}}\gamma)(1-\sin(\alpha))+(\ddot{\mathbf{x}}^{\mathrm{T}}(\ddot{\gamma}-\ddot{\lambda})-\dot{\mathbf{x}}^{\mathrm{T}}(\dot{\gamma}-\dot{\lambda}))(1-\cos(\alpha))^2\\
&-(\dot{\mathbf{x}}^{\mathrm{T}}(\ddot{\gamma}-\ddot{\lambda})+\ddot{\mathbf{x}}^{\mathrm{T}}(\dot{\gamma}-\dot{\lambda}))\sin(\alpha)(1-\cos(\alpha))
\end{aligned}\tag{13.104}$$

$$
\begin{aligned}
&\leq (\mathbf{y}^{\mathrm{T}}\lambda + \mathbf{z}^{\mathrm{T}}\gamma)(1-\sin(\alpha)) + (\ddot{\mathbf{x}}^{\mathrm{T}}\mathbf{H}\ddot{\mathbf{x}} - \dot{\mathbf{x}}^{\mathrm{T}}\mathbf{H}\dot{\mathbf{x}})(1-\cos(\alpha))^2 \\
&\quad + \dot{\mathbf{x}}^{\mathrm{T}}\mathbf{H}\dot{\mathbf{x}}(1-\cos(\alpha))^2 + \ddot{\mathbf{x}}^{\mathrm{T}}\mathbf{H}\ddot{\mathbf{x}}\sin^2(\alpha) \\
&= (\mathbf{y}^{\mathrm{T}}\lambda + \mathbf{z}^{\mathrm{T}}\gamma)(1-\sin(\alpha)) + \ddot{\mathbf{x}}^{\mathrm{T}}\mathbf{H}\ddot{\mathbf{x}}(1-\cos(\alpha))^2 + \ddot{\mathbf{x}}^{\mathrm{T}}\mathbf{H}\ddot{\mathbf{x}}\sin^2(\alpha)
\end{aligned}
$$

Dividing both the sides by $2n$ proves the second inequality of the lemma. Combining (13.104) and (13.45) proves the first inequality of the lemma. ■

Proof of Lemma 13.12:
From the second inequality of (13.54), it must have

$$
\mu(\alpha) - \mu \leq \mu \sin(\alpha)\left(-1 + \frac{\ddot{\mathbf{x}}^{\mathrm{T}}\mathbf{H}\ddot{\mathbf{x}}}{2n\mu}\sin(\alpha) + \frac{\ddot{\mathbf{x}}^{\mathrm{T}}\mathbf{H}\ddot{\mathbf{x}}}{2n\mu}\sin^3(\alpha)\right)
$$

Clearly, if $\frac{\ddot{\mathbf{x}}^{\mathrm{T}}\mathbf{H}\ddot{\mathbf{x}}}{2n\mu} \leq \frac{1}{2}$, for any $\alpha \in [0, \frac{\pi}{2}]$, the function

$$
f(\alpha) := \left(-1 + \frac{\ddot{\mathbf{x}}^{\mathrm{T}}\mathbf{H}\ddot{\mathbf{x}}}{2n\mu}\sin(\alpha) + \frac{\ddot{\mathbf{x}}^{\mathrm{T}}\mathbf{H}\ddot{\mathbf{x}}}{2n\mu}\sin^3(\alpha)\right) \leq 0
$$

and $\mu(\alpha) \leq \mu$. If $\frac{\ddot{\mathbf{x}}^{\mathrm{T}}\mathbf{H}\ddot{\mathbf{x}}}{2n\mu} > \frac{1}{2}$, using Lemma 13.5, the function f has one real solution $\sin(\alpha) \in (0,1)$. The solution is given as

$$
\sin(\hat{\alpha}) = \sqrt[3]{\frac{n\mu}{\ddot{\mathbf{x}}^{\mathrm{T}}\mathbf{H}\ddot{\mathbf{x}}} + \sqrt{\left(\frac{n\mu}{\ddot{\mathbf{x}}^{\mathrm{T}}\mathbf{H}\ddot{\mathbf{x}}}\right)^2 + \left(\frac{1}{3}\right)^3}} + \sqrt[3]{\frac{n\mu}{\ddot{\mathbf{x}}^{\mathrm{T}}\mathbf{H}\ddot{\mathbf{x}}} - \sqrt{\left(\frac{n\mu}{\ddot{\mathbf{x}}^{\mathrm{T}}\mathbf{H}\ddot{\mathbf{x}}}\right)^2 + \left(\frac{1}{3}\right)^3}}
$$

This proves the Lemma. ■

Proof of Lemma 13.13:
Since $\sin(\tilde{\alpha})$ is the only positive real solution of (13.56) in $[0,1]$ and $q(0) < 0$, substituting a_0, a_1, a_2, a_3 and a_4 into (13.56) yields, for all $\sin(\alpha) \leq \sin(\tilde{\alpha})$,

$$
\begin{aligned}
&\left(\left\|\ddot{\mathbf{p}} \circ \ddot{\omega} - \dot{\omega} \circ \dot{\mathbf{p}} - \frac{1}{2n}(\ddot{\mathbf{p}}^{\mathrm{T}}\ddot{\omega} - \dot{\omega}^{\mathrm{T}}\dot{\mathbf{p}})\mathbf{e}\right\|\right)\sin^4(\alpha) \\
&+ \left(\left\|\dot{\mathbf{p}} \circ \ddot{\omega} + \dot{\omega} \circ \ddot{\mathbf{p}} - \frac{1}{2n}(\dot{\mathbf{p}}^{\mathrm{T}}\ddot{\omega} + \dot{\omega}^{\mathrm{T}}\ddot{\mathbf{p}})\mathbf{e}\right\|\right)\sin^3(\alpha) \\
&\leq -\left(2\theta\frac{\dot{\mathbf{p}}^{\mathrm{T}}\dot{\omega}}{2n}\right)\sin^4(\alpha) - \left(2\theta\frac{\dot{\mathbf{p}}^{\mathrm{T}}\dot{\omega}}{2n}\right)\sin^2(\alpha) + \theta\mu(1-\sin(\alpha)) \qquad (13.105)
\end{aligned}
$$

Using (13.38), (13.39), (13.31), (13.32), (13.58), Lemma 13.2, (13.105), and the first inequality of (13.54) yields

$$
\left\|\mathbf{p}(\alpha) \circ \omega(\alpha) - \mu(\alpha)\mathbf{e}\right\|
$$

$$
\begin{aligned}
&= \left\| \left(\mathbf{p} - \dot{\mathbf{p}}\sin(\alpha) + \ddot{\mathbf{p}}(1-\cos(\alpha))\right) \circ \left(\omega - \dot{\omega}\sin(\alpha) + \ddot{\omega}(1-\cos(\alpha))\right) - \mu(\alpha)\mathbf{e} \right\| \\
&= \left\| (\mathbf{p}\circ\omega - \mu\mathbf{e})(1-\sin(\alpha)) + \left(\ddot{\mathbf{p}}\circ\ddot{\omega} - \dot{\mathbf{p}}\circ\dot{\omega} - \frac{1}{2n}(\ddot{\mathbf{p}}^{\mathrm{T}}\ddot{\omega} - \dot{\mathbf{p}}^{\mathrm{T}}\dot{\omega})\mathbf{e} \right)(1-\cos(\alpha))^2 \right. \\
&\quad \left. - \left(\dot{\mathbf{p}}\circ\ddot{\omega} + \dot{\omega}\circ\ddot{\mathbf{p}} - \frac{1}{2n}(\dot{\mathbf{p}}^{\mathrm{T}}\ddot{\omega} + \ddot{\mathbf{p}}^{\mathrm{T}}\dot{\omega})\mathbf{e} \right)\sin(\alpha)(1-\cos(\alpha)) \right\| \\
&\leq (1-\sin(\alpha))\left\|\mathbf{p}\circ\omega - \mu\mathbf{e}\right\| + \left\| (\ddot{\mathbf{p}}\circ\ddot{\omega} - \dot{\mathbf{p}}\circ\dot{\omega} - \frac{1}{2n}(\ddot{\mathbf{p}}^{\mathrm{T}}\ddot{\omega} - \dot{\mathbf{p}}^{\mathrm{T}}\dot{\omega}))\mathbf{e} \right\| (1-\cos(\alpha))^2 \\
&\quad + \left\| (\dot{\mathbf{p}}\circ\ddot{\omega} + \dot{\omega}\circ\ddot{\mathbf{p}} - \frac{1}{2n}(\dot{\mathbf{p}}^{\mathrm{T}}\ddot{\omega} + \ddot{\mathbf{p}}^{\mathrm{T}}\dot{\omega})\mathbf{e} \right\| \sin(\alpha)(1-\cos(\alpha)) \qquad (13.106) \\
&\leq \theta\mu(1-\sin(\alpha)) + \left\| (\ddot{\mathbf{p}}\circ\ddot{\omega} - \dot{\mathbf{p}}\circ\dot{\omega} - \frac{1}{2n}(\ddot{\mathbf{p}}^{\mathrm{T}}\ddot{\omega} - \dot{\mathbf{p}}^{\mathrm{T}}\dot{\omega}))\mathbf{e} \right\| \sin^4(\alpha) + a_3\sin^3(\alpha) \\
&\leq 2\theta\mu(1-\sin(\alpha)) - \left(2\theta\frac{\dot{\mathbf{p}}^{\mathrm{T}}\dot{\omega}}{2n} \right)(\sin^4(\alpha) + \sin^2(\alpha)) \\
&\leq 2\theta\left(\mu(1-\sin(\alpha)) - \left(\frac{\dot{\mathbf{x}}^{\mathrm{T}}\mathbf{H}\dot{\mathbf{x}}}{2n}\right)\left((1-\cos(\alpha))^2 + \sin^2(\alpha)\right) \right) \\
&\leq 2\theta\mu(\alpha) \qquad (13.107)
\end{aligned}
$$

Hence, the point $(\mathbf{x}(\alpha), \mathbf{p}(\alpha), \omega(\alpha))$ satisfies the proximity condition for $\mathcal{N}_2(2\theta)$. To check the positivity condition $(\mathbf{p}(\alpha), \omega(\alpha)) > 0$, in view of the initial condition $(\mathbf{p}, \omega) > 0$, it follows from (13.107) and Corollary 13.1 that, for $\sin(\alpha) \leq \sin(\bar{\alpha})$ and $\theta < 0.5$,

$$p_i(\alpha)\omega_i(\alpha) \geq (1-2\theta)\mu(\alpha) > 0 \qquad (13.108)$$

Therefore, it cannot have $p_i(\alpha) = 0$ or $\omega_i(\alpha) = 0$ for any index i when $\alpha \in [0, \sin^{-1}(\bar{\alpha})]$. This proves $(\mathbf{p}(\alpha), \omega(\alpha)) > 0$ ■

Remark 13.5 It is worthwhile to note, by examining the proof of Lemma 13.13, that $\sin(\tilde{\alpha})$ is selected for the proximity condition (13.107) to hold, and $\sin(\bar{\alpha})$ is selected for $\mu(\alpha) > 0$, thereby assuring the positivity condition (13.108) to hold. ■

Proof of Lemma 13.14:
Since,

$$\left\|\frac{\dot{\mathbf{p}}}{\mathbf{p}}\right\|^2 = \sum_{i=1}^{2n}\left(\frac{\dot{p}_i}{p_i}\right)^2, \quad \left\|\frac{\dot{\omega}}{\omega}\right\|^2 = \sum_{i=1}^{2n}\left(\frac{\dot{\omega}_i}{\omega_i}\right)^2$$

from Lemma 13.9 and (13.25), it must have

$$
\begin{aligned}
&\left(\frac{n}{1-\theta}\right)^2 \\
\geq\ & \left\|\frac{\dot{\mathbf{p}}}{\mathbf{p}}\right\|^2 \left\|\frac{\dot{\omega}}{\omega}\right\|^2 = \left(\sum_{i=1}^{2n}\left(\frac{\dot{p}_i}{p_i}\right)^2\right)\left(\sum_{i=1}^{2n}\left(\frac{\dot{\omega}_i}{\omega_i}\right)^2\right)
\end{aligned}
$$

$$\geq \sum_{i=1}^{2n}\left(\frac{\dot{p}_i}{p_i}\frac{\dot{\omega}_i}{\omega_i}\right)^2 = \left\|\frac{\dot{\mathbf{p}}}{\mathbf{p}}\circ\frac{\dot{\omega}}{\omega}\right\|^2$$

$$\geq \sum_{i=1}^{2n}\left(\frac{\dot{p}_i\dot{\omega}_i}{(1+\theta)\mu}\right)^2 = \frac{1}{(1+\theta)^2\mu^2}\left\|\dot{\mathbf{p}}\circ\dot{\omega}\right\|^2$$

i.e.,

$$\left\|\dot{\mathbf{p}}\circ\dot{\omega}\right\|^2 \leq \left(\frac{1+\theta}{1-\theta}n\mu\right)^2$$

This proves (13.59). Using

$$\left\|\frac{\ddot{\mathbf{p}}}{\mathbf{p}}\right\|^2 = \sum_{i=1}^{2n}\left(\frac{\ddot{p}_i}{p_i}\right)^2, \quad \left\|\frac{\ddot{\omega}}{\omega}\right\|^2 = \sum_{i=1}^{2n}\left(\frac{\ddot{\omega}_i}{\omega_i}\right)^2$$

and Lemma 13.10, then following the same procedure, it is easy to verify (13.60). From (13.47) and (13.50), one obtains

$$\begin{aligned}
\left(\frac{2n}{(1-\theta)}\right)\left(\frac{4(1+\theta)n^2}{(1-\theta)^3}\right) &\geq \left(\left\|\frac{\dot{\mathbf{p}}}{\mathbf{p}}\right\|^2+\left\|\frac{\dot{\omega}}{\omega}\right\|^2\right)\left(\left\|\frac{\ddot{\mathbf{p}}}{\mathbf{p}}\right\|^2+\left\|\frac{\ddot{\omega}}{\omega}\right\|^2\right) \\
&\geq \left\|\frac{\ddot{\mathbf{p}}}{\mathbf{p}}\right\|^2\left\|\frac{\dot{\omega}}{\omega}\right\|^2+\left\|\frac{\dot{\mathbf{p}}}{\mathbf{p}}\right\|^2\left\|\frac{\ddot{\omega}}{\omega}\right\|^2 \\
&= \left(\sum_{i=1}^{2n}\left(\frac{\ddot{p}_i}{p_i}\right)^2\right)\left(\sum_{i=1}^{2n}\left(\frac{\dot{\omega}_i}{\omega_i}\right)^2\right)+\left(\sum_{i=1}^{2n}\left(\frac{\dot{p}_i}{p_i}\right)^2\right)\left(\sum_{i=1}^{2n}\left(\frac{\ddot{\omega}_i}{\omega_i}\right)^2\right) \\
&\geq \sum_{i=1}^{2n}\left(\frac{\ddot{p}_i\dot{\omega}_i}{p_i\omega_i}\right)^2+\sum_{i=1}^{2n}\left(\frac{\dot{p}_i\ddot{\omega}_i}{p_i\omega_i}\right)^2 \\
&\geq \sum_{i=1}^{2n}\left(\frac{\ddot{p}_i\dot{\omega}_i}{(1+\theta)\mu}\right)^2+\sum_{i=1}^{2n}\left(\frac{\dot{p}_i\ddot{\omega}_i}{(1+\theta)\mu}\right)^2 \\
&= \frac{1}{(1+\theta)^2\mu^2}\left(\left\|\ddot{\mathbf{p}}\circ\dot{\omega}\right\|^2+\left\|\dot{\mathbf{p}}\circ\ddot{\omega}\right\|^2\right)
\end{aligned} \tag{13.109}$$

i.e.,

$$\left\|\ddot{\mathbf{p}}\circ\dot{\omega}\right\|^2+\left\|\dot{\mathbf{p}}\circ\ddot{\omega}\right\|^2 \leq \frac{(2n)^3(1+\theta)^3}{(1-\theta)^4}\mu^2$$

This proves the lemma. ■

Proof of Lemma 13.15:
First notice that $q(\sin(\alpha))$ is a monotonic increasing function of $\sin(\alpha)$ for $\alpha \in [0, \frac{\pi}{2}]$ and $q(\sin(0)) < 0$, therefore, one only needs to show that $q(\frac{\theta}{\sqrt{n}}) < 0$ for $\theta \leq 0.22$. Using Lemma 13.6 yields

$$\left\|\dot{\mathbf{p}}\circ\ddot{\omega}+\dot{\omega}\circ\ddot{\mathbf{p}}-\frac{1}{2n}(\dot{\mathbf{p}}^{\mathrm{T}}\ddot{\omega}+\dot{\omega}^{\mathrm{T}}\ddot{\mathbf{p}})\mathbf{e}\right\| \leq \left\|\dot{\mathbf{p}}\circ\ddot{\omega}\right\|+\left\|\dot{\omega}\circ\ddot{\mathbf{p}}\right\|$$

$$\left\|\ddot{\mathbf{p}}\circ\ddot{\omega}-\dot{\omega}\circ\dot{\mathbf{p}}-\frac{1}{2n}(\ddot{\mathbf{p}}^{\mathrm{T}}\ddot{\omega}-\dot{\omega}^{\mathrm{T}}\dot{\mathbf{p}})\mathbf{e}\right\|\leq\left\|\ddot{\mathbf{p}}\circ\ddot{\omega}\right\|+\left\|\dot{\omega}\circ\dot{\mathbf{p}}\right\|$$

In view of Lemmas 13.14, 13.9, and 13.10, from (13.56), it must have, for $\alpha\in[0,\frac{\pi}{2}]$,

$$\begin{aligned}q(\sin(\alpha))\leq&\left(\left\|\ddot{\mathbf{p}}\circ\ddot{\omega}\right\|+\left\|\dot{\omega}\circ\dot{\mathbf{p}}\right\|+2\theta\frac{\dot{\mathbf{p}}^{\mathrm{T}}\dot{\omega}}{2n}\right)\sin^4(\alpha)\\&+\left(\left\|\dot{\mathbf{p}}\circ\ddot{\omega}\right\|+\left\|\dot{\omega}\circ\ddot{\mathbf{p}}\right\|\right)\sin^3(\alpha)\\&+2\theta\frac{\dot{\mathbf{p}}^{\mathrm{T}}\dot{\omega}}{2n}\sin^2(\alpha)+\theta\mu\sin(\alpha)-\theta\mu\\\leq&\mu\Bigg(\left(\frac{2(1+\theta)^2}{(1-\theta)^3}n^2+\frac{n(1+\theta)}{(1-\theta)}+\frac{\theta(1+\theta)}{(1-\theta)}\right)\sin^4(\alpha)\\&+4\sqrt{2}\frac{(1+\theta)^{\frac{3}{2}}}{(1-\theta)^2}n^{\frac{3}{2}}\sin^3(\alpha)\\&+\frac{\theta(1+\theta)}{(1-\theta)}\sin^2(\alpha)+\theta\sin(\alpha)-\theta\Bigg)\end{aligned}$$

Since $n\geq 1$ and $\theta>0$, substituting $\sin(\alpha)=\frac{\theta}{\sqrt{n}}$ gives

$$\begin{aligned}q\left(\frac{\theta}{\sqrt{n}}\right)\leq&\mu\Bigg(\left(\frac{2(1+\theta)^2}{(1-\theta)^3}n^2+\frac{n(1+\theta)}{(1-\theta)}+\frac{\theta(1+\theta)}{(1-\theta)}\right)\frac{\theta^4}{n^2}\\&+4\sqrt{2}\frac{(1+\theta)^{\frac{3}{2}}n^{\frac{3}{2}}}{(1-\theta)^2}\frac{\theta^3}{n^{\frac{3}{2}}}+\frac{\theta(1+\theta)}{(1-\theta)}\frac{\theta^2}{n}+\theta\frac{\theta}{\sqrt{n}}-\theta\Bigg)\\=&\theta\mu\Bigg(\frac{2\theta^3(1+\theta)^2}{(1-\theta)^3}+\frac{\theta^3(1+\theta)}{n(1-\theta)}+\frac{\theta^4(1+\theta)}{(1-\theta)n^2}\\&+\frac{4\sqrt{2}\theta^2(1+\theta)^{\frac{3}{2}}}{(1-\theta)^2}+\frac{\theta^2(1+\theta)}{n(1-\theta)}+\frac{\theta}{\sqrt{n}}-1\Bigg)\\\leq&\theta\mu\Bigg(\frac{2\theta^3(1+\theta)^2}{(1-\theta)^3}+\frac{\theta^3(1+\theta)}{(1-\theta)}+\frac{\theta^4(1+\theta)}{(1-\theta)}\\&+\frac{4\sqrt{2}\theta^2(1+\theta)^{\frac{3}{2}}}{(1-\theta)^2}+\frac{\theta^2(1+\theta)}{(1-\theta)}+\theta-1\Bigg):=\theta\mu p(\theta)\end{aligned}\tag{13.110}$$

Since $p(\theta)$ is monotonic increasing function of $\theta\in[0,1)$, $p(0)<0$, and it is easy to verify that $p(0.22)<0$, this proves the lemma. ■

Proof of Lemma 13.16:
Using Lemma 13.6 yields

$$0\leq\left\|\Delta\mathbf{p}\circ\Delta\omega-\frac{1}{2n}(\Delta\mathbf{p}^{\mathrm{T}}\Delta\omega)\mathbf{e}\right\|^2\leq\|\Delta\mathbf{p}\circ\Delta\omega\|^2\tag{13.111}$$

Pre-multiplying $\left(\mathbf{P}(\alpha)\Omega(\alpha)\right)^{-\frac{1}{2}}$ on the both sides of (13.66) yields

$$\mathbf{D}\Delta\omega+\mathbf{D}^{-1}\Delta\mathbf{p}=\left(\mathbf{P}(\alpha)\Omega(\alpha)\right)^{-\frac{1}{2}}\left(\mu(\alpha)\mathbf{e}-\mathbf{P}(\alpha)\Omega(\alpha)\mathbf{e}\right)$$

Let $\mathbf{u}=\mathbf{D}\Delta\omega$, $\mathbf{v}=\mathbf{D}^{-1}\Delta\mathbf{p}$, from (13.63), it must have

$$\mathbf{u}^{\mathrm{T}}\mathbf{v}=\Delta\mathbf{p}^{\mathrm{T}}\Delta\omega=\Delta\mathbf{y}^{\mathrm{T}}\Delta\lambda+\Delta\mathbf{z}^{\mathrm{T}}\Delta\gamma=\Delta\mathbf{x}^{\mathrm{T}}(\Delta\gamma-\Delta\lambda)=\Delta\mathbf{x}^{\mathrm{T}}\mathbf{H}\Delta\mathbf{x}\geq 0 \quad (13.112)$$

Using Lemma 13.4 and the assumption of $(\mathbf{x}(\alpha),\mathbf{p}(\alpha),\omega(\alpha))\in\mathcal{N}_2(2\theta)$ yields

$$\begin{aligned}\left\|\Delta\mathbf{p}\circ\Delta\omega\right\| &= \left\|\mathbf{u}\circ\mathbf{v}\right\|\leq 2^{-\frac{3}{2}}\left\|\left(\mathbf{P}(\alpha)\Omega(\alpha)\right)^{-\frac{1}{2}}\left(\mu(\alpha)\mathbf{e}-\mathbf{P}(\alpha)\Omega(\alpha)\mathbf{e}\right)\right\|^2\\ &= 2^{-\frac{3}{2}}\sum_{i=1}^{2n}\frac{(\mu(\alpha)-p_i(\alpha)\omega_i(\alpha))^2}{p_i(\alpha)\omega_i(\alpha)}\\ &\leq 2^{-\frac{3}{2}}\frac{\|\mu(\alpha)\mathbf{e}-\mathbf{p}(\alpha)\circ\omega(\alpha)\|^2}{\min_i p_i(\alpha)\omega_i(\alpha)}\\ &\leq 2^{-\frac{3}{2}}\frac{(2\theta)^2\mu(\alpha)^2}{(1-2\theta)\mu(\alpha)}=2^{\frac{1}{2}}\frac{\theta^2\mu(\alpha)}{(1-2\theta)} \end{aligned}\quad (13.113)$$

Define $(\mathbf{p}^{k+1}(t),\omega^{k+1}(t))=(\mathbf{p}(\alpha),\omega(\alpha))+t(\Delta\mathbf{p},\Delta\omega)$. From (13.66) and (13.40), one gets

$$\mathbf{p}(\alpha)^{\mathrm{T}}\Delta\omega+\omega(\alpha)^{\mathrm{T}}\Delta\mathbf{p}=2n\mu-\sum_{i=1}^{2n}p_i(\alpha)\omega_i(\alpha)=0 \quad (13.114)$$

Therefore,

$$\begin{aligned}\mu^{k+1}(t) &= \frac{\left(\mathbf{p}(\alpha)+t\Delta\mathbf{p}\right)^{\mathrm{T}}\left(\omega(\alpha)+t\Delta\omega\right)}{2n}\\ &= \frac{\mathbf{p}(\alpha)^{\mathrm{T}}\omega(\alpha)+t^2\Delta\mathbf{p}^{\mathrm{T}}\Delta\omega}{2n}=\mu(\alpha)+t^2\frac{\Delta\mathbf{p}^{\mathrm{T}}\Delta\omega}{2n}\end{aligned}\quad (13.115)$$

Since $\Delta\mathbf{p}^{\mathrm{T}}\Delta\omega=\Delta\mathbf{x}^{\mathrm{T}}\mathbf{H}\Delta\mathbf{x}\geq 0$, it must have $\mu^{k+1}(t)\geq\mu(\alpha)$. Using (13.115), (13.66), (13.111), and (13.113) yields

$$\begin{aligned}&\left\|\mathbf{p}^{k+1}(t)\circ\omega^{k+1}(t)-\mu^{k+1}(t)\mathbf{e}\right\|\\ =&\left\|(\mathbf{p}(\alpha)+t\Delta\mathbf{p})\circ(\omega(\alpha)+t\Delta\omega)-\mu(\alpha)\mathbf{e}-\frac{t^2}{2n}\left(\Delta\mathbf{p}^{\mathrm{T}}\Delta\omega\right)\mathbf{e}\right\|\\ =&\left\|\mathbf{p}(\alpha)\circ\omega(\alpha)+t[\omega(\alpha)\circ\Delta\mathbf{p}+\mathbf{p}(\alpha)\circ\Delta\omega]+t^2\Delta\mathbf{p}\circ\Delta\omega-\mu(\alpha)\mathbf{e}-\frac{t^2}{2n}\left(\Delta\mathbf{p}^{\mathrm{T}}\Delta\omega\right)\mathbf{e}\right\|\\ =&\left\|\mathbf{p}(\alpha)\circ\omega(\alpha)+t[\mu(\alpha)\mathbf{e}-\mathbf{p}(\alpha)\circ\omega(\alpha)]+t^2\Delta\mathbf{p}\circ\Delta\omega-\mu(\alpha)\mathbf{e}-\frac{t^2}{2n}\left(\Delta\mathbf{p}^{\mathrm{T}}\Delta\omega\right)\mathbf{e}\right\|\end{aligned}$$

$$
\begin{aligned}
&= \left\| (1-t)\left[\mathbf{p}(\alpha)\circ\omega(\alpha)-\mu(\alpha)\mathbf{e}\right]+t^2\left(\Delta\mathbf{p}\circ\Delta\omega-\frac{1}{2n}\left(\Delta\mathbf{p}^{\mathrm{T}}\Delta\omega\right)\mathbf{e}\right)\right\| \\
&\leq (1-t)(2\theta)\mu(\alpha)+t^2\frac{2^{\frac{1}{2}}\theta^2}{(1-2\theta)}\mu(\alpha) \\
&\leq \left((1-t)(2\theta)+t^2\frac{2^{\frac{1}{2}}\theta^2}{(1-2\theta)}\right)\mu^{k+1} := f(t,\theta)\mu^{k+1} \qquad (13.116)
\end{aligned}
$$

Therefore, taking $t=1$ gives $\left\|\mathbf{p}^{k+1}\circ\omega^{k+1}-\mu^{k+1}\mathbf{e}\right\| \leq \frac{2^{\frac{1}{2}}\theta^2}{(1-2\theta)}\mu^{k+1}$. It is easy to see that, for $\theta \leq 0.29$,

$$\frac{2^{\frac{1}{2}}\theta^2}{(1-2\theta)} = 0.2832 < \theta$$

For $\theta \leq 0.29$ and $t \in [0,1]$, noticing

$$0 \leq f(t,\theta) \leq f(t,0.29) \leq 0.58(1-t)+0.2832t^2 < 1$$

and using Corollary 13.1, one gets, for an additional condition $\sin(\alpha) \leq \sin^{-1}(\bar{\alpha})$;

$$
\begin{aligned}
p_i^{k+1}(t)\omega_i^{k+1}(t) &\geq (1-f(t,\theta))\,\mu^{k+1}(t) \\
&= (1-f(t,\theta))\left(\mu(\alpha)+\frac{t^2}{n}\Delta p^{\mathbf{T}}\Delta\omega\right) \\
&\geq (1-f(t,\theta))\,\mu(\alpha) \\
&> 0 \qquad (13.117)
\end{aligned}
$$

Therefore, $(\mathbf{p}^{k+1}(t),\omega^{k+1}(t)) > 0$ for $t \in [0,1]$, i.e., $(\mathbf{p}^{k+1},\omega^{k+1}) > 0$. This finishes the proof. ∎

Proof of Lemma 13.17:
The first inequality of (13.67) follows from (13.112). Pre-multiplying both sides of (13.66) by $\mathbf{P}^{-\frac{1}{2}}(\alpha)\Omega^{-\frac{1}{2}}(\alpha)$ gives

$$\mathbf{P}^{-\frac{1}{2}}(\alpha)\Omega^{\frac{1}{2}}(\alpha)\Delta\mathbf{p}+\mathbf{P}^{\frac{1}{2}}(\alpha)\Omega^{-\frac{1}{2}}(\alpha)\Delta\omega=\mathbf{P}^{-\frac{1}{2}}(\alpha)\Omega^{-\frac{1}{2}}(\alpha)\left(\mu(\alpha)\mathbf{e}-\mathbf{P}(\alpha)\Omega(\alpha)\mathbf{e}\right)$$

Let

$$
\begin{aligned}
\mathbf{u} &= \mathbf{P}^{-\frac{1}{2}}(\alpha)\Omega^{\frac{1}{2}}(\alpha)\Delta\mathbf{p} \\
\mathbf{v} &= \mathbf{P}^{\frac{1}{2}}(\alpha)\Omega^{-\frac{1}{2}}(\alpha)\Delta\omega
\end{aligned}
$$

and

$$\mathbf{w} = \mathbf{P}^{-\frac{1}{2}}(\alpha)\Omega^{-\frac{1}{2}}(\alpha)\left(\mu(\alpha)\mathbf{e}-\mathbf{P}(\alpha)\Omega(\alpha)\mathbf{e}\right)$$

in view of (13.112), it must have

$$\mathbf{u}^{\mathrm{T}}\mathbf{v} = \Delta\mathbf{p}^{\mathrm{T}}\Delta\omega \geq 0$$

Using Lemma 13.3 and the assumption of $(\mathbf{x}(\alpha), \mathbf{p}(\alpha), \omega(\alpha)) \in \mathcal{N}_2(2\theta)$ yields

$$\begin{aligned}
\|\mathbf{u}\|^2 + \|\mathbf{v}\|^2 = & \sum_{i=1}^{2n} \left(\frac{(\Delta p_i)^2 \omega_i(\alpha)}{p_i(\alpha)} + \frac{(\Delta \omega_i)^2 p_i(\alpha)}{\omega_i(\alpha)} \right) \\
\leq & \|\mathbf{w}\|^2 = \sum_{i=1}^{2n} \frac{(\mu(\alpha) - p_i(\alpha)\omega_i(\alpha))^2}{p_i(\alpha)\omega_i(\alpha)} \\
\leq & \frac{\sum_{i=1}^{2n} (\mu(\alpha) - p_i(\alpha)\omega_i(\alpha))^2}{\min_i p_i(\alpha)\omega_i(\alpha)} \\
\leq & \frac{(2\theta)^2 \mu^2(\alpha)}{(1-2\theta)\mu(\alpha)} = \frac{(2\theta)^2 \mu(\alpha)}{(1-2\theta)}
\end{aligned} \tag{13.118}$$

Dividing both sides by $\mu(\alpha)$ and using $p_i(\alpha)\omega_i(\alpha) \geq \mu(\alpha)(1-2\theta)$ yields

$$\begin{aligned}
& \sum_{i=1}^{2n} (1-2\theta) \left(\frac{(\Delta p_i)^2}{p_i^2(\alpha)} + \frac{(\Delta \omega_i)^2}{\omega_i^2(\alpha)} \right) \\
= & (1-2\theta) \left(\left\| \frac{\Delta \mathbf{p}}{\mathbf{p}(\alpha)} \right\|^2 + \left\| \frac{\Delta \omega}{\omega(\alpha)} \right\|^2 \right) \\
\leq & \frac{(2\theta)^2}{(1-2\theta)}
\end{aligned} \tag{13.119}$$

i.e.,

$$\left\| \frac{\Delta \mathbf{p}}{\mathbf{p}(\alpha)} \right\|^2 + \left\| \frac{\Delta \omega}{\omega(\alpha)} \right\|^2 \leq \left(\frac{2\theta}{1-2\theta} \right)^2 \tag{13.120}$$

Invoking Lemma 13.1, one gets

$$\left\| \frac{\Delta \mathbf{p}}{\mathbf{p}(\alpha)} \right\|^2 \cdot \left\| \frac{\Delta \omega}{\omega(\alpha)} \right\|^2 \leq \frac{1}{4} \left(\frac{2\theta}{1-2\theta} \right)^4 \tag{13.121}$$

This gives

$$\left\| \frac{\Delta \mathbf{p}}{\mathbf{p}(\alpha)} \right\| \cdot \left\| \frac{\Delta \omega}{\omega(\alpha)} \right\| \leq \frac{2\theta^2}{(1-2\theta)^2} \tag{13.122}$$

Using Cauchy-Schwarz inequality leads to

$$\frac{(\Delta \mathbf{p})^{\mathrm{T}}(\Delta \omega)}{\mu(\alpha)}$$

$$\leq \sum_{i=1}^{2n} \frac{|\Delta p_i||\Delta \omega_i|}{\mu(\alpha)}$$

$$\leq (1+2\theta)\sum_{i=1}^{2n} \frac{|\Delta p_i|}{p_i(\alpha)} \frac{|\Delta \omega_i|}{\omega_i(\alpha)}$$

$$=(1+2\theta)\left|\frac{\Delta \mathbf{p}}{\mathbf{p}(\alpha)}\right|^{\mathrm{T}}\left|\frac{\Delta \omega}{\omega(\alpha)}\right|$$

$$\leq (1+2\theta)\left\|\frac{\Delta \mathbf{p}}{\mathbf{p}(\alpha)}\right\| \cdot \left\|\frac{\Delta \omega}{\omega(\alpha)}\right\|$$

$$\leq \frac{2\theta^2(1+2\theta)}{(1-2\theta)^2} \tag{13.123}$$

Therefore,

$$\frac{(\Delta \mathbf{p})^{\mathrm{T}}(\Delta \omega)}{2n} \leq \frac{\theta^2(1+2\theta)}{n(1-2\theta)^2}\mu(\alpha) \tag{13.124}$$

This proves the lemma. ■

Proof of Lemma 13.19:
Using Lemmas 13.18, 13.11, 13.2, 13.8, 13.9, and 13.10, and noticing $\ddot{\mathbf{p}}^{\mathrm{T}}\ddot{\omega} \geq 0$ and $\dot{\mathbf{p}}^{\mathrm{T}}\dot{\omega} \geq 0$ yields

$$\mu^{k+1} \leq \mu(\alpha)\left(1+\frac{\theta^2(1+2\theta)}{n(1-2\theta)^2}\right) = \mu(\alpha)\left(1+\frac{\delta_0}{n}\right) \tag{13.125a}$$

$$=\mu^k\left[1-\sin(\alpha)+\left(\frac{\ddot{\mathbf{p}}^{\mathrm{T}}\ddot{\omega}}{2n\mu}-\frac{\dot{\mathbf{p}}^{\mathrm{T}}\dot{\omega}}{2n\mu}\right)(1-\cos(\alpha))^2\right.$$
$$\left.-\left(\frac{\dot{\mathbf{p}}^{\mathrm{T}}\ddot{\omega}}{2n\mu}+\frac{\dot{\omega}^{\mathrm{T}}\ddot{\mathbf{p}}}{2n\mu}\right)\sin(\alpha)(1-\cos(\alpha))\right]\left(1+\frac{\delta_0}{n}\right)$$

$$\leq \mu^k\left(1-\sin(\alpha)+\frac{\ddot{\mathbf{p}}^{\mathrm{T}}\ddot{\omega}}{2n\mu}\sin^4(\alpha)+\left(\left|\frac{\dot{\mathbf{p}}^{\mathrm{T}}\ddot{\omega}}{2n\mu}\right|+\left|\frac{\dot{\omega}^{\mathrm{T}}\ddot{\mathbf{p}}}{2n\mu}\right|\right)\sin^3(\alpha)\right)\left(1+\frac{\delta_0}{n}\right)$$

$$\leq \mu^k\left(1-\sin(\alpha)+\frac{n(1+\theta)^2}{(1-\theta)^3}\sin^4(\alpha)+\frac{2(2n)^{\frac{1}{2}}(1+\theta)^{\frac{3}{2}}}{(1-\theta)^2}\sin^3(\alpha)\right)\left(1+\frac{\delta_0}{n}\right) \tag{13.125b}$$

Substituting $\sin(\alpha)=\frac{\theta}{\sqrt{n}}$ into (13.125b) gives

$$\mu^{k+1} \leq \mu^k\left(1-\frac{\theta}{\sqrt{n}}+\frac{n(1+\theta)^2}{(1-\theta)^3}\frac{\theta^4}{n^2}+\frac{2(2n)^{\frac{1}{2}}(1+\theta)^{\frac{3}{2}}}{(1-\theta)^2}\frac{\theta^3}{n^{\frac{3}{2}}}\right)\left(1+\frac{\delta_0}{n}\right)$$

$$=\mu^k\left(1-\frac{\theta}{\sqrt{n}}+\frac{\theta^4(1+\theta)^2}{n(1-\theta)^3}+\frac{2^{\frac{3}{2}}\theta^3(1+\theta)^{\frac{3}{2}}}{n(1-\theta)^2}\right)\left(1+\frac{\delta_0}{n}\right)$$

$$=\mu^k\left(1-\frac{\theta}{\sqrt{n}}+\frac{\delta_0}{n}+\frac{\theta^4(1+\theta)^2}{n(1-\theta)^3}+\frac{2^{\frac{3}{2}}\theta^3(1+\theta)^{\frac{3}{2}}}{n(1-\theta)^2}-\frac{\theta\delta_0}{n^{\frac{3}{2}}}\right.$$
$$\left.+\frac{\delta_0}{n}\left[\frac{\theta^4(1+\theta)^2}{n(1-\theta)^3}+\frac{2^{\frac{3}{2}}\theta^3(1+\theta)^{\frac{3}{2}}}{n(1-\theta)^2}\right]\right)$$
$$=\mu^k\left(1-\frac{\theta}{\sqrt{n}}\left[1-\frac{\delta_0}{\sqrt{n}\theta}-\frac{\theta^3(1+\theta)^2}{\sqrt{n}(1-\theta)^3}-\frac{2^{\frac{3}{2}}\theta^2(1+\theta)^{\frac{3}{2}}}{\sqrt{n}(1-\theta)^2}\right]\right.$$
$$\left.-\frac{\theta\delta_0}{n^{\frac{3}{2}}}\left[1-\frac{\theta^3(1+\theta)^2}{\sqrt{n}(1-\theta)^3}-\frac{2^{\frac{3}{2}}\theta^2(1+\theta)^{\frac{3}{2}}}{\sqrt{n}(1-\theta)^2}\right]\right)$$

Since

$$1-\frac{\theta^3(1+\theta)^2}{\sqrt{n}(1-\theta)^3}-\frac{2^{\frac{3}{2}}\theta^2(1+\theta)^{\frac{3}{2}}}{\sqrt{n}(1-\theta)^2}$$
$$\geq 1-\frac{\theta^3(1+\theta)^2}{(1-\theta)^3}-\frac{2^{\frac{3}{2}}\theta^2(1+\theta)^{\frac{3}{2}}}{(1-\theta)^2}:=f(\theta)$$

where $f(\theta)$ is a monotonic decreasing function of θ, and for $\theta \leq 0.37$, $f(\theta) > 0$. Therefore, for $\theta \leq 0.37$, the following relation holds.

$$\mu^{k+1}\leq\mu^k\left(1-\frac{\theta}{\sqrt{n}}\left[1-\frac{\delta_0}{\sqrt{n}\theta}-\frac{\theta^3(1+\theta)^2}{\sqrt{n}(1-\theta)^3}-\frac{2^{\frac{3}{2}}\theta^2(1+\theta)^{\frac{3}{2}}}{\sqrt{n}(1-\theta)^2}\right]\right)$$
$$=\mu^k\left(1-\frac{\theta}{\sqrt{n}}\left[1-\frac{\theta(1+2\theta)}{\sqrt{n}(1-2\theta)^2}-\frac{\theta^3(1+\theta)^2}{\sqrt{n}(1-\theta)^3}-\frac{2^{\frac{3}{2}}\theta^2(1+\theta)^{\frac{3}{2}}}{\sqrt{n}(1-\theta)^2}\right]\right) \tag{13.126}$$

Since

$$1-\frac{\theta(1+2\theta)}{\sqrt{n}(1-2\theta)^2}-\frac{\theta^3(1+\theta)^2}{\sqrt{n}(1-\theta)^3}-\frac{2^{\frac{3}{2}}\theta^2(1+\theta)^{\frac{3}{2}}}{\sqrt{n}(1-\theta)^2}$$
$$\geq 1-\frac{\theta(1+2\theta)}{(1-2\theta)^2}-\frac{\theta^3(1+\theta)^2}{(1-\theta)^3}-\frac{2^{\frac{3}{2}}\theta^2(1+\theta)^{\frac{3}{2}}}{(1-\theta)^2}:=g(\theta) \tag{13.127}$$

where $g(\theta)$ is a monotonic decreasing function of θ, one can conclude, for $\theta \leq 0.19$, $g(\theta) > 0.0976 > 0$. For $\theta = 0.19$, it must have $\theta g(\theta) > 0.0185$ and

$$\mu^{k+1}\leq\mu^k\left(1-\frac{0.0185}{\sqrt{n}}\right)$$

This proves (13.70). ∎

Chapter 14

Spacecraft Control Using CMG

Control Moment Gyros (CMGs) are important type of actuators used in spacecraft control because of their well-known torque amplification property [100]. The conventional use of CMG keeps the flywheel spinning in a constant speed, while torques of the CMG are produced by changing the gimbal's rotational speed [83]. A more complicated operational concept is the so-called variable-speed control moment gyros (VSCMG) in which the flywheel's speed of the CMG is allowed to be changed too. This idea was first proposed by Ford in his Ph.D dissertation [50] where he derived a mathematical model for VSCMGs which is now widely used in literatures. Because of the extra freedom of VSCMG, it can generate torques on a plane perpendicular to the gimbal axis while the conventional CMG can only generate a torque in a single direction at any instant of time [263].

The existing designs of spacecraft control system using CMG or VSCMG rely on the calculation of the desired torques and then determines the VSCMG's gimbal speed and flywheel speed. These designs have a fundamental problem because there are *singular points* where the gimbals speed and flywheel speed cannot be found given the desired torques. Extensive literatures focus on this difficulty of implementation in the last few decades, for example, Oh and Vadali [143] proposed singularity-robust steering law which avoids failure but produces an errant torque; Junkins and Kim [84] enhanced the pseudo-inverse technical using singular value decomposition (SVD); Ford and Hall [52] extended SVD analysis to singular direction avoidance; Zhang et al. [269] formulated the singularity avoidance problem as a nonlinear optimization problem. Gui et al. [62]

adopted a modified direct-inverse steering law. There are good survey papers [101, 223] that include extensive references.

Another difficulty associated with the control system design using CMG or VSCMG is that the nonlinear dynamical models for these type of actuators are much more complicated than other types of actuators used for spacecraft attitude control systems. Most proposed designs, for example [51, 52, 62, 80, 83, 120, 175, 261, 263], use Lyapunov stability theory for nonlinear systems. There are two shortcomings of this design method: first, there is no systematic way to find the desired Lyapunov function, and second, the design does not consider the system performance but only stability.

In this chapter, we propose a different operational concept for VSCMG: the flywheels of the cluster of the VSCMG does not always spin at high speed, they spin at high speed only when they need to. The same is true for the gimbals. This operational strategy makes the origin (the state variables at zero) an equilibrium point of the nonlinear system which can be regarded as an equivalent linear time-varying (LTV) system. Therefore, some mature linear system design methods can be used and system performance can be part of the design by using these linear system design methods. Additional advantages of the proposed operational concept are: (a) energy saving due to normally reduced spin speed of flywheels and gimbals, (b) singularity-free because the control of the spacecraft is always achievable by accelerating or decelerating the flywheels and gimbals, therefore, there is no inverse from desired torques to the speeds of the gimbals and flywheels.

It is worthwhile to point out that the nonlinear model can be viewed as a linear time-varying (LTV) system. The design methods for linear time-invariant (LTI) systems may be repeatedly applied to LTV systems. A popular design method for LTV system is the so-called gain scheduling design method, which has been discussed in several decades, for example, [105, 168, 170, 177]. The basic idea is to fix the time-varying model in a number of 'frozen' models and using linear system design method for each of these 'frozen' linear time-invariant systems. When the parameters of the LTV system are not in these 'frozen' points, interpolation is used to calculate the feedback gain matrix.

Although, gain scheduling design has been proved to be effective for many applications of LTV systems, it has an intrinsic limitation for some time-varying systems that have many independent time-varying variables, which is the case for spacecraft control using VSCMGs. As we will see that if a control system model has many independent time-varying parameters, then, the computation for the gain scheduling design will be too expensive to be feasible. Therefore, another popular control system design method will be considered, the so-called model predictive control (MPC) [7]. To meet some required stability conditions imposed on the LTV system [169], it is proposed to use the robust pole assignment design [201, 258] for the MPC design and establish the condition of uniformly exponential stability. The content of this chapter is based on [255].

14.1 Spacecraft Model Using Variable-speed CMG

Assuming that there are N variable-speed CMGs installed in a spacecraft, following the notations of [50], we define a matrix $\mathbf{A}_s = [\mathbf{s}_1, \mathbf{s}_2, \ldots, \mathbf{s}_N]$ such that the columns of $\mathbf{A}_s$, $\mathbf{s}_j$ ($j = 1, \ldots, N$), specify the unit spin axes of the flywheels in the spacecraft body frame. Similarly, we define $\mathbf{A}_g = [\mathbf{g}_1, \mathbf{g}_2, \ldots, \mathbf{g}_N]$ the matrix whose columns are the unit gimbal axes and $\mathbf{A}_t = [\mathbf{t}_1, \mathbf{t}_2, \ldots, \mathbf{t}_N]$ the matrix whose columns are the unit axes of the transverse (torque) directions, both are represented in the spacecraft body frame. Whereas $\mathbf{A}_g$ is a constant matrix, the matrices $\mathbf{A}_s$ and $\mathbf{A}_t$ depend on the gimbal angles. Let $\gamma = [\gamma_1, \ldots, \gamma_N]^{\mathrm{T}} \in [0, 2\pi] \times \cdots \times [0, 2\pi] := \Pi$ be the vector of N gimbal angles, and

$$[\dot{\gamma}_1, \ldots, \dot{\gamma}_N]^{\mathrm{T}} = \dot{\gamma} := \omega_g = [\omega_{g_1}, \ldots, \omega_{g_N}]^{\mathrm{T}} \tag{14.1}$$

be the vector of N gimbal speed, then the following relations hold [262] (see Figure 14.1).

$$\dot{\mathbf{s}}_i = \dot{\gamma}_i \mathbf{t}_i = \omega_{g_i} \mathbf{t}_i, \quad \dot{\mathbf{t}}_i = -\dot{\gamma}_i \mathbf{s}_i = -\omega_{g_i} \mathbf{s}_i, \quad \dot{\mathbf{g}}_i = 0 \tag{14.2}$$

Denote

$$\Gamma^c = \mathrm{diag}(\cos(\gamma)), \quad \Gamma^s = \mathrm{diag}(\sin(\gamma)) \tag{14.3}$$

A different but related expression is given in [50][1] Let $\mathbf{A}_{s_0}$ and $\mathbf{A}_{t_0}$ be initial spin axes and gimbal axes matrices at $\gamma_0 = \mathbf{0}$, then

$$\mathbf{A}_s(\gamma) = \mathbf{A}_{s_0} \Gamma^c + \mathbf{A}_{t_0} \Gamma^s \tag{14.4a}$$
$$\mathbf{A}_t(\gamma) = \mathbf{A}_{t_0} \Gamma^c - \mathbf{A}_{s_0} \Gamma^s \tag{14.4b}$$

This gives

$$\dot{\mathbf{A}}_s = \mathbf{A}_t \mathrm{diag}(\dot{\gamma}) = \mathbf{A}_t \mathrm{diag}(\omega_g) \tag{14.5a}$$
$$\dot{\mathbf{A}}_t = -\mathbf{A}_s \mathrm{diag}(\dot{\gamma}) = -\mathbf{A}_s \mathrm{diag}(\omega_g) \tag{14.5b}$$

which are identical to the formulas of (14.2). Let J_{s_j}, J_{g_j}, and J_{t_j} be the wheel spin axis inertia, the gimbal axis inertia, and the transverse axis inertia of the j-th CMG, let three $N \times N$ matrices be defined as

$$\mathbf{J}_s = \mathrm{diag}(J_{s_j}), \quad \mathbf{J}_g = \mathrm{diag}(J_{g_j}), \quad \mathbf{J}_t = \mathrm{diag}(J_{t_j}) \tag{14.6}$$

Let $\omega = [\omega_1, \omega_2, \omega_3]^{\mathrm{T}}$ be the spacecraft body angular rate with respect to the inertial frame, $\beta = [\beta_1, \ldots, \beta_N]^{\mathrm{T}}$ be the vector of N flywheel angles, and

$$[\dot{\beta}_1, \ldots, \dot{\beta}_N]^{\mathrm{T}} = \dot{\beta} := \omega_s = [\omega_{s_1}, \ldots, \omega_{s_N}]^{\mathrm{T}} \tag{14.7}$$

be the vector of N flywheel speeds. Denote

$$\mathbf{h}_s = [J_{s_1} \dot{\beta}_1, \ldots, J_{s_N} \dot{\beta}_N]^{\mathrm{T}} = \mathbf{J}_s \omega_s \tag{14.8}$$

[1] There are some typos in the signs in [50] which are corrected in (14.4) and (14.5).

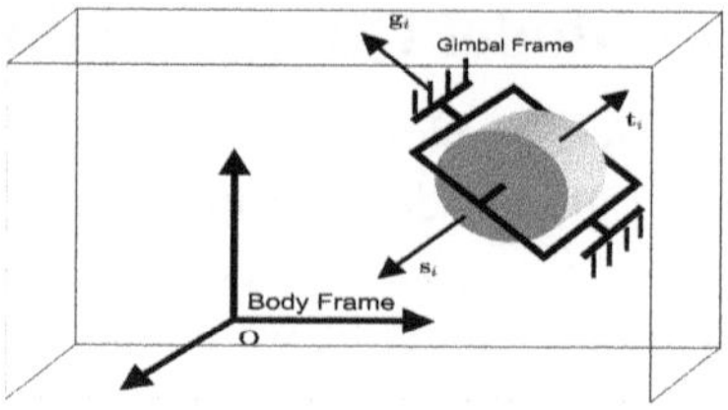

Figure 14.1: Spacecraft body with a single VSCMG.

$$\mathbf{h}_g = [J_{g_1}\dot{\gamma}_1, \dots, J_{g_N}\dot{\gamma}_N]^{\mathrm{T}} = \mathbf{J}_g\omega_g \tag{14.9}$$

and $\mathbf{h}_t$ be the N-dimensional vectors representing the components of an absolute angular momentum of the VSCMGs about their spin axes, gimbal axes, and transverse axes, respectively. Note that the angular momentum generated by the ith flywheel represented in the body frame is given by $\mathbf{s}_i J_{s_i}\dot{\beta}_i$ and the angular momentum generated by the ith gimbal represented in the body frame is given by $\mathbf{g}_i J_{g_i}\dot{\gamma}_i$, the total angular momentum of the spacecraft with a cluster of VSCMGs represented in the body frame is given as

$$\begin{aligned} \mathbf{h} &= \mathbf{J}_b\omega + \sum_{i=1}^{N}\mathbf{s}_i J_{s_i}\dot{\beta}_i + \sum_{i=1}^{N}\mathbf{g}_i J_{g_i}\dot{\gamma}_i = \mathbf{J}_b\omega + \mathbf{A}_s\mathbf{h}_s + \mathbf{A}_g\mathbf{h}_g \\ &= \mathbf{J}_b\omega + \mathbf{A}_s\mathbf{J}_s\omega_s + \mathbf{A}_g\mathbf{J}_g\omega_g \end{aligned} \tag{14.10}$$

Taking derivative of (14.10) and using (14.2) and $\dot{\mathbf{J}} = 0$, noticing that gimbal axes are fixed, we have

$$\begin{aligned} \dot{\mathbf{h}} &= \mathbf{J}_b\dot{\omega} + \sum_{i=1}^{N}\left(\dot{\mathbf{s}}_i J_{s_i}\dot{\beta}_i + \mathbf{s}_i J_{s_i}\ddot{\beta}_i\right) + \sum_{i=1}^{N}(\dot{\mathbf{g}}_i J_{g_i}\dot{\gamma}_i + \mathbf{g}_i J_{g_i}\ddot{\gamma}_i) \\ &= \mathbf{J}_b\dot{\omega} + \sum_{i=1}^{N}\left(\dot{\gamma}_i\mathbf{t}_i J_{s_i}\dot{\beta}_i + \mathbf{s}_i J_{s_i}\ddot{\beta}_i\right) + \sum_{i=1}^{N}\mathbf{g}_i J_{g_i}\ddot{\gamma}_i \\ &= -\omega\times\mathbf{h} + \mathbf{t}_e \end{aligned} \tag{14.11}$$

where $\mathbf{t}_e$ is the external torque. Denote $\Omega_s = \mathrm{diag}(\omega_s)$ and $\Omega_g = \mathrm{diag}(\omega_g)$. This equation can be written as a compact form as follows.

$$\begin{aligned} & \mathbf{J}_b\dot{\omega} + \mathbf{A}_t\mathbf{J}_s\Omega_s\omega_g + \mathbf{A}_s\mathbf{J}_s\dot{\omega}_s + \mathbf{A}_g\mathbf{J}_g\dot{\omega}_g \\ = \; & -\omega\times(\mathbf{J}_b\omega + \mathbf{A}_s\mathbf{J}_s\omega_s + \mathbf{A}_g\mathbf{J}_g\omega_g) + \mathbf{t}_e \end{aligned} \tag{14.12}$$

Note that the torques generated by wheel acceleration or deceleration in the directions defined by $\mathbf{A}_s$ are given by

$$\mathbf{t}_s = -\mathbf{J}_s\dot{\omega}_s = [t_{s_1}, \dots, t_{s_N}]^{\mathrm{T}} \tag{14.13}$$

(note that vectors $\mathbf{t}_i$ in $\mathbf{A}_t$ are axes and scalars t_{s_i} in $\mathbf{t}_s$ are torques) and the torques generated by gimbals' acceleration or deceleration in the directions defined by $\mathbf{A}_g$ are given by

$$\mathbf{t}_g = -\mathbf{J}_g \dot{\omega}_g = [t_{g1}, \ldots, t_{gN}]^{\mathrm{T}} \tag{14.14}$$

the dynamical equation can be expressed as

$$\mathbf{J}_b\dot{\omega} + \mathbf{A}_t\mathbf{J}_s\Omega_s\omega_g + \omega \times (\mathbf{J}_b\omega + \mathbf{A}_s\mathbf{J}_s\omega_s + \mathbf{A}_g\mathbf{J}_g\omega_g) = \mathbf{A}_s\mathbf{t}_s + \mathbf{A}_g\mathbf{t}_g + \mathbf{t}_e \tag{14.15}$$

Let

$$\bar{\mathbf{q}} = [q_0, q_1, q_2, q_3]^{\mathrm{T}} = [q_0, \mathbf{q}^{\mathrm{T}}]^{\mathrm{T}} = \left[\cos(\frac{\alpha}{2}), \hat{\mathbf{e}}^{\mathrm{T}}\sin(\frac{\alpha}{2})\right]^{\mathrm{T}} \tag{14.16}$$

be the quaternion representing the rotation of the body frame relative to the inertial frame, where $\hat{\mathbf{e}}$ is the unit length rotational axis and α is the rotation angle about $\hat{\mathbf{e}}$. Therefore, in view of (4.9), the reduced kinematics equation becomes

$$\begin{aligned}
\begin{bmatrix} \dot{q}_1 \\ \dot{q}_2 \\ \dot{q}_3 \end{bmatrix} &= \frac{1}{2}\begin{bmatrix} f & -q_3 & q_2 \\ q_3 & f & -q_1 \\ -q_2 & q_1 & f \end{bmatrix}\begin{bmatrix} \omega_1 \\ \omega_2 \\ \omega_3 \end{bmatrix} \\
&= \mathbf{g}(q_1, q_2, q_3, \omega) = \frac{1}{2}\left(\sqrt{1-q_1^2-q_2^2-q_3^2}\mathbf{I}_3 + \mathbf{q}^{\times}\right)\omega
\end{aligned}$$

where $f = \sqrt{1-q_1^2-q_2^2-q_3^2}$, or simply

$$\dot{\mathbf{q}} = \mathbf{g}(\mathbf{q}, \omega) \tag{14.17}$$

The nonlinear time-varying spacecraft control system model can be written as follows:

$$\begin{aligned}
\begin{bmatrix} \dot{\omega}_g \\ \dot{\omega}_s \\ \dot{\omega} \\ \dot{\mathbf{q}} \end{bmatrix} &= \begin{bmatrix} \mathbf{0} \\ \mathbf{0} \\ -\mathbf{J}_b^{-1}\left[\mathbf{A}_t\mathbf{J}_s\Omega_s\omega_g + \omega \times (\mathbf{J}_b\omega + \mathbf{A}_s\mathbf{J}_s\omega_s + \mathbf{A}_g\mathbf{J}_g\omega_g)\right] \\ \mathbf{g}(\mathbf{q}, \omega) \end{bmatrix} \\
&\quad + \begin{bmatrix} -\mathbf{J}_g^{-1}\mathbf{t}_g \\ -\mathbf{J}_s^{-1}\mathbf{t}_s \\ \mathbf{J}_b^{-1}(\mathbf{A}_s\mathbf{t}_s + \mathbf{A}_g\mathbf{t}_g + \mathbf{t}_e) \\ \mathbf{0} \end{bmatrix} \\
&= \mathbf{F}(\omega, \omega_g, \omega_s, \mathbf{q}, t) + \mathbf{G}(\mathbf{t}_s, \mathbf{t}_g, \mathbf{t}_e, t)
\end{aligned} \tag{14.18}$$

or simply

$$\dot{\mathbf{x}} = \mathbf{F}(\mathbf{x}, \gamma(t)) + \mathbf{G}(\mathbf{u}, \mathbf{t}_e, \gamma(t)) \tag{14.19}$$

where the state variable vector is $\mathbf{x} = [\omega_g^{\mathrm{T}}, \omega_s^{\mathrm{T}}, \omega^{\mathrm{T}}, \mathbf{q}^{\mathrm{T}}]^{\mathrm{T}}$, the control variable vector is $\mathbf{u} = [\mathbf{t}_g^{\mathrm{T}}, \mathbf{t}_s^{\mathrm{T}}]^{\mathrm{T}}$, disturbance torque vector is $\mathbf{t}_e$, and $\mathbf{F}$ and $\mathbf{G}$ are functions of time t because the parameters of ω, ω_s, ω_g, $\mathbf{q}$, $\mathbf{A}_s$ and $\mathbf{A}_t$ are functions of time t.

The system dimension is $n = 2N + 6$. The control input dimension is $2N$. Clearly, an equilibrium of (14.18) is $\mathbf{x}_e = \mathbf{0} = [\omega^{\mathrm{T}}, \omega_s^{\mathrm{T}}, \omega_g^{\mathrm{T}}, \mathbf{q}^{\mathrm{T}}]^{\mathrm{T}}$. Notice that

$$\mathbf{A}_t\mathbf{J}_s\Omega_s\omega_g = \frac{1}{2}(\mathbf{A}_t\mathbf{J}_s\Omega_s\omega_g + \mathbf{A}_t\mathbf{J}_s\Omega_g\omega_s) \tag{14.20}$$

and

$$\begin{aligned} & \omega \times (\mathbf{J}_b\omega + \mathbf{A}_s\mathbf{J}_s\omega_s + \mathbf{A}_g\mathbf{J}_g\omega_g) \\ = \; & (\omega \times \mathbf{J}_b)\omega + \frac{1}{2}[(\omega \times \mathbf{A}_s\mathbf{J}_s)\omega_s + (\omega \times \mathbf{A}_g\mathbf{J}_g)\omega_g \\ & -(\mathbf{A}_s\mathbf{J}_s\omega_s + \mathbf{A}_g\mathbf{J}_g\omega_g)^{\times}\omega] \end{aligned} \tag{14.21}$$

Let

$$\mathbf{F}_{31} = -\frac{1}{2}\mathbf{J}_b^{-1}[\mathbf{A}_t\mathbf{J}_s\Omega_s + \omega \times \mathbf{A}_g\mathbf{J}_g] \tag{14.22}$$

$$\mathbf{F}_{32} = -\frac{1}{2}\mathbf{J}_b^{-1}[\mathbf{A}_t\mathbf{J}_s\Omega_g + \omega \times \mathbf{A}_s\mathbf{J}_s] \tag{14.23}$$

$$\mathbf{F}_{33} = \mathbf{J}_b^{-1}\left[(\mathbf{J}_b\omega)^{\times} + \frac{1}{2}(\mathbf{A}_s\mathbf{J}_s\omega_s + \mathbf{A}_g\mathbf{J}_g\omega_g)^{\times}\right] \tag{14.24}$$

and

$$\mathbf{F}_{43} = \frac{1}{2}\left(\sqrt{1 - q_1^2 - q_2^2 - q_3^2}\mathbf{I}_3 + \mathbf{q}^{\times}\right) \tag{14.25}$$

Then, Eq. (14.18) can be written as the following linear time-varying model

$$\begin{aligned} \begin{bmatrix} \dot{\omega}_g \\ \dot{\omega}_s \\ \dot{\omega} \\ \dot{\mathbf{q}} \end{bmatrix} & = \begin{bmatrix} \mathbf{0} & \mathbf{0} & \mathbf{0} & \mathbf{0} \\ \mathbf{0} & \mathbf{0} & \mathbf{0} & \mathbf{0} \\ \mathbf{F}_{31} & \mathbf{F}_{32} & \mathbf{F}_{33} & \mathbf{0} \\ \mathbf{0} & \mathbf{0} & \mathbf{F}_{43} & \mathbf{0} \end{bmatrix} \begin{bmatrix} \omega_g \\ \omega_s \\ \omega \\ \mathbf{q} \end{bmatrix} \\ & + \begin{bmatrix} -\mathbf{J}_g^{-1} & \mathbf{0} \\ \mathbf{0} & -\mathbf{J}_s^{-1} \\ \mathbf{J}_b^{-1}\mathbf{A}_g & \mathbf{J}_b^{-1}\mathbf{A}_s \\ \mathbf{0} & \mathbf{0} \end{bmatrix} \begin{bmatrix} \mathbf{t}_g \\ \mathbf{t}_s \end{bmatrix} + \begin{bmatrix} \mathbf{0} \\ \mathbf{0} \\ \mathbf{J}_b^{-1} \\ \mathbf{0} \end{bmatrix} \mathbf{t}_e \\ & = \mathbf{A}(t)\mathbf{x} + \mathbf{B}(t)\mathbf{u} + \mathbf{C}\mathbf{t}_e \end{aligned} \tag{14.26}$$

where $\mathbf{C}$ is a time-invariant matrix. The linear system is time-varying because ω, ω_s, ω_g, $\mathbf{q}$, $\mathbf{A}_s$ and $\mathbf{A}_t$ in $\mathbf{A}$ and $\mathbf{B}$ are all functions of t.

Given $\mathbf{A}_{s0}$, $\mathbf{A}_{t0}$, and ω_g, then, $\mathbf{A}_s$ and $\mathbf{A}_t$ can be calculated by the integration of (14.5). But using (14.3) and (14.4) is a better method because it ensures that the columns of $\mathbf{A}_s$ and $\mathbf{A}_t$ are unit vectors as required. Notice that the ith column of $\mathbf{A}_s$ and the ith column of $\mathbf{A}_t$, $i = 1, \ldots, n$, must be perpendicular to each other, an even better method to update $\mathbf{A}_t$ is to use the cross product

$$\mathbf{t}_i = \mathbf{g}_i \times \mathbf{s}_i, \quad i = 1, \ldots, n \tag{14.27}$$

to prevent $\mathbf{t}_i$ and $\mathbf{s}_i$ from losing perpendicularity due to the numerical error accumulation. In simulation, integration of (14.1) can be used to obtain γ which is needed in the computation of (14.3), but in engineering practice, the encoder measurement should be used to get γ.

14.2 Spacecraft Attitude Control Using VSCMG

Assuming that the closed-loop linear time-varying system is given by

$$\dot{\mathbf{x}} = \bar{\mathbf{A}}(t)\mathbf{x}(t), \quad \mathbf{x}(t_0) = \mathbf{x}_0 \tag{14.28}$$

It is well-known that even if all the eigenvalues of $\bar{\mathbf{A}}(t)$, denoted by $\mathcal{R}_e[\lambda(t)]$, are in the left half complex plane for all t, the system may not be stable [169, pages 113–114]. But the following theorem (cf. [169, pages 117–119]) provides a nice stability criterion for the closed-loop system (14.28).

Theorem 14.1
Suppose for the linear time-varying system (14.28) with $\bar{\mathbf{A}}(t)$ continuously differentiable there exist finite positive constants α, μ such that, for all t, $\|\bar{\mathbf{A}}(t)\| \leq \alpha$ and every point-wise eigenvalue of $\bar{\mathbf{A}}(t)$ satisfies $\mathcal{R}_e[\lambda(t)] \leq -\mu$. Then there exists a positive constant β such that if the time derivative of $\bar{\mathbf{A}}(t)$ satisfies $\|\dot{\bar{\mathbf{A}}}(t)\| \leq \beta$ for all t, the state equation is uniformly exponentially stable.

This theorem is the theoretical base for the linear time-varying control system design. We need at least that $\mathcal{R}_e[\lambda(t)] \leq -\mu$ holds for $t \geq 0$, which is the design criterion in this section.

14.2.1 Gain Scheduling Control

Gain scheduling control design is fully discussed in [168] and it seems to be applicable to this LTV system. The main idea of gain scheduling is: (a) select a set of fixed parameters' values, which represent the range of the plant dynamics, each member in the fixed parameter set is called a 'frozen model', for each frozen model, the gain is designed by a linear time-invariant design method, and all gains are installed in the computer on-board; (b) when spacecraft flies on orbit, in between operating points, the gain is interpolated using the designs for the fixed parameters' values that cover the operating points. As an example, for $i = 1,\ldots,N$, let $\gamma_i \in \{2\pi/p_\gamma, 4\pi/p_\gamma, \cdots, 2\pi\}$ be a set of p_γ fixed points equally spread in $[0, 2\pi]$. Then, for N VSCMGs, there are p_γ^N possible fixed parameters' combinations. For example, if $N = 4$ and $p_\gamma = 8$, we can represent the grid composed of these fixed points in a matrix form as follows:

$$\begin{bmatrix} \pi/4 & \pi/2 & 3\pi/4 & \pi & 5\pi/4 & 3\pi/2 & 7\pi/4 & 2\pi \\ \pi/4 & \pi/2 & 3\pi/4 & \pi & 5\pi/4 & 3\pi/2 & 7\pi/4 & 2\pi \\ \pi/4 & \pi/2 & 3\pi/4 & \pi & 5\pi/4 & 3\pi/2 & 7\pi/4 & 2\pi \\ \pi/4 & \pi/2 & 3\pi/4 & \pi & 5\pi/4 & 3\pi/2 & 7\pi/4 & 2\pi \end{bmatrix} \tag{14.29}$$

and each fixed γ is a vector composed of γ_i ($i = 1,2,3,4$) which can be any element of ith row. If γ is not one of those fixed points, we have $\gamma_i \in [\kappa(i), \kappa(i)+1]$ for all $i \in [1,\cdots,N]$. Assume that γ_i is in the interior of $(\kappa(i), \kappa(i)+1)$ for all $i \in [1,\cdots,N]$. Then, γ meets the following conditions:

$$\gamma = \begin{bmatrix} \gamma_1 \in (\kappa(1), \kappa(1)+1) \\ \vdots \\ \gamma_N \in (\kappa(N), \kappa(N)+1) \end{bmatrix} \tag{14.30}$$

Using the example of (14.29), if $\gamma = \left[\frac{5\pi}{8}, \frac{3\pi}{8}, \frac{7\pi}{16}, \frac{15\pi}{8}\right]^{\mathrm{T}}$, then

$$\gamma \in \left[\left(\frac{\pi}{2}, \frac{3\pi}{4}\right), \left(\frac{\pi}{4}, \frac{\pi}{2}\right), \left(\frac{\pi}{4}, \frac{\pi}{2}\right), \left(\frac{7\pi}{4}, 2\pi\right)\right]^{\mathrm{T}}$$

To use gain scheduling control, we need also to consider fixed points for ω, ω_s, ω_g, and $\mathbf{q}$ in their possible operational ranges. Let p_w, p_{w_s}, p_{w_g}, and p_q be the number of the fixed points for ω, ω_s, ω_g, and $\mathbf{q}$. The total vertices for the entire polytope (including a grid of all possible time-varying parameters) will be $p_\gamma^N p_w^3 p_{w_s}^N p_{w_g}^N p_q^3$.

For each of these ($p_\gamma^N p_w^3 p_{w_s}^N p_{w_g}^N p_q^3$) fixed models, we need to conduct a control design to calculate the feedback gain matrix for each 'frozen' model. If the system (14.26) at time t happens to have all parameters equal to some fixed point, we can use a 'frozen' feedback gain to control the system (14.26). Otherwise, we need to construct a gain matrix based on these 'frozen' gain matrices. Assuming that each parameter has some moderate number of fixed points, say 8, and the control system has $N = 4$ gimbals, the total number of the fixed models will be 8^{18}, each needs to compute a feedback matrix, an impossibly computational task.

14.2.2 Model Predictive Control

Unlike the gain scheduling control design in which most computation is done off-line, model predictive control computes the feedback gain matrix on-line for the LTV system (14.26) in which $\mathbf{A}$ and $\mathbf{B}$ matrices are updated in every sampling period. It is straightforward to verify that for any given γ, if $\mathbf{x} \neq \mathbf{x}_e$, the frozen linear system (14.26) is controllable. In theory, one can use either robust pole assignment [258, 201], or LQR design [108], or $\mathbf{H}_\infty$ design [272] for the on-line design, but $\mathbf{H}_\infty$ design costs significant more computational time and should not be considered for this on-line design problem. Since LTV system design should meet the condition of $\mathcal{R}_e[\lambda(t)] \leq -\mu$ required in Theorem 14.1, robust pole assignment design is clearly a better choice than LQR design for this purpose. Another attractive feature of the robust pole assignment design is that the perturbation of the closed-loop eigenvalues between sampling period are expected to be small. It is worthwhile to note that a robust pole assignment design [201] minimizes an upper bound of $\mathbf{H}_\infty$ norm which means that the design is robust to the modeling error and reduces the impact of disturbance torques on the system output [241, 250]. Additional merits about this method, such as computational speed which is important for the on-line design, is discussed in [150]. Therefore, the method of [201] is used in the proposed design.

The proposed design algorithm is given as follows:

Algorithm 14.1

Data: $\mathbf{J}_b$, $\mathbf{J}_s$, $\mathbf{J}_g$, *and* $\mathbf{A}_g$
Initial condition: $\mathbf{x} = \mathbf{x}_0$, $\gamma = \gamma_0$, $\mathbf{A}_{s0}$, *and* $\mathbf{A}_{t0}$

Step 1: Update $\mathbf{A}$ *and* $\mathbf{B}$ *based on the latest* γ *and* $\mathbf{x}$.

Step 2: Calculate the gain $\mathbf{K}$ *using robust pole assignment algorithm* `robpole` *(cf. Appendix C or [201]).*

Step 3: Apply feedback $\mathbf{u} = \mathbf{K}\mathbf{x}$ *to (14.18) or (14.26).*

Step 4: Update γ *and* $\mathbf{x} = [\omega^{\mathrm{T}}, \omega_s^{\mathrm{T}}, \omega_g^{\mathrm{T}}, \mathbf{q}^{\mathrm{T}}]^{\mathrm{T}}$. *Go back to Step 1.*

14.2.3 Robust Pole Assignment

Although `robpole` developed in [201] is the most efficient robust pole assignment algorithm [150], the efficiency of `robpole` in on-line application should be further improved by exploring the system structure of $\mathbf{A}$ and the fact that $\mathbf{J}_g$, $\mathbf{J}_b$, and $\mathbf{A}_g$ are constant matrices in (14.26). Let $\Lambda = \mathrm{diag}(\lambda_i)$ and $\mathbf{X} = [\mathbf{x}_1, \ldots, \mathbf{x}_n]$ with $\|\mathbf{x}_i\| = 1$ such that

$$(\mathbf{A} + \mathbf{B}\mathbf{K})\mathbf{X} = \mathbf{X}\Lambda \tag{14.31}$$

The algorithm of `robpole` can be summarized as follows (for details, see Appendix C):

Algorithm 14.2
`robpole`
Data: $\mathbf{A}$, $\mathbf{B}$, *and diagonal matrix* $\Lambda = \mathrm{diag}(\lambda_i)$ *with* λ_i *being the desired closed-loop poles.*

Step 1: QR decomposition for $\mathbf{B}$ *yields orthogonal* $\mathbf{Q} = [\mathbf{Q}_0 \;\; \mathbf{Q}_1]$ *and triangular* $\mathbf{R}$ *such that*

$$\mathbf{B} = [\mathbf{Q}_0 \;\; \mathbf{Q}_1] \begin{bmatrix} \mathbf{R} \\ \mathbf{0} \end{bmatrix} \tag{14.32}$$

Step 2: QR decomposition for $(\mathbf{A}^{\mathrm{T}} - \lambda_i \mathbf{I})\mathbf{Q}_1$ *yields orthogonal* $\mathbf{V} = [\mathbf{V}_{0i} \;\; \mathbf{V}_{1i}]$ *and triangular* $\mathbf{Y}$ *such that*

$$(\mathbf{A}^{\mathrm{T}} - \lambda_i \mathbf{I})\mathbf{Q}_1 = [\mathbf{V}_{0i} \;\; \mathbf{V}_{1i}] \begin{bmatrix} \mathbf{Y} \\ \mathbf{0} \end{bmatrix}, \quad i = 1, \ldots, n \tag{14.33}$$

Step 3: Cyclically select one real or a pair of (real or complex conjugate) unit length eigenvectors such that $\mathbf{x}_i \in \mathcal{S}_i = span(\mathbf{V}_{1i})$ *and the robustness measure* $\det(\mathbf{X})$ *is maximized.*

Step 4: The feedback matrix is given by

$$\mathbf{K} = \mathbf{R}^{-1}\mathbf{Q}_0^{\mathrm{T}}(\mathbf{X}\Lambda\mathbf{X}^{-1} - \mathbf{A}) \tag{14.34}$$

Step 3 in Algorithm 14.2 looks very complex but it turns out, by some careful investigation, that this step mainly involves two rank-one QR decomposition updates and a rank-two singular value decomposition (SVD). The rank-two SVD admits an analytical solution [201]. Since $\mathbf{A}$ in (14.26) has a lot of zeros, the calculation in parentheses in Steps 2 and 4 can save substantial flops, especially in Step 2 which is done for $i = 1, \ldots, n$. Another major saving can be achieved in Step 1 by using the fact that the first three columns of $\mathbf{B}$ are constant (not time-varying). Assume that

$$\mathbf{B} = \begin{bmatrix} -\mathbf{J}_g^{-1} & \mathbf{0} \\ \mathbf{0} & -\mathbf{J}_s^{-1} \\ \mathbf{J}_b^{-1}\mathbf{A}_g & \mathbf{J}_b^{-1}\mathbf{A}_s \\ \mathbf{0} & \mathbf{0} \end{bmatrix} = [\mathbf{Q}_0 \;\; \mathbf{Q}_1] \begin{bmatrix} \mathbf{R} \\ \mathbf{0} \end{bmatrix} \tag{14.35}$$

or equivalently

$$\mathbf{Q}^{\mathrm{T}}\mathbf{B} = \begin{bmatrix} \mathbf{Q}_0^{\mathrm{T}} \\ \mathbf{Q}_1^{\mathrm{T}} \end{bmatrix} \mathbf{B} = [\mathbf{R}_1 \;\; \mathbf{R}_2] \tag{14.36}$$

As time evolves and $\mathbf{A}_s$ changes, $\mathbf{R}_2 = \mathbf{Q}^{\mathrm{T}} \begin{bmatrix} \mathbf{0} & -\mathbf{J}_s^{-\mathrm{T}} & \mathbf{A}_s^{\mathrm{T}}\mathbf{J}_b^{-\mathrm{T}} & \mathbf{0} \end{bmatrix}^{\mathrm{T}}$ changes but $\mathbf{R}_1$ is constant and triangular. Therefore, the QR decomposition needs only to zero a few non-zeros in $\mathbf{R}_2$ to make $\mathbf{R}$ triangular. This reduces significant amount of flops in every sampling time.

14.3 Simulation Test

The proposed design method is simulated using the model and data in [83, 262, 261]. We assume that the four VSCMGs are mounted in pyramid configuration[2]

[2]Pyramid configuration was extensively studied because four CMGs are the minimum having one degree of redundancy [100]. But a detailed study [100] showed that CMG control using Pyramid configuration and inverse from torque to flywheel speed cannot avoid singularity.

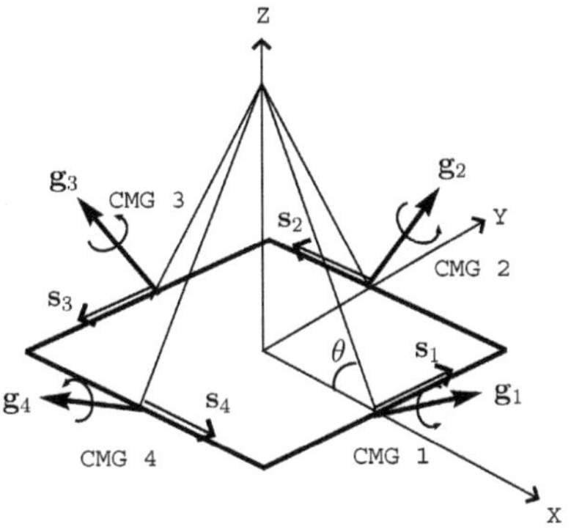

Figure 14.2: VSCMG system with pyramid configuration concept.

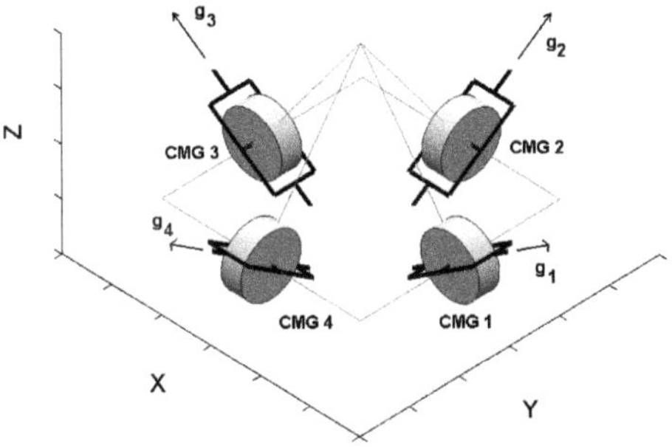

Figure 14.3: VSCMG system with pyramid configuration.

as shown in Figures 14.2 and 14.3. The angle of each pyramid side to its base is $\theta = 54.75$ degree; the inertia matrix of the spacecraft is given by [261] as

$$\mathbf{J}_b = \begin{bmatrix} 15053 & 3000 & -1000 \\ 3000 & 6510 & 2000 \\ -1000 & 2000 & 11122 \end{bmatrix} \quad \text{kg} \cdot m^2 \tag{14.37}$$

The spin axis inertial matrix is given by $\mathbf{J}_s = \text{diag}(0.7, 0.7, 0.7, 0.7)$ kg $\cdot\, m^2$ and the gimbal axis inertia matrix is given by $\mathbf{J}_g = \text{diag}(0.1, 0.1, 0.1, 0.1)$ kg $\cdot\, m^2$. The initial wheel speeds are 2π radians per second for all wheels. The initial gimbal speeds are all zeros. The initial spacecraft body rate vector is randomly generated by MATLAB® using $rand(3,1) * 10^{-3}$ and the initial spacecraft attitude vector is a reduced quaternion randomly generated by MATLAB using $rand(3,1) * 10^{-1}$. The gimbal axis matrix is fixed and given by [262] (cf. Figures 14.2 and 14.3.)

$$\mathbf{A}_g = \begin{bmatrix} \sin(\theta) & 0 & -\sin(\theta) & 0 \\ 0 & \sin(\theta) & 0 & -\sin(\theta) \\ \cos(\theta) & \cos(\theta) & \cos(\theta) & \cos(\theta) \end{bmatrix} \tag{14.38}$$

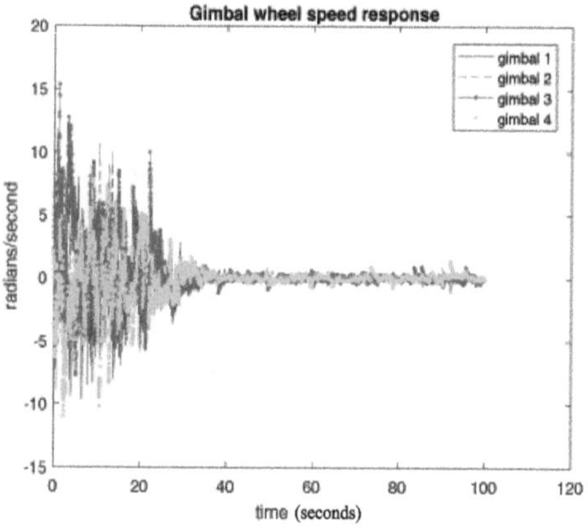

Figure 14.4: Gimbal wheel ω_g response.

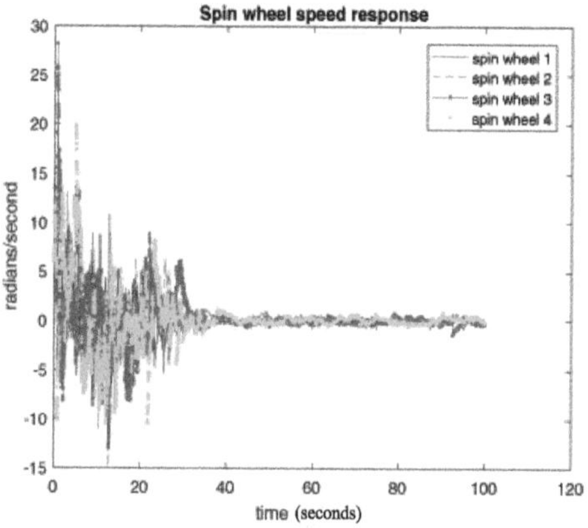

Figure 14.5: Spin wheel ω_s response.

The initial flywheel axis matrix can be obtained using Figures 14.2 and 14.3 and is given by

$$\mathbf{A}_s = \begin{bmatrix} 0 & -1 & 0 & 1 \\ 1 & 0 & -1 & 0 \\ 0 & 0 & 0 & 0 \end{bmatrix} \tag{14.39}$$

The initial transverse matrix $\mathbf{A}_t$ can be obtained by the formula of (14.27). The desired or designed closed-loop poles are selected as

$$\{-3.0, -3.1, -2.9, -3.2, -2.1, -2.2, -2.0, -1.9, -3.4, -3.5, -3.3, -2.7, -2.6, -2.8\}$$

The simulation test results for (14.26) using control Algorithm 3.1 are given in Figures 14.4–14.7.

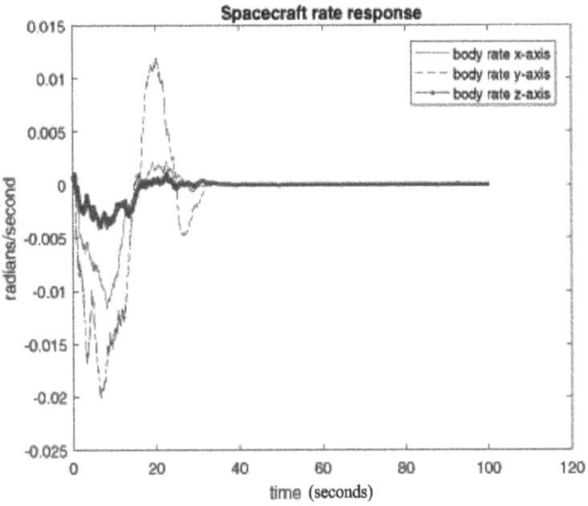

Figure 14.6: Spacecraft body rate ω response.

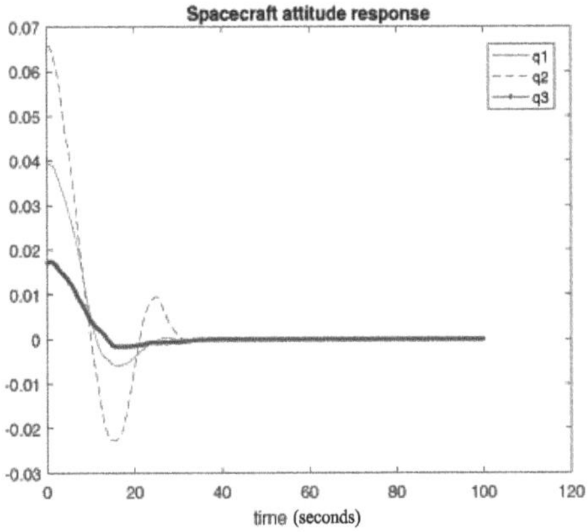

Figure 14.7: Attitude q_0, q_1, q_2, and q_3 response.

Remark 14.1 The simulation shows that the computational time for robust pole assignment design is very efficient. But if this algorithm does not meet the on-line computational requirement, a faster but not a robust pole assignment algorithm proposed by Misrikhanov and Ryabchenko is available [131], which is discussed in Appendix C. ■

Chapter 15

Spacecraft Rendezvous and Docking

15.1 Introduction

Spacecraft rendezvous is an important operation in many space missions. There has been extensive research in this field and hundreds of successful rendezvous missions, see, for example, the survey paper [116] and references therein. The entire rendezvous process can be divided into several phases, including phasing, close-range rendezvous, final approaching, and docking. In the early phase, the chaser flies to the target with the aid from the ground station and orbital translation control is the main concern. For this purpose, the well-known Hill [72] or Clohessy and Wiltshire [32] equations are adequate for the control system design if the orbit is circular. But in the final approaching and docking phase, coupled orbital and attitude control may be required. Moreover, it is desired to consider the case that the orbit of the target spacecraft is not circular. To achieve this requirement, more complex models introduced in [96, 148, 215] should be considered. Although these models are developed for more general purpose, they can be easily tailored for spacecraft rendezvous and docking control.

The research of spacecraft rendezvous has attracted renewed interest in recent years as a result of new development in control theory and increased space missions involving rendezvous and soft docking. Various design methods have been considered for this control system design problem. For example, reachability was considered in [264]; an adaptive output feedback control was proposed for this purpose in [185]; a multi-objective robust H_∞ control method was investigated in [54]; a Lyapunov differential equation approach was studied for elliptical orbital rendezvous with constrained controls [270]; a gain scheduled control of linear systems was applied to spacecraft rendezvous problem subject to actuator saturation [271]; and various control design methods were considered for 6 degree of freedom (DOF) spacecraft proximity operations [97, 110, 121, 128, 191, 193, 195, 196, 235, 268]. NASA is working on some concept validation flight test [167]. All these methods have their merits in solving the challenging problem under various conditions, but none of them address a fundamental issue, i.e., to achieve the soft docking.

In this chapter, a recently proposed model in [96] is carefully examined. The measurable variables and controllable inputs in the mission of the final approaching and docking phase are then determined. Some reasonable assumptions that normally hold via engineering design are made clear. Because of the merits discussed in [243, 250], a reduced quaternion concept proposed in [243] is adopted, which slightly simplifies the model of [96]. To make the general model useful for the control system design, a thruster configuration is considered and modeled in the control system model. This control system model can be viewed either as a nonlinear model or a linear time-varying (LTV) model. Using the linear time-varying model is preferred because a linear system is easier to handle than a nonlinear system and the corresponding design methods are capable to consider the system performance which is very important as *soft docking does not allow oscillation crossing the horizontal line for the relative position and relative attitude (between the target and the chaser) in the spacecraft rendezvous and docking phase*.

There are two popular methods that deal with time-varying control system design with the consideration of system performance. The first one is gaining scheduling [170] and the second one is model predictive control (MPC) [7]. A simple analysis in the previous chapter shows that the former is the most efficient when all time-varying parameters explicitly depend on time; and the later is more appropriate when many parameters depend implicitly on time. The rendezvous and docking control falls into the second category. Therefore, an MPC-based method is proposed to design the rendezvous and docking control. Although several LTI design methods, such as LQR, $\mathbf{H}_\infty$, and robust pole assignment, take the performance into the design consideration and can be used in the MPC-based design, only robust pole assignment method can directly take system oscillation into the design consideration because oscillation is directly related to the closed-loop pole positions [41]. In addition, robust pole assignment guarantees that the closed-loop poles are not sensitive to the parameter changes in the system [150] that is important given the system is time-varying. Moreover, robust pole assignment design minimizes an upper bound of $\mathbf{H}_\infty$ norm which means that the design is robust to the modeling error and reduces the impact of disturbance forces on the system output (see Chapter 9 and [250]). Among many robust pole assignment algorithms, we suggest a globally convergent algorithm [201] because of its fast on-line computation and other merits [150]. Two design examples and simulation are used to show the efficiency and effectiveness of the proposed method.

This chapter is mainly based on [257]. Section 2 summarizes the complete rendezvous model and its implication for rendezvous and docking control. Section 3 discusses the MPC-based method for spacecraft control using robust pole assignment. Section 4 provides some design examples and simulation results.

15.2 Spacecraft Model for Rendezvous

This section first presents the model developed by Kristiansen et al. in [96]. It then discusses the assumptions derived from the application of final approaching and docking phase in the rendezvous process and present a simplified version to be used in this chapter. For the sake of simplicity, the scalar notation a is used for the magnitude of $\|\mathbf{a}\|$. The following assumption is made throughout the chapter.

Assumption 1 *Chaser and target can exchange position, attitude and rotational rate information in real time.*

This assumption can be achieved by engineering design. But this assumption is not essential because extensive research, for relative pose determination techniques, has been performed and many of these techniques are expected to be used in the future missions (see a survey paper [145]).

15.2.1 The Model for Translation Dynamics

As shown in Figure 15.1, the inertial frame is defined by standard earth-centered inertial (ECI) frame $\mathcal{F}_i$ with $\mathbf{i}_x$, $\mathbf{i}_y$, and $\mathbf{i}_z$ being the coordinate axes. Let $\mathbf{r}_t$ be the vector from the Earth's center to the center of the mass of the target. Let the angular momentum vector of the target orbit be denoted by $\mathbf{h} = \mathbf{r}_t \times \dot{\mathbf{r}}_t$. The target orbital frame $\mathcal{F}_{to}$ is the spacecraft RSW frame discussed in Chapter 3 with the origin at center of the mass of the target. The coordinate vectors of the RSW frame are

$$\mathbf{e}_r = \mathbf{r}_t / r_t \tag{15.1a}$$

$$\mathbf{e}_w = \mathbf{h} / h \tag{15.1b}$$

$$\mathbf{e}_s = \mathbf{e}_w \times \mathbf{e}_r \tag{15.1c}$$

Several other vectors are defined in RSW frame $\mathcal{F}_{to}$: $\mathbf{e}_v$ is the vector in the spacecraft velocity direction. $\mathbf{e}_n$ is defined to be orthogonal to $\mathbf{e}_v$ and $\mathbf{e}_w$ as $\mathbf{e}_n = \mathbf{e}_v \times \mathbf{e}_w$. If the spacecraft *orbit is circular*, then $\mathbf{e}_v = \mathbf{e}_s$ and $\mathbf{e}_n = \mathbf{e}_r$. The transformation from ECI frame to the RSW frame (the target orbit frame) is given in (3.16). The body frames of the target and chaser, $\mathcal{F}_{tb}$ and $\mathcal{F}_{cb}$, have their origins at their centers of mass and their coordinate vectors are their principal axes of the inertia.

The relative position vector between target and chaser is defined by

$$\mathbf{p} = \mathbf{r}_c - \mathbf{r}_t = x\mathbf{e}_r + y\mathbf{e}_s + z\mathbf{e}_w \tag{15.2}$$

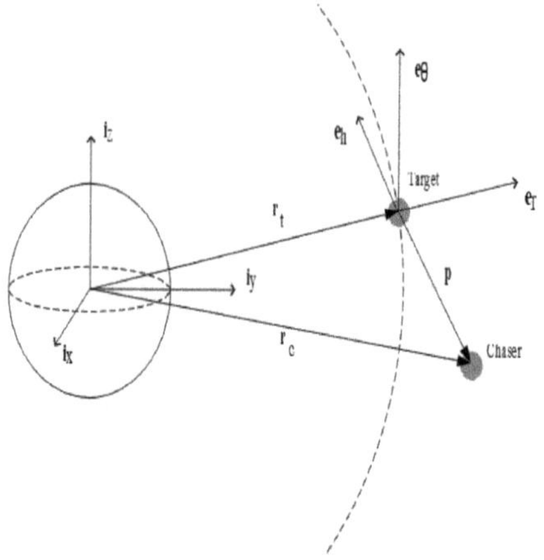

Figure 15.1: Spacecraft coordinate frame.

p is available in real time if GPS is installed in both spacecraft and Assumption 1 holds. Spacecraft acceleration can be written as

$$\mathbf{a} = a_r\mathbf{e}_r + a_s\mathbf{e}_s + a_w\mathbf{e}_w = a_n\mathbf{e}_n + a_v\mathbf{e}_v + a_w\mathbf{e}_w \tag{15.3}$$

The spacecraft velocity vector can be derived according to Figure 15.2 as follows. Let v_r and v_s be the velocity components in $\mathbf{e}_r$ and $\mathbf{e}_s$. Than, $v_r = \dot{r}_t$, and $v_s = r_t\dot{\theta}$, where θ is the *true anomaly*. Equations (2.51), (2.14), and (2.29) will be used, which are listed below for easy reference:

$$r_t = \frac{p}{1+e\cos(\theta)} = \frac{a(1-e^2)}{1+e\cos(\theta)} \tag{15.4}$$

where e is the eccentricity of the spacecraft orbit, a is the semi-major axis of the orbit, and p is semi-latus rectum,

$$h = r_t^2\frac{d\theta}{dt} \tag{15.5}$$

and

$$r_t = \frac{h^2/\mu}{1+e\cos(\theta)} \tag{15.6}$$

where μ is the geocentric gravitational constant of the Earth. From aforementioned equations, the following relations follow:

$$\begin{aligned} v_r &= \dot{r}_t = \frac{a(1-e^2)e\sin(\theta)\dot{\theta}}{(1+e\cos(\theta))^2} \\ &= \frac{a(1-e^2)eh\sin(\theta)}{r_t^2(1+e\cos(\theta))^2} \end{aligned}$$

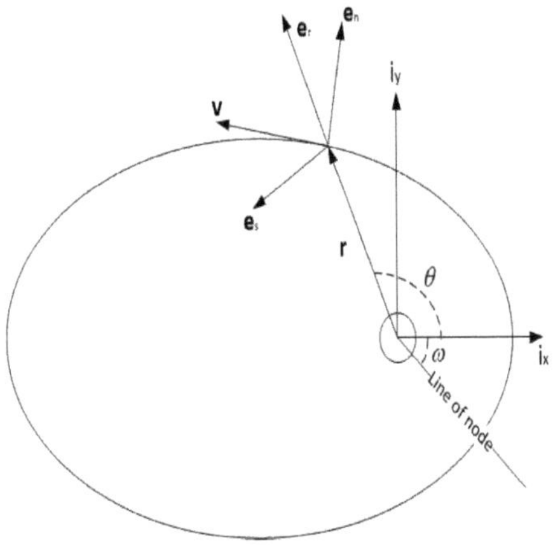

Figure 15.2: Spacecraft coordinate in orbital plan.

$$
\begin{aligned}
&= \frac{eh\sin(\theta)}{r_t(1+e\cos(\theta))} \\
&= \frac{eh\sin(\theta)}{h^2/\mu} \\
&= \frac{\mu}{h}e\sin(\theta) \qquad (15.7)
\end{aligned}
$$

Using (2.30) $p = \frac{h^2}{\mu}$ yields

$$v_s = r_t\dot{\theta} = r_t h/r_t^2 = h/r_t = \frac{h^2\mu}{hr_t\mu} = \frac{p\mu}{hr_t} \qquad (15.8)$$

Combining (15.7) and (15.8) gives

$$\mathbf{v} = \dot{\mathbf{r}}_t = \frac{\mu}{h}\left(e\sin(\theta)\mathbf{e}_r + \frac{p}{r_t}\mathbf{e}_s\right) \qquad (15.9)$$

Since $\mathbf{e}_v$ is pointing to the velocity vector,

$$\mathbf{e}_v = \frac{\mathbf{v}}{v} = \frac{h}{pv}\left(e\sin(\theta)\mathbf{e}_r + \frac{p}{r_t}\mathbf{e}_s\right) \qquad (15.10)$$

Since $\mathbf{e}_n$ is perpendicular to $\mathbf{e}_v$ and $\mathbf{e}_w$,

$$\mathbf{e}_n = \mathbf{e}_v \times \mathbf{e}_w = \frac{h}{pv}\left(\frac{p}{r_t}\mathbf{e}_r - e\sin(\theta)\mathbf{e}_s\right) \qquad (15.11)$$

The coordinate transformation between the orbit plane acceleration vector components can be found from above equations as

$$\begin{bmatrix} a_r \\ a_s \end{bmatrix} = \frac{h}{pv}\begin{bmatrix} \frac{p}{r_t} & e\sin(\theta) \\ -e\sin(\theta) & \frac{p}{r_t} \end{bmatrix}\begin{bmatrix} a_n \\ a_v \end{bmatrix} \qquad (15.12)$$

so that

$$\mathbf{C}_a^l = \frac{h}{pv} \begin{bmatrix} \frac{p}{r_t} & e\sin(\theta) & 0 \\ -e\sin(\theta) & \frac{p}{r_t} & 0 \\ 0 & 0 & \frac{pv}{h} \end{bmatrix} \tag{15.13}$$

Note that $\mathbf{C}_a^l$ is not in general a proper rotation matrix since $\det(\mathbf{C}_a^l) = 1 + e^2 + 2e\cos(\theta)$. When $e = 0$, $\mathbf{C}_a^l$ is a rotational matrix.

For the two-body problem, using equation (2.2) $\mathbf{f} = \frac{Gm_1m_2\mathbf{r}}{r^3}$ (m_1 is the mass of the Earth and m_2 is the mass of the spacecraft) and $\mathbf{a} = \frac{d\mathbf{r}^2}{dt^2}$, the fundamental differential equation can be found as

$$\frac{d\mathbf{r}^2}{dt^2} + \frac{\mu}{r^3}\mathbf{r} = 0 \tag{15.14}$$

where $\mu = G(m_1 + m_2) \approx Gm_1$, $G = 6.669 * 10^{-11} m^3/kg - s^2$ is the universal constant of gravitation. This equation can be generalized to include force terms due to aerodynamic disturbances, gravitational forces from other bodies, solar radiation, magnetic fields, and so on. In addition, it can be augmented to include control input vectors from on-board actuators. Accordingly, (15.14) should be expressed for the target and chaser spacecraft as

$$\frac{d\mathbf{r}_t^2}{dt^2} = -\frac{\mu}{r_t^3}\mathbf{r}_t + \frac{\mathbf{f}_{dt}}{m_t} + \frac{\mathbf{f}_{at}}{m_t} \tag{15.15}$$

$$\frac{d\mathbf{r}_c^2}{dt^2} = -\frac{\mu}{r_c^3}\mathbf{r}_c + \frac{\mathbf{f}_{dc}}{m_c} + \frac{\mathbf{f}_{ac}}{m_c} \tag{15.16}$$

where $\mathbf{f}_{dt}$ and $\mathbf{f}_{dc}$ are the disturbance actions due to external effects; $\mathbf{f}_{at}$ and $\mathbf{f}_{ac}$ are the actuator forces of the target and chaser spacecraft, respectively. In addition, spacecraft masses are assumed to be small relative to the mass of the Earth. The second order derivative of the relative position vector is given by

$$\ddot{\mathbf{p}} = \ddot{\mathbf{r}}_c - \ddot{\mathbf{r}}_t = -\frac{\mu}{r_c^3}\mathbf{r}_c + \frac{\mathbf{f}_{dc}}{m_c} + \frac{\mathbf{f}_{ac}}{m_c} + \frac{\mu}{r_t^3}\mathbf{r}_t - \frac{\mathbf{f}_{dt}}{m_t} - \frac{\mathbf{f}_{at}}{m_t} \tag{15.17}$$

Simple manipulating on the formula gives

$$m_c\ddot{\mathbf{p}} = -m_c\mu\left(\frac{\mathbf{r}_t + \mathbf{p}}{(r_t + p)^3} - \frac{\mathbf{r}_t}{r_t^3}\right) + \mathbf{f}_{ac} + \mathbf{f}_{dc} - \frac{m_c}{m_t}(\mathbf{f}_{at} + \mathbf{f}_{dt}) \tag{15.18}$$

In view of (15.2), the dynamics of the chaser spacecraft relative to the target spacecraft, referenced in the target orbit frame $\mathcal{F}_{to}$, can be expressed as

$$\mathbf{r}_c = \mathbf{r}_t + \mathbf{p} = (r_t + x)\mathbf{e}_r + y\mathbf{e}_s + z\mathbf{e}_w \tag{15.19}$$

Taking derivative on this equation twice with respect to time yields

$$\ddot{\mathbf{r}}_c = (\ddot{r}_t + \ddot{x})\mathbf{e}_r + 2(\dot{r}_t + \dot{x})\dot{\mathbf{e}}_r + (r_t + x)\ddot{\mathbf{e}}_r + \ddot{y}\mathbf{e}_s$$

$$+2\dot{y}\dot{\mathbf{e}}_s + y\ddot{\mathbf{e}}_s + \ddot{z}\mathbf{e}_w + 2\dot{z}\dot{\mathbf{e}}_w + z\ddot{\mathbf{e}}_w \tag{15.20}$$

By using the true anomaly θ of the target spacecraft, the following relationships hold.

$$\dot{\mathbf{e}}_r = \dot{\theta}\mathbf{e}_s,\ \dot{\mathbf{e}}_s = -\dot{\theta}\mathbf{e}_r,\ \ddot{\mathbf{e}}_r = \ddot{\theta}\mathbf{e}_s - \dot{\theta}^2\mathbf{e}_r,\ \ddot{\mathbf{e}}_s = -\ddot{\theta}\mathbf{e}_r - \dot{\theta}^2\mathbf{e}_s \tag{15.21}$$

Substituting of (15.21) into (15.20), while recognizing that no out-of-plane motion exists in the ideal case, and hence $\dot{\mathbf{e}}_w = \ddot{\mathbf{e}}_w = 0$, yields

$$\begin{aligned} \ddot{\mathbf{r}}_c &= \left[\ddot{r}_t + \ddot{x} - 2\dot{y}\dot{\theta} - \dot{\theta}^2(r_t + x) - y\ddot{\theta}\right]\mathbf{e}_r \\ &\quad + \left[\ddot{y} + 2\dot{\theta}(\dot{r}_t + \dot{x}) + \ddot{\theta}(r_t + x) - y\dot{\theta}^2\right]\mathbf{e}_s + \ddot{z}\mathbf{e}_w \end{aligned} \tag{15.22}$$

Moreover, the position of the target spacecraft can be expressed as $\mathbf{r}_t = r_t\mathbf{e}_r$, and taking derivative for this expression twice with respect to time and inserting (15.21), result in

$$\ddot{\mathbf{r}}_t = \ddot{r}_t\mathbf{e}_r + 2\dot{r}_t\dot{\mathbf{e}}_r + r_t\ddot{\mathbf{e}}_r = (\ddot{r}_t - r_t\dot{\theta}^2)\mathbf{e}_r + (2\dot{r}_t\dot{\theta} + r_t\ddot{\theta})\mathbf{e}_s \tag{15.23}$$

Subtracting (15.23) and (15.22) into (15.17) results in the formulation of the position vector acceleration represented in the $\mathcal{F}_{to}$ frame:

$$\ddot{\mathbf{p}} = \ddot{\mathbf{r}}_c - \ddot{\mathbf{r}}_t = (\ddot{x} - 2\dot{y}\dot{\theta} - \dot{\theta}^2 x - \ddot{\theta}y)\mathbf{e}_r + (\ddot{y} + 2\dot{\theta}\dot{x} + \ddot{\theta}x - \dot{\theta}^2 y)\mathbf{e}_s + \ddot{z}\mathbf{e}_w \tag{15.24}$$

Substituting (15.24), (15.19), and (15.1) into (15.18) gives

$$\begin{aligned} m_c\ddot{\mathbf{p}} &= m_c\left((\ddot{x} - 2\dot{y}\dot{\theta} - \dot{\theta}^2 x - \ddot{\theta}y)\mathbf{e}_r + (\ddot{y} + 2\dot{\theta}\dot{x} + \ddot{\theta}x - \dot{\theta}^2 y)\mathbf{e}_s + \ddot{z}\mathbf{e}_w\right) \\ &= -m_c\mu\left(\frac{\mathbf{r}_c}{r_c^3} - \frac{\mathbf{r}_t}{r_t^3}\right) + \mathbf{f}_a + \mathbf{f}_d \\ &= -m_c\mu\left(\frac{(r_t + x)}{r_c^3}\mathbf{e}_r + \frac{y}{r_c^3}\mathbf{e}_s + \frac{z}{r_c^3}\mathbf{e}_w - \frac{1}{r_t^2}\mathbf{e}_r\right) + \mathbf{f}_a + \mathbf{f}_d \end{aligned} \tag{15.25}$$

where $\mathbf{f}_a = \mathbf{f}_{ac}$ and $\mathbf{f}_d = \mathbf{f}_{dc}$ and forces on target spacecraft is omitted. Denoting

$$\mathbf{d} = \begin{bmatrix} x \\ y \\ z \end{bmatrix},\quad \dot{\mathbf{d}} = \begin{bmatrix} \dot{x} \\ \dot{y} \\ \dot{z} \end{bmatrix},\quad \ddot{\mathbf{d}} = \begin{bmatrix} \ddot{x} \\ \ddot{y} \\ \ddot{z} \end{bmatrix}$$

as in [238], the nonlinear model (15.25) of spacecraft translation dynamics can be rewritten as follows:

$$m_c\ddot{\mathbf{d}} + \mathbf{C}_t(\dot{\theta})\dot{\mathbf{d}} + \mathbf{D}_t(\dot{\theta}, \ddot{\theta}, r_c)\mathbf{d} + \mathbf{n}_t(r_c, r_t) = \mathbf{f}_a + \mathbf{f}_d \tag{15.26}$$

where

$$\mathbf{C}_t(\dot{\theta}) = 2m_c\dot{\theta}\begin{bmatrix} 0 & -1 & 0 \\ 1 & 0 & 0 \\ 0 & 0 & 0 \end{bmatrix} \tag{15.27}$$

$$\mathbf{D}_t(\dot{\theta},\ddot{\theta},r_c) = m_c \begin{bmatrix} \frac{\mu}{r_c^3} - \dot{\theta}^2 & -\ddot{\theta} & 0 \\ \ddot{\theta} & \frac{\mu}{r_c^3} - \dot{\theta}^2 & 0 \\ 0 & 0 & \frac{\mu}{r_c^3} \end{bmatrix} \tag{15.28}$$

$$\mathbf{n}_t(r_c,r_t) = m_c\mu \begin{bmatrix} r_t/r_c^3 - 1/r_t^2 \\ 0 \\ 0 \end{bmatrix} \tag{15.29}$$

$\mathbf{f}_a$ is the control force vector, and $\mathbf{f}_d$ is the disturbance force vector, both are applied in chaser's body frame. It is worthwhile to note that

$$\mathbf{n}_t(r_c,r_t)\big|_{r_c=r_t} = \mathbf{0} \tag{15.30}$$

The calculation of $\dot{\theta}$ is given by (15.5)

$$\dot{\theta} = \frac{h}{r_t^2}$$

where h is a constant depending on the specific orbit, and r_t is provided by GPS.

Case 1: If the orbit is circular, $\dot{\theta}$ is a constant because both h and r_t are constants. Hence, $\ddot{\theta} = 0$. Noticing that, during the docking phase, $r_c \approx r_t$ and the latter is a constant, therefore, $\mathbf{C}_t(\dot{\theta})$ and $\mathbf{D}_t(\dot{\theta},\ddot{\theta},r_c)$ are constants.

Case 2: If the orbit is elliptic, using (15.4) gives

$$\dot{\theta} = \frac{h}{r_t^2} = \frac{h(1+e\cos(\theta))^2}{p^2} = \frac{h(1+e\cos(\theta))^2}{a^2(1-e^2)^2} \tag{15.31}$$

where e, a, and p are all constants. Taking derivative for both sides of $r_t^2\dot{\theta} = h$ and noticing that h is a constant yields

$$2r_t\dot{r}_t\dot{\theta} + r_t^2\ddot{\theta} = 0$$

Substituting (15.4) and (15.7) into this equation gives

$$\ddot{\theta} = -\frac{2\dot{r}_t\dot{\theta}}{r_t} = -\frac{2\mu e\sin(\theta)\dot{\theta}}{hr_t} = \frac{2\mu e\dot{\theta}\sin(\theta)(1+e\cos(\theta))}{ha(1-e^2)} \tag{15.32}$$

According to (15.31) and (15.32), to calculate $\dot{r}_t(t)$, $\dot{\theta}$ and $\ddot{\theta}$, one needs to know θ. Let $t=0$ be the time that the spacecraft passing from the perigee. A function of $\theta(t)$ can be found as follows: from (2.61)

$$M = \frac{2\pi t}{T} = \psi - e\sin(\psi)$$

where T is the spacecraft orbital period, t is the time elapsed since the spacecraft passes the perigee, M is the mean anomaly, ψ is the eccentric anomaly. Therefore, given t, one can calculate M. Given M and e, one can calculate ψ by using Newton's method. Given ψ, one can calculate θ by using (2.50) which is given as follows:

$$\tan\left(\frac{\theta}{2}\right) = \sqrt{\frac{1+e}{1-e}}\tan\left(\frac{\psi}{2}\right) \tag{15.33}$$

Therefore, according to Assumption 1, $\mathbf{C}_t(\dot{\theta})$, $\mathbf{D}_t(\dot{\theta},\ddot{\theta},r_c)$ and $\mathbf{n}_t(r_c,r_t)$ are known but in general are time-varying since r_c, r_t, θ, $\dot{\theta}$, and $\ddot{\theta}$ are all time-varying.

15.2.2 The Model for Attitude Dynamics

Let the unit quaternion $\bar{\mathbf{q}} = \left[q_0, \mathbf{q}^{\mathrm{T}}\right]^{\mathrm{T}}$ be the relative attitude of the target and chaser, where

$$\mathbf{q}^{\mathrm{T}} = [q_1, q_2, q_3] \tag{15.34}$$

The inverse of the quaternion is defined in (3.49) as $\bar{\mathbf{q}}^{-1} = \left[q_0, -\mathbf{q}^{\mathrm{T}}\right]^{\mathrm{T}}$. Let $\bar{\mathbf{q}}_{i,cb} = [q_{c0}, q_{c1}, q_{c2}, q_{c3}]$ be the relative quaternion from chaser's body frame to the inertial frame, and $\bar{\mathbf{q}}_{i,tb} = [q_{t0}, q_{t1}, q_{t2}, q_{t3}]$ be the relative quaternion from the target's body frame to the inertial frame. Notice that $\bar{\mathbf{q}}_{i,cb}$ is measurable from the chaser and $\bar{\mathbf{q}}_{i,tb}$ is measurable from the target. Using the Assumption 1, equations (3.49) and (3.63), we have

$$\bar{\mathbf{q}} = \bar{\mathbf{q}}_{i,cb}^{-1}\bar{\mathbf{q}}_{i,tb} = \begin{bmatrix} q_{t0} & -q_{t1} & -q_{t2} & -q_{t3} \\ q_{t1} & q_{t0} & q_{t3} & -q_{t2} \\ q_{t2} & -q_{t3} & q_{t0} & q_{t1} \\ q_{t3} & q_{t2} & -q_{t1} & q_{t0} \end{bmatrix} \begin{bmatrix} q_{c0} \\ -q_{c1} \\ -q_{c2} \\ -q_{c3} \end{bmatrix} \tag{15.35}$$

which, according to Assumption 1, is measurable. The relative angular velocity between frames $\mathcal{F}_{cb}$ and $\mathcal{F}_{tb}$ expressed in frame $\mathcal{F}_{cb}$ is given by

$$\omega = \omega_{i,cb}^{cb} - \mathbf{R}_{tb}^{cb}\omega_{i,tb}^{tb} = [\omega_1, \omega_2, \omega_3]^{\mathrm{T}} \tag{15.36}$$

where $\omega_{i,cb}^{cb}$ is the angular velocity of the chaser spacecraft body frame relative to the inertial frame, expressed in the chaser spacecraft body frame, $\omega_{i,cb}^{cb}$ is measurable from chaser; $\omega_{i,tb}^{tb}$ is the angular velocity of the target spacecraft body frame relative to the inertial frame, expressed in the target spacecraft body frame, $\omega_{i,tb}^{tb}$ is measurable from target; $\mathbf{R}_{tb}^{cb}$ is the rotational matrix from $\mathcal{F}_{tb}$ to $\mathcal{F}_{cb}$ which is an equivalent rotation of $\bar{\mathbf{q}}$ and is given by (3.55)

$$\mathbf{R}_{tb}^{cb} = (q_0^2 - \mathbf{q}^{\mathrm{T}}\mathbf{q})\mathbf{I} + 2\mathbf{q}\mathbf{q}^{\mathrm{T}} - 2q_0\mathbf{S}(\mathbf{q}) \tag{15.37}$$

where $\mathbf{S}(\mathbf{q}) = \mathbf{q}^{\times}$ is the cross product operator. Using Assumption 1 again, we conclude that ω is available from measurements. Let $\mathbf{J}_c$ and $\mathbf{J}_t$ be the inertia matrices of the chaser and target, respectively.

Assume that a quaternion $\bar{\mathbf{q}}$ rotates frame a to frame b, then the corresponding direction cosine matrix is given by (3.60) which is provided below for easy reference.

$$\mathbf{R}_a^b = \begin{bmatrix} 2q_0^2-1+2q_1^2 & 2q_1q_2+2q_0q_3 & 2q_1q_3-2q_0q_2 \\ 2q_1q_2-2q_0q_3 & 2q_0^2-1+2q_2^2 & 2q_2q_3+2q_0q_1 \\ 2q_1q_3+2q_0q_2 & 2q_2q_3-2q_0q_1 & 2q_0^2-1+2q_3^2 \end{bmatrix} \tag{15.38}$$

Let $\omega_{i,sb}^{sb}$ be the angular velocity of the spacecraft relative to the inertial frame, expressed in the spacecraft body frame, where $s \in \{c,t\}$. In view of (4.2), the spacecraft dynamical model can be written as

$$J_s\dot{\omega}_{i,sb}^{sb} = -\mathbf{S}(\omega_{i,sb}^{sb})J_s\omega_{i,sb}^{sb} + \mathbf{t}_{ds} + \mathbf{t}_{as} \tag{15.39}$$

where $\mathbf{t}_{ds}$ is the disturbance torque applied to the spacecraft body and expressed in the body frame, and $\mathbf{t}_{as}$ is the control torque applied to the spacecraft body and expressed in the body frame. In view of (3.13), the derivative of the rotational matrix$\mathbf{R}_b^a$ that rotates from b frame to a frame is given by

$$\dot{\mathbf{R}}_b^a = -\mathbf{S}(\omega_{a,b}^b)\mathbf{R}_b^a = \mathbf{S}(\omega_{a,b}^a)\mathbf{R}_b^a \tag{15.40}$$

Using definition of ω in (15.36), (15.40), and $\mathbf{a}\times\mathbf{b} = -\mathbf{b}\times\mathbf{a}$, the relative attitude dynamics can be expressed as

$$\begin{aligned} \mathbf{J}_c\dot{\omega} &= \mathbf{J}_c\left(\dot{\omega}_{i,cb}^{cb} - \dot{\mathbf{R}}_{tb}^{cb}\omega_{i,tb}^{tb} - \mathbf{R}_{tb}^{cb}\dot{\omega}_{i,tb}^{tb}\right) \\ &= \mathbf{J}_c\dot{\omega}_{i,cb}^{cb} - \mathbf{J}_c\mathbf{S}(\omega_{cb,tb}^{cb})\mathbf{R}_{tb}^{cb}\omega_{i,tb}^{tb} - \mathbf{J}_c\mathbf{R}_{tb}^{cb}\dot{\omega}_{i,tb}^{tb} \\ &= \mathbf{J}_c\dot{\omega}_{i,cb}^{cb} - \mathbf{J}_c\mathbf{S}(\omega_{cb,tb}^{cb})\omega_{i,tb}^{cb} - \mathbf{J}_c\mathbf{R}_{tb}^{cb}\dot{\omega}_{i,tb}^{tb} \\ &= \mathbf{J}_c\dot{\omega}_{i,cb}^{cb} + \mathbf{J}_c\mathbf{S}(\omega_{i,tb}^{cb})\omega_{cb,tb}^{cb} - \mathbf{J}_c\mathbf{R}_{tb}^{cb}\dot{\omega}_{i,tb}^{tb} \\ &= \mathbf{J}_c\dot{\omega}_{i,cb}^{cb} - \mathbf{J}_c\mathbf{S}(\omega_{i,tb}^{cb})\omega_{tb,cb}^{cb} - \mathbf{J}_c\mathbf{R}_{tb}^{cb}\dot{\omega}_{i,tb}^{tb} \\ &= \mathbf{J}_c\dot{\omega}_{i,cb}^{cb} - \mathbf{J}_c\mathbf{S}(\omega_{i,tb}^{cb})\omega - \mathbf{J}_c\mathbf{R}_{tb}^{cb}\dot{\omega}_{i,tb}^{tb} \end{aligned} \tag{15.41}$$

where $\omega_{cb,tb}^{cb} = -\omega_{tb,cb}^{cb}$ and $\omega_{tb,cb}^{cb} = \omega$ are used in the last two equalities. Using $\omega_{i,cb}^{cb} = \omega + \mathbf{R}_{tb}^{cb}\omega_{i,tb}^{tb}$ and applying (15.39) to $\mathbf{J}_c\dot{\omega}_{i,cb}^{cb}$ yield

$$\begin{aligned} \mathbf{J}_c\dot{\omega}_{i,cb}^{cb} &= -\mathbf{S}(\omega + \mathbf{R}_{tb}^{cb}\omega_{i,tb}^{tb})\mathbf{J}_c(\omega + \mathbf{R}_{tb}^{cb}\omega_{i,tb}^{tb}) + \mathbf{t}_{dc} + \mathbf{t}_{ac} \\ &= -\mathbf{S}(\omega)\mathbf{J}_c(\omega + \mathbf{R}_{tb}^{cb}\omega_{i,tb}^{tb}) - \mathbf{S}\left(\mathbf{R}_{tb}^{cb}\omega_{i,tb}^{tb}\right)\mathbf{J}_c\left(\omega + \mathbf{R}_{tb}^{cb}\omega_{i,tb}^{tb}\right) + \mathbf{t}_{dc} + \mathbf{t}_{ac} \\ &= \mathbf{S}\left(\mathbf{J}_c(\omega + \mathbf{R}_{tb}^{cb}\omega_{i,tb}^{tb})\right)\omega - \mathbf{S}(\mathbf{R}_{tb}^{cb}\omega_{i,tb}^{tb})\mathbf{J}_c\left(\omega + \mathbf{R}_{tb}^{cb}\omega_{i,tb}^{tb}\right) + \mathbf{t}_{dc} + \mathbf{t}_{ac} \\ &= \left[\mathbf{S}\left(\mathbf{J}_c(\omega + \mathbf{R}_{tb}^{cb}\omega_{i,tb}^{tb})\right) - \mathbf{S}\left(\mathbf{R}_{tb}^{cb}\omega_{i,tb}^{tb}\right)\mathbf{J}_c\right]\omega \\ &\quad -\mathbf{S}\left(\mathbf{R}_{tb}^{cb}\omega_{i,tb}^{tb}\right)\mathbf{J}_c\left(\mathbf{R}_{tb}^{cb}\omega_{i,tb}^{tb}\right) + \mathbf{t}_{dc} + \mathbf{t}_{ac} \end{aligned} \tag{15.42}$$

It is straightforward to see that

$$\mathbf{J}_c\mathbf{S}\left(\omega_{i,tb}^{cb}\right)\omega = \mathbf{J}_c\mathbf{S}\left(\mathbf{R}_{tb}^{cb}\omega_{i,tb}^{tb}\right)\omega \tag{15.43}$$

Using (15.39) again gives

$$
\begin{aligned}
& \mathbf{J}_c \mathbf{R}_{tb}^{cb} \dot{\omega}_{i,tb}^{tb} \\
= \; & \mathbf{J}_c \mathbf{R}_{tb}^{cb} \mathbf{J}_t^{-1} \mathbf{J}_t \dot{\omega}_{i,tb}^{tb} \\
= \; & -\mathbf{J}_c \mathbf{R}_{tb}^{cb} \mathbf{J}_t^{-1} \mathbf{S}\left(\omega_{i,tb}^{tb}\right) \mathbf{J}_t \omega_{i,tb}^{tb} + \mathbf{J}_c \mathbf{R}_{tb}^{cb} \mathbf{J}_t^{-1} \mathbf{t}_{dt} + \mathbf{J}_c \mathbf{R}_{tb}^{cb} \mathbf{J}_t^{-1} \mathbf{t}_{at} \quad (15.44)
\end{aligned}
$$

Substituting (15.42), (15.43), and (15.44) into (15.41), we get the model of relative attitude dynamics which is given in chaser's frame as follows (see also [96, 268]):

$$
\mathbf{J}_c \dot{\omega} + \mathbf{C}_r(\omega, \mathbf{q}) \omega + \mathbf{n}_r(\omega, \mathbf{q}) = \mathbf{t}_c + \mathbf{t}_d \quad (15.45)
$$

where $\mathbf{t}_c = \mathbf{t}_{ac} - \mathbf{J}_c \mathbf{R}_{tb}^{cb} \mathbf{J}_t^{-1} \mathbf{t}_{at}$ and $\mathbf{t}_d = \mathbf{t}_{dc} - \mathbf{J}_c \mathbf{R}_{tb}^{cb} \mathbf{J}_t^{-1} \mathbf{t}_{dt}$ are control torque and disturbance torque, respectively, $\mathbf{C}_r(\omega)$ and $\mathbf{n}_r(\omega)$ are given as follows:

$$
\mathbf{C}_r(\omega, \mathbf{q}) = \mathbf{J}_c \mathbf{S}(\mathbf{R}_{tb}^{cb} \omega_{i,tb}^{tb}) + \mathbf{S}(\mathbf{R}_{tb}^{cb} \omega_{i,tb}^{tb}) \mathbf{J}_c - \mathbf{S}(\mathbf{J}_c(\omega + \mathbf{R}_{tb}^{cb} \omega_{i,tb}^{tb})) \quad (15.46)
$$

$$
\mathbf{n}_r(\omega, \mathbf{q}) = \mathbf{S}(\mathbf{R}_{tb}^{cb} \omega_{i,tb}^{tb}) \mathbf{J}_c \mathbf{R}_{tb}^{cb} \omega_{i,tb}^{tb} - \mathbf{J}_c \mathbf{R}_{tb}^{cb} \mathbf{J}_t^{-1} \mathbf{S}(\omega_{i,tb}^{tb}) \mathbf{J}_t \omega_{i,tb}^{tb} \quad (15.47)
$$

In the rest discussion, we consider the rendezvous and soft docking by controlling the chaser spacecraft. Therefore, $\mathbf{t}_c = \mathbf{t}_{ac}$ and $\mathbf{t}_d = \mathbf{t}_{dc}$. At the end of the docking phase, the rotation matrix satisfies $\mathbf{R}_{tb}^{cb} = \mathbf{I}$. If target spacecraft is aligned with the inertial frame, then $\omega_{i,tb}^{tb} = 0$, $\mathbf{C}_r(\omega, \mathbf{q}) = -\mathbf{S}(\mathbf{J}_c \omega)$, and $\mathbf{n}_r(\omega, \mathbf{q}) = 0$.

In the final approaching and docking phase, using reduced quaternion dynamics equation as proposed in [243] can easily be justified because of the small attitude error. The attitude dynamics is given as follows:

$$
\begin{aligned}
\dot{\mathbf{q}} &= \begin{bmatrix} \dot{q}_1 \\ \dot{q}_2 \\ \dot{q}_3 \end{bmatrix} \\
&= \frac{1}{2} \begin{bmatrix} \sqrt{1 - q_1^2 - q_2^2 - q_3^2} & -q_3 & q_2 \\ q_3 & \sqrt{1 - q_1^2 - q_2^2 - q_3^2} & -q_1 \\ -q_2 & q_1 & \sqrt{1 - q_1^2 - q_2^2 - q_3^2} \end{bmatrix} \begin{bmatrix} \omega_1 \\ \omega_2 \\ \omega_3 \end{bmatrix} \\
&= \frac{1}{2} \mathbf{T} \omega \quad (15.48)
\end{aligned}
$$

15.2.3 A Complete Model for Rendezvous and Docking

Let

$$
\mathbf{v} = \dot{\mathbf{d}} \quad (15.49)
$$

which can be obtained by $\dot{\mathbf{d}} \approx \Delta \mathbf{d} / \Delta t$. Now, the result can be summarized by combining equations (15.49), (15.26), (15.45), and (15.48), which yields the complete model for rendezvous and docking:

$$\dot{\mathbf{x}} = \begin{bmatrix} \dot{\mathbf{d}} \\ \dot{\mathbf{v}} \\ \dot{\mathbf{q}} \\ \dot{\omega} \end{bmatrix} = \begin{bmatrix} \mathbf{0} & \mathbf{I} & \mathbf{0} & \mathbf{0} \\ -\frac{1}{m_c}\mathbf{D}_t & -\frac{1}{m_c}\mathbf{C}_t & \mathbf{0} & \mathbf{0} \\ \mathbf{0} & \mathbf{0} & \mathbf{0} & \frac{1}{2}\mathbf{T} \\ \mathbf{0} & \mathbf{0} & \mathbf{0} & -\mathbf{J}_c^{-1}\mathbf{C}_r \end{bmatrix} \begin{bmatrix} \mathbf{d} \\ \mathbf{v} \\ \mathbf{q} \\ \omega \end{bmatrix} - \begin{bmatrix} \mathbf{0} \\ \frac{1}{m_c}\mathbf{n}_t \\ \mathbf{0} \\ \mathbf{J}_c^{-1}\mathbf{n}_r \end{bmatrix} + \begin{bmatrix} \mathbf{0} \\ \frac{1}{m_c}\mathbf{f}_c \\ \mathbf{0} \\ \mathbf{J}_c^{-1}\mathbf{t}_c \end{bmatrix} \tag{15.50}$$

Since $\mathbf{D}_t$, $\mathbf{C}_t$, $\mathbf{T}$, $\mathbf{C}_r$, $\mathbf{n}_t$, and $\mathbf{n}_r$ depend on $\mathbf{q}$, ω, r_c, r_t, θ which are all time-varying, equation (15.50) can be treated as a linear time-varying system.

It is well-known that the control force vector and control torque vector depend on the thruster configurations and many configurations are reported in different systems, for example, [36, 236, 249]. Let $\mathbf{F}_a$ and $\mathbf{T}_a$ be the thruster configuration related matrices that define the control force vector and control torque vector, i.e.,

$$\mathbf{f}_c = \mathbf{F}_a\mathbf{f}_a, \quad \mathbf{t}_c = \mathbf{T}_a\mathbf{f}_a \tag{15.51}$$

where $\mathbf{f}_a$ is the vector of forces generated by thrusters. Substituting (15.51) into (15.50), we have

$$\begin{aligned} \dot{\mathbf{x}} &= \begin{bmatrix} \dot{\mathbf{d}} \\ \dot{\mathbf{v}} \\ \dot{\mathbf{q}} \\ \dot{\omega} \end{bmatrix} = \begin{bmatrix} \mathbf{0} & \mathbf{I} & \mathbf{0} & \mathbf{0} \\ -\frac{1}{m_c}\mathbf{D}_t & -\frac{1}{m_c}\mathbf{C}_t & \mathbf{0} & \mathbf{0} \\ \mathbf{0} & \mathbf{0} & \mathbf{0} & \frac{1}{2}\mathbf{T} \\ \mathbf{0} & \mathbf{0} & \mathbf{0} & -\mathbf{J}_c^{-1}\mathbf{C}_r \end{bmatrix} \begin{bmatrix} \mathbf{d} \\ \mathbf{v} \\ \mathbf{q} \\ \omega \end{bmatrix} - \begin{bmatrix} \mathbf{0} \\ \frac{1}{m_c}\mathbf{n}_t \\ \mathbf{0} \\ \mathbf{J}_c^{-1}\mathbf{n}_r \end{bmatrix} + \begin{bmatrix} \mathbf{0} \\ \frac{1}{m_c}\mathbf{F}_a \\ \mathbf{0} \\ \mathbf{J}_c^{-1}\mathbf{T}_a \end{bmatrix} \mathbf{f}_a \\ &= \mathbf{A}(t)\mathbf{x} - \mathbf{n}_d(t) + \mathbf{B}\mathbf{f}_a \end{aligned} \tag{15.52}$$

Assuming that the chaser's mass change due to fuel consumption is negligible, the matrix $\mathbf{B}$ is then time-invariant. For illustrative purpose, in the rest of the discussion, it is assumed that the thrusters have the configuration considered in [268] which is described in Figure 15.3. But the same idea can be used in other thruster configurations. Therefore, the following relations are easily obtained from Figure 15.3.

$$\mathbf{F}_a = \begin{bmatrix} 0 & 0 & 1 & -1 & 0 & 0 \\ 0 & 0 & 0 & 0 & 1 & -1 \\ 1 & -1 & 0 & 0 & 0 & 0 \end{bmatrix} \tag{15.53}$$

and

$$\mathbf{T}_a = \begin{bmatrix} \frac{L_2}{2} & \frac{L_2}{2} & 0 & 0 & \frac{L_3}{2} & \frac{L_3}{2} \\ -\frac{L_1}{2} & -\frac{L_1}{2} & \frac{L_3}{2} & \frac{L_3}{2} & 0 & 0 \\ 0 & 0 & -\frac{L_2}{2} & -\frac{L_2}{2} & \frac{L_1}{2} & \frac{L_1}{2} \end{bmatrix} \tag{15.54}$$

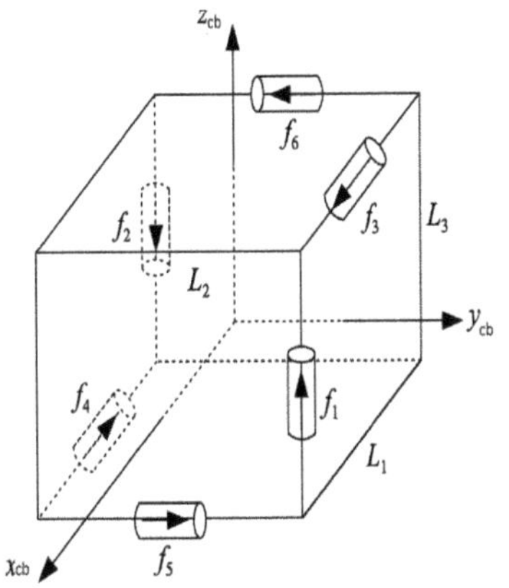

Figure 15.3: Coordinate Frame.

It is easy to check that the following matrix

$$\mathbf{G} := \begin{bmatrix} \mathbf{F}_a \\ \mathbf{T}_a \end{bmatrix} \tag{15.55}$$

is full row rank matrix. As a matter of fact, in engineer practice, thruster configuration should always be designed to be able to fully control the translation and attitude operations. Therefore, we may make the following assumption in the rest of the chapter:

Assumption 2 *The configuration matrix* **G** *is always a full row rank matrix.*

15.3 Model Predictive Control System Design

Although it is difficult to analyze the close-loop stability for MPC control system designs, Theorem 14.1 (see also [169, pages 117–119]) provides a nice sufficient stability criterion for the linear time-varying system. This theorem is the theoretical base for us to use the MPC design for the linear time-varying system. One of the main conditions of the theorem requires that the closed-loop system at every fixed time satisfies $\mathcal{R}_e[\lambda(t)] \leq -\mu$. Clearly, robust pole assignment design guarantees that this condition holds at all sampling time. For any time between the fixed sampling time, robust pole assignment design minimizes the sensitivity of the closed-loop poles to the parameter changes. This is another reason that we select robust pole assignment design over LQR design. The last and the most important reason we select robust pole assignment design is that prescribed pole places are directly related to the closed-loop system performance. In rendezvous and soft docking control, we do not want the relative position and relative attitude response to have any oscillation crossing the horizontal line to avoid collision.

Among various pole assignment algorithms, the one proposed in [201, 258] is chosen because it converges faster than other popular algorithms [150], a critical requirement in MPC design.

We will divide the control force into two parts. The first part is used to cancel $\mathbf{n}_d(t)$ in (15.52). This can be achieved simply by solving the following linear system of equations.

$$\begin{bmatrix} \mathbf{F}_a \\ \mathbf{T}_a \end{bmatrix} \mathbf{u}_1 = \mathbf{G}\mathbf{u}_1 = \begin{bmatrix} \mathbf{n}_t(t) \\ \mathbf{n}_r(t) \end{bmatrix} \tag{15.56}$$

which gives

$$\mathbf{u}_1 = \mathbf{G}^{\dagger} \begin{bmatrix} \mathbf{n}_t(t) \\ \mathbf{n}_r(t) \end{bmatrix} := \mathbf{G}^{\dagger}\mathbf{n} \tag{15.57}$$

where $\mathbf{G}^{\dagger}$ is pseudo-inverse of $\mathbf{G}$. In the example, equations (15.53) and (15.54) implies $\mathbf{G}^{\dagger} = \mathbf{G}^{-1}$.

The design of second part of the thruster force $\mathbf{u}_2$ is based on the following linear time-varying system:

$$\dot{\mathbf{x}} = \mathbf{A}(t)\mathbf{x} + \mathbf{B}\mathbf{u}_2 \tag{15.58}$$

where $\mathbf{x}$, $\mathbf{A}(t)$, and $\mathbf{B}$ are defined as in (15.52). At every sampling time t, $\mathbf{A}(t)$ is evaluated based on the measurable variables. The robust pole assignment algorithm of [201] is called to get the feedback matrix

$$\mathbf{u}_2 = \mathbf{K}(t)\mathbf{x}$$

The feedback force $\mathbf{f}_a = \mathbf{u}_1 + \mathbf{u}_2$ is applied to the linear time-varying system (15.52). The new variables are measured and the next $\mathbf{A}(t)$ is evaluated in the next sampling time, and the process is repeated. *To avoid the oscillation crossing the horizontal line for relative position and relative attitude in the rendezvous and docking process, i.e., to achieve soft docking, the closed-loop poles should be assigned on the negative real axis, i.e., all the poles should be negative and real.*

The MPC algorithm using robust pole assignment is summarized as follows:

Algorithm 15.1

Data: μ, m_c, L_1, L_2, L_3, $\mathbf{J}_c$, $\mathbf{J}_t$, $\mathbf{F}_a$, $\mathbf{T}_a$, *and* $\mathbf{B}$.
Initial condition: At time t_0, *take the measurements* $\theta = \theta_0$, $\mathbf{r}_c$, $\mathbf{r}_t$, $\bar{\mathbf{q}}_{i,tb}$, $\bar{\mathbf{q}}_{i,cb}$, $\omega_{i,cb}^{cb}$, $\omega_{i,tb}^{tb}$, *calculate* $\mathbf{d}$, $\mathbf{r}$, $\mathbf{q}$, $\mathbf{R}_{tb}^{cb}$, ω, *which gives* $\mathbf{x} = \mathbf{x}_0$.

Step 1: Update $\mathbf{n}_t(r_c, r_t)$, $\mathbf{n}_r(\omega, \mathbf{q})$ *which gives* $\mathbf{n}_d(t)$*; update* $\mathbf{A}(t)$ *using* $\mathbf{D}_t(\dot{\theta}, \ddot{\theta}, \mathbf{r}_c)$, $\mathbf{C}_t(\dot{\theta})$, $\mathbf{C}_r(\omega, \mathbf{q})$, *and* $\mathbf{T}(\mathbf{q})$.

Step 2: Calculate the gain $\mathbf{K}$ *for the linear time-varying system (15.58) using robust pole assignment algorithm implemented as* `robpole` *(cf. Appendix C or [201]).*

Step 3: Apply the controlled thruster force $\mathbf{f}_a = \mathbf{u}_1 + \mathbf{u}_2 = \mathbf{G}^{\dagger}\mathbf{n} + \mathbf{K}\mathbf{x}$ *to (15.52).*

Step 4: Take the measurements θ, $\mathbf{r}_c$, $\mathbf{r}_t$, $\bar{\mathbf{q}}_{i,tb}$, $\bar{\mathbf{q}}_{i,cb}$, $\omega_{i,cb}^{cb}$, $\omega_{i,tb}^{tb}$, *calculate* $\mathbf{d}$, $\mathbf{r}$, $\mathbf{q}$, $\mathbf{R}_{tb}^{cb}$, ω, *which gives* $\mathbf{x}$. *Go back to Step 1.*

Remark 15.1 It is worthwhile to emphasize that $\mathbf{B}$ in (15.58) is a constant matrix. This information can be used in `robpole` to reduce the computational burden for the MPC control scheme. ■

15.4 Simulation Test

In this section, two simulation test examples are presented to support the design idea. The simulation examples of [268, 270] and their parameters are used. The simulation results are compared to other designs to demonstrate the superiority of the proposed design.

The first simulation test example is borrowed from [268]. The physics constants, such as gravitational constant $\mu = 3.986004418 * 10^{14} m^3/(kg \cdot s^2)$, Earth's radius 6,371,000 m, are taken from [219]. The rest parameters are taken from [268]: the target spacecraft orbit is circular and the altitude is 250 km, $L_1 = L_2 = L_3 = 2$ m, the mass of the chaser is 10 kg and its inertia matrix is $\mathbf{J}_c = \text{diag}[10,\ 10,\ 10] kg \cdot m^2$, the mass of the target is 10 kg and its inertia matrix is given as

$$\mathbf{J}_t = \begin{bmatrix} 10 & 2.5 & 3.5 \\ 2.5 & 10 & 4.5 \\ 3.5 & 4.5 & 10 \end{bmatrix} kg \cdot m^2$$

$\mathbf{F}_a$ is given in (15.53), $\mathbf{T}_a$ is given in (15.54). The initial condition is set as

$$\mathbf{p}(0) = [10,\ -10,\ 10]^{\mathrm{T}} m$$

$$\mathbf{d}(0) = [0,\ 0,\ 0]^{\mathrm{T}} m/s$$

$$\bar{\mathbf{q}}(0) = [0.3772,\ -0.4329,\ 0.6645,\ 0.4783]^{\mathrm{T}}$$

$$\omega(0) = [0,\ 0,\ 0]^{\mathrm{T}} rad/s$$

To avoid the oscillation of relative distance and relative attitude to guarantee the soft docking, all closed-loop eigenvalues are assigned in negative real axis. Therefore, the proposed closed-loop poles are set to

$$\begin{array}{l} -0.0410, -0.0411, -0.0412, -0.0413, -0.0414, -0.0415, \\ -0.0416, -0.0417, -0.0418, -0.0419, -0.0420, -0.0421. \end{array} \tag{15.59}$$

Applying the on-line Algorithm 15.1 to this problem, the simulation is performed and the closed-loop responses are shown in Figures 15.4–15.6. Figure 15.4 is the response of relative position between the chaser and the target and Figure 15.5 is the response of relative attitude between the chaser and the target. These figures show that the design successfully avoid the oscillation crossing the horizontal line during the docking process and achieved the soft docking. Figure 15.6 depicts the forces in six thrusters used in this docking process, the maximum forces is about 0.17 Newton, which is much smaller than the maximum forces[1] used in the design of [268], which is in the range of 30 Newton.

Comparing to the simulation tests in [7,8,9,10,11,12,13,14,15], the simulation using the proposed method is the only one that does not have oscillation crossing the horizontal line in relative position and attitude responses, which is a clear indication that the design achieves soft docking. The on-line computational time for each call of `robpole` is about 0.1 second on a Dell PC with Intel Core i5-4440 CPU @ 3.10GHz and installed memory of 12GB. Since `robpole` is a MATLAB® code which is an interpreted code. Computational experience shows that a compiled code can be magnitude faster than interpreted code. Therefore, the algorithm will be fast enough in real-time application.

The second simulation test example uses the same spacecraft parameters as described in the first example but uses an elliptical orbit described in Table 1 of [270], where the semi-major axis $a = 2.4616 \times 10^7$ meters, the eccentricity $e = 0.73074$, the specific angular momentum $h = 6.762 \times 10^{10} m^2/s$, and the period of the orbit is 38,436 seconds. To show that the proposed method can achieve the performance of no oscillation crossing the horizontal line for the relative position and relative attitude between the target and chaser spacecraft with measurement error, control error, and external effect, the simulation is performed as follows: the $\mathbf{x}(t)$ applied in feedback is up to 5% deviation from calculated true $\mathbf{x}(t)$. This deviation can be the result of either measurement error or control error or disturbance force. The performance of the position responses and attitude responses in this simulation are provided in Figures 15.7 and 15.8, the required force is given in Figure 15.9. Clearly, the performance of relative position and relative attitude responses meets the design requirement, i.e., there is no oscillation crossing the horizontal line. Also it has been seen that the design is insensitive to measurement error, control error, and external disturbance effect.

[1]but much longer time is used than the design of [268].

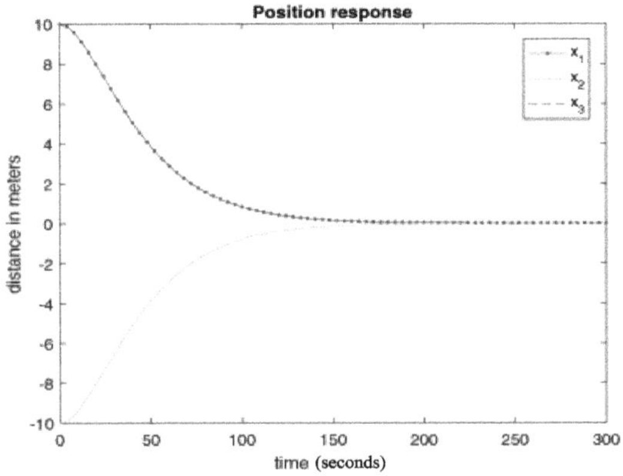

Figure 15.4: Position response for the circular orbit.

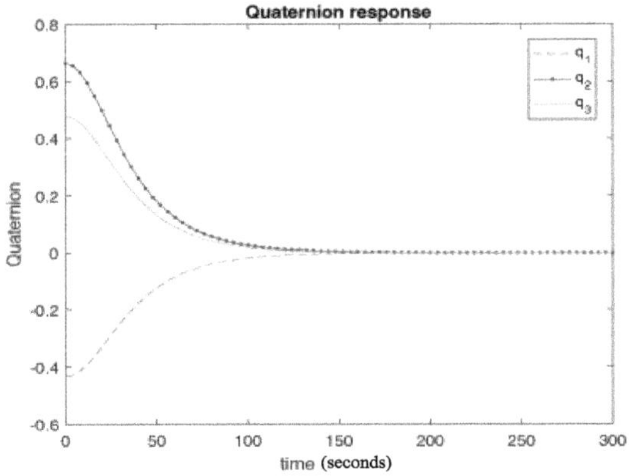

Figure 15.5: Attitude response for the circular orbit.

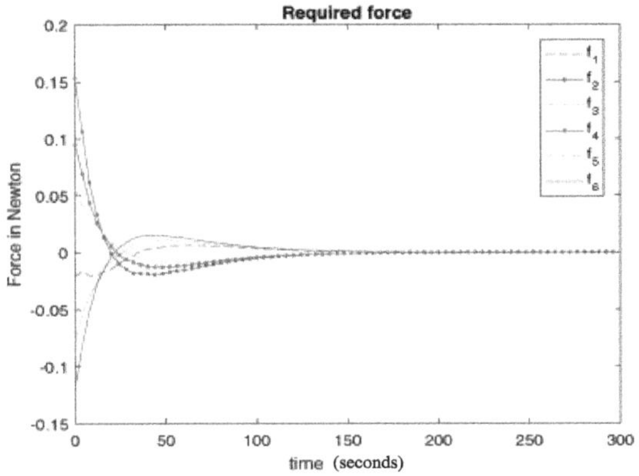

Figure 15.6: Required forces for the circular orbit.

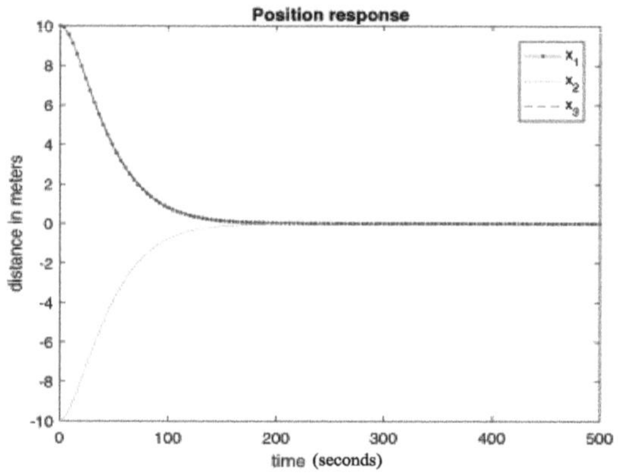

Figure 15.7: Position response for the elliptical orbit.

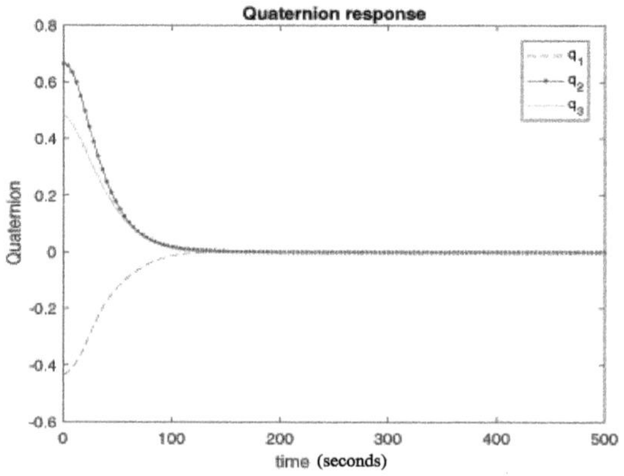

Figure 15.8: Attitude response for the elliptical orbit.

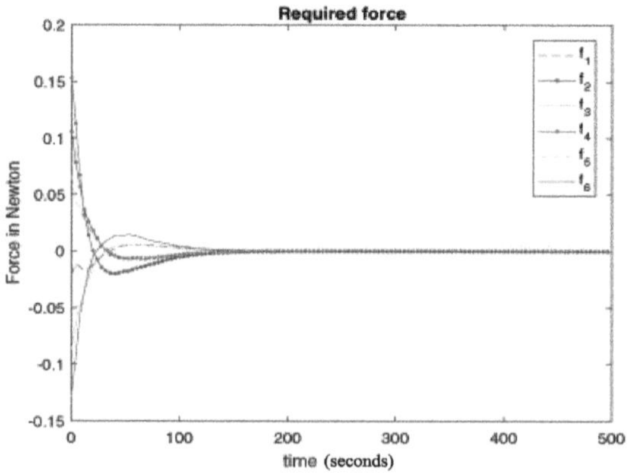

Figure 15.9: Required forces for the elliptical orbit.

Appendix A

First Order Optimality Conditions

In this Appendix, we present the first order optimality conditions for the general constrained optimization problems. These conditions are applicable to linear optimization problem which has linear objective function and linear constraints, convex quadratic optimization problem which has a convex quadratic objective function and linear constraints, and general nonlinear optimization problem which has general nonlinear objective function and nonlinear constraints. Although the first order optimality conditions for the general constrained optimization problems are necessary conditions, these conditions are necessary and sufficient conditions for both linear optimization problem and convex quadratic optimization problem which are considered extensively in this book.

A.1 Problem Introduction

Consider the general optimization problem:

$$\begin{array}{lll} \min_{\mathbf{x}\in\mathbf{R}^n} & f(\mathbf{x}) & \\ \text{subject to} & c_i(\mathbf{x}) = \mathbf{0}, & i \in \mathcal{E} \\ & c_i(\mathbf{x}) \geq \mathbf{0}, & i \in \mathcal{I} \end{array}$$

where f is the objective function and c_i are the constraint functions; these functions are all smooth, real-valued on a subset of $\mathbf{R}^n$, and $\mathcal{E}$ and $\mathcal{I}$ are two finite sets of indices for equality constraints and inequality constraints, respectively.

The feasible set Ω is defined as the set of all points $\mathbf{x}$ that satisfy all the constraints; i.e.,

$$\Omega = \{\mathbf{x} | c_i(\mathbf{x}) = \mathbf{0}, \quad i \in \mathcal{E}; \quad c_i(\mathbf{x}) \geq \mathbf{0}, \quad i \in \mathcal{I}\} \tag{A.1}$$

so that one can rewrite (A.1) as

$$\min_{\Omega} \quad f(\mathbf{x}) \tag{A.2}$$

A vector $\mathbf{x}^*$ is a local solution of the problem (A.1) if $\mathbf{x}^* \in \Omega$ and there is a neighborhood $\mathcal{N}$ of $\mathbf{x}^*$ such that $f(\mathbf{x}) \geq f(\mathbf{x}^*)$ for $\mathbf{x} \in \mathcal{N} \cap \Omega$. A vector $\mathbf{x}^*$ is a strict local solution of the problem (A.1) if $\mathbf{x}^* \in \Omega$ and there is a neighborhood $\mathcal{N}$ of $\mathbf{x}^*$ such that $f(\mathbf{x}) > f(\mathbf{x}^*)$ for $\mathbf{x} \in \mathcal{N} \cap \Omega$ with $\mathbf{x} \neq \mathbf{x}^*$. A vector $\mathbf{x}^*$ is a global solution of the problem (A.1) if $\mathbf{x}^* \in \Omega$ such that $f(\mathbf{x}) \geq f(\mathbf{x}^*)$. A vector $\mathbf{x}^*$ is a strict global solution of the problem (A.1) if $\mathbf{x}^* \in \Omega$ such that $f(\mathbf{x}) > f(\mathbf{x}^*)$ for $\mathbf{x} \in \Omega$ with $\mathbf{x} \neq \mathbf{x}^*$.

A.2 Karush-Kuhn-Tucker Conditions

To state the first order optimality conditions, we introduce the Lagrangian function for the constrained optimization problem (A.1) which is defined as

$$\mathcal{L}(\mathbf{x}, \lambda) = f(\mathbf{x}) - \sum_{i \in \mathcal{E} \cup \mathcal{I}} \lambda_i c_i(\mathbf{x}) \tag{A.3}$$

The active set at any feasible $\mathbf{x}$ is the union of the set $\mathcal{E}$ and the indices of the active inequality constraints given by

$$\mathcal{A}(\mathbf{x}) = \mathcal{E} \cup \{i \in \mathcal{I} | c_i(\mathbf{x}) = \mathbf{0}\} \tag{A.4}$$

The first order optimality conditions are directly related to the linearly independent constraint qualification (LICQ) which is defined as follows:

Definition A.1 Given the point $\mathbf{x}^*$ and the active set $\mathcal{A}(\mathbf{x}^*)$ defined by (A.4), the linear independent constraint qualification is said to be held if the set of active constraint gradients $\{\nabla_i c_i(\mathbf{x}^*), i \in \mathcal{A}(\mathbf{x}^*)\}$ is linearly independent.

Note that if this condition holds, none of the active constraint gradients can be zero. Now we are ready to state the first-order necessary conditions.

Theorem A.1
Suppose that $\mathbf{x}^*$ *is a local solution of (A.1) and that the LICQ holds at* $\mathbf{x}^*$. *Then, there is a Lagrange multiplier vector* λ^*, *with components* $\lambda_i, i \in \mathcal{E} \cup \mathcal{I}$ *such that the following conditions are satisfied at* $(\mathbf{x}^*, \lambda^*)$

$$\nabla_x \mathcal{L}(\mathbf{x}^*, \lambda^*) = \mathbf{0} \tag{A.5a}$$
$$c_i(\mathbf{x}^*) = \mathbf{0}, \quad \forall i \in \mathcal{E} \tag{A.5b}$$
$$c_i(\mathbf{x}^*) \geq \mathbf{0}, \quad \forall i \in \mathcal{I} \tag{A.5c}$$
$$\lambda_i^* \geq \mathbf{0}, \quad \forall i \in \mathcal{I} \tag{A.5d}$$
$$\lambda_i^* c_i(\mathbf{x}^*) = \mathbf{0}, \quad \forall i \in \mathcal{E} \cup \mathcal{I} \tag{A.5e}$$

The proof of Theorem A.1 is very technical, therefore, is omitted. Readers who are interested in the proof are referred to [142]. The conditions of (A.5) are widely known as the Karush-Kuhn-Tucker conditions or KKT conditions for short. The KKT conditions were first proved by Karush in his master thesis in 1939 [89] and rediscovered by Kuhn and Tucker in 1951 [98]. A special solution is important and deserve its own definition:

Definition A.2 Given a local solution $\mathbf{x}^*$ of (A.1) and a vector λ^* satisfying (A.5), we say that the solution is strict complementary if exactly one of λ_i^* and $c_i(\mathbf{x}^*)$ is zero for each index $i \in \mathcal{I}$. In other words, $\lambda_i^* > 0$ for each $i \in \mathcal{I} \cap \mathcal{A}(\mathbf{x}^*)$.

Appendix B

Optimal Control

This appendix provides a brief review of optimal control with focus on discrete-time linear system. The reasons behind the choice of the materials are (a) most computer controlled systems are based on the discrete-time system, (b) the non-linear systems are normally reduced to linear systems so that the complexity is manageable in system design, and (c) we have included the minimum materials in the appendices that will be necessary to understand the main body of the book.

B.1 General Discrete-time Optimal Control Problem

Let the nonlinear system be described by the general discrete-time dynamical equations:

$$\mathbf{x}_{k+1} = \mathbf{f}_k(\mathbf{x}_k, \mathbf{u}_k) \tag{B.1}$$

where $\mathbf{x}_k \in \mathbf{R}^n$ is the state of the system, $\mathbf{u} \in \mathbf{R}^m$ is the control input, and the initial condition is $\mathbf{x}_0$. The subscript k indicates that in general the system and its model can be time-varying. Let the cost function of the system be given as:

$$J = \phi(N, \mathbf{x}_N) + \sum_{k=0}^{N-1} L_k(\mathbf{x}_k, \mathbf{u}_k) \tag{B.2}$$

where $k \in [0, N]$ is the time interval on a discrete scale with a fixed sample step, ϕ is the cost of the final state deviation from zero, and $L_k(\mathbf{x}_k, \mathbf{u}_k)$ is the cost of state and control input at each intermediate time $k \in [0, N-1]$. The optimal control problem is to find an optimal solution $\mathbf{u}_k^*$ on the interval $[0, N-1]$ that minimizes the cost function (B.2) along the trajectory $\mathbf{x}_k^*$ defined by (B.1).

It is worthwhile to note that the discrete-time optimal control problem is a nonlinear constrained optimization problem with its constraint defined by (B.1). This problem is a special case discussed in Appendix A (with only equality constraints (B.1) and objective function of (B.2)) and the solution should satisfy the KKT conditions. Thus, let $\lambda_k \in \mathbf{R}^n$ be the Lagrange multiplier vector, and we define an augmented cost function by

$$J' = \phi(N, \mathbf{x}_N) + \sum_{k=0}^{N-1} \left[L_k(\mathbf{x}_k, \mathbf{u}_k) + \lambda_{k+1}^{\mathrm{T}} (\mathbf{f}_k(\mathbf{x}_k, \mathbf{u}_k) - \mathbf{x}_{k+1}) \right] \tag{B.3}$$

Let the Hamiltonian function be

$$H_k(\mathbf{x}_k, \mathbf{u}_k) = L_k(\mathbf{x}_k, \mathbf{u}_k) + \lambda_{k+1}^{\mathrm{T}} \mathbf{f}_k(\mathbf{x}_k, \mathbf{u}_k) \tag{B.4}$$

then, by rearranging the terms in (B.3), we have

$$J' = \phi(N, \mathbf{x}_N) + \lambda_N^{\mathrm{T}} \mathbf{x}_N + H_0(\mathbf{x}_0, \mathbf{u}_0) + \sum_{k=1}^{N-1} \left[H_k(\mathbf{x}_k, \mathbf{u}_k) - \lambda_k^{\mathrm{T}} \mathbf{x}_k \right] \tag{B.5}$$

The first order necessary optimal conditions are

$$\frac{\partial J'}{\partial \lambda_{k+1}} = \mathbf{0} \quad \Rightarrow \quad \mathbf{x}_{k+1} = \mathbf{f}_k(\mathbf{x}_k, \mathbf{u}_k) \tag{B.6a}$$

$$\frac{\partial J'}{\partial \mathbf{x}_k} = \mathbf{0} \quad \Rightarrow \quad \lambda_k = \frac{\partial H_k}{\partial \mathbf{x}_k} = \left(\frac{\partial \mathbf{f}_k}{\partial \mathbf{x}_k} \right)^{\mathrm{T}} \lambda_{k+1} + \frac{\partial L_k}{\partial \mathbf{x}_k} \tag{B.6b}$$

$$\mathbf{0} = \frac{\partial H_k}{\partial \mathbf{u}_k} = \left(\frac{\partial \mathbf{f}_k}{\partial \mathbf{u}_k} \right)^{\mathrm{T}} \lambda_{k+1} + \frac{\partial L_k}{\partial \mathbf{u}_k} \tag{B.6c}$$

$$0 = \frac{\partial H_0}{\partial \mathbf{x}_0} d\mathbf{x}_0 \tag{B.6d}$$

$$0 = \left(\frac{\partial \phi}{\partial \mathbf{x}_N} - \lambda_N \right)^{\mathrm{T}} d\mathbf{x}_N \tag{B.6e}$$

Since in our problem, $\mathbf{x}_0$ is given, $d\mathbf{x}_0$ is zero, the equation (B.6d) can be omitted. If $\mathbf{x}_N$ is fixed, then we can omit (B.6e). But if $\mathbf{x}_N$ is a free state, then

$$\lambda_N = \frac{\partial \phi}{\partial \mathbf{x}_N} \tag{B.7}$$

is a valid equation. In summary, solving the nonlinear system of equations (B.6) will find the optimal control input $\mathbf{u}_k^*$.

B.2 Solution of Discrete-time LQR Control Problem

In theory, the solution of (B.6) provides the the solution of the general discrete-time optimal control problem. But it is in general very difficult to find the solution of (B.6). In engineering practice, engineers normally reduces the nonlinear system to a linearized system, design the control system for the linear system, and then verify the design actually works for the nonlinear system. Therefore, the solution of discrete-time linear quadratic optimal control problem has been extensively studied. In this case, the system dynamics is reduced to

$$\mathbf{x}_{k+1} = \mathbf{A}_k\mathbf{x}_k + \mathbf{B}_k\mathbf{u}_k \tag{B.8}$$

with the initial condition $\mathbf{x}_0$. The cost function of the system is reduced to:

$$J = \frac{1}{2}\mathbf{x}_N^{\mathrm{T}}\mathbf{Q}_N\mathbf{x}_N + \frac{1}{2}\sum_{k=0}^{N-1}\left(\mathbf{x}_k^{\mathrm{T}}\mathbf{Q}_k\mathbf{x}_k + \mathbf{u}_k^{\mathrm{T}}\mathbf{R}_k\mathbf{u}_k\right) \tag{B.9}$$

where $\mathbf{Q}_k$ and $\mathbf{R}_k$ are positive semi-definite. This problem is referred to as the Linear Quadratic Regulator (LQR) problem. This is a convex quadratic optimization problem discussed in Appendix A. Its Hamiltonian function is given as:

$$H_k(\mathbf{x}_k, \mathbf{u}_k) = \frac{1}{2}\left(\mathbf{x}_k^{\mathrm{T}}\mathbf{Q}_k\mathbf{x}_k + \mathbf{u}_k^{\mathrm{T}}\mathbf{R}_k\mathbf{u}_k\right) + \lambda_{k+1}^{\mathrm{T}}\left(\mathbf{A}_k\mathbf{x}_k + \mathbf{B}_k\mathbf{u}_k\right) \tag{B.10}$$

The first order necessary optimal conditions (B.6) are then reduced to

$$\mathbf{x}_{k+1} = \frac{\partial H_k}{\partial \lambda_{k+1}} = \mathbf{A}_k\mathbf{x}_k + \mathbf{B}_k\mathbf{u}_k \tag{B.11a}$$

$$\lambda_k = \frac{\partial H_k}{\partial \mathbf{x}_k} = \mathbf{Q}_k\mathbf{x}_k + \mathbf{A}_k^{\mathrm{T}}\lambda_{k+1} \tag{B.11b}$$

$$\mathbf{0} = \frac{\partial H_k}{\partial \mathbf{u}_k} = \mathbf{R}_k\mathbf{u}_k + \mathbf{B}_k^{\mathrm{T}}\lambda_{k+1} \tag{B.11c}$$

$$\mathbf{0} = \mathbf{Q}_N\mathbf{x}_N - \lambda_N \tag{B.11d}$$

with $\mathbf{x}_0$ being given. If $\mathbf{x}_N$ is known, we can find the optimal solution by (a) using (B.11d) to get λ_N, (b) using (B.11c) to get $\mathbf{u}_{N-1} = -\mathbf{R}_{N-1}^{-1}\mathbf{B}_{N-1}^{\mathrm{T}}\lambda_N$, (c) using (B.11a) to get $\mathbf{x}_{N-1}$, and (d) using (B.11b) to get λ_{N-1}. Repeating Steps (b), (c), and (d), we should find $\mathbf{x}_0$ as expected. The main problem is that $\mathbf{x}_N$ is not known at the very beginning if $\mathbf{x}_N$ is a fee variable. From (B.11c), we have

$$\mathbf{u}_k = -\mathbf{R}_k^{-1}\mathbf{B}_k^{\mathrm{T}}\lambda_{k+1} \tag{B.12}$$

A very important assumption in the so-called sweep method [24] is about the relation between λ_k and $\mathbf{x}_k$ which is given as follows:

$$\lambda_k = \mathbf{P}_k\mathbf{x}_k \tag{B.13}$$

where $\mathbf{P}_k \in \mathbf{R}^{n\times n}$ is a matrix to be determined. Substituting this relation into (B.11a) yields

$$\mathbf{x}_{k+1} = \mathbf{A}\mathbf{x}_k - \mathbf{B}_k\mathbf{R}_k^{-1}\mathbf{B}_k^{\mathrm{T}}\mathbf{P}_{k+1}\mathbf{x}_{k+1} \tag{B.14}$$

Solving for $\mathbf{x}_{k+1}$ gives

$$\mathbf{x}_{k+1} = (\mathbf{I} + \mathbf{B}_k \mathbf{R}_k^{-1} \mathbf{B}_k^{\mathrm{T}} \mathbf{P}_{k+1})^{-1} \mathbf{A}_k \mathbf{x}_k \tag{B.15}$$

Now substituting (B.13) into (B.11b) yields

$$\mathbf{P}_k \mathbf{x}_k = \mathbf{Q}_k \mathbf{x}_k + \mathbf{A}_k^{\mathrm{T}} \mathbf{P}_{k+1} \mathbf{x}_{k+1} \tag{B.16}$$

Substituting (B.15) into (B.16) yields

$$\mathbf{P}_k \mathbf{x}_k = \mathbf{Q}_k \mathbf{x}_k + \mathbf{A}_k^{\mathrm{T}} \mathbf{P}_{k+1} (\mathbf{I} + \mathbf{B}_k \mathbf{R}_k^{-1} \mathbf{B}_k^{\mathrm{T}} \mathbf{P}_{k+1})^{-1} \mathbf{A}_k \mathbf{x}_k \tag{B.17}$$

Since this equation holds for all possible $\mathbf{x}_k$, we must have the following Riccati matrix equation:

$$\mathbf{P}_k = \mathbf{Q}_k + \mathbf{A}_k^{\mathrm{T}} \mathbf{P}_{k+1} (\mathbf{I} + \mathbf{B}_k \mathbf{R}_k^{-1} \mathbf{B}_k^{\mathrm{T}} \mathbf{P}_{k+1})^{-1} \mathbf{A}_k \tag{B.18}$$

Using the Woodbury identity [155], the Riccati matrix equation can be written as follows:

$$\mathbf{P}_k = \mathbf{Q}_k + \mathbf{A}_k^{\mathrm{T}} \left[\mathbf{P}_{k+1} - \mathbf{P}_{k+1} \mathbf{B}_k (\mathbf{R}_k + \mathbf{B}_k^{\mathrm{T}} \mathbf{P}_{k+1} \mathbf{B}_k)^{-1} \mathbf{B}_k^{\mathrm{T}} \mathbf{P}_{k+1} \right] \mathbf{A}_k \tag{B.19}$$

From (B.11d), we have $\boldsymbol{\lambda}_N = \mathbf{Q}_N \mathbf{x}_N$; therefore,

$$\mathbf{P}_N = \mathbf{Q}_N \tag{B.20}$$

Substituting backward in equation (B.19), we can calculate the solution of the discrete Riccati equation. From (B.12), we have

$$\begin{aligned} \mathbf{u}_k &= -\mathbf{R}_k^{-1} \mathbf{B}_k^{\mathrm{T}} \boldsymbol{\lambda}_{k+1} \\ &= -\mathbf{R}_k^{-1} \mathbf{B}_k^{\mathrm{T}} \mathbf{P}_{k+1} (\mathbf{A}_k \mathbf{x}_k + \mathbf{B}_k \mathbf{u}_k) \\ &= -(\mathbf{R}_k + \mathbf{B}_k^{\mathrm{T}} \mathbf{P}_{k+1} \mathbf{B}_k)^{-1} \mathbf{B}_k^{\mathrm{T}} \mathbf{P}_{k+1} \mathbf{A}_k \mathbf{x}_k \end{aligned} \tag{B.21}$$

Finally, using (B.8), we can obtain the entire state response trajectory.

B.3 LQR Control for Discrete-time LTI System

This section considers a more specific problem: the linear quadratic regulator control for discrete-time linear time-invariant system. There are several reasons that we pay special attention to this problem. First, for linear time-varying system, the Riccati equation solution needs a lot of storage space, especially when N is large. Second, many engineer system can be approximately modeled by linear time-invariant system. For computer-controlled system, the model is in discrete-time. The LQR control for discrete-time LTI system is described as follows:

$$\mathbf{x}_{k+1} = \mathbf{A}\mathbf{x}_k + \mathbf{B}\mathbf{u}_k \tag{B.22}$$

where $\mathbf{A}$ and $\mathbf{B}$ are constant matrices and the initial condition $\mathbf{x}_0$ is given. The cost function of the system is given as:

$$J = \lim_{N\to\infty} \left[\frac{1}{2}\mathbf{x}_N^{\mathrm{T}}\mathbf{Q}\mathbf{x}_N + \frac{1}{2}\sum_{k=0}^{N-1}\left(\mathbf{x}_k^{\mathrm{T}}\mathbf{Q}\mathbf{x}_k + \mathbf{u}_k^{\mathrm{T}}\mathbf{R}\mathbf{u}_k\right)\right] \tag{B.23}$$

where $\mathbf{Q}$ and $\mathbf{R}$ are constant matrices. A key idea to solve this problem is to consider a linear system of equations involving both the state variable $\mathbf{x}_k$ and the co-state variable λ_k. Combining the relation of (B.11) and (B.12) gives the following (see also [208]):

$$\begin{bmatrix} \mathbf{x}_{k+1} \\ \lambda_k \end{bmatrix} = \begin{bmatrix} \mathbf{A} & -\mathbf{B}\mathbf{R}^{-1}\mathbf{B}^{\mathrm{T}} \\ \mathbf{Q} & \mathbf{A}^{\mathrm{T}} \end{bmatrix} \begin{bmatrix} \mathbf{x}_k \\ \lambda_{k+1} \end{bmatrix} \tag{B.24}$$

If $\mathbf{A}$ is invertible, then

$$\mathbf{x}_k = \mathbf{A}^{-1}\mathbf{x}_{k+1} + \mathbf{A}^{-1}\mathbf{B}\mathbf{R}^{-1}\mathbf{B}^{\mathrm{T}}\lambda_{k+1} \tag{B.25}$$

This allows us to have a different expression of (B.24)

$$\begin{bmatrix} \mathbf{x}_k \\ \lambda_k \end{bmatrix} = \begin{bmatrix} \mathbf{A}^{-1} & \mathbf{A}^{-1}\mathbf{B}\mathbf{R}^{-1}\mathbf{B}^{\mathrm{T}} \\ \mathbf{Q}\mathbf{A}^{-1} & \mathbf{A}^{\mathrm{T}} + \mathbf{Q}\mathbf{A}^{-1}\mathbf{B}\mathbf{R}^{-1}\mathbf{B}^{\mathrm{T}} \end{bmatrix} \begin{bmatrix} \mathbf{x}_{k+1} \\ \lambda_{k+1} \end{bmatrix} := \mathbf{H} \begin{bmatrix} \mathbf{x}_{k+1} \\ \lambda_{k+1} \end{bmatrix} \tag{B.26}$$

It is straightforward to verify that

$$\mathbf{H}^{-1} = \begin{bmatrix} \mathbf{A} + \mathbf{B}\mathbf{R}^{-1}\mathbf{B}^{\mathrm{T}}\mathbf{A}^{-\mathrm{T}}\mathbf{Q} & -\mathbf{B}\mathbf{R}^{-1}\mathbf{B}^{\mathrm{T}}\mathbf{A}^{-\mathrm{T}} \\ -\mathbf{A}^{-\mathrm{T}}\mathbf{Q} & \mathbf{A}^{-\mathrm{T}} \end{bmatrix} \tag{B.27}$$

Similar to the assumption for the linear time-varying system, we make a very important assumption as follows:

$$\lambda_k = \mathbf{P}\mathbf{x}_k \tag{B.28}$$

where $\mathbf{P} \in \mathbf{R}^{n\times n}$ is a constant matrix. We expect that the matrix $\mathbf{P}$ is the solution of the Riccati equation (B.19) with $\mathbf{A}_k = \mathbf{A}$, $\mathbf{B}_k = \mathbf{B}$, $\mathbf{Q}_k = \mathbf{Q}$, and $\mathbf{R}_k = \mathbf{R}$ as $k \to \infty$, i.e.,

$$\mathbf{P} = \mathbf{Q} + \mathbf{A}^{\mathrm{T}}\left[\mathbf{P} - \mathbf{P}\mathbf{B}(\mathbf{R} + \mathbf{B}^{\mathrm{T}}\mathbf{P}\mathbf{B})^{-1}\mathbf{B}^{\mathrm{T}}\mathbf{P}\right]\mathbf{A} \tag{B.29}$$

To find $\mathbf{P}$ satisfying (B.28), first we show that there is a matrix $\mathbf{W}$ such that

$$\mathbf{W}^{-1}\mathbf{H}\mathbf{W} = \mathbf{D} \tag{B.30}$$

where $\mathbf{D}$ is a diagonal matrix. Moreover, if μ is an eigenvalue of $\mathbf{D}$, then $\frac{1}{\mu}$ is also an eigenvalue of $\mathbf{D}$ with the same multiplicity. Let $[\mathbf{f}^{\mathrm{T}}, \mathbf{g}^{\mathrm{T}}]^{\mathrm{T}}$ be the eigenvector corresponding to the eigenvalue of μ. Then,

$$\begin{bmatrix} \mathbf{A}^{-1} & \mathbf{A}^{-1}\mathbf{B}\mathbf{R}^{-1}\mathbf{B}^{\mathrm{T}} \\ \mathbf{Q}\mathbf{A}^{-1} & \mathbf{A}^{\mathrm{T}} + \mathbf{Q}\mathbf{A}^{-1}\mathbf{B}\mathbf{R}^{-1}\mathbf{B}^{\mathrm{T}} \end{bmatrix} \begin{bmatrix} \mathbf{f} \\ \mathbf{g} \end{bmatrix} = \mu \begin{bmatrix} \mathbf{f} \\ \mathbf{g} \end{bmatrix} \tag{B.31}$$

This can be rearranged as

$$\begin{bmatrix} (\mathbf{A}+\mathbf{BR}^{-1}\mathbf{B}^{\mathrm{T}}\mathbf{A}^{-\mathrm{T}}\mathbf{Q})^{\mathrm{T}} & -(\mathbf{A}^{-\mathrm{T}}\mathbf{Q})^{\mathrm{T}} \\ -(\mathbf{BR}^{-1}\mathbf{B}^{\mathrm{T}}\mathbf{A}^{-\mathrm{T}})^{\mathrm{T}} & \mathbf{A}^{-1} \end{bmatrix} \begin{bmatrix} \mathbf{g} \\ -\mathbf{f} \end{bmatrix} = \mu \begin{bmatrix} \mathbf{g} \\ -\mathbf{f} \end{bmatrix} \tag{B.32}$$

Since $\mathbf{Q}$ and $\mathbf{R}$ are symmetric, the last equation is equivalent to

$$\mathbf{H}^{-\mathrm{T}} \begin{bmatrix} \mathbf{g} \\ -\mathbf{f} \end{bmatrix} = \mu \begin{bmatrix} \mathbf{g} \\ -\mathbf{f} \end{bmatrix}$$

i.e., μ is an eigenvalue of $\mathbf{H}^{-\mathrm{T}}$; therefore, μ is an eigenvalue of $\mathbf{H}^{-1}$. This proves that $\frac{1}{\mu}$ is also an eigenvalue of $\mathbf{H}$ and there is an invertible matrix $\mathbf{W}$ and a diagonal matrix $\mathbf{D}$ such that (B.30) holds. Now, we consider $[\mathbf{w}_k^{\mathrm{T}}, \mathbf{z}_k^{\mathrm{T}}]^{\mathrm{T}}$ which satisfies the following relation

$$\begin{bmatrix} \mathbf{x}_k \\ \lambda_k \end{bmatrix} = \mathbf{W} \begin{bmatrix} \mathbf{w}_k \\ \mathbf{z}_k \end{bmatrix} = \begin{bmatrix} \mathbf{W}_{11} & \mathbf{W}_{12} \\ \mathbf{W}_{21} & \mathbf{W}_{22} \end{bmatrix} \begin{bmatrix} \mathbf{w}_k \\ \mathbf{z}_k \end{bmatrix} \tag{B.33}$$

Combining (B.26), (B.30), and (B.33), and using the fact that both μ and $\frac{1}{\mu}$ are eigenvalues of $\mathbf{H}$, we have

$$\begin{bmatrix} \mathbf{w}_k \\ \mathbf{z}_k \end{bmatrix} = \mathbf{D} \begin{bmatrix} \mathbf{w}_{k+1} \\ \mathbf{z}_{k+1} \end{bmatrix} := \begin{bmatrix} \mathbf{M} & \mathbf{0} \\ \mathbf{0} & \mathbf{M}^{-1} \end{bmatrix} \begin{bmatrix} \mathbf{w}_{k+1} \\ \mathbf{z}_{k+1} \end{bmatrix} \tag{B.34}$$

where $\mathbf{M}$ is a diagonal matrix and all diagonal elements are outside the unit circle. Repeatedly using (B.34), we have

$$\begin{bmatrix} \mathbf{w}_k \\ \mathbf{z}_k \end{bmatrix} = \begin{bmatrix} \mathbf{M}^{N-k} & \mathbf{0} \\ \mathbf{0} & \mathbf{M}^{-(N-k)} \end{bmatrix} \begin{bmatrix} \mathbf{w}_N \\ \mathbf{z}_N \end{bmatrix} \tag{B.35}$$

Since we want to let $N \to \infty$ for the steady-state solution to the infinite time problem, and $\mathbf{M}$ is not stable, we rewrite (B.35) as

$$\begin{bmatrix} \mathbf{w}_N \\ \mathbf{z}_k \end{bmatrix} = \begin{bmatrix} \mathbf{M}^{-(N-k)} & \mathbf{0} \\ \mathbf{0} & \mathbf{M}^{-(N-k)} \end{bmatrix} \begin{bmatrix} \mathbf{w}_k \\ \mathbf{z}_N \end{bmatrix} \tag{B.36}$$

Now we consider the relations between $\mathbf{x}_k$ and λ_k to determine $\mathbf{P}$. From (B.33), we have

$$\lambda_N = \mathbf{W}_{21}\mathbf{w}_N + \mathbf{W}_{22}\mathbf{z}_N = \mathbf{P}\mathbf{x}_N = \mathbf{P}(\mathbf{W}_{11}\mathbf{w}_N + \mathbf{W}_{12}\mathbf{z}_N) \tag{B.37}$$

Solving $\mathbf{z}_N$ in terms of $\mathbf{w}_N$ yields

$$\mathbf{z}_N = -(\mathbf{W}_{22} - \mathbf{P}\mathbf{W}_{12})^{-1}(\mathbf{W}_{21} - \mathbf{P}\mathbf{W}_{11})\mathbf{w}_N := \mathbf{T}\mathbf{w}_N \tag{B.38}$$

From (B.36) and (B.38), we have

$$\mathbf{z}_k = \mathbf{M}^{-(N-k)}\mathbf{z}_N = \mathbf{M}^{-(N-k)}\mathbf{T}\mathbf{w}_N = \mathbf{M}^{-(N-k)}\mathbf{T}\mathbf{M}^{-(N-k)}\mathbf{w}_k := \mathbf{T}_k\mathbf{w}_k \tag{B.39}$$

Using (B.33) again,

$$\lambda_k = \mathbf{W}_{21}\mathbf{w}_k + \mathbf{W}_{22}\mathbf{z}_k = \mathbf{P}\mathbf{x}_k = \mathbf{P}(\mathbf{W}_{11}\mathbf{w}_k + \mathbf{W}_{12}\mathbf{z}_k) \tag{B.40}$$

Substituting (B.39) into (B.40) yields

$$(\mathbf{W}_{21} + \mathbf{W}_{22}\mathbf{T}_k)\mathbf{w}_k = \mathbf{P}(\mathbf{W}_{11} + \mathbf{W}_{12}\mathbf{T}_k)\mathbf{w}_k \tag{B.41}$$

Since this must hold for all $\mathbf{w}_k$, we have

$$\mathbf{P} = (\mathbf{W}_{21} + \mathbf{W}_{22}\mathbf{T}_k)(\mathbf{W}_{11} + \mathbf{W}_{12}\mathbf{T}_k)^{-1} \tag{B.42}$$

As $N \to \infty$, we have, $\mathbf{T}_k = \mathbf{M}^{-(N-k)}\mathbf{T}\mathbf{M}^{-(N-k)} \to \mathbf{0}$, therefore,

$$\mathbf{P} = \mathbf{W}_{21}\mathbf{W}_{11}^{-1} \tag{B.43}$$

Now we prove that $\mathbf{P}$ is the solution of the Riccati equation (B.29). Note that

$$\begin{aligned} \mathbf{H} &= \begin{bmatrix} \mathbf{A}^{-1} & \mathbf{A}^{-1}\mathbf{B}\mathbf{R}^{-1}\mathbf{B}^{\mathrm{T}} \\ \mathbf{Q}\mathbf{A}^{-1} & \mathbf{A}^{\mathrm{T}} + \mathbf{Q}\mathbf{A}^{-1}\mathbf{B}\mathbf{R}^{-1}\mathbf{B}^{\mathrm{T}} \end{bmatrix} \\ &= \begin{bmatrix} \mathbf{A} & \mathbf{0} \\ -\mathbf{Q} & \mathbf{I} \end{bmatrix}^{-1} \begin{bmatrix} \mathbf{I} & \mathbf{B}\mathbf{R}^{-1}\mathbf{B}^{\mathrm{T}} \\ \mathbf{0} & \mathbf{A}^{\mathrm{T}} \end{bmatrix} := \mathbf{E}^{-1}\mathbf{F} \end{aligned} \tag{B.44}$$

From (B.30) and (B.34), we have

$$\begin{aligned} & \mathbf{H}\mathbf{W} = \mathbf{W}\mathbf{D} \\ \Longleftrightarrow \quad & \mathbf{F}\mathbf{W} = \mathbf{E}\mathbf{W}\mathbf{D} \\ \Longleftrightarrow \quad & \begin{bmatrix} \mathbf{I} & \mathbf{B}\mathbf{R}^{-1}\mathbf{B}^{\mathrm{T}} \\ \mathbf{0} & \mathbf{A}^{\mathrm{T}} \end{bmatrix} \begin{bmatrix} \mathbf{W}_{11} \\ \mathbf{W}_{21} \end{bmatrix} = \begin{bmatrix} \mathbf{A} & \mathbf{0} \\ -\mathbf{Q} & \mathbf{I} \end{bmatrix} \begin{bmatrix} \mathbf{W}_{11}\mathbf{M} \\ \mathbf{W}_{21}\mathbf{M} \end{bmatrix} \end{aligned} \tag{B.45}$$

The first row of (B.45) gives

$$\mathbf{W}_{11} + \mathbf{B}\mathbf{R}^{-1}\mathbf{B}^{\mathrm{T}}\mathbf{W}_{21} = \mathbf{A}\mathbf{W}_{11}\mathbf{M}$$

$$\Longleftrightarrow \quad \mathbf{A} = \mathbf{W}_{11}\mathbf{M}^{-1}\mathbf{W}_{11}^{-1} + \mathbf{B}\mathbf{R}^{-1}\mathbf{B}^{\mathrm{T}}\mathbf{W}_{21}\mathbf{M}^{-1}\mathbf{W}_{11}^{-1} \tag{B.46}$$

$$\Longleftrightarrow \quad \mathbf{A} = \left[\mathbf{W}_{11} + \mathbf{B}\mathbf{R}^{-1}\mathbf{B}^{\mathrm{T}}\mathbf{W}_{21}\right]\mathbf{M}^{-1}\mathbf{W}_{11}^{-1} \tag{B.47}$$

The second row of (B.45) gives

$$\mathbf{A}^{\mathrm{T}}\mathbf{W}_{21} = -\mathbf{Q}\mathbf{W}_{11}\mathbf{M} + \mathbf{W}_{21}\mathbf{M}$$

$$\Longleftrightarrow \quad \mathbf{Q} = \mathbf{W}_{21}\mathbf{W}_{11}^{-1} - \mathbf{A}^{\mathrm{T}}\mathbf{W}_{21}\mathbf{M}^{-1}\mathbf{W}_{11}^{-1} \tag{B.48}$$

Denote $\mathbf{G} = \mathbf{B}^{\mathrm{T}}\mathbf{P}\mathbf{B}$. Substituting (B.43), (B.46), (B.47), and (B.48) into (B.29) yields

$$\begin{aligned} & -\mathbf{P} + \mathbf{A}^{\mathrm{T}}\left[\mathbf{P} - \mathbf{P}\mathbf{B}(\mathbf{R} + \mathbf{B}^{\mathrm{T}}\mathbf{P}\mathbf{B})^{-1}\mathbf{B}^{\mathrm{T}}\mathbf{P}\right]\mathbf{A} + \mathbf{Q} \\ = \quad & -\mathbf{W}_{21}\mathbf{W}_{11}^{-1} + \mathbf{A}^{\mathrm{T}}\mathbf{W}_{21}\mathbf{W}_{11}^{-1}\left(\mathbf{W}_{11}\mathbf{M}^{-1}\mathbf{W}_{11}^{-1} + \mathbf{B}\mathbf{R}^{-1}\mathbf{B}^{\mathrm{T}}\mathbf{W}_{21}\mathbf{M}^{-1}\mathbf{W}_{11}^{-1}\right) \end{aligned}$$

$$
\begin{aligned}
& -\mathbf{A}^{\mathrm{T}}\mathbf{PB}(\mathbf{R}+\mathbf{B}^{\mathrm{T}}\mathbf{PB})^{-1}\mathbf{B}^{\mathrm{T}}\mathbf{PA}+\mathbf{W}_{21}\mathbf{W}_{11}^{-1}-\mathbf{A}^{\mathrm{T}}\mathbf{W}_{21}\mathbf{M}^{-1}\mathbf{W}_{11}^{-1} \\
= \; & \mathbf{A}^{\mathrm{T}}\mathbf{W}_{21}\mathbf{W}_{11}^{-1}\mathbf{BR}^{-1}\mathbf{B}^{\mathrm{T}}\mathbf{W}_{21}\mathbf{M}^{-1}\mathbf{W}_{11}^{-1}-\mathbf{A}^{\mathrm{T}}\mathbf{PB}(\mathbf{R}+\mathbf{B}^{\mathrm{T}}\mathbf{PB})^{-1}\mathbf{B}^{\mathrm{T}}\mathbf{PA} \\
= \; & \mathbf{A}^{\mathrm{T}}\mathbf{PB}\left[\mathbf{R}^{-1}\mathbf{B}^{\mathrm{T}}\mathbf{W}_{21}\mathbf{M}^{-1}\mathbf{W}_{11}^{-1}-(\mathbf{R}+\mathbf{B}^{\mathrm{T}}\mathbf{PB})^{-1}\mathbf{B}^{\mathrm{T}}\mathbf{PA}\right] \\
= \; & \mathbf{A}^{\mathrm{T}}\mathbf{PB}\left[\mathbf{R}^{-1}\mathbf{B}^{\mathrm{T}}\mathbf{W}_{21}-(\mathbf{R}+\mathbf{G})^{-1}\mathbf{B}^{\mathrm{T}}\mathbf{W}_{21}\mathbf{W}_{11}^{-1}\left(\mathbf{W}_{11}+\mathbf{BR}^{-1}\mathbf{B}^{\mathrm{T}}\mathbf{W}_{21}\right)\right] \\
& \mathbf{M}^{-1}\mathbf{W}_{11}^{-1} \\
= \; & \mathbf{A}^{\mathrm{T}}\mathbf{PB}\left[\mathbf{R}^{-1}-(\mathbf{R}+\mathbf{G})^{-1}-(\mathbf{R}+\mathbf{G})^{-1}\mathbf{GR}^{-1}\right]\mathbf{B}^{\mathrm{T}}\mathbf{W}_{21}\mathbf{M}^{-1}\mathbf{W}_{11}^{-1} \\
= \; & \mathbf{A}^{\mathrm{T}}\mathbf{PB}(\mathbf{R}+\mathbf{G})^{-1}\left[(\mathbf{R}+\mathbf{G})\mathbf{R}^{-1}-\mathbf{I}-\mathbf{GR}^{-1}\right]\mathbf{B}^{\mathrm{T}}\mathbf{W}_{21}\mathbf{M}^{-1}\mathbf{W}_{11}^{-1} \\
= \; & \mathbf{0} \qquad \text{(since } (\mathbf{R}+\mathbf{G})\mathbf{R}^{-1}-\mathbf{I}-\mathbf{GR}^{-1}=\mathbf{0}\text{)}
\end{aligned} \tag{B.49}
$$

This proves that $\mathbf{P}=\mathbf{W}_{21}\mathbf{W}_{11}^{-1}$ is indeed the solution of the discrete Riccati equation.

The optimal feedback is given by

$$
\mathbf{u}_k = -(\mathbf{R}+\mathbf{B}^{\mathrm{T}}\mathbf{PB})^{-1}\mathbf{B}^{\mathrm{T}}\mathbf{PA}\mathbf{x}_k = -\mathbf{K}\mathbf{x}_k \tag{B.50}
$$

Appendix C

Robust Pole Assignment

This appendix provides a brief review of robust pole assignment with focus on continuous-time linear system. The reasons behind the choice of the materials are (a) most existing literatures discuss continuous-time linear system, (b) extension to the discrete-time system is straightforward, and (c) we have included the minimum materials in the appendices that will be necessary to understand the main body of the book. In this appendix, we will consider the following linear time-invariant system:

$$\dot{\mathbf{x}} = \mathbf{A}\mathbf{x} + \mathbf{B}\mathbf{u} \tag{C.1}$$

where $\mathbf{x} \in \mathbf{R}^n$, $\mathbf{u} \in \mathbf{R}^m$, $\mathbf{A} \in \mathbf{R}^{n\times n}$, and $\mathbf{B} \in \mathbf{R}^{n\times m}$. We assume that $(\mathbf{A}, \mathbf{B})$ is controllable, and $rank(\mathbf{B}) = m > 1$. Under this assumption, the pole assignment design is not unique. Therefore, we can use the extra degrees of freedom to achieve more desired features than the required performance. One of the important desired features is system robustness to the modeling error. A design with this feature is called the robust pole assignment, which can be defined as follows:

Robust pole assignment: For system given in (C.1) with $(\mathbf{A}, \mathbf{B})$ controllable and $rank(\mathbf{B}) > 1$, and a given set of desired close-loop poles $\{\lambda_1, \dots, \lambda_n\}$, robust pole assignment design is to find a feedback control $\mathbf{u} = -\mathbf{K}\mathbf{x} = \mathbf{F}\mathbf{x}$ such that the closed-loop poles are as much insensitive to the system parameter perturbation as possible.

C.1 Eigenvalue Sensitivity to the Perturbation

For square matrices, a variety of robustness measures have been proposed to measure the robustness of their eigen-structure. When all the eigenvalues are simple, the first order sensitivity of each individual λ_i to uncertainty is given by the *eigenvalue condition number* [227]

$$c_i := \frac{\|\mathbf{y}_i\|_2 \|\mathbf{x}_i\|_2}{|\mathbf{y}_i^{\mathrm{T}} \mathbf{x}_i|} \tag{C.2}$$

where $\mathbf{y}_i$ and $\mathbf{x}_i$ are the left and right eigenvectors associated with λ_i; c_i is the Frobenius norm of the gradient of $\lambda_i(\mathbf{X})$ with respect to $\mathbf{X}$ under the (natural) trace inner product. We use

$$c_\infty := \max_i \; c_i \tag{C.3}$$

to denote the worst-case eigenvalue condition number. For the case where $\mathbf{X}$ is non-defective but has repeated eigenvalues, see [192] for a definition of the corresponding condition numbers.

The Bauer-Fike theorem [58] established that c_∞ is upper-bounded by the *spectral condition number* of the matrix of eigenvectors

$$\kappa_2(\mathbf{X}) := \|\mathbf{X}\|_2 \|\mathbf{X}^{-1}\|_2 \tag{C.4}$$

and hence this is often used as a robustness measure. The *Frobenius condition number* of $\mathbf{X}$ is given by

$$\kappa_{fro}(\mathbf{X}) := \|\mathbf{X}\|_{fro} \|\mathbf{X}^{-1}\|_{fro} \tag{C.5}$$

Since $\kappa_2(\mathbf{X}) \leq \kappa_{fro}(\mathbf{X})$, the Frobenius condition number provides a more conservative bound on the eigenvalue sensitivity than $\kappa_2(\mathbf{X})$, but enjoys the virtue of being differentiable, and hence is often used as a robustness measure.

Minimizing the measures c_∞, $\kappa_2(\mathbf{X})$ and $\kappa_{fro}(\mathbf{X})$ corresponds to superior robustness, with perfect robustness being achieved only when the eigenvector matrix is unitary, i.e., when $\mathbf{M}$ is normal.

Another robustness measure was proposed in [239] which is given as follows:

$$|\det(\mathbf{X})| := \sqrt{\det(\mathbf{X}\mathbf{X}^*)} \tag{C.6}$$

where all columns of $\mathbf{X}$ are unit length and $\mathbf{X}^*$ is the complex conjugate of $\mathbf{X}$. This robustness measure is the volume of the box spanned by unit length column vectors of $\mathbf{X}$ and is clearly a good measure of orthogonality, and hence it may be used as the robustness measure. Again, the perfect robustness being achieved for this metric is only when the eigenvector matrix is unitary.

Let $\sigma_i, i = 1, 2, \ldots, n$ be the singular values of $\mathbf{X}$ with σ_1 the largest singular value and σ_n the smallest singular values. It is well known that

$$\kappa_2(\mathbf{X}) = \sigma_1 / \sigma_n \tag{C.7}$$

Because

$$\sigma_1^2 \leq \sum_{i=1}^{n} \sigma_i^2 = \text{trace}(\mathbf{X}^{\mathrm{T}}\mathbf{X}) = n \tag{C.8}$$

we have $\sigma_1 \leq \sqrt{n}$, i.e., σ_1 is bounded above. On the other hand, if $\sigma_n \to 0$, then $\kappa_2 \to \infty$. Therefore, σ_n is the dominant factor of $\kappa_2(\mathbf{X})$. The rest of this section is to estimate the low bound of σ_n. First, we introduce a Lemma [79].

Lemma C.1

Suppose that the real coefficient polynomial $f(x) = a_n x^n + a_{n-1}x^{n-1} + \ldots + a_1 x + a_0$ with $a_n > 0$ and a_{n-k} being the first negative coefficient, and B is the greatest value among all the absolute values of the negative coefficients. Then

$$N = 1 + \sqrt[k]{B/a_n} \tag{C.9}$$

is a upper bound of positive root of $f(x)$.

Proof C.1 Assume that

$$x > 1 + \sqrt[k]{B/a_n} \tag{C.10}$$

Since $a_{n-1}, a_{n-2}, \ldots, a_{n-k+1} \geq 0$, and $a_{n-k}, a_{n-k-1}, \ldots, a_0 \geq -B$, we have

$$\begin{aligned}
f(x) &= a_n x^n + a_{n-1}x^{n-1} + \ldots + a_1 x + a_0 \\
&\geq a_n x^n - B(x^{n-k} + x^{n-k-1} + \ldots + x + 1) \\
&= a_n x^n - B\frac{x^{n-k+1} - 1}{x-1} \\
&> a_n x^n - B\frac{x^{n-k+1}}{x-1} \\
&= \frac{x^{n-k+1}}{x-1}\left[a_n x^{k-1}(x-1) - B\right] \\
&> \frac{x^{n-k+1}}{x-1}\left[a_n (x-1)^k - B\right]
\end{aligned} \tag{C.11}$$

For any x satisfying (C.10), it must have $f(x) > 0$. Therefore, a upper bound of positive solution of $f(x) = 0$ is given by (C.9). ■

Theorem C.1

Suppose that $\mathbf{X}$ is a matrix composed of standardized eigenvectors and generalized eigenvectors of matrix $\mathbf{A}$. Denote $\Delta = \det(\mathbf{X}^{\mathrm{T}}\mathbf{X})$. Then, we have

$$\sigma_n^2 \geq \frac{1}{1 + \frac{1}{\Delta}\left(\frac{n}{n-1}\right)^{n-1}} \tag{C.12}$$

Proof C.2 Let $\lambda_i, i = 1, \ldots, n$ be the eigenvalues of $\mathbf{X}^{\mathrm{T}}\mathbf{X}$, note that $\lambda_i = \sigma_i^2$, we have a matrix $\mathbf{Y}$ such that

$$(\mathbf{X}^{\mathrm{T}}\mathbf{X})\mathbf{Y} = \mathbf{Y}\mathrm{diag}(\sigma_1^2, \ldots, \sigma_n^2) \tag{C.13}$$

moreover (C.8) and the following relation

$$\prod_{i=1}^{n} \sigma_i^2 = \prod_{i=1}^{n} \lambda_i(\mathbf{X}^{\mathrm{T}}\mathbf{X}) = \det(\mathbf{X}^{\mathrm{T}}\mathbf{X}) = \Delta \tag{C.14}$$

hold. Denote the sets

$$K_1 = \{\sigma_i | \sigma_i \text{ satisfy } (C.13), (C.8), (C.14)\}$$

$$K_2 = \{\sigma_i | \sigma_i \text{ satisfy } (C.8), (C.14)\}$$

We have $\min_{\sigma_i \in K_1} \sigma_n^2 \geq \min_{\sigma_i \in K_2} \sigma_n^2$. The lower bound is established based on $\min_{\sigma_i \in K_2} \sigma_n^2$. The Lagrangian for this minimization problem is given by

$$L = \sigma_n^2 + \beta_0 \left(\prod_{i=1}^{n} \sigma_i^2 - \Delta \right) + \beta_1 \left(\sum_{i=1}^{n} \sigma_i^2 - n \right)$$

Therefore, we have

$$\frac{\partial L}{\partial \sigma_i} = 2\beta_0 \prod_{j \neq i} \sigma_j^2 \sigma_i + 2\beta_1 \sigma_i = 0, \quad i \neq n \tag{C.15a}$$

$$\frac{\partial L}{\partial \sigma_n} = 2\sigma_n + 2\beta_0 \prod_{j \neq n} \sigma_j^2 \sigma_n + 2\beta_1 \sigma_n = 0 \tag{C.15b}$$

$$\frac{\partial L}{\partial \beta_0} = \prod_{i=1}^{n} \sigma_i^2 - \Delta = 0 \tag{C.15c}$$

$$\frac{\partial L}{\partial \beta_1} = \sum_{i=1}^{n} \sigma_i^2 - n = 0 \tag{C.15d}$$

and

$$\begin{aligned} & \frac{\partial L}{\partial \sigma_n} - \frac{\partial L}{\partial \sigma_i} \\ = \; & 2\sigma_n + 2\beta_0 \prod_{j \neq i,n} \sigma_j^2 \left(\sigma_n \sigma_i^2 - \sigma_n^2 \sigma_i \right) + 2\beta_1 (\sigma_n - \sigma_i) \\ = \; & 2\sigma_n + 2 \left[\beta_0 \prod_{j \neq i,n} \sigma_j^2 \sigma_n \sigma_i - \beta_1 \right] (\sigma_i - \sigma_n) = 0 \end{aligned} \tag{C.16}$$

Since $\det(\mathbf{X}^{\mathrm{T}}\mathbf{X}) \neq 0$, we have $\sigma_n \neq 0$, which means

$$\sigma_i \neq \sigma_n, \quad \forall i \neq n \tag{C.17}$$

Since

$$\frac{\partial L}{\partial \sigma_i} - \frac{\partial L}{\partial \sigma_j} = \left[2\beta_0 \left(\prod_{k \neq i,j} \sigma_k^2 \sigma_j \sigma_i \right) - 2\beta_1 \right] (\sigma_j - \sigma_i) = 0$$

we have either one or both of the following relations hold.

$$\sigma_j = \sigma_i, \quad i \neq j, \quad \forall i, j \neq n \tag{C.18a}$$

$$\beta_1/\beta_0 = \prod_{k \neq i,j} \sigma_k^2 \sigma_j \sigma_i, \quad \forall i, j, k \neq n, \quad i \neq j \neq k \tag{C.18b}$$

By symmetry, the second relation is equivalent to the first one. Substituting (C.17) and (C.18a) into (C.15c) and (C.15d) and denoting $\lambda = \lambda_i, i \neq n$, we have

$$\lambda^{(n-1)}\lambda_n = \Delta, \quad (n-1)\lambda + \lambda_n = n \tag{C.19a}$$

$$f(\lambda_n) = (n - \lambda_n)^{(n-1)}\lambda_n - \Delta(n-1)^{(n-1)} = 0 \tag{C.19b}$$

Since λ_n is positive and is a solution of (C.19b), λ_n must be greater than or equal to the smallest positive root of (C.19b). It is hard to get the analytic solution of the smallest positive root. But we can estimate the lower bound of the smallest positive root. Considering $\phi(\lambda_n) = \lambda_n^n f(\frac{1}{\lambda_n})$, if α is an arbitrary positive root of $f(\lambda_n)$, then $\frac{1}{\alpha}$ is a positive root of $\phi(\lambda_n)$. If N is an upper bound of the positive roots of $\phi(\lambda_n)$, then $\frac{1}{N}$ is a lower bound of the positive roots of $f(\lambda_n)$. Note that

$$\begin{aligned}
\phi(\lambda_n) &= \lambda_n^n \left[\frac{1}{\lambda_n}\left(n - \frac{1}{\lambda_n}\right)^{n-1} - \Delta(n-1)^{(n-1)} \right] \\
&= (n\lambda_n - 1)^{n-1} - \Delta(n-1)^{n-1}\lambda_n^n \\
&= \Delta(n-1)^{n-1}\lambda_n^n - n^{n-1}\lambda_n^{n-1} + n^{n-2}(n-1)\lambda_n^{n-1} + \ldots + (-1)^{n-1} = 0
\end{aligned} \tag{C.20}$$

According to the Lemma, we have

$$N = 1 + \frac{1}{\Delta}\left(\frac{n}{n-1}\right)^{(n-1)} \tag{C.21}$$

Therefore, the smallest positive solution λ_n^* of $f(\lambda_n)$ satisfies

$$\lambda_n^* \geq \frac{1}{1 + \frac{1}{\Delta}\left(\frac{n}{n-1}\right)^{(n-1)}} \tag{C.22}$$

This finishes the proof. ■

Unlike $\kappa_2(\mathbf{X})$, which depends on the largest and the smallest singular values (C.7), $|\det(\mathbf{X})| = \prod_{i=1}^n \sigma_i$ depends on all singular values of $\mathbf{X}$, which may be a better robustness measure than $\kappa_2(\mathbf{X})$. Because the merits of c_∞ and $|\det(\mathbf{X})|$, these two robustness measures were used in [150] to compare different robust pole assignment algorithms. To make the range of $|\det(\mathbf{X})|$ similar to other robustness metrics, [150] introduces

$$\Gamma(\mathbf{X}) := 1 - \log(|\det(\mathbf{X})|) \tag{C.23}$$

as an equivalent alternative to the maximization measure $|\det(\mathbf{X})|$. The use of $\Gamma(\mathbf{X})$ is preferred for consistency with c_∞ in that smaller values correspond to superior robustness. Since computation of $\log(|\det(\mathbf{X})|)$ is not numerically reliable in MATLAB®, an equivalent alternative is proposed as follows:

$$\Gamma(\mathbf{X}) = 1 - \sum_{i}^{n} \log(\sigma_i) \tag{C.24}$$

C.2 Robust Pole Assignment Algorithms

The rest of this appendix discusses two algorithms because of their speed. The speed is very important as we will use these algorithms in Model Predictive Control (MPC) which involves extensive on-line computations.

The first algorithm is an efficient algorithm proposed in [201] which is an extension of [90] and probably the most efficient among all robust pole assignment algorithms. A critical observation is based on the following two theorems given in [90].

Theorem C.2
Given $\Lambda = \mathrm{diag}\{\lambda_1, \lambda_2, \ldots, \lambda_n\}$, *the prescribed closed-loop eigenvalues, and* $\mathbf{X}$, *a non-singular matrix composed of closed-loop eigenvectors, then there exists* $\mathbf{F}$, *a solution to*

$$(\mathbf{A} + \mathbf{BF})\mathbf{X} = \mathbf{X}\Lambda \tag{C.25}$$

if and only if

$$\mathbf{U}_1^{\mathrm{T}}(\mathbf{AX} - \mathbf{X}\Lambda) = \mathbf{0} \tag{C.26}$$

where

$$\mathbf{B} = \begin{bmatrix}\mathbf{U}_0 & \mathbf{U}_1\end{bmatrix}\begin{bmatrix}\mathbf{Z}\\ \mathbf{0}\end{bmatrix} \tag{C.27}$$

with $\mathbf{U} := \begin{bmatrix}\mathbf{U}_0 & \mathbf{U}_1\end{bmatrix}$ *orthogonal and* $\mathbf{Z}$ *nonsingular. Then* $\mathbf{F}$ *is given by*

$$\mathbf{F} := \mathbf{Z}^{-1}\mathbf{U}_0^{\mathrm{T}}(\mathbf{X}\Lambda\mathbf{X}^{-1} - \mathbf{A}) \tag{C.28}$$

Proof C.3 The assumption that $\mathbf{B}$ is full rank implies the existence of decomposition (C.27). From (C.25), $\mathbf{F}$ must satisfy

$$\mathbf{BF} = \mathbf{X}\Lambda\mathbf{X}^{-1} - \mathbf{A} \tag{C.29}$$

and pre-multiplying $\mathbf{U}^{\mathrm{T}}$ gives the following equations

$$\mathbf{ZF} = \mathbf{U}_0^{\mathrm{T}}(\mathbf{X}\Lambda\mathbf{X}^{-1} - \mathbf{A}) \tag{C.30a}$$

$$\mathbf{0} = \mathbf{U}_1^{\mathrm{T}}(\mathbf{X}\Lambda\mathbf{X}^{-1} - \mathbf{A}) \tag{C.30b}$$

Since $\mathbf{X}$ and $\mathbf{Z}$ are invertible, equations (C.30) implies (C.26) and (C.28). ∎

Also, the columns $\mathbf{x}_i$, $i = 1,\ldots,n$, of $\mathbf{X}$ must satisfy the following constraint.

Theorem C.3
The eigenvector $\mathbf{x}_i$ of $\mathbf{A}+\mathbf{BF}$ corresponding to the assigned eigenvalue $\lambda_i \in \mathcal{L}$ must belong to the space

$$\mathcal{S}_i := \ker(\mathbf{U}_1^{\mathrm{T}}(\mathbf{A}-\lambda_i\mathbf{I})) \tag{C.31}$$

Proof C.4 From (C.26), we have $\mathbf{U}_1^{\mathrm{T}}(\mathbf{A}-\lambda_i\mathbf{I})\mathbf{x}_i = 0$, for $\forall i$. This proves the theorem. ∎

The main idea of the algorithm is to select $\mathbf{X} = [\mathbf{x}_1,\ldots,\mathbf{x}_n]$ nonsingular such that $\mathbf{x}_i \in \mathcal{S}_i$ and $\|\mathbf{x}_i\|_2 = 1$, $i = 1,\ldots,n$, such that $|\det(\mathbf{X})|$ is maximized. For some $\mathbf{X}$, if $\det(\mathbf{X})$ is minimized, let $\hat{\mathbf{X}}$ be a matrix that is equal to $\mathbf{X}$ except the sign of one column is changed, then, $\det\hat{\mathbf{X}}$ is maximized. Therefore, the robust pole assignment problem is reduced to solve the following optimization problem:

$$\max\ \det(\mathbf{X})\ \text{ s.t. } \mathbf{x}_i \in \mathcal{S}_i,\ \|\mathbf{x}_i\| = 1,\ i = 1,\ldots,n,\ \det(\mathbf{X}) \neq 0 \tag{C.32}$$

Once the optimal $\mathbf{X}$ is obtained, the feedback matrix is given by (C.28). First, $\mathcal{S}_i$ can be obtained by the following steps:

Step 1: Using a QR decomposition of $\mathbf{B}$, we can get $\mathbf{U}_0$, $\mathbf{U}_1$, and $\mathbf{Z}$ as given in (C.27).

Step 2: Using QR decompositions for every prescribed closed-loop pole λ_i

$$(\mathbf{U}_1^{\mathrm{T}}(\mathbf{A}-\lambda_i\mathbf{I}))^{\mathrm{T}} = [\mathbf{V}_{0i}\ \ \mathbf{V}_i]\begin{bmatrix}\mathbf{Y}\\ \mathbf{0}\end{bmatrix}$$

The m columns of $\mathbf{V}_i$ is the orthonormal base of $\mathcal{S}_i$.

Starting from the initial point $\mathbf{X}^0 = [\mathbf{x}_i^0,\ldots,\mathbf{x}_n^0]$ with $\mathbf{x}_i^0 \in \mathbf{V}_i$, the main trick of the algorithm is to select one or at most two $\mathbf{x}_i$ at a time to increase the robustness measurement $\det(\mathbf{X})$ so that every iteration becomes extremely efficient. This strategy is based on several useful theorems [201, 258]. The first one is a method of updating one column of $\mathbf{X}$ at a time.

Let $\mathbf{x}_j \in \mathcal{S}_j$ and $\|\mathbf{x}_j\|_2 = 1$ for $\forall j \in \{1,\ldots,n\}$. Let i be any index in $\{1,\ldots,n\}$ such that $\mathbf{x}_i$ is a real eigenvector corresponding to a real eigenvalue, $\mathbf{X}(\xi) = [\mathbf{x}_1,\ldots,\mathbf{x}_{i-1},\xi,\mathbf{x}_{i+1},\ldots,\mathbf{x}_n]$, and $\mathbf{X}_- = [\mathbf{x}_1,\ldots,\mathbf{x}_{i-1},\mathbf{x}_{i+1},\ldots,\mathbf{x}_n]$. Let $\mathbf{u}_i(\mathbf{X}_-)$ be the unit length vector orthogonal to $\mathbf{X}_-$ such that the inner product of $\langle\mathbf{u}_i(\mathbf{X}_-),\mathbf{x}_i\rangle > 0$. Then, we can replace $\mathbf{x}_i$ by ξ, a new eigenvector of λ_i, such that $\det(\mathbf{X}(\xi))$ is maximized. This is an optimization problem, its mathematical formula and the corresponding solution is given as the following theorem (see also [240]).

Theorem C.4
Let $\mathbf{x}_i \in \mathcal{S}_i$ be a real eigenvector corresponding to a prescribed real closed-loop pole λ_i. Consider the following optimization problem:

$$\max \det(\mathbf{X}(\xi)) \quad s.t. \quad \|\xi\| = 1, \quad \xi \in \mathcal{S}_i \tag{C.33}$$

we have

$$\det(\mathbf{X}(\xi)) = \langle \xi, \mathbf{u}_i(\mathbf{X}_-) \rangle \sqrt{\det(\mathbf{X}_-^{\mathrm{T}}\mathbf{X}_-)} \tag{C.34}$$

the optimization problem (C.33) is reduced to

$$\max \langle \xi, \mathbf{u}_i(\mathbf{X}_-) \rangle \quad \text{s.t.} \quad \|\xi\| = 1, \quad \xi \in \mathcal{S}_i \tag{C.35}$$

The optimal solution of (C.35) is given by

$$\xi = \frac{\mathbf{V}_i \mathbf{V}_i^{\mathrm{T}} \mathbf{u}_i(\mathbf{X}_-)}{\|\mathbf{V}_i^{\mathrm{T}} \mathbf{u}_i(\mathbf{X}_-)\|} \tag{C.36}$$

Proof C.5 Let $\mathbf{P}$ be the permutation matrix such that

$$\mathbf{XP} = [\xi, \mathbf{X}_-] \tag{C.37}$$

and let $\mathbf{Q} \in \mathbf{R}^{n\times(n-1)}$ and $\mathbf{R} \in \mathbf{R}^{(n-1)\times(n-1)}$ be any two matrices such that $\mathbf{Q}^{\mathrm{T}}\mathbf{Q} = \mathbf{I}$ and

$$\mathbf{X}_- = \mathbf{QR} \tag{C.38}$$

Using (C.37) and (C.38), we obtain

$$\begin{aligned}
(\det(\mathbf{X}))^2 &= \det(\mathbf{X}^{\mathrm{T}}\mathbf{X}) \\
&= \det(\mathbf{P}^{\mathrm{T}}\mathbf{X}^{\mathrm{T}}\mathbf{X}\mathbf{P}) = \det([\xi, \mathbf{X}_-]^{\mathrm{T}}[\xi, \mathbf{X}_-]) \\
&= \det\left(\begin{bmatrix} \xi^{\mathrm{T}}\xi & \xi^{\mathrm{T}}\mathbf{X}_- \\ \mathbf{X}_-^{\mathrm{T}}\xi & \mathbf{X}_-^{\mathrm{T}}\mathbf{X}_- \end{bmatrix}\right) \\
&= (\xi^{\mathrm{T}}\xi - \xi^{\mathrm{T}}\mathbf{X}_-(\mathbf{X}_-^{\mathrm{T}}\mathbf{X}_-)^{-1}\mathbf{X}_-^{\mathrm{T}}\xi)\det(\mathbf{X}_-^{\mathrm{T}}\mathbf{X}_-) \\
&= (\xi^{\mathrm{T}}\xi - \xi^{\mathrm{T}}\mathbf{QR}(\mathbf{R}^{\mathrm{T}}\mathbf{R})^{-1}\mathbf{R}^{\mathrm{T}}\mathbf{Q}^{\mathrm{T}}\xi)\det(\mathbf{X}_-^{\mathrm{T}}\mathbf{X}_-) \\
&= (\xi^{\mathrm{T}}\xi - \xi^{\mathrm{T}}\mathbf{Q}\mathbf{Q}^{\mathrm{T}}\xi)\det(\mathbf{X}_-^{\mathrm{T}}\mathbf{X}_-)
\end{aligned} \tag{C.39}$$

Now note that $[\mathbf{Q}, \mathbf{u}_i(\mathbf{X}_-)]$ is an orthogonal matrix so that

$$\mathbf{Q}\mathbf{Q}^{\mathrm{T}} + \mathbf{u}_i(X_-)\mathbf{u}_i(\mathbf{X}_-)^{\mathrm{T}} = \mathbf{I}$$

Thus

$$\begin{aligned}(\det(\mathbf{X}))^2 &= \xi^{\mathrm{T}}\mathbf{u}_i(X_-)\mathbf{u}_i(\mathbf{X}_-)^{\mathrm{T}}\xi\det(\mathbf{X}_-^{\mathrm{T}}\mathbf{X}_-)\\ &= \langle\xi,\mathbf{u}_i(\mathbf{X}_-)\rangle^2\det(\mathbf{X}_-^{\mathrm{T}}\mathbf{X}_-)\end{aligned} \tag{C.40}$$

Note that $\langle\xi,\mathbf{u}_i(\mathbf{X}_-)\rangle$ and $\det(\mathbf{X})$ have the same sign since (i) they are both linear in ξ, (ii) in view of (C.40) they vanish simultaneously and (iii) they are both positive at $\xi = \mathbf{x}_i$. The first claim follows. To prove the second part, notice that $\xi \in S_i$ implies $\xi = \mathbf{V}_i\mathbf{z}$ for some $\mathbf{z}$. Since $\|\xi\| = 1$, it is easy to see that maximizing $\mathbf{z}^{\mathrm{T}}\mathbf{V}_i^{\mathrm{T}}\mathbf{u}_i(\mathbf{X}_-)$ with $\|\xi\| = 1$ is given by (C.36). ∎

If all prescribed closed-loop poles are real, then by cyclically updating the eigenvector one by one, we will continuously improve the robustness measure of $\det(\mathbf{X})$. This is the strategy proposed by [90]. Computation involved in Theorem C.4, given $\mathbf{X}$, is inexpensive. The main task, computation of $\mathbf{u}_i(\mathbf{X}_-)$, can be effected by means of a QR factorization of $\mathbf{X}_-$. Kautsky et al. note that the QR factorization to be carried out at iteration $k > 0$ can be obtained by a rank-one update of that computed at iteration $k-1$, requiring only $O(n^2)$ operations; and that the subsequent projection on ξ that yields the ith column of the new $\mathbf{X}$ requires $O(nm)$ operations, for a total of $O(n^2)+O(mn)$ operations per iteration.

Although optimizing one eigenvector at a time makes the problem simple and computationally attractive, there are two reasons to consider optimizing more eigenvectors at a time if we can still maintain the low cost in each iteration. First, we may achieve better convergence rate because more eigenvectors are optimized at each iteration. Second, the method proposed above cannot be directly applied to problems which have prescribed complex conjugate eigenvalues, i.e., have complex conjugate eigenvectors. We now consider a updating method for two eigenvectors at a time.

For simplicity of exposition, suppose that n is even, say $n = 2p$. Given $\mathbf{X} = [\mathbf{x}_1,\ldots,\mathbf{x}_n]$, let

$$\mathbf{X}_= = [\mathbf{x}_1,\cdots,\mathbf{x}_{2i-2},\mathbf{x}_{2i+1},\cdots,\mathbf{x}_n]$$

and

$$\mathbf{X}(\xi,\eta) = [\mathbf{x}_1,\ldots,\mathbf{x}_{2i-2},\xi,\eta,\mathbf{x}_{2i+1},\ldots,\mathbf{x}_n]$$

Let $\mathbf{U}_i(\mathbf{X}_=) \in \mathbf{R}^{n\times n}$, $i = 1,\ldots,p$, be defined by

$$\mathbf{U}_i(\mathbf{X}_=) = (\mathbf{u}\mathbf{v}^{\mathrm{T}} - \mathbf{v}\mathbf{u}^{\mathrm{T}})$$

where $\mathbf{u}$, $\mathbf{v} \in \mathbf{R}^n$ form an orthonormal basis for the orthogonal complement of the set

$$\{\mathbf{x}_1,\ldots,\mathbf{x}_{2(i-1)},\mathbf{x}_{2i+1},\ldots,\mathbf{x}_n\}$$

and satisfy the inequality

$$\langle\mathbf{x}_{2i-1},\mathbf{u}\rangle\langle\mathbf{x}_{2i},\mathbf{v}\rangle \geq \langle\mathbf{x}_{2i-1},\mathbf{v}\rangle\langle\mathbf{x}_{2i},\mathbf{u}\rangle \tag{C.41}$$

(note that the latter can be achieved by proper choice of the orientation of **u** and **v**). It is readily checked that $\mathbf{U}_i(\mathbf{X}_=)$ is thus uniquely determined (although **u** and **v** are not) and is continuous as a function of **X**.

Theorem C.5
Let $\mathbf{x}_{2i-1} \in \mathcal{S}_{2i-1}$ and $\mathbf{x}_{2i} \in \mathcal{S}_{2i}$ be two real eigenvectors corresponding to two prescribed real closed-loop poles λ_{2i-1} and λ_{2i}. Consider the following optimization problem:

$$\max \det(\mathbf{X}(\xi,\eta)) \quad \text{s.t.} \quad \|\xi\| = 1, \quad \xi \in \mathcal{S}_{2i-1}, \quad \|\eta\| = 1, \quad \eta \in \mathcal{S}_{2i} \tag{C.42}$$

we have

$$\det(\mathbf{X}) = \langle \xi, \mathbf{U}_i(\mathbf{X}_=)\eta \rangle \sqrt{\det(\mathbf{X}_=^{\mathrm{T}}\mathbf{X}_=)} \tag{C.43}$$

the optimization problem (C.42) is reduced to

$$\text{maximize } \langle \xi, \mathbf{U}_i(\mathbf{X}_=)\eta \rangle \quad \text{s.t.} \quad \|\xi\| = 1, \quad \xi \in \mathcal{S}_{2i-1}, \quad \|\eta\| = 1, \quad \eta \in \mathcal{S}_{2i} \tag{C.44}$$

Proof C.6 Let **P** be a permutation matrix such that

$$\mathbf{XP} = [\xi, \eta, \mathbf{X}_=] \tag{C.45}$$

and let $\mathbf{Q} \in \mathbf{R}^{n\times(n-2)}$ and $\mathbf{R} \in \mathbf{R}^{(n-2)\times(n-2)}$ be any two matrices such that $\mathbf{Q}^{\mathrm{T}}\mathbf{Q} = \mathbf{I}$ and

$$\mathbf{X}_= = \mathbf{QR} \tag{C.46}$$

Using (C.45) and (C.46) we have

$$\begin{aligned}
(\det(\mathbf{X}))^2 &= \det(\mathbf{X}^{\mathrm{T}}\mathbf{X}) \\
&= \det(\mathbf{P}^{\mathrm{T}}\mathbf{X}^{\mathrm{T}}\mathbf{XP}) = \det([\xi,\eta,\mathbf{X}_=]^{\mathrm{T}}[\xi,\eta,\mathbf{X}_=]) \\
&= \det\begin{pmatrix} \begin{bmatrix} \xi^{\mathrm{T}} \\ \eta^{\mathrm{T}} \end{bmatrix}[\xi,\eta] & \begin{bmatrix} \xi^{\mathrm{T}} \\ \eta^{\mathrm{T}} \end{bmatrix}\mathbf{X}_= \\ \mathbf{X}_=^{\mathrm{T}}[\xi,\eta] & \mathbf{X}_=^{\mathrm{T}}\mathbf{X}_= \end{pmatrix} \\
&= \det\left(\begin{bmatrix} \xi^{\mathrm{T}} \\ \eta^{\mathrm{T}} \end{bmatrix}[\xi,\eta] - \begin{bmatrix} \xi^{\mathrm{T}} \\ \eta^{\mathrm{T}} \end{bmatrix}\mathbf{X}_=(\mathbf{X}_=^{\mathrm{T}}\mathbf{X}_=)^{-1}\mathbf{X}_=^{\mathrm{T}}[\xi,\eta]\right)\det(\mathbf{X}_=^{\mathrm{T}}\mathbf{X}_=) \\
&= \det\left(\begin{bmatrix} \xi^{\mathrm{T}} \\ \eta^{\mathrm{T}} \end{bmatrix}[\xi,\eta] - \begin{bmatrix} \xi^{\mathrm{T}} \\ \eta^{\mathrm{T}} \end{bmatrix}\mathbf{QR}(\mathbf{R}^{\mathrm{T}}\mathbf{R})^{-1}\mathbf{R}^{\mathrm{T}}\mathbf{Q}^{\mathrm{T}}[\xi,\eta]\right)\det(\mathbf{X}_=^{\mathrm{T}}\mathbf{X}_=) \\
&= \det\left(\begin{bmatrix} \xi^{\mathrm{T}} \\ \eta^{\mathrm{T}} \end{bmatrix}[\xi,\eta] - \begin{bmatrix} \xi^{\mathrm{T}} \\ \eta^{\mathrm{T}} \end{bmatrix}\mathbf{QQ}^{\mathrm{T}}[\xi,\eta]\right)\det(\mathbf{X}_=^{\mathrm{T}}\mathbf{X}_=)
\end{aligned} \tag{C.47}$$

Now, $\mathbf{U}_i(\mathbf{X}_=) = (\mathbf{uv}^{\mathrm{T}} - \mathbf{vu}^{\mathrm{T}})$ with $\{\mathbf{u},\mathbf{v}\}$ an orthonormal basis for the null space of $\mathbf{X}_=^{\mathrm{T}}$ satisfying (C.41). Thus $[\mathbf{Q},\mathbf{u},\mathbf{v}]$ is orthogonal, therefore

$$\mathbf{QQ}^{\mathrm{T}} + \begin{bmatrix} \mathbf{u} & \mathbf{v} \end{bmatrix}\begin{bmatrix} \mathbf{u}^{\mathrm{T}} \\ \mathbf{v}^{\mathrm{T}} \end{bmatrix} = \mathbf{I}$$

and

$$(\det(\mathbf{X}))^2 = \det\left(\begin{bmatrix} \xi^{\mathrm{T}} \\ \eta^{\mathrm{T}} \end{bmatrix}[\mathbf{u},\mathbf{v}]\begin{bmatrix} \mathbf{u}^{\mathrm{T}} \\ \mathbf{v}^{\mathrm{T}} \end{bmatrix}[\xi,\eta]\right)\det(\mathbf{X}_{=}^{\mathrm{T}}\mathbf{X}_{=})$$
$$= \left(\det\left(\begin{bmatrix} \xi^{\mathrm{T}} \\ \eta^{\mathrm{T}} \end{bmatrix}[\mathbf{u},\mathbf{v}]\right)\right)^2 \det(\mathbf{X}_{=}^{\mathrm{T}}\mathbf{X}_{=})$$
$$= (\langle\xi,\mathbf{u}\rangle\langle\eta,\mathbf{v}\rangle - \langle\xi,\mathbf{v}\rangle\langle\eta,\mathbf{u}\rangle)^2 \det(\mathbf{X}_{=}^{\mathrm{T}}\mathbf{X}_{=}) \tag{C.48}$$

Next,

$$\mathrm{sgn}(\det(\mathbf{X})) = \mathrm{sgn}(\langle\xi,\mathbf{u}\rangle\langle\eta,\mathbf{v}\rangle - \langle\xi,\mathbf{v}\rangle\langle\eta,\mathbf{u}\rangle)$$

since (i) the arguments in both sides are quadratic in ξ,η, (ii) in view of (C.48) they vanish simultaneously and (iii) in view of (C.41) and since $\det(\mathbf{X}) > 0$ they are both positive at $\xi = \mathbf{x}_{2i-1}$, and $\eta = \mathbf{x}_{2i}$. Thus,

$$\det(\mathbf{X}) = (\langle\xi,\mathbf{u}\rangle\langle\eta,\mathbf{v}\rangle - \langle\xi,\mathbf{v}\rangle\langle\eta,\mathbf{u}\rangle)\sqrt{\det(\mathbf{X}_{=}^{\mathrm{T}}\mathbf{X}_{=})}$$
$$= \langle\xi,\mathbf{U}_i(\mathbf{X}_{=})\eta\rangle\sqrt{\det(\mathbf{X}_{=}^{\mathrm{T}}\mathbf{X}_{=})} \tag{C.49}$$

The claim follows. ■

Since $\xi \in \mathcal{S}_{2i-1}$ and $\eta \in \mathcal{S}_{2i}$, noticing that $\mathbf{V}_{2i-1}$ and $\mathbf{V}_{2i}$ are the bases of $\mathcal{S}_{2i-1}$ and $\mathcal{S}_{2i}$, we can reduce the problem a little further.

Proposition C.1
The optimization (C.44) is equivalent to the following optimization problem:

$$\text{maximize } \langle\mu,\mathbf{V}_{2i-1}^{\mathrm{T}}\mathbf{U}_i(\mathbf{X}_{=})\mathbf{V}_{2i}\nu\rangle \text{ s.t. } \|\mu\| = 1, \quad \|\nu\| = 1 \tag{C.50}$$

Noticing that $\mathbf{U}_i(\mathbf{X}_{=})$ is a rank 2 matrix, we can solve the optimization problem (C.50) very efficiently.

Theorem C.6
Let $\mathbf{X} = [\mathbf{x}_1,\ldots,\mathbf{x}_n] \in \mathbf{X}$, let $i \in \{1,\ldots,p\}$, let $\sigma_1 \geq \sigma_2$ be the top two singular values of $\mathbf{V}_{2i-1}^{\mathrm{T}}\mathbf{U}_i(\mathbf{X}_{=})\mathbf{V}_{2i}$, and for $j = 1,2$, let μ_j, $\nu_j \in \mathbf{R}^m$ form a left-right singular vector pair associated with σ_j with the property that $\langle\mu_1,\mu_2\rangle = 0$ and $\langle\nu_1,\nu_2\rangle = 0$. Then for $i = 1,\ldots,p$, the optimal update defined by (C.44) or (C.50) is given by:

$$\mathbf{X} = [\mathbf{x}_1,\ldots,\mathbf{x}_{2i-2},\xi,\eta,\mathbf{x}_{2i+1},\ldots,\mathbf{x}_n],$$

where $\begin{bmatrix} \xi \\ \eta \end{bmatrix} = \sqrt{2}\zeta/\|\zeta\|$ *and*

(i) if $\sigma_1 > \sigma_2$, then, ζ is the orthogonal projection of $\begin{bmatrix} \mathbf{x}_{2i-1} \\ \mathbf{x}_{2i} \end{bmatrix}$ *on the span of*

$\left\{\begin{bmatrix} \mathbf{V}_{2i-1}\mu_1 \\ \mathbf{V}_{2i}\nu_1 \end{bmatrix}\right\}$, *i.e.,*

$$\begin{aligned}\zeta &= \begin{bmatrix} \mathbf{V}_{2i-1}\mu_1 \\ \mathbf{V}_{2i}\nu_1 \end{bmatrix}\begin{bmatrix} \mathbf{V}_{2i-1}\mu_1 \\ \mathbf{V}_{2i}\nu_1 \end{bmatrix}^{\mathrm{T}}\begin{bmatrix} \mathbf{x}_{2i-1} \\ \mathbf{x}_{2i} \end{bmatrix} \\ &= (\langle \mathbf{x}_{2i-1}, \mathbf{V}_{2i-1}\mu_1\rangle + \langle \mathbf{x}_{2i}, \mathbf{V}_{2i}\nu_1\rangle)\begin{bmatrix} \mathbf{V}_{2i-1}\mu_1 \\ \mathbf{V}_{2i}\nu_1 \end{bmatrix} \end{aligned} \tag{C.51}$$

(ii) if $\sigma_1 = \sigma_2$, then, ζ is the orthogonal projection of $\begin{bmatrix} \mathbf{x}_{2i-1} \\ \mathbf{x}_{2i} \end{bmatrix}$ *on the span of* $\left\{\begin{bmatrix} \mathbf{V}_{2i-1}\mu_1 \\ \mathbf{V}_{2i}\nu_1 \end{bmatrix}, \begin{bmatrix} \mathbf{V}_{2i-1}\mu_2 \\ \mathbf{V}_{2i}\nu_2 \end{bmatrix}\right\}$, *i.e.,*

$$\zeta = \begin{bmatrix} \mathbf{V}_{2i-1} & \mathbf{0} \\ \mathbf{0} & \mathbf{V}_{2i} \end{bmatrix}\begin{bmatrix} \mu_1 & \mu_2 \\ \nu_1 & \nu_2 \end{bmatrix}\begin{bmatrix} \mu_1 & \mu_2 \\ \nu_1 & \nu_2 \end{bmatrix}^{\mathrm{T}}\begin{bmatrix} \mathbf{V}_{2i-1} & \mathbf{0} \\ \mathbf{0} & \mathbf{V}_{2i} \end{bmatrix}^{\mathrm{T}}\begin{bmatrix} \mathbf{x}_{2i-1} \\ \mathbf{x}_{2i} \end{bmatrix} \tag{C.52}$$

Proof C.7 The proof is straightforward by using (C.50) and therefore omitted. ■

Consider now the case where the set of desired poles includes a number of complex conjugate pairs. Let $\lambda_1, \ldots, \lambda_n$ be the eigenvalues to be assigned. For the sake of simplicity of exposition, again assume that n is even. Moreover assume that $\{\lambda_1, \ldots, \lambda_n\} \cap \mathbf{R} = \{\lambda_1, \ldots, \lambda_{2p}\}$, i.e., $\lambda_1, \ldots, \lambda_{2p}$ are real and $\lambda_{2p+1}, \ldots, \lambda_n$ are complex; let c be the number of complex pairs, i.e., $c = n/2 - p$, and assume that $\lambda_{2i} = \overline{\lambda}_{2i-1}, i = p+1, \ldots, p+c$. Clearly, candidate sets of eigenvectors of the closed loop matrix $\mathbf{A} + \mathbf{BF}$ must include c complex conjugate pairs. Moreover, as in the real case, they must satisfy additional conditions. The next theorem extends Theorem C.3.

Theorem C.7
Let $\mathbf{X} = [\mathbf{x}_1, \ldots, \mathbf{x}_n]$, nonsingular, and $\Lambda = \mathrm{diag}(\lambda_1, \ldots, \lambda_n)$ be two complex $n \times n$ matrices such that (i) for $j = 1, \ldots, 2p$, $\lambda_j \in \mathbf{R}$ and $\mathbf{x}_j \in \mathbf{R}^n$, and (ii) for $i = p+1, \ldots, p+c$, $\lambda_{2i} = \overline{\lambda}_{2i-1}$ and $\mathbf{x}_{2i} = \overline{\mathbf{x}}_{2i-1}$. Then

(i) $\mathbf{X}\Lambda\mathbf{X}^{-1}$ *is real, and*

(ii) $(\mathbf{A} + \mathbf{BF})\mathbf{X} = \mathbf{X}\Lambda$ *for some real matrix* $\mathbf{F}$ *if and only if* $\mathbf{x}_j \in \mathcal{S}_j,\ j = 1, \ldots, n.$

Proof C.8 Let $\mathbf{P} \in \mathbf{R}^{n\times n}$ be the permutation matrix that exchanges columns $2i-1$ and $2i$, $i = p+1, \ldots, p+c$, of the matrix it post-multiplies. Thus, $\mathbf{X} = \overline{\mathbf{X}}\mathbf{P}$, $\overline{\Lambda} = \mathbf{P}\Lambda\mathbf{P}$, and $\mathbf{P}^{-1} = \mathbf{P}$. Then

$$\overline{\mathbf{X}\Lambda\mathbf{X}^{-1}} = \overline{\mathbf{X}}\mathbf{P}\Lambda\mathbf{P}\overline{\mathbf{X}}^{-1} = \overline{\mathbf{X}}\mathbf{P}\Lambda(\overline{\mathbf{X}}\mathbf{P})^{-1} = \mathbf{X}\Lambda\mathbf{X}^{-1}$$

proving the first claim. Now suppose that for some real matrix $\mathbf{F}$, $(\mathbf{A}+\mathbf{BF})\mathbf{X} = \mathbf{X}\Lambda$. Then

$$\mathbf{AX} - \mathbf{X}\Lambda = -\mathbf{BFX}$$

implying that

$$(\mathbf{A} - \lambda_j \mathbf{I})\mathbf{x}_j \in \mathbf{R_C}(\mathbf{B}), \ \ j = 1, \ldots, n \tag{C.53}$$

where $\mathbf{R_C}(\mathbf{B}) = \{\mathbf{By} : \mathbf{y} \in \mathbf{C}^m\}$, i.e., $\mathbf{x}_j \in \mathcal{S}_j$ for $j = 1, \ldots, n$. Finally, suppose that (C.53) holds, i.e., for some $\mathbf{y}_j \in \mathbf{C}^m$, $j = 1, \ldots, n$

$$(\mathbf{A} - \lambda_j \mathbf{I})\mathbf{x}_j = \mathbf{By}_j$$

Thus,

$$\mathbf{AX} - \mathbf{X}\Lambda = \mathbf{BY}$$

with $\mathbf{Y} := [\mathbf{y}_1, \ldots, \mathbf{y}_n] \in \mathbf{C}^{m \times n}$, i.e.,

$$\mathbf{A} - \mathbf{X}\Lambda\mathbf{X}^{-1} = \mathbf{BYX}^{-1}$$

The left-hand side is a real matrix. Thus,

$$\mathbf{BYX}^{-1} = \mathcal{R}_e(\mathbf{BYX}^{-1}) = \mathbf{B}\mathcal{R}_e(\mathbf{YX}^{-1})$$

The last claim then follows by setting $\mathbf{F} = \mathcal{R}_e(\mathbf{YX}^{-1})$. ■

In view of this result, we will focus on a modification of problem (C.32) given by

$$\begin{aligned} \max \ \det(\mathbf{X}) \ \ \text{s.t.} \ \ & \mathbf{x}_i \in \mathcal{S}_i, \ \ \|\mathbf{x}_i\| = 1, \ \ i = 1, \ldots, n \\ & \mathbf{x}_{2i-1} = \bar{\mathbf{x}}_{2i}, i = p+1, \ldots, p+c \\ & \mathbf{x}_i \in \mathbf{R}^n, i = 1, \ldots, 2p; \ \det(\mathbf{X}) \neq 0 \end{aligned} \tag{C.54}$$

Since we consider only the case of updating a pair of complex conjugate eigenvectors, we can reduce the problem quite a bit and solve the reduced problem efficiently. Let

$$\tilde{\mathbf{U}}_i(\mathbf{X}_=) = \mathbf{u}\bar{\mathbf{u}}^{\mathrm{T}} - \bar{\mathbf{u}}\mathbf{u}^{\mathrm{T}}$$

where $\mathbf{u} = \mathbf{u}_R + \mathbf{j}\mathbf{u}_I$ is such that $\sqrt{2}\mathbf{u}_R$, $\sqrt{2}\mathbf{u}_I \in \mathbf{R}^n$ form an orthonormal basis for the orthogonal complement of the set

$$\{\mathbf{x}_1, \ldots, \mathbf{x}_{2p}\} \cup \{\mathcal{R}_e(\mathbf{x}_{2j}), \mathcal{I}_m(\mathbf{x}_{2j}), j = p+1, \ldots, p+c, j \neq i\}$$

and satisfy the inequality

$$\langle \mathcal{R}_e(\mathbf{x}_{2i}), \mathbf{u}_R \rangle \langle \mathcal{I}_m(\mathbf{x}_{2i}), \mathbf{u}_I \rangle \geq \langle \mathcal{R}_e(\mathbf{x}_{2i}), \mathbf{u}_I \rangle \langle \mathcal{I}_m(\mathbf{x}_{2i}), \mathbf{u}_R \rangle \tag{C.55}$$

Again, it is readily checked that $\tilde{\mathbf{U}}_i(\mathbf{X}_=)$ is thus uniquely determined and is continuous as a function of $\mathbf{X}_=$.

Lemma C.2
Let $i \in \{p+1,\ldots,p+c\}$, and let $\mathbf{u} = \mathbf{u}_R + \mathbf{j}\mathbf{u}_I$ be such that $\sqrt{2}\mathbf{u}_R$, $\sqrt{2}\mathbf{u}_I \in \mathbf{R}^n$ form an orthonormal basis for the orthogonal complement of the set

$$\{\mathbf{x}_1,\ldots,\mathbf{x}_{2p}\} \cup \{\mathcal{R}_e(\mathbf{x}_{2j}), \mathcal{I}_m(\mathbf{x}_{2j}), j = p+1,\ldots,p+c, j \neq i\}$$

Then $\{\mathbf{u},\overline{\mathbf{u}}\}$ is an orthonormal basis for the null space of $[\mathbf{x}_1,\cdots,\mathbf{x}_{2i-2},\mathbf{x}_{2i+1},\cdots,\mathbf{x}_n]^*$.

The following theorem is critical to provide an efficient method to update two complex conjugate eigenvectors.

Theorem C.8
Let $\mathbf{X} = [\mathbf{x}_1,\ldots,\mathbf{x}_n] \in \mathbf{C}^{n\times n}$ with $\mathbf{x}_1,\ldots,\mathbf{x}_{2p} \in \mathbf{R}^n$, and $\mathbf{x}_{2i-1} = \overline{\mathbf{x}}_{2i}$, $i = p+1,\ldots,p+c$ be such that $\det(\mathbf{X}) \neq 0$, let $\eta \in \mathbf{C}^n$, let $i \in \{p+1,\ldots,p+c\}$, and let

$$\mathbf{X}(\eta) = [\mathbf{x}_1,\cdots,\mathbf{x}_{2i-2},\overline{\eta},\eta,\mathbf{x}_{2i+1},\cdots,\mathbf{x}_n] \tag{C.56}$$

and

$$\mathbf{X}_= = [\mathbf{x}_1,\cdots,\mathbf{x}_{2i-2},\mathbf{x}_{2i+1},\cdots,\mathbf{x}_n]$$

Then $\mathbf{X}_=^*\mathbf{X}_=$ is nonsingular, and

$$|\det(\mathbf{X})| = |\langle \eta, \tilde{\mathbf{U}}_i(\mathbf{X}_=)\eta\rangle| \sqrt{\det(\mathbf{X}_=^*\mathbf{X}_=)} \tag{C.57}$$

Proof C.9 Let $\mathbf{P}$ be the permutation matrix such that

$$\mathbf{XP} = (\overline{\eta},\eta,\mathbf{X}_=) \tag{C.58}$$

and let $\mathbf{Q} \in \mathbf{C}^{n\times(n-2)}$ and $\mathbf{R} \in \mathbf{C}^{(n-2)\times(n-2)}$ be any two matrices such that $\mathbf{Q}^*\mathbf{Q} = \mathbf{I}$ and

$$\mathbf{X}_= = \mathbf{QR} \tag{C.59}$$

Using (C.58) and (C.59) we have

$$\begin{aligned}
|\det(\mathbf{X})|^2 &= \det(\mathbf{X}^*\mathbf{X}) \\
&= \det(\mathbf{P}^{\mathrm{T}}\mathbf{X}^*\mathbf{X}\mathbf{P}) = \det((\overline{\eta},\eta,\mathbf{X}_=)^*(\overline{\eta},\eta,\mathbf{X}_=)) \\
&= \det\begin{pmatrix} \begin{bmatrix} \overline{\eta}^* \\ \eta^* \end{bmatrix} [\overline{\eta},\eta] & \begin{bmatrix} \overline{\eta}^* \\ \eta^* \end{bmatrix} \mathbf{X}_= \\ \mathbf{X}_=^*[\overline{\eta},\eta] & \mathbf{X}_=^*\mathbf{X}_= \end{pmatrix} \\
&= \det\left(\begin{bmatrix} \overline{\eta}^* \\ \eta^* \end{bmatrix} [\overline{\eta},\eta] - \begin{bmatrix} \overline{\eta}^* \\ \eta^* \end{bmatrix} \mathbf{X}_=(\mathbf{X}_=^*\mathbf{X}_=)^{-1}\mathbf{X}_=^*[\overline{\eta},\eta]\right)\det(\mathbf{X}_=^*\mathbf{X}_=) \\
&= \det\left(\begin{bmatrix} \overline{\eta}^* \\ \eta^* \end{bmatrix} [\overline{\eta},\eta] - \begin{bmatrix} \overline{\eta}^* \\ \eta^* \end{bmatrix} \mathbf{QR}(\mathbf{R}^*\mathbf{R})^{-1}\mathbf{R}^*\mathbf{Q}^*[\overline{\eta},\eta]\right)\det(\mathbf{X}_=^*\mathbf{X}_=)
\end{aligned}$$

$$= \det\left(\begin{bmatrix} \overline{\eta}^* \\ \eta^* \end{bmatrix}[\overline{\eta},\eta] - \begin{bmatrix} \overline{\eta}^* \\ \eta^* \end{bmatrix}\mathbf{Q}\mathbf{Q}^*[\overline{\eta},\eta]\right)\det(\mathbf{X}_=^*\mathbf{X}_=) \qquad \text{(C.60)}$$

Since by Lemma C.2, $\{\mathbf{u},\overline{\mathbf{u}}\}$ is an orthonormal basis for the null space of $\mathbf{X}_=^*$, thus $[\mathbf{Q},\mathbf{u},\overline{\mathbf{u}}]$ is unitary, therefore $\mathbf{Q}\mathbf{Q}^* + \begin{bmatrix} \mathbf{u},\overline{\mathbf{u}} \end{bmatrix}\begin{bmatrix} \mathbf{u}^* \\ \overline{\mathbf{u}}^* \end{bmatrix} = \mathbf{I}$ and

$$\begin{aligned}
|\det(\mathbf{X})|^2 &= \det\left(\begin{bmatrix} \overline{\eta}^* \\ \eta^* \end{bmatrix}[\mathbf{u},\overline{\mathbf{u}}]\begin{bmatrix} \mathbf{u}^* \\ \overline{\mathbf{u}}^* \end{bmatrix}[\overline{\eta},\eta]\right)\det(\mathbf{X}_=^*\mathbf{X}_=) \\
&= \left|\det\left(\begin{bmatrix} \mathbf{u}^* \\ \overline{\mathbf{u}}^* \end{bmatrix}[\overline{\eta},\eta]\right)\right|^2\det(\mathbf{X}_=^*\mathbf{X}_=) \\
&= \left|\det\left[\begin{bmatrix} \overline{\mathbf{u}}^{\mathrm{T}} \\ \mathbf{u}^{\mathrm{T}} \end{bmatrix}[\overline{\eta},\eta]\right]\right|^2\det(\mathbf{X}_=^*\mathbf{X}_=) \\
&= \left|((\mathbf{u}^{\mathrm{T}}\eta)(\overline{\mathbf{u}}^{\mathrm{T}}\overline{\eta}) - (\overline{\mathbf{u}}^{\mathrm{T}}\eta)(\mathbf{u}^{\mathrm{T}}\overline{\eta})\right|^2\det(\mathbf{X}_=^*\mathbf{X}_=) \\
&= \left|\langle\eta,(\overline{\mathbf{u}}\mathbf{u}^{\mathrm{T}} - \mathbf{u}\overline{\mathbf{u}}^{\mathrm{T}})\eta\rangle\right|^2\det(\mathbf{X}_=^*\mathbf{X}_=)
\end{aligned}$$

The claim follows by taking the square root on both the sides. ∎

This theorem suggests that: when updating a pair of complex conjugate eigenvectors $\overline{\mathbf{x}}_{2i}$ and $\mathbf{x}_{2i}$, we can reduce the problem (C.54) to the following problem:

$$\max\ \det(\mathbf{X}(\eta)) \quad \text{s.t.} \quad \overline{\eta} \in \mathcal{S}_{2i-1},\ \ \eta \in \mathcal{S}_{2i},\ \ \|\overline{\eta}\| = 1,\ \ \|\eta\| = 1 \qquad \text{(C.61)}$$

Theorem C.9
Let $\mathbf{X} = [\mathbf{x}_1,\ldots,\mathbf{x}_n] \in \mathbf{X}$, let $i \in \{p+1,\ldots,p+c\}$, let σ_1 and σ_2, with $\sigma_1 \geq \sigma_2$, be the two nonzero singular values of $\mathbf{V}_{2i}^*\tilde{\mathbf{U}}_i(\mathbf{X}_=)\mathbf{V}_{2i}$, and let μ_ℓ, $\ell = 1,2$, denote unit-length singular vectors[1] associated with σ_ℓ with the property that $\langle\mu_1,\ \mu_2\rangle = 0$. Then for $i = p+1,\ldots,p+c$, the optimal update defined by (C.61) is given by

$$\mathbf{X} = [\mathbf{x}_1,\ldots,\mathbf{x}_{2i-2},\overline{\eta},\eta,\mathbf{x}_{2i+1},\ldots,\mathbf{x}_n]$$

where $\eta = \zeta/\|\zeta\|$ and

(i) if $\sigma_1 > \sigma_2$, ζ is the orthogonal projection of $\mathbf{x}_{2i}$ on the span $\mathbf{V}_{2i}\mu_1$, i.e.,

$$\zeta = \mathbf{V}_{2i}\mu_1\mu_1^*\mathbf{V}_{2i}^*\mathbf{x}_{2i} \qquad \text{(C.62)}$$

[1] In fact, since $\mathbf{V}_{2i}^*\tilde{\mathbf{U}}_i(\mathbf{X})\mathbf{V}_{2i}$ is Hermitian, left and right singular vectors have the same direction (but opposite orientation when the corresponding eigenvalue is negative).

(ii) if $\sigma_1 = \sigma_2$, ζ *is the orthogonal projection of* $\mathbf{x}_{2i}$ *on the span* $\{\mathbf{V}_{2i}\mu_1, \mathbf{V}_{2i}\mu_2\}$, *i.e.,*

$$\zeta = \mathbf{V}_{2i}[\mu_1,\ \mu_2][\mu_1,\ \mu_2]^* \mathbf{V}_{2i}^* \mathbf{x}_{2i} \tag{C.63}$$

Proof C.10 The proof is straightforward and therefore omitted. ∎

Thus, for a given $\mathbf{X} = [\mathbf{x}_1, \ldots, \mathbf{x}_n]$, the updated $\mathbf{X}$ can be computed as follows: For $i = 1, \ldots, p$, i.e., two real eigenvectors,

Step (1). Compute an orthonormal basis $\{\mathbf{u}, \mathbf{v}\} \subset \mathbf{R}^n$ for the orthogonal complement of

$$\{\mathbf{x}_1, \ldots, \mathbf{x}_{2i-2}, \mathbf{x}_{2i+1}, \ldots, \mathbf{x}_n\}$$

Step (2). Evaluate $\mathbf{a}_1 = \mathbf{V}_{2i-1}^{\mathrm{T}}\mathbf{u}$, $\mathbf{a}_2 = \mathbf{V}_{2i-1}^{\mathrm{T}}\mathbf{v}$, $\mathbf{b}_1 = \mathbf{V}_{2i}^{\mathrm{T}}\mathbf{v}$, and $\mathbf{b}_2 = \mathbf{V}_{2i}^{\mathrm{T}}\mathbf{u}$.

Step (3). Compute a singular value decomposition of

$$[\mathbf{a}_1,\ \mathbf{a}_2][\mathbf{b}_1,\ -\mathbf{b}_2]^{\mathrm{T}}\ (= \mathbf{V}_{2i-1}^{\mathrm{T}}\mathbf{U}_i(\mathbf{X}_=)\mathbf{V}_{2i})$$

(If the nonzero singular values are distinct, the singular vectors corresponding to the second singular value need not be computed.)

Step (4). Compute (ξ, η) as per Theorem C.6 to obtain the updated $\mathbf{X}$.

For $i = p+1, \ldots, p+c$, i.e., a pair of complex conjugate eigenvectors,

Step (1). Compute an orthonormal basis $\{\sqrt{2}\mathbf{u}_R,\ \sqrt{2}\mathbf{u}_I\} \subset \mathbf{R}^n$ for the orthogonal complement of

$$\{\mathbf{x}_1, \ldots, \mathbf{x}_{2p}\} \cup \{\mathcal{R}_e(\mathbf{x}_{2j}), \mathcal{I}_m(\mathbf{x}_{2j}), j = p+1, \ldots, p+c, j \neq i\}$$

Step (2). Evaluate $\mathbf{a}_1 = \mathbf{V}_{2i}^*(\sqrt{2}\mathbf{u})$, and $\mathbf{a}_2 = \mathbf{V}_{2i}^*(\sqrt{2}\bar{\mathbf{u}})$, where $\mathbf{u} = \mathbf{u}_R + \mathbf{j}\mathbf{u}_I$.

Step (3). Compute a singular value decomposition of

$$[\mathbf{a}_1,\ \mathbf{a}_2][\mathbf{a}_1,\ -\mathbf{a}_2]^*\ (= 2\mathbf{V}_{2i}^*\tilde{U}_i(\mathbf{X}_=)\mathbf{V}_{2i})$$

(If the nonzero singular values are distinct, the singular vector corresponding to the second singular value need not be computed.)

Step (4). Compute η as per Theorem C.9 to obtain the updated $\mathbf{X}$.

In both cases, the bulk of the computation takes place in Step (1). Note that no attention is paid to satisfying (C.41) and (C.55). It is readily checked that the only effect of enforcing these inequalities is to possibly change the orientation of some columns of **X**.

Several efficient algorithms are proposed in [201] based on the theory described in this section. These algorithms are proved to be globally convergent. Many details on efficient implementation are discussed in [201, 241]. A MATLAB implementation for one of the algorithms is available in MathWorks' file exchange website:

https://www.mathworks.com/matlabcentral/fileexchange/53969-robpole

C.3 Misrikhanov and Ryabchenko Algorithm

Strictly speaking, Misrikhanov and Ryabchenko Algorithm [131] is not designed for robust pole assignment but simply a pole assignment algorithm. But this design is likely related to the minimum gain pole assignment design. One of the main merits of this design is that the algorithm is probably the fastest one among all pole assignment algorithms in this author's opinion. This feature is important for Model Predictive Control (MPC) which requires on-line computation for pole assignment gain matrix.

For a set of prescribed closed-loop poles, assuming that the system $(\mathbf{A}, \mathbf{B})$ is controllable, Misrikhanov and Ryabchenko proposed a slightly different decomposition of (C.25) given as follows:

$$\mathbf{A} + \mathbf{BF} = \mathbf{X}\Lambda\mathbf{X}^{-1} \tag{C.64}$$

where Λ is a block diagonal matrix. In Λ, for each ith real closed-loop pole λ_i, the corresponding diagonal cell block is 1×1; for each pair of complex conjugate closed-loop poles, the corresponding diagonal cell block is 2×2 of the form:

$$\begin{bmatrix} \mathcal{R}_e(\lambda_i) & \mathcal{I}_m(\lambda_i) \\ -\mathcal{I}_m(\lambda_i) & \mathcal{R}_e(\lambda_i) \end{bmatrix} \tag{C.65}$$

Let $\mathbf{B}^{\perp^{\mathrm{T}}} = \mathrm{null}(\mathbf{B}^{\mathrm{T}})$ be an orthonormal matrix satisfying conditions:

$$\mathbf{B}^{\perp}\mathbf{B} = \mathbf{0}_{(n-m)\times m} \tag{C.66a}$$

$$\mathbf{B}^{\perp}\mathbf{B}^{\perp^{\mathrm{T}}} = \mathbf{I}_{n-m} \tag{C.66b}$$

Let $\mathbf{B}^{+} = (\mathbf{B}^{\mathrm{T}}\mathbf{B})^{-1}\mathbf{B}^{\mathrm{T}}$ be the Moore-Penrose pseudo-inverse of $\mathbf{B}$ matrix. The following lemma is important in Misrikhanov and Ryabchenko Algorithm.

Lemma C.3
Let $\mathbf{X} \in \mathbf{R}^{m\times(n-m)}$, $\mathbf{Y} \in \mathbf{R}^{m\times m}$, *and* $\mathbf{F} = \mathbf{X}\mathbf{B}^{\perp} + \mathbf{Y}\mathbf{B}^{+} - \mathbf{B}^{+}\mathbf{A}$. *Then,* $\mathbf{A}+\mathbf{B}\mathbf{F}$ *is similar to the following matrix:*

$$\begin{bmatrix} \mathbf{B}^{\perp}\mathbf{A}\mathbf{B}^{\perp^{\mathrm{T}}} & \mathbf{B}^{\perp}\mathbf{A}\mathbf{B} \\ \mathbf{X} & \mathbf{Y} \end{bmatrix} \tag{C.67}$$

Proof C.11 The proof here is due to Tits [200]. Consider the invertible matrix

$$\mathbf{T} = \begin{bmatrix} \mathbf{B}^{\perp} \\ \mathbf{B}^{+} \end{bmatrix}$$

It is easy to verify that the inverse of $\mathbf{T}$ is given by

$$\mathbf{T}^{-1} = \left[\mathbf{B}^{\perp^{\mathrm{T}}}, \mathbf{B}\right]$$

because

$$\left[\mathbf{B}^{\perp^{\mathrm{T}}}, \mathbf{B}\right] \begin{bmatrix} \mathbf{B}^{\perp} \\ \mathbf{B}^{+} \end{bmatrix} = \mathbf{I}_n$$

and

$$\begin{bmatrix} \mathbf{B}^{\perp} \\ \mathbf{B}^{+} \end{bmatrix} \left[\mathbf{B}^{\perp^{\mathrm{T}}}, \mathbf{B}\right] = \begin{bmatrix} \mathbf{I}_{n-m} & \mathbf{0} \\ \mathbf{0} & \mathbf{I}_m \end{bmatrix}$$

Therefore, the following relations hold.

$$\mathbf{F}\mathbf{B}^{\perp^{\mathrm{T}}} = \mathbf{X}\mathbf{B}^{\perp}\mathbf{B}^{\perp^{\mathrm{T}}} + \mathbf{Y}\mathbf{B}^{+}\mathbf{B}^{\perp^{\mathrm{T}}} - \mathbf{B}^{+}\mathbf{A}\mathbf{B}^{\perp^{\mathrm{T}}} = \mathbf{X} - \mathbf{B}^{+}\mathbf{A}\mathbf{B}^{\perp^{\mathrm{T}}} \tag{C.68a}$$

$$\mathbf{F}\mathbf{B} = \mathbf{X}\mathbf{B}^{\perp}\mathbf{B} + \mathbf{Y}\mathbf{B}^{+}\mathbf{B} - \mathbf{B}^{+}\mathbf{A}\mathbf{B} = \mathbf{Y} - \mathbf{B}^{+}\mathbf{A}\mathbf{B} \tag{C.68b}$$

In view of (C.68) and (C.66), one can write the similar transformation as follows:

$$\begin{aligned} & \mathbf{T}(\mathbf{A}+\mathbf{B}\mathbf{F})\mathbf{T}^{-1} \\ = & \begin{bmatrix} \mathbf{B}^{\perp} \\ \mathbf{B}^{+} \end{bmatrix} (\mathbf{A}+\mathbf{B}\mathbf{F}) \left[\mathbf{B}^{\perp^{\mathrm{T}}}, \mathbf{B}\right] \\ = & \begin{bmatrix} \mathbf{B}^{\perp}(\mathbf{A}+\mathbf{B}\mathbf{F})\mathbf{B}^{\perp^{\mathrm{T}}} & \mathbf{B}^{\perp}(\mathbf{A}+\mathbf{B}\mathbf{F})\mathbf{B} \\ \mathbf{B}^{+}(\mathbf{A}+\mathbf{B}\mathbf{F})\mathbf{B}^{\perp^{\mathrm{T}}} & \mathbf{B}^{+}(\mathbf{A}+\mathbf{B}\mathbf{F})\mathbf{B} \end{bmatrix} \\ = & \begin{bmatrix} \mathbf{B}^{\perp}\mathbf{A}\mathbf{B}^{\perp^{\mathrm{T}}} & \mathbf{B}^{\perp}\mathbf{A}\mathbf{B} \\ \mathbf{B}^{+}\mathbf{A}\mathbf{B}^{\perp^{\mathrm{T}}} + \mathbf{F}\mathbf{B}^{\perp^{\mathrm{T}}} & \mathbf{B}^{+}\mathbf{A}\mathbf{B} + \mathbf{F}\mathbf{B} \end{bmatrix} \\ = & \begin{bmatrix} \mathbf{B}^{\perp}\mathbf{A}\mathbf{B}^{\perp^{\mathrm{T}}} & \mathbf{B}^{\perp}\mathbf{A}\mathbf{B} \\ \mathbf{X} & \mathbf{Y} \end{bmatrix} \end{aligned} \tag{C.69}$$

This proves the lemma. ■

Taking $\mathbf{X} = \mathbf{0}$ and $\mathbf{Y} = \Phi$, we have the following:

Lemma C.4
Let $\mathbf{F} = \Phi\mathbf{B}^{+} - \mathbf{B}^{+}\mathbf{A}$. *Then,* $\mathbf{A}+\mathbf{BF}$ *is similar to the following matrix:*

$$\begin{bmatrix} \mathbf{B}^{\perp}\mathbf{A}\mathbf{B}^{\perp^{\mathrm{T}}} & \mathbf{B}^{\perp}\mathbf{A}\mathbf{B} \\ \mathbf{0} & \Phi \end{bmatrix} \tag{C.70}$$

Therefore, if $\mathbf{B}^{\perp}\mathbf{A}\mathbf{B}^{\perp^{\mathrm{T}}}$ *is asymptotically stable, then* $\mathbf{A}+\mathbf{BF}$ *is asymptotically stable. Moreover,* $eig(\mathbf{A}+\mathbf{BF}) = eig(\mathbf{B}^{\perp}\mathbf{A}\mathbf{B}^{\perp^{\mathrm{T}}}) \cup eig(\Phi)$.

Lemma C.4 is the fundamental idea of Misrikhanov and Ryabchenko Algorithm. Let

$$\begin{array}{lcc}
\text{Level 0:} & \mathbf{A}_0 = \mathbf{A} & \mathbf{B}_0 = \mathbf{B}, \\
\text{Level 1:} & \mathbf{A}_1 = \mathbf{B}_0^{\perp}\mathbf{A}_0\mathbf{B}_0^{\perp^{\mathrm{T}}} & \mathbf{B}_1 = \mathbf{B}_0^{\perp}\mathbf{A}_0\mathbf{B}_0 \\
 & \cdots & \cdots \\
\text{Level k:} & \mathbf{A}_k = \mathbf{B}_{k-1}^{\perp}\mathbf{A}_{k-1}\mathbf{B}_{k-1}^{\perp^{\mathrm{T}}} & \mathbf{B}_k = \mathbf{B}_{k-1}^{\perp}\mathbf{A}_{k-1}\mathbf{B}_{k-1} \\
 & \cdots & \cdots \\
\text{Level L:} & \mathbf{A}_L = \mathbf{B}_{L-1}^{\perp}\mathbf{A}_{L-1}\mathbf{B}_{L-1}^{\perp^{\mathrm{T}}} & \mathbf{B}_L = \mathbf{B}_{L-1}^{\perp}\mathbf{A}_{L-1}\mathbf{B}_{L-1}
\end{array} \tag{C.71}$$

where $L = ceil(n/m) - 1$. The technical base of Misrikhanov and Ryabchenko Algorithm is the following theorem.

Theorem C.10
Let linear system $(\mathbf{A},\mathbf{B})$ *is fully controllable and the matrix* $\mathbf{F} \in \mathbf{R}^{m\times r}$ *satisfies*

$$\begin{array}{cc}
\mathbf{F} = \mathbf{F}_0 = \Phi_0\mathbf{B}_0^{-} - \mathbf{B}_0^{-}\mathbf{A} & \mathbf{B}_0^{-} = \mathbf{B}_0^{+} - \mathbf{F}_1\mathbf{B}_0^{\perp} \\
\mathbf{F}_1 = \Phi_1\mathbf{B}_1^{-} - \mathbf{B}_1^{-}\mathbf{A}_1 & \mathbf{B}_1^{-} = \mathbf{B}_1^{+} - \mathbf{F}_2\mathbf{B}_0^{\perp} \\
\cdots & \cdots \\
\mathbf{F}_k = \Phi_k\mathbf{B}_k^{-} - \mathbf{B}_k^{-}\mathbf{A}_{k+1} & \mathbf{B}_k^{-} = \mathbf{B}_k^{+} - \mathbf{F}_{k+1}\mathbf{B}_k^{\perp} \\
\cdots & \cdots \\
\mathbf{F}_L = \mathbf{B}_L^{+}(\Phi_L - \mathbf{A}_L) &
\end{array} \tag{C.72}$$

Then,

$$eig(\mathbf{A}+\mathbf{BF}) = \cup_{i=1}^{L+1} eig(\Phi_{i-1}) \tag{C.73}$$

The proof of this theorem is omitted. Interested readers are referred to [131]. The Misrikhanov and Ryabchenko algorithm is given as follows:

Algorithm C.1

Step 0: *Select* $\Phi_k \in \mathbf{R}^{m\times m}$, $k = 0,\ldots,L-1$, *and* $\Phi_L \in \mathbf{R}^{L\times L}$, *all diagonal matrices, satisfying (C.73). Let* $\mathbf{A}_0 = \mathbf{A}$, $\mathbf{B}_0 = \mathbf{B}$, *and* $\mathbf{F}_L = 0$. *Calculate* $\mathbf{B}_0^{+}$.

Step 1: For $k = 1,\ldots,L-1$, *calculate*

$$\mathbf{B}_{k-1}^{\perp}, \quad \mathbf{A}_k = \mathbf{B}_{k-1}^{\perp}\mathbf{A}_{k-1}\mathbf{B}_{k-1}^{\perp^T}, \quad \mathbf{B}_k = \mathbf{B}_{k-1}^{\perp}\mathbf{A}_{k-1}\mathbf{B}_{k-1}, \quad \mathbf{B}_k^{+} \tag{C.74}$$

Step 2: For $k = L-1,\ldots,0$, *calculate* $\mathbf{B}_k^{-} = \mathbf{B}_k^{+} - \mathbf{F}_{k+1}\mathbf{B}_k^{\perp}$ *and* $\mathbf{F}_k = \Phi_k\mathbf{B}_k^{-} - \mathbf{B}_k^{-}\mathbf{A}_k$.

Step 3: Set $\mathbf{F} = \mathbf{F}_0$.

The easiest way to select Φ_k is to have block diagonal Φ_k such that (C.73) holds. A more complicated but attractive way is to select $\Phi_k = \mathbf{X}\Lambda\mathbf{X}^{\mathrm{T}}$ such that $||\mathbf{F}_k||_f^2 = ||\Phi_k\mathbf{B}_k^{-} - \mathbf{B}_k^{-}\mathbf{A}_k||_f^2$ is minimized, where $\mathbf{X}$ is an orthogonal matrix with $\mathbf{X}^{\mathrm{T}}\mathbf{X} = \mathbf{I}$, and Λ is a real block diagonal matrix with 1×1 blocks for real poles and 2×2 blocks for complex poles. Clearly minimizing $||\mathbf{F}_k||_f^2$ is a minimum gain pole assignment design [149].

Sine $\mathbf{X}$ is an orthogonal matrix, this problem is equivalent to

$$\min \quad ||\Lambda\mathbf{X}^{\mathrm{T}}\mathbf{B}_k^{-} - \mathbf{X}^{\mathrm{T}}\mathbf{B}_k^{-}\mathbf{A}_k||_f^2 \tag{C.75a}$$

$$s.t. \quad \mathbf{X}^{\mathrm{T}}\mathbf{X} = \mathbf{I} \tag{C.75b}$$

or

$$\min \quad Tr[(\mathbf{B}_k^{-})^{\mathrm{T}}\mathbf{X}\Lambda^{\mathrm{T}}\Lambda\mathbf{X}\mathbf{B}_k^{-} - (\mathbf{B}_k^{-})^{\mathrm{T}}\mathbf{X}\Lambda^{\mathrm{T}}\mathbf{X}^{\mathrm{T}}\mathbf{B}_k^{-}\mathbf{A}_k - \mathbf{A}_k^{\mathrm{T}}(\mathbf{B}_k^{-})^{\mathrm{T}}\mathbf{X}\Lambda\mathbf{X}^{\mathrm{T}}\mathbf{B}_k^{-}] \tag{C.76a}$$

$$s.t. \quad \mathbf{X}^{\mathrm{T}}\mathbf{X} = \mathbf{I} \tag{C.76b}$$

where $Tr[\cdot]$ is the trace of the matrix. The optimization problem can be efficiently solved using a conjugate gradient method on Riemannian manifold [1]. The Matlab code is available in

`http://www.mathworks.com/matlabcentral/fileexchange/47591-unit-opt-zip.`

Handreds randowly generated problems are tested with starting point $\mathbf{X}_0 = \mathbf{I}$, and the optimization solutions stay in $\mathbf{X}^* = \mathbf{I}$. Therefore, Misrikhanov and Ryabchenko Algorithm is likely a minimum gain pole assignment. Hence, there is no need to solve (C.76) in Step 2, selecting diagonal Φ_k is good enough.

References

[1] Abrudan, T. 2008. Advanced Optimization Algorithms for Sensor Arrays and Multi-antenna Communications, Department of Signal Processing and Acoustics. Aalto University, Finland.

[2] Alessio, A. and A. Bemporad. 2009. A survey of explicit model predictive control. pp. 345–369. *In*: L. Magni and D.M. Raimondo and F. Allgower [eds.]. Nonlinear Model Predictive Control: Toward New Challenging Applications. Springer-Verlag, Berlin.

[3] Alfriend, K.T. 1975. Magnetic attitude control system for dual-spin satellites. AIAA Journal. 13: 817–822.

[4] Anderson, B.D.O. and J.B.I. Moore. 1979. Optimal Filtering. Prentice-Hall, Inc., Englewood Cliffs, N.J.

[5] de Angelis, E.L., F. Giulietti, A.H.J. de Ruiter and G. Avanzini. 2016. Spacecraft attitude control using magnetic and mechanical actuation. Journal of Guidance, Control, and Dynamics. 39(3): 564–573.

[6] Arnold, W.F. III and A.J. Laub. 1984. Generalized eigenproblem algorithms and software for algebraic Riccati equations. Proceedings of the IEEE. 72: 1746–1754.

[7] Astrom, K.J. and B. Wittenmark. 2013. Computer-controlled Systems: Theory and Design. Dover Publications, Inc., Mineola, NY.

[8] Athans, M. and P.L. Falb. 1966. Optimal Control: An Introduction to the Theory and Its Applications. McGraw-Hill, Inc, New York.

[9] Bar-Itzhack, I.Y. and Y. Oshman. 1985. Attitude determination from vector observations: quaternion estimation. IEEE Transactions on Aerospace and Electronic Systems. 21: 128–136.

[10] Bastow, J.G. (ed.). 1965. Proceedings of the Magnetic Workshop, JPL TM33-216, Jet Propulsion Laboratory, Pasadena, Calif., September.

[11] Beletskii, V.V. 1966. Motion of an artificial satellite about its center of mass. Technical Report, NASA TT-F429.

[12] Bemporad, A. and M. Morari. 1999. Robust model predictive control: a survey. Robustness in Identification and Control. 245: 207–226.

[13] Bemporad, A., M. Morari, V. Dua and E. Pistikopoulos. 2002. The explicit linear quadratic regulator for constrained systems. Automatica. 38: 3-20.

[14] Berkelaar, A.B., K. Roos and T. Terlaky. 1997. The optimal set and optimal partition approach to linear and quadratic programming. *In*: H. Greenberg and T. Gal [eds.]. Recent Advances in Sensitivity Analysis and Parametric Programming, Kluwer Publishers, Berlin.

[15] Bertsekas, D.P. 1982. Projected Newton methods for optimization problems with simple constraints. SIAM J. on Control and Optimization. 20: 221–246.

[16] Bhat, S.P. 2005. Controllability of nonlinear time-varying systems: application to spacecraft attitude control using magnetic actuation. IEEE Transactions on Automatic Control. 50(11): 1725–1735.

[17] Bhat, S.P. and D.S. Bernstein. 2000. A topological obstruction to continuous global stabilization of rotational motion and the unwinding phenomenon. Systems & Control Letters. 39: 63–70.

[18] Bierman, G.J. 1976. Measurement updating using the U-D factorization. Automatica. 12(4): 375–382.

[19] Bittanti, S. 1991. Periodic Riccati equation. pp. 127–162. *In*: S. Bittanti et al. eds.. Springer, Berlin.

[20] Bittanti, S., P. Colaneri and G. Guardabassi. 1986. Analysis of the periodic Lyapunov and Riccati equations via canonical decomposition. SIAM J. Control and Optimization. 24(6): 1138–1149.

[21] Bittanti, S., P. Colaneri and G.D. Nicolao. 1989. A note on the maximal solution of the periodic Riccati equation. IEEE Transactions on Automatic Control. 15(12): 1316–1319.

[22] Black, H.D. 1964. A passive system for determining the attitude of a satellite. AIAA Journal. 2: 1350–1351.

[23] Boskovic, J., S. Li and R. Mehra. 2000. Robust adaptive variable structure control of spacecraft under control input saturation. Journal of Guidance, Control and Dynamics. 24: 14–22.

[24] Bryson, A.E. and Ho, Y.C. 1975. Applied Optimal Control: Optimization, Estimation, and Control. Taylor & Francis, New York.

[25] Di Cairano, S., H. Park and I. Kolmanovsky. 2012. Model predictive control approach for guidance of spacecraft rendezvous and proximity maneuvering. International Journal of Robust and Nonlinear Control. 22(12): 1398–1427.

[26] Carlson, N.A. 1973. Fast triangular formulation of the square root filter. AIAA Journal. 11(9): 1259–1264.

[27] Cheng, Y. and M.D. Shuster. 2014. Improvement to the implementation of the QUEST Algorithm. Journal of Guidance, Control, and Dynamics. 37(1): 301–305.

[28] do Carmo, M.P. 1976. Differential Geometry of Curves and Surfaces. Prentice-Hall, New Jersey.

[29] Chen, X., W.H. Steyn, S. Hodgart and Y. Hashida. 1999. Optimal combined reaction-wheel momentum management for earth-pointing satellites. Journal of Guidance, Control, and Dynamics. 22(4): 543–550.

[30] Chen, X. and X. Wu. 2010. Model predictive control of cube satellite with magnet-torque. pp. 997–1002. *In*: Proceedings of the 2010 IEEE International Conference on Information and Automation, Harbin, China.

[31] Cheon, Y. and J. Kim. 2007. Unscented filtering in a unit quaternion space for spacecraft attitude estimation. pp. 66–71. *In*: Proceedings of IEEE International Symposium on Industrial Electronics.

[32] Clohessy, W.H. and R.S. Wiltshire. 1960. Terminal guidance system for satellite rendezvous. Journal of Aerospace Sciences. 27(9): 653–658.

[33] Cochrane, C.J., J. Blacksberg, M.A. Anders and P.M. Lenahan. 2016. Vectorized magnetometer for space applications using electrical readout of atomic scale defects in silicon carbide. Scientific Reports. 6: 370–377.

[34] Crassidis, J.L. and F.L. Markley. 2003. Unscented filtering for spacecraft attitude estimation. Journal of Guidance, Control, and Dynamics. 26(4): 536–542.

[35] Crassidis, J.L., F.L. Markley and Y. Cheng. 2007. A survey of nonlinear attitude estimation methods. Journal of Guidance, Control and Dynamics. 30(1): 12–28.

[36] Curti, F., M. Romano and R. Bevilacqua. 2010. Lyapunov-based thrusters selection for spacecraft control: analysis and experimentation. Journal of Guidance, Control, and Dynamics. 33(4): 198–219.

[37] Curtis, H.D. 2005. Orbital Mechanics for Engineering Students. Elsevier Butterworth-Heinemann, Burlington, MA.

[38] Davenport, P. 1965. A vector approach to the algebra of rotations with applications. Technical Report, NASA, X-546-65-437.

[39] Davis, J. 2004. Mathematical modeling of Earth's magnetic field, TN, Virginia Polytechnic Institute and State University, Blacksburg, VA.

[40] Dieci, L. and T. Eirola. 1999. On smooth decompositions of matrices. SIAM Journal on Matrix Analysis and Applications. 20: 800–819.

[41] Dorf, R.C. and R.H. Bishop. 2008. Modern Control Systems. Pearson Prentice Hall, Upper Saddle River, NJ.

[42] Doyle, J., K. Glover, P. Khargonekar and B. Francis. 1989. State-space solutions to standard $\mathbf{H}_2$ and $\mathbf{H}_\infty$ control problems. IEEE Transactions on Automatic Control. 34: 831–847.

[43] Droll, P.W. and E.J. Iuler. 1967. Magnetic properties of selected spacecraft materials. Proceedings of symposium on space magnetic exploration and technology. Engineering Report. 9: 189–197.

[44] Dzielsk, J., E. Bergmann and J. Paradiso. 1990. A computational algorithm for spacecraft control and momentum management. pp. 1320–1325. *In*: Proceedings of the American Control Conference.

[45] Egeland, O. and J.T. Gravdahl. 2002. Modeling and simulation for automatic control. Marine Cybernetics, Trondheim, Norway.

[46] Finlay, C.C., S. Maus, C.D. Beggan, T.N. Bondar, A. Chambodut, T.A. Chernova, A. Chulliat, V.P. Golovkov, B. Hamilton, M. Hamoudi, R. Holme, G. Hulot, W. Kuang, B. Langlais, V. Lesur, F.J. Lowes, H. Lhr, S. Macmillan, M. Mandea, S. McLean, C. Manoj, M. Menvielle, I. Michaelis, N. Olsen, J. Rauberg, M. Rother, T.J. Sabaka, A. Tangborn, L. Tffner-Clausen, E. Thbault , A.W.P. Thomson, I. Wardinski, Z. Wei and T.I. Zvereva. 2010. International geomagnetic reference field: the eleventh generation. Geophysical Journal International 183: 1216–1230.

[47] Finlay, C.C., E. Thbault and H. Toh. 2015. Special issue of International geomagnetic reference field-the twelfth generation, Earth, Planets and Space. pp. 67–158.

[48] Flamm, D.S. 1991. A new shift-invariant representation for periodic linear system. Systems and Control Letters. 17(1): 9-14.

[49] Forbes, J.R., A.H.J. de Ruiter and D.E. Zlotnik. 2014. Continuous-time norm-constrained Kalman filter. Automatica. 50(10): 2546–2554.

[50] Ford, K.A. 1997. Reorientations of flexible spacecraft using momentum exchange devices. Ph.D dissertation, Dept. of Aeronautics and Astronautics, Air Force Institute of of Technology, OH.

[51] Ford, K.A. and C.D. Hall. 1997. Flexible spacecraft reorientations using gimbal momentum wheels. pp. 1895–1914. *In*: F. Hoots, B. Kaufman, P.J. Cefola and D.B. Spencer [eds.]. Advances in the Astronautical Sciences, Astrodynamics. Univelt, San Diego, 97.

[52] Ford, K.A. and C.D. Hall. 2000. Singular direction avoidance steering for control moment gyros. Journal of Guidance, Control, and Dynamics. 23(4): 648–656.

[53] Fortescue, P., J. Stark and G. Swinerd. 2008. Spacecraft Systems Engineering. John Wiley & Sons, West Hoboken, NJ.

[54] Gao, H., X. Yang and P. Shi. 2009. Multi-Objective Robust H_∞ Control of Spacecraft Rendezvous. IEEE Transactions on Control Systems Technology. 17(4): 794–802.

[55] Gelb, A. (ed.). 1974. Applied Optimal Estimation. The MIT press, Cambridge, MA, USA.

[56] Giulietti, F., A.A. Quarta and P. Tortora. 2006. Optimal control laws for momentum-wheel desaturation using magnetorquers. Journal of Guidance, Control, and Dynamics. 29(6): 1464–1468.

[57] Goldman, A.J. and A.W. Tucker. 1956. Theory of linear programming. pp. 53–97. *In*: H.W. Kuhn and Tucker [eds.]. Linear Equalities and Related Systems. Princeton University Press, Princeton.

[58] Golub, G.H. and C.F. Van Loan. 1989. Matrix Computations. The Johns Hopkins University press, Baltimore.

[59] Grace, A., A.J. Laub, J.N. Little and C. Thompson. 1990. Control System Toolbox for Use with MATLAB: User's Guide. The MathWorks, Inc., South Natick, MA.

[60] Grasselli, O.M. and S. Longhi. 1991. Pole-placement for non-reachable periodic discrete-time system. Mathematics of Control, Signals and Systems. 4: 439–455.

[61] Greenwood, D.T. 1988. Principles of Dynamics. Prentice-Hall Inc., Englewood Cliffs, New Jersey, 2nd edition.

[62] Gui, H., G. Vukovich and S. Xu. 2015. Attitude tracking of a rigid spacecraft using two internal torques. IEEE Transactions on Aerospace and Electronic Systems. 51(4): 2900–2913.

[63] Guler, O. and Y. Ye. 1993. Convergence behavior of interior-point algorithms. Mathematical Programming. 60: 215–228.

[64] Hall, A.C. 1943. The Analysis and Synthesis of Linear Servomechanisms. The Technology Press, Cambridge, MA.

[65] Harris, M. and R. Lyle. 1969. Magnetic fields-earth and extraterrestrial. NASA Report, SP-8017.

[66] Harris, M. and R. Lyle. 1969. Spacecraft gravitational torques. Technical Report, NASA SP-8024.

[67] Hartley, E.N., P.A. Trodden, A.G. Richards and J.M. Maciejowski. 2012. Model predictive control system design and implementation for spacecraft rendezvous. Control Engineering Practice. 20(7): 695–713.

[68] Haseltine, E.L. and J.B. Rawlings. 2005. Critical evaluation of extended Kalman filtering and moving-horizon estimation. Industrial and Engineering Chemistry Research. 44(8): 2451–2460.

[69] Hegrenæs, O., J.T. Gravdahl and P. Tøndel. 2005. Spacecraft attitude control using explicit model predictive control. Automatica. 41(12): 2107–2114.

[70] Hench, J.J. and A.J. Laub. 1994. Numerical solution of the discrete-time periodic Riccati equation. IEEE Transactions on Automatic Control. 39(6): 1197–1210.

[71] Herbison-Evans, D. 1994. Solving quartics and cubics for graphics. TR94-487, University of Sydney, Sydney, Australia.

[72] Hill, G.W. 1878. Researches in lunar theory. American Journal of Mathematics. 1(1): 5–26.

[73] Horn, R.A. and C.R. Johnson. 1985. Matrix Analysis. Cambridge University Press, Cambridge.

[74] Hughes, P.C. 1986. Spacecraft Attitude Dynamics. Wiley, New York, USA.

[75] Available at: http://www.ngdc.noaa.gov/IAGA/vmod/igrf.html

[76] Available at: https://www.mathworks.com/matlabcentral/fileexchange/54255-quartic-roots-m.

[77] Available at: https://www.mathworks.com/matlabcentral/fileexchange/54499-newtonrattitude-m.

[78] Available at: http://en.wikipedia.org/wiki/Magnetometer.

[79] Jiang, E.X. 1978. Linear Algebra. People Educational Press (in Chinese).

[80] Jin, J. and I. Hwang. 2011. Attitude control of a spacecraft with single variable-speed control moment gyroscope. Journal of Guidance, Control, and Dynamics. 34(6): 1920–1925.

[81] Juang, J.N., H.Y. Kim and J.L. Junkins. 2003. An efficient and robust singular value method for star pattern recognition and attitude determination. Technical Report, NASA/TM-2003-212142, Langley Research Center, NASA, Hampton, Virginia.

[82] Julier, S., J. Uhlmann and H.F. Durrant-Whyte. 2000. A new method for the nonlinear transformation of means and covariances in filters and estimators. IEEE Transactions on Automatic Control. 45(3): 477–482.

[83] Jung, D. and P. Tsiotras. 2004. An experimental comparison of CMG steering control laws. AIAA/AAS Astrodynamics Specialist Conference and Exhibit, Guidance, Navigation, and Control. Providence, Rhode Island.

[84] Junkins, J.L. and Y. Kim. 1993. Introduction to dynamics and control of flexible structures. Washington, DC: AIAA, pp. 9-64.

[85] Jury, E.I. and F.J. Mullin. 1958. A note on the operational solution of linear difference equations. J. Franklin Inst. 266: 189–205.

[86] Jury, E.I. and F.J. Mullin. 1959. The analysis of sampled data control systems with a periodically time-varying sampling rate. IRE Transactions on Automatic Control. AC-4: 15–21.

[87] Kalman, R.E. 1960. A new approach to linear filtering and prediction problem. Trans. ASME J. Basic Engineering. 82D: 35–45.

[88] Karush, W. 1939. Minima of Functions of Several Variables with Inequalities as Side Constraints, M.Sc. Dissertation. Dept. of Mathematics, Univ. of Chicago, Chicago, Illinois.

[89] Kalman, R.E. 1960. A new approach to linear filtering and prediction problem. Trans. ASME J. Basic Engineering. 82D: 35–45.

[90] Kautsky, J., N.K. Nichols and P. Van Dooren. 1985. Robust pole assignment in linear state feedback. International Journal of Control 41: 1129–1155.

[91] Keat, J. 1977. Analysis of least-squares attitude determination, Routine DOAOP. Technical Report, Comp. Sc. Corp., CSC/TM-77/6034.

[92] Khalil, H.K. 1992. Nonlinear System. Macmillan Publishing Company, New York.

[93] Khan, N., S. Fekri, R. Ahmad and Dawei Gu. 2011. New results on robust state estimation in spacecraft attitude control. pp. 90–95. The 50th IEEE Conference on Decision and Control and European Control Conference (CDC-ECC), Orlando, FL, USA.

[94] Khargonekar, P.P., K. Poolla and A. Tanenbaum. 1985. Robust control of linear time-invariant plants using periodic compensation. IEEE Transactions on Automatic Control. 30(11): 1088–1096.

[95] Kleinman, D.L. 1968. On an iterative technique for Riccati equation computation. IEEE IEEE Transactions on Automatic Control. 13(1): 114–115.

[96] Kristiansen, R., E.I. Grotli, P.J. Nicklasso and J.T. Gravdahl. 2007. A model of relative translation and rotation in leader-follower spacecraft formations. Modeling, Identification and Control. 28(1): 3–13.

[97] Kristiansena, R., P.J. Nicklasso, J.T. Gravdahlb. 2008. Spacecraft coordination control in 6DOF: Integrator back-stepping vs passivity-based control. Automatica. 44(11): 2896–2901.

[98] Kuhn, H.W. and A.W. Tucker. 1951. Nonlinear programming. pp. 481–492. *In*: Proceedings of 2nd Berkeley Symposium, Berkeley, University of California Press.

[99] Kuipers, J.B. 1998. Quaternions and rotation sequences: A primer with applications to orbits, aerospace, and virtual reality. Princeton University Press, Princeton and Oxford.

[100] Kurokawa, H. 1998. A geometric study of single gimbal control moment gyros. Technical Report No. 175, National Institute of Advanced Industrial Science and Technology, Tsukuba, Japan.

[101] Kurokawa, H. 2007. Survey of theory and steering laws of single-gimbal control moment gyros. Journal of Guidance, Control, and Dynamics. 30(5): 1331–1340.

[102] Laub, A.J. 1972. Canonical forms for σ-symplectic matrices. M.S. thesis, School of Mathematics, Univ. of Minnesota, MN.

[103] Laub, A.J. 1979. A Schur method for solving algebraic Riccati equations. IEEE Transactions on Automatic Control. 24(6): 913–921.

[104] LaViola Jr., J.J. 2003. A comparison of unscented and extended Kalman filtering for estimating quaternion motion. pp. 2435–2440. *In*: Proceedings of the American Control Conference Denver, Colorado.

[105] Lawrence, D.A. and W.J. Rugh. 1990. On a stability theorem for nonlinear systems with slowly varying inputs. IEEE Transactions on Automatic Control. 35: 860–864.

[106] Lefferts, E.J., F.L. Markley and M.D. Shuster. 1982. Kalman filtering for spacecraft attitude estimation. Journal of Guidance, Control and Dynamics. 5(5): 417–429.

[107] Lenz, J.E. 1990. A review of magnetic sensors. Proceedings of the IEEE. 78: 973–989.

[108] Lewis, F.L., D. Vrabie and V.L. Syrmos. 2012. Optimal Control, 3rd Edition. John Wiley & Sons, Inc., New York, USA.

[109] Ley, W., K. Wittmann and W. Hallmann. 2009. Handbook of Space Technology. Wiley, Munich, Germany.

[110] Li, Q., J. Yuan., B. Zhang and C. Gao. 2017. Model predictive control for autonomous rendezvous and docking with a tumbling target. Aerospace Science and Technology. 69: 700–711.

[111] Liebe, C.C. and S. Mobasser. 2001. MEMS based sun sensor. pp. 1565 -1572. IEEE Proceedings of Aerospace Conference.

[112] Lou, G.Y. 1973. Models of Earth's atmosphere (90 to 2500 km). Technical Report, NASA SP-8021.

[113] Lovera, M. and A. Astolfi. 2004. Spacecraft attitude control using magnetic actuators. Automatica. 40: 1405–1414.

[114] Lovera, M. and A. Astolfi. 2006. Global magnetic attitude control of spacecraft in the presence of gravity gradient. IEEE Transactions on Aerospace and Electronic System. 42(3): 796–805.

[115] Lovera, M., E. Marchi and S. Bittanti. 2002. Periodic attitude control techniques for small satellites with magnetic actuators. IEEE Transactions on Control System Technology. 10(1): 90–95.

[116] Luo, Y., J. Zhang and G. Tang. 2014. Survey of orbital dynamics and control of space rendezvous. Chinese Journal of Aeronautics 27(1): 1–11.

[117] Lyle, R. and P. Stabekis. 1971. Spacecraft aerodynamic torques. Technical Report, NASA SP-8058.

[118] Macfarlane, A.G.J. 1963. An eigenvector solution of the optimal linear regulator problem. Journal of Electronics and Control. 14(6): 643–654.

[119] Magnus, J.R. 1985. On differentiating eigenvalues and eigenvectors. Econometric Theory. 1: 179-191.

[120] Malik, M.S.I. and S. Asghar. 2013. Inverse free steering law for small satellite attitude control and power tracking with VSCMGs. Advances in Space Research. 53: 97–109.

[121] Malladi, B.P., R.G. Sanfelice, E. Butcher and J. Wang. 2016. Robust hybrid supervisory control for rendezvous and docking of a spacecraft. IEEE 55th Conference on Decision and Control.

[122] Morf, M. and T. Kailath. 1975. Square root algorithms for least squares estimation. IEEE transactions on Automatic Control. 20(4): 487–497.

[123] Markley, F.L. 1993. Attitude determination using vector observations: a fast optimal matrix algorithm. Journal of Astronautical Sciences. 41(2): 261–280.

[124] Markley, F.L. 2002. Fast quaternion attitude estimation from two vector measurements. Journal of Guidance, Control and Dynamics. 25: 411–414.

[125] Markley, F.L. 2003. Attitude error representations for Kalman filtering. Journal of Guidance and Control. 26(2): 311–317.

[126] Markley, F.L. 2004. Multiplicative vs. additive filtering for spacecraft attitude determination. *In*: Proceedings of the 6th Conference on Dynamics and Control Systems and Structures in Space (DCSSS). D22. Riomaggiore, Italy.

[127] Markley, F.L. and D. Mortari. 2000. Quaternion attitude estimation using vector observations. Journal of Astronautical Sciences. 48(2): 359–380.

[128] Massari, M. and M. Zamaro. 2014. Application of SDRE technique to orbital and attitude control of spacecraft formation flying. Acta Astronautica. 94(1): 409–420.

[129] Mayne, D.Q., J.R. Rawlings, C.V. Rao and P.O.M. Scokaert. 2000. Constrained model predictive control: stability and optimality. Automatica. 36(6): 789–814.

[130] Meyer, R.A. and C.S. Burrus. 1975. A unified analysis of multirate and periodically time-varying digital filters. IEEE Transactions on Circuits and System. 22(3): 162–168.

[131] Misrikhanov, M.S. and V.N. Ryabchenko. 2011. Pole placement for controlling a large scale power system. Automation and Remote Control. 72(10): 2123–2146.

[132] McGee, I.A. and S.F. Schmidt. 1985. Discovery of the Kalman filter as a practical tool for aerospace industry. Technical Report, NASA-TM-86847, NASA.

[133] Mizuno, S., M. Todd and Y. Ye. 1993. On adaptive step primal-dual interior-point algorithms for linear programming. Mathematics of Operations Research. 18: 964–981.

[134] Monteiro, R. and I. Adler. 1989. Interior path following primal-dual algorithms. Part I: linear programming. Mathematical Programming. 44: 27–41.

[135] Morgan, D., S.J. Chung and F.Y. Hadaegh. 2014. Model predictive control of swarms of spacecraft using sequential convex programming. Journal of Guidance, Control, and Dynamics. 37(6): 1725–1740.

[136] Mortari, D. 1997. ESOQ: A closed-form solution to the Wahba problem. Journal of Astronautical Sciences. 45(2): 195–205.

[137] Murnaghan, F.D. and A. Wintner. 1931. A canonical form for real matrices under orthogonal transformations. Proc. Nat. Acad. Sci. 17: 417–420.

[138] Musser, K.L. and W.L. Ebert. 1989. Autonomous spacecraft attitude control using magnetic torquing only. pp. 23–38. *In*: Proceedings of the Flight Mechanics and Estimation Theory Symposium, NASA Goddard Space Flight Center, Greenbelt, MD.

[139] Nakamura, Y. and H. Hanafusa. 1986. Inverse kinematic solution with singularity robustness for robot manipulator control. Journal of Dynamic Systems, Measurement, and Control. 108(3): 163–171.

[140] Naumann, R.J. 1961. Recent information gained from satellite orientation measurements. Planetary and Space Sciences. 7: 445–453.

[141] Navabi, M. and M. Barati. 2017. Mathematical modeling and simulation of the Earth's magnetic field: A comparative study of the models on the spacecraft attitude control application. Applied Mathematical Modeling. 46: 365–381.

[142] Nocedal, J. and S.J. Wright. 1999. Numerical Optimization. Springer-Verlag, New York.

[143] Oh, H.S. and S.R. Vadali. 1991. Feedback control and steering laws for spacecraft using single gimbal control moment gyros. Journal of the Astronautical Sciences. 39(2): 183–203.

[144] Oppenheim, A.V., R.W. Schafer and J.R. Buck. 1999. Discrete-Time Signal Processing. Prentice Hall, New York.

[145] Opromolla, R., G. Fasano, G. Rufino and M. Grassi. 2017. A review of cooperative and uncooperative spacecraft pose determination techniques for close-proximity operations. Progress in Aerospace Sciences. 93: 53–72.

[146] Paghis, I., C.A. Franklin and J. Mar. 1967. Alouette I: the first three years in orbit, Part III, DRTE Report 1159. Department of Defence (Canada).

[147] Paielli, R. and R. Bach. 1993. Control with realization of linear error dynamics. Journal of Guidance, Control and Dynamics. 16: 182–189.

[148] Pan, H. and V. Kapila. 2001. Adaptive Nonlinear Control for Spacecraft Formation Flying with Coupled Translational and Attitude Dynamics. *In*: Proceedings of the Conference on Decision and Control, Orlando, FL.

[149] Pandey, A., R. Schmid and T. Nguyen. 2015. Performance survey of minimum gain exact pole placement methods. pp. 1808–1812. *In*: Proceedings of 2015 European Control Conference.

[150] Pandey, A., R. Schmid, T. Nguyen, Yaguang Yang, V. Sima and A.L. Tits. 2014. Performance survey of robust pole placement methods. pp. 3186–3191. *In*: Proceedings of 53rd Conference on Decision and Control. Los Angeles, California, USA.

[151] Pappas, T., A.J. Laub and N.R. Sandell. 1980. On the numerical solution of the discrete-time algebraic Riccati equation. IEEE Transactions on Automatic Control. 25(4): 631–641.

[152] Patera, R.P. 2018. Attitude estimation based on observation vector inertia. Advances in Space Research. 62: 383–397.

[153] Perea, L., J. How, L. Breger and P. Elosegui. 2007. Nonlinearity in Sensor Fusion: Divergence Issues in EKF, modified truncated SOF, and UKF, AIAA Guidance, Navigation and Control Conference and Exhibit, South Carolina.

[154] Persson, S.M. and I. Sharf. 2013. Invariant trapezoidal Kalman filter for application to attitude estimation. Journal of Guidance and Control. 36(3): 721–733.

[155] Petersen, K.B. and M.S. Pedersen. 2012. The Matrix Cookbook. Available online at: http://matrixcookbook.com.

[156] Pittelkau, M.E. 1993. Optimal periodic control for spacecraft pointing and attitude determination. Journal of Guidance, Control, and Dynamics. 16(6): 1078–1084.

[157] Polyanin, A.D. and A.V. Manzhirov. 2007. Handbook of Mathematics For Engineers and Scientists. Chapman & Hall/CRC, Noca Raton, FL.

[158] Potter, J.E. and R.G. Stern. 1963. Statistial filtering of space navigation measurements. In Proceedings 1963 AIAA Guidance and Control Conference, Cambridge, MA.

[159] Psiaki, M.L. 2001. Magnetic torque attitude control via asymptotic periodic linear quadratic regulator. Journal of Guidance, Control, and Dynamics. 24(2): 386–394.

[160] Pulecchi, T., M. Lovera and A. Varga. 2010. Optimal discrete-time design of three-axis magnetic attitude control law. IEEE Transactions on Control System Technology. 18(3): 714–722.

[161] Qin, S.J. and T.A. Badgwell. 2003. A survey of industrial model predictive control technology. Control Engineering Practice. 11: 733–764.

[162] Rao, C.V., S.J. Wright and J.B. Rawling. 1998. Application of interior-point methods to model predictive control. Journal of Optimization Theory and Applications. 99: 723–757.

[163] Reyhanoglu, M. and J.R. Hervas. 2011. Three-axis magnetic attitude control algorithm for small satellites. *In*: Proceedings of 5th International Conference on Recent Advance Technologies, Istanbul.

[164] Reynolds, R.G. 1998. Quaternion parameterization and a simple algorithm for global attitude estimation. Journal of Guidance, Control, and Dynamics. 21: 669–671.

[165] Rodriguez-Vazquez, A.L., M.A. Martin-Prats and F. Bernelli-Zazzera. 2012. Full magnetic satellite attitude control using ASRE method. The first IAA Conference on Dynamics and Control of Space Systems, DyCoSS 2012, Porto, Portugal.

[166] Rodriquez-Vazouez, A., M.A. Martin-Prats and F. Bernelli-Zazzera. 2015. Spacecraft magnetic attitude control using approximating sequence Riccati equations. IEEE transactions on Aerospace and Electronic Systems. 51(4): 3374–3385.

[167] Roscoe, C.W.T., J.J. Westphal and E. Mosleh. 2018. Overview and GNC design of the CubeSat Proximity Operations Demonstration (CPOD) mission. Acta Astronautica. Doi.org/10.1016/j.actaastro.2018.03.033.

[168] Rugh, W.J. 1990. Analytical framework for gain scheduling. pp. 1688–1694. *In*: American Control Conference, San Diego, CA, USA.

[169] Rugh, W.J. 1993. Linear System Theory. pp. 4–5. Prentice-Hall, Inc., Englewood Cliffs, New Jersey.

[170] Rugh, W.J. and J.S. Shamma. 2000. Research on gain scheduling. Automatica. 36: 1401–1425.

[171] Samaan, M.A., C. Bruccoleri, D. Mortari and J.L. Junkins. 2003. Novel techniques for creating nearly uniform star catalog. Advances in the Astronautical Sciences. 116: 1961–1704.

[172] Samaan, M. and S. Theil. 2012. Development of a low cost star tracker for the SHEFEX mission. Aerospace Science and Technology. 23(1): 469–478.

[173] Schalknowsky, S. and M. Harris. 1969. Spacecraft magnetic torques. Technical Report, NASA SP-8018.

[174] Schaub, H. and J.L. Junkins. 2003. Analytical Mechanics of Space Systems. AIAA Education Series, Reston, VA.

[175] Schaub, H., S. Vadali and J.L. Junkins. 1998. Feedback control law for variable speed control moment gyroscopes. Journal of the Astronautical Sciences. 46(3): 307–328.

[176] Serway, R.A. and J.W. Jewett. 2004. Physics for Scientists and Engineers. Books/Cole Thomson Learing, Belmont, CA.

[177] Shahruz, S.M. and S. Behtash. 1992. Design of controllers for linear parameter-varying systems by the gain scheduling technique. Journal of Mathematical Analysis and Applications. 168(1): 195–217.

[178] Shmakov, S.L. 2011. A universal method of solving quartic equations. International Journal of Pure and Applied Mathematics. 71(2): 251–259.

[179] Shuster, M.D. 1993. A survey of attitude presentation. Journal of the Astronautical Sciences. 27: 439–517.

[180] Shuster, M.D. and S.D. Oh. 1981. Three-axis attitude determination form vector observations. Journal of Guidance and Control. 4: 70–77.

[181] Sidi, M. 1997. Spacecraft Dynamics and Control: A Practical Engineering Approach. Cambridge University Press, Cambridge, UK.

[182] Siliani, E. and M. Lovera. 2006. Magnetic spacecraft attitude control: a survey and some new results. Control Engineering Practice. 13: 357–371.

[183] Sima, V., A.L. Tits and Y. Yang. 2006. Computational experience with robust pole assignment algorithms. IEEE Conference on Computer Aided Control Systems Design, Munich, Germany.

[184] Singla, P., J.L Crassida and J.L. Junkins. 2003. Spacecraft angular rate estimation algorithms for star tracker-based attitude determination. Advances in the Astronautical Sciences. 114: 1303–1316.

[185] Singla, P., K. Subbarao, J.L. Junkins. 2006. Adaptive output feedback control for spacecraft rendezvous and docking under measurement uncertainty. Journal of Guidance Control and Dynamics. 29(4): 892–902.

[186] Smith, S.T. 1994. Optimization techniques on Riemannian manifolds. Fields Institute Communications. 3: 113–136.

[187] Smith, G.L., S.G. Schmidt and L.A. McGee. 1962. Application of statistical filter theory to the optimal estimation of position and velocity on board a circumlunar vehicle. Technical Report, NASA Technical Report R-135, NASA.

[188] Spratling, B.B. and D. Mortari. 2009. A survey on star identification algorithms. Algorithms. 2: 93–107.

[189] Stewart, G.W. and J.G. Sun. 1990. Matrix Perturbation Theory. Academic Press, San Diego.

[190] Stoltz, P.M., S. Sivapiragasam and T. Anthony. 1998. Satellite orbit-raising using LQR control with fixed thrusters. Advances in the Astronautical Sciences. 98: 109–120.

[191] Sun, H., S. Lia and S. Fei. 2011. A composite control scheme for 6DOF spacecraft formation control. Acta Astronautica, Vol. 69, No. 7-8, 2011, pp. 595-611.

[192] Sun, J. 1995. On worst-case condition numbers of a nondefective multiple eigenvalue. Numerische Mathematik. 68: 373–382.

[193] Sun, L. and W. Huo. 2015. 6-DOF integrated adaptive back-stepping control for spacecraft proximity operations. IEEE Transactions on Aerospace and Electronic Systems. 51(3): 2433–2443.

[194] Sun, L., W. Huo and Z. Jiao. 2016. Robust Nonlinear Adaptive Relative Pose Control for Cooperative Spacecraft During Rendezvous and Proximity Operations. IEEE Transactions on Control Systems Technology. 25: 1840–1847.

[195] Sun, L., W. Huo and Z. Jiao. 2017. Adaptive Back stepping Control of Spacecraft Rendezvous and Proximity Operations With Input Saturation and Full-State Constraint. IEEE Transactions on Industrial Electronics. 64: 480–492.

[196] Sun, L. and Z. Zheng. 2018. Adaptive relative pose control of spacecraft with model couplings and uncertainties. Acta Astronautica. 143: 29–36.

[197] Sutherland, R., I. Kolmanovsky and A.R. Girard. Attitude control of a 2U cubesat by magnetic and air drag torques. IEEE Transactions on Control Systems Technology, to appear. DOI: 10.1109/TCST.2018.2791979.

[198] Tayebi, A. 2008. Unit quaternion-based output feedback for the attitude tracking problem. IEEE Transactions on Automatic Control. 53(6): 1516–1520.

[199] Thebault et al. 2015. International geomagnetic reference field: the 12th generation, Earth, Planets and Space. 67: 79–98.

[200] Tits, A.L. 2014. Private communication.

[201] Tits, A.L. and Y. Yang. 1996. Globally convergent algorithms for robust pole assignment by state feedback. IEEE Transactions on Automatic Control. 41: 1432–1452.

[202] Tøndel, P., A. Johansen and A. Bemporad. 2001. An algorithm for multiparametric quadratic programming and explicit MPS solution. pp. 1199–1204. *In*: IEEE conference on Decision and Control.

[203] Treqouet, J.-F., D. Arzelier, D. Peaucelle, C. Pittet and L. Zaccarian. 2015. Reaction wheel desaturation using magnetorquers and and static input allocation. IEEE Transactions on Control System Technology. 23(2): 525–539.

[204] US Nautical Almanac Office. 2001. The astronomical almanac for the year 2001, data for astronomy, space science, geodesy, surveying, navigation and other applications.

[205] Vallado, D.A. 2004. Fundamentals of astrodynamics and applications. Microcosm Press, El Segundo, CA.

[206] Varga, A. 2008. On solving periodic Riccati equations. Numerical Linear Algebra with Applications. 15(12): 809–835.

[207] Varga, A. 2013. Computational issues for linear periodic system: paradigms, algorithms, open problems. International Journal of Control. 86(7): 1227–1239.

[208] Vaughan, D.R. 1970. A nonrecursive algebraic solution for the discrete Riccati equation. IEEE Transactions on Automatic Control. 15(5): 597–599.

[209] Verhaegen, M. and P. Van Dooren. 1978. Numerical aspects of different Kalman filter. IEEE transactions on Automatic Control. 31(10): 907–917.

[210] Volland, H. 1969. A theory of thermospheric dynamics—I, diurnal and solar cycle variations. Planet Space Sciences. 17: 1581–1597.

[211] Volland, H. 1969. A theory of thermospheric dynamics—II, geomagnetic activities effect, 27-day variation and semiannual variation. Planet Space Sciences. 17: 1709–1724.

[212] Votel, R. and D. Sinclair. 2012. Comparison of control moment gyros and reaction wheels for small earth-observing satellites, 26th Annual AIAA/USU Conference on Small Satellites.

[213] Wahba, G. 1965. A least squares estimate of spacecraft attitude. SIAM Review. 7: 409.

[214] Wallsgrove, R. and M. Akella. 2005. Globally stabilizing saturated control in the presence of bounded unknown disturbances. Journal of Guidance, Control and Dynamics. 28: 957–963.

[215] Wang, P.K.C. and F.Y. Hadaegh. 1996. Coordination and control of multiple micro-spacecraft moving in formation. Journal of Astronautical Sciences. 44(3): 315–355.

[216] Wang, Y. and S. Boyd. 2010. Fast Model Predictive Control Using On line Optimization. IEEE Transactions on control systems technology. 18: 267–278.

[217] de Weck, O.L. 2001. Attitude determination and control: (ADCS). Department of Aeronautics and Astronautics, MIT.

[218] Wen, J. and K. Kreutz-Delgado. 1991. The attitude control problem. IEEE Transactions On Automatic Control. 36: 1148–1161.

[219] Wertz, J. 1978. Spacecraft Attitude Determination and Control. Kluwer Academic Publishers, Dordrecht, Holland.

[220] Wie, B. 1998. Vehicle Dynamics and Control. AIAA Education Series, Reston, VA.

[221] Wie, B. 2003. New singularity escape/avoidance logic for control moment gyro systems. AIAA Guidance, Navigation, and Control Conference, Austin, TX.

[222] Wie, B. 2004. Solar sail attitude control and dynamics, Part I. Journal of Guidance, Control, and Dynamics. 27: 526–535.

[223] Wie, B., D. Bailey and C. Heigerg. 2001. Singularity robust steering logic for redundant single-gimbal control moment gyros. Journal of Guidance, Control, and Dynamics. 24(5): 865–872.

[224] Wie, B., H. Weiss and A. Arapostathis. 1989. Quaternion feedback regulator for spacecraft eigenaxis rotations. Journal of Guidance, Control and Dynamics 12: 375–380.

[225] Wiener, N. 1949. Extrapolation, Interpolation, and Smoothing of Stationary Time Series. MIT Press, Cambridge, MA.

[226] Wiesel, N. 1989. Spaceflight Dynamics, McGraw-Hill, New York.

[227] Wilkinson, J.H. 1965. The Algebraic eigenvalue problem. Charendon Press, Oxford.

[228] Wisniewski, R. 1997. Linear time varying approach to satellite attitude control using only electromagnetic actuation. pp. 243–251. *In*: Proceedings of the AIAA Guidance, Navigation, and Control Conference, New Orleans.

[229] Won, C.H. 1999. Comparative study of various control methods for attitude control of a LEO satellite. Aerospace Science and Technology. 3: 323–333.

[230] Wonham, W.M. 1968. On the separation theorem of stochastic control. SIAM Journal on Control. 6: 312–326.

[231] Wood, M., W.H. Chen and D. Fertin. 2006. Model predictive control of low earth orbiting spacecraft with magneto- torquers. pp. 2908–2913. *In*: Computer Aided Control System Design. The IEEE International Conference on Control Applications.

[232] Wright, S.J. 1993. Interior point methods for optimal control of discrete time systems. Journal of Optimization Theory and Applications. 77: 161–187.

[233] Wright, S.J. 1997. Primal-Dual Interior-Point Methods. SIAM, Philadelphia.

[234] Wu, J., Z. Zhou, B. Gao, R. Li, Y. Cheng and H. Fourati. 2018. Fast linear quaternion attitude estimator using vector observations. IEEE Transactions on Automation Science and Engineering. 15: 307–319.

[235] Xu, R., H. Ji, K. Li, Y. Kang and K. Yang. 2015. Relative position and attitude coupled control with finite-time convergence for spacecraft rendezvous and docking. IEEE 54th Annual Conference on Decision and Control (CDC0, Osaka, Japan.

[236] Xu, Y., A. Tatsch, N.G. Fitz-Coy. 2005. Chattering free sliding mode control for a 6 DOF formation flying mission. pp. 2005–6464. *In*: Proceedings of AIAA Guidance, Navigation, and Control Conference and Exhibit, San Francisco, USA, AIAA.

[237] Yan, H., I.M. Ross and K.T. Alfriend. 2007. Pseudo-spectral feedback control for three-axes magnetic attitude stabilization in elliptic orbits. Journal of Guidance, Control, and Dynamics. 30(4): 1107–1115.

[238] Yan, Q., G. Yang, V. Kapila and M. de Queiroz. 2000. Nonlinear dynamics and output feedback control of multiple spacecraft in elliptical orbits. pp. 839–843. *In*: Proceedings of the American Control Conference, Chicago, Illinois.

[239] Yang, Y. 1989. A new condition number of eigenvalue and its application in control theory. Journal of Computational Mathematics. 7(1): 15–21.

[240] Yang, Y. 1992. A globally convergent algorithm for robust pole assignment. Science in China (Series A). 23(12): 1126–1131. (In Chinese).

[241] Yang, Y. 1996. Robust system design: pole assignment approach. University of Maryland at College Park, College Park, MD.

[242] Yang, Y. 2007. Globally convergent optimization algorithms on Riemannian manifolds: uniform framework for unconstrained and constrained optimization. Journal of Optimization Theory and Applications. 132(2): 245–265.

[243] Yang, Y. 2010. Quaternion based model for momentum biased nadir pointing spacecraft. Aerospace Science and Technology. 14: 199–202.

[244] Yang, Y. 2011. A polynomial arc-search interior-point algorithm for convex quadratic programming. European Journal of Operational Research. 215: 25–38.

[245] Yang, Y. 2012. Analytic LQR design for spacecraft control system based on quaternion model. Journal of Aerospace Engineering. 25: 448–453.

[246] Yang, Y. 2012. Spacecraft attitude determination and control: quaternion based method. Annual Reviews in Control. 36(2): 198–219.

[247] Yang, Y. 2013. A polynomial arc-search interior-point algorithm for linear programming. Journal of Optimization Theory and Applications. 158: 859–873.

[248] Yang, Y. 2013. Constrained LQR design using interior-point arc-search method for convex quadratic programming with box constraints. arXiv: 1304.4685.

[249] Yang, Y. 2014. Quaternion based LQR spacecraft control design is a robust pole assignment design. Journal of Aerospace Engineering. 27(1): 168–176.

[250] Yang, Y. 2014. Attitude control in spacecraft orbit-raising using a reduced quaternion model. Advances in Aircraft and Spacecraft Science. 1(4): 427–421.

[251] Yang, Y. 2015. Attitude determination using Newton's method on Riemannian manifold. Proceedings of the IMechE, Part G: Journal of Aerospace Engineering. 229(14): 2737–2742.

[252] Yang, Y. 2016. Controllability of spacecraft using only magnetic torques. IEEE Transactions on Aerospace and Electronic Systems. 52(2): 954–961.

[253] Yang, Y. 2017. Spacecraft attitude and reaction wheel desaturation combined control method. IEEE Transactions on Aerospace and Electronic Systems. 53(1): 286–295.

[254] Yang, Y. 2017. An efficient algorithm for periodic Riccati equation with periodically time-varying input matrix. Automatica. 78(4): 103–109.

[255] Yang, Y. 2018. Singularity-free spacecraft attitude control using variable-speed control Moment gyroscopes. IEEE Transactions on Aerospace and Electronic Systems 54(3): 1511–1518.

[256] Yang, Y. 2018. An efficient LQR design for discrete-time linear periodic system based on a novel lifting method. Automatica. 87: 383–388.

[257] Yang, Y. 2019. Coupled orbital and attitude control in spacecraft rendezvous and soft docking. Proc. IMechE Part G: Journal of Aerospace Engineering. DOI: 10.1177/0954410018792991.

[258] Yang, Y. and A.L. Tits. 1993. On robust pole assignment by state feedback. pp. 2765–2766. *In*: Proceedings of the American Control Conference, San Francisco.

[259] Yang, Y. and Z. Zhou. 2013. An analytic solution to Wahba's problem. Aerospace Science and Technology. 30: 46–49.

[260] Yang, Y. and Z. Zhou. 2016. Spacecraft dynamics should be considered in Kalman filter based attitude estimation 26th AAS/AIAA Space Flight Mechanics Meeting, Napa, CA.

[261] Yoon, H. and P. Tsiotras. 2002. Spacecraft adaptive attitude and power tracking with variable speed control moment gyroscopes. Journal of Guidance, Control, and Dynamics. 25(6): 1081–1090.

[262] Yoon, H. and P. Tsiotras. 2004. Singularity analysis of variable-speed control moment gyros. Journal of Guidance, Control, and Dynamics. 27(3): 374–386.

[263] Yoon, H. and P. Tsiotras. 2006. Spacecraft line-of-sight control using a single variable-speed control moment gyro. Journal of Guidance, Control, and Dynamics. 29(6): 1295–1308.

[264] Zagaris, C. and M. Romano. 2018. Reachability analysis of planar spacecraft docking with rotating body in close proximity. Journal of Guidance, Control, and Dynamics. 41: 1416–1422.

[265] Zanchttin, A.M. and M. Lovera. 2011. H_∞ attitude control of magnetically actuated satellites. Proceedings of 18th IFAC World Congress, Milano, Italy.

[266] Zanetti, R., M. Majji, R.H. Bishop and D. Mortari. 2009. Norm-constrained Kalman filter. Journal of Guidance and Control. 32(5): 1458–1465.

[267] Zhang, Y. 1996. Solving large-scale linear programs by interior-point methods under the MATLAB environment. Technical Report, Department of Mathematics and Statistics, University of Maryland, Baltimore County, Maryland.

[268] Zhang, F. and G. Duan. 2012. Integrated Relative Position and Attitude Control of Spacecraft in Proximity Operation Missions. International Journal of Automation and Computing.9(4): 342–351.

[269] Zhang, J., K. Ma, G. Meng and S. Tian. 2015. Spacecraft maneuvers via singularity-avoidance of control moment gyros based on dual-mode model predictive control. IEEE Transactions on Aerospace and Electronic Systems. 51: 2546–2559.

[270] Zhou, B., Z. Lin and G. Duan. 2011. Lyapunov differential equation approach to elliptical orbital rendezvous with constrained controls. Journal of Guidance Control and Dynamics. 34(2): 892–902.

[271] Zhou, B., Q. Wang, Z. Lin and G. Duan. 2014. Gain scheduled control of linear systems subject to actuator saturation with application to spacecraft rendezvous. IEEE Transactions on Control Systems Technology. 22(5): 2031–2038.

[272] Zhou, K., J.C. Doyle and K. Glover. 1996. Robust and Optimal Control. Prentice Hall, Inc., New Jersey.

[273] Zhou, Z. and R. Colgren. 2005. Nonlinear spacecraft attitude tracking controller for large non-constant rate commands. International Journal of Control. 78: 311–325.

Index

Milton Keynes UK
Ingram Content Group UK Ltd.
UKHW051014071024
449327UK00012B/237

9 780367 780357